**Fundamentals of Spacecraft Thermal Design**

**Progress in Astronautics and Aeronautics**

Martin Summerfield,
Series Editor
PRINCETON UNIVERSITY

VOLUMES

1. **Solid Propellant Rocket Research. 1960**

   EDITORS
   Martin Summerfield
   PRINCETON UNIVERSITY

2. **Liquid Rockets and Propellants. 1960**

   Loren E. Bollinger
   THE OHIO STATE UNIVERSITY
   Martin Goldsmith
   THE RAND CORPORATION
   Alexis W. Lemmon Jr.
   BATTELLE MEMORIAL INSTITUTE

3. **Energy Conversion for Space Power. 1961**

   Nathan W. Snyder
   INSTITUTE FOR DEFENSE ANALYSES

4. **Space Power Systems. 1961**

   Nathan W. Snyder
   INSTITUTE FOR DEFENSE ANALYSES

5. **Electrostatic Propulsion. 1961**

   David B. Langmuir
   SPACE TECHNOLOGY LABORATORIES, INC.
   Ernst Stuhlinger
   NASA GEORGE C. MARSHALL SPACE FLIGHT CENTER
   J. M. Sellen Jr.
   SPACE TECHNOLOGY LABORATORIES

6. **Detonation and Two-Phase Flow. 1962**

   S. S. Penner
   CALIFORNIA INSTITUTE OF TECHNOLOGY
   F. A. Williams
   HARVARD UNIVERSITY

7. **Hypersonic Flow Research. 1962**

   Frederick R. Riddell
   AVCO CORPORATION

8. **Guidance and Control. 1962**

   Robert E. Roberson
   CONSULTANT
   James S. Farrior
   LOCKHEED MISSILES AND SPACE COMPANY

9. **Electric Propulsion Development. 1963**

   Ernst Stuhlinger
   NASA GEORGE C. MARSHALL SPACE FLIGHT CENTER

10. **Technology of Lunar Exploration. 1963**

    Clifford I. Cummings and
    Harold R. Lawrence
    JET PROPULSION LABORATORY

11. **Power Systems for Space Flight. 1963**
Morris A. Zipkin and
Russell N. Edwards
GENERAL ELECTRIC COMPANY

12. **Ionization in High-Temperature Gases. 1963**
Kurt E. Shuler, Editor
NATIONAL BUREAU OF STANDARDS
John B. Fenn, Associate Editor
PRINCETON UNIVERSITY

13. **Guidance and Control — II. 1964**
Robert C. Langford
GENERAL PRECISION INC.
Charles J. Mundo
INSTITUTE OF NAVAL STUDIES

14. **Celestial Mechanics and Astrodynamics. 1964**
Victor G. Szebehely
YALE UNIVERSITY OBSERVATORY

15. **Heterogeneous Combustion. 1964**
Hans G. Wolfhard
INSTITUTE FOR DEFENSE ANALYSES
Irvin Glassman
PRINCETON UNIVERSITY
Leo Green Jr.
AIR FORCE SYSTEMS COMMAND

16. **Space Power Systems Engineering. 1966**
George C. Szego
INSTITUTE FOR DEFENSE ANALYSES
J. Edward Taylor
TRW INC.

17. **Methods in Astrodynamics and Celestial Mechanics. 1966**
Raynor L. Duncombe
U.S. NAVAL OBSERVATORY
Victor G. Szebehely
YALE UNIVERSITY OBSERVATORY

18. **Thermophysics and Temperature Control of Spacecraft and Entry Vehicles. 1966**
Gerhard B. Heller
NASA GEORGE C. MARSHALL SPACE FLIGHT CENTER

19. **Communication Satellite Systems Technology. 1966**
Richard B. Marsten
RADIO CORPORATION OF AMERICA

20. **Thermophysics of Spacecraft and Planetary Bodies**
Radiation Properties of Solids and the Electromagnetic Radiation Environment in Space. 1967
Gerhard B. Heller
NASA GEORGE C. MARSHALL SPACE FLIGHT CENTER

21. Thermal Design Principles of Spacecraft and Entry Bodies. 1969
Jerry T. Bevans
TRW SYSTEMS

22. Stratospheric Circulation. 1969
Willis L. Webb
ATMOSPHERIC SCIENCES LABORATORY, WHITE SANDS, AND UNIVERSITY OF TEXAS AT EL PASO

23. Thermophysics: Applications to Thermal Design of Spacecraft. 1970
Jerry T. Bevans
TRW SYSTEMS

24. Heat Transfer and Spacecraft Thermal Control. 1971
John W. Lucas
JET PROPULSION LABORATORY

25. Communication Satellites for the 70's: Technology. 1971
Nathaniel E. Feldman
THE RAND CORPORATION
Charles M. Kelly
THE AEROSPACE CORPORATION

26. Communication Satellites for the 70's: Systems. 1971
Nathaniel E. Feldman
THE RAND CORPORATION
Charles M. Kelly
THE AEROSPACE CORPORATION

27. Thermospheric Circulation. 1972
Willis L. Webb
ATMOSPHERIC SCIENCES LABORATORY, WHITE SANDS, AND UNIVERSITY OF TEXAS AT EL PASO

28. Thermal Characteristics of the Moon. 1972
John W. Lucas
JET PROPULSION LABORATORY

29. Fundamentals of Spacecraft Thermal Design. 1972
John W. Lucas
JET PROPULSION LABORATORY

(Other volumes are planned.)

The MIT Press
Cambridge, Massachusetts,
and London, England

Progress in
Astronautics and Aeronautics

An American Institute of Aeronautics
and Astronautics Series

Martin Summerfield, Series Editor

Volume 29

**Fundamentals of Spacecraft
Thermal Design**

Technical papers selected from the AIAA
9th Aerospace Sciences Meeting, January
1971, and the AIAA 6th Thermophysics
Conference, April 1971, subsequently
revised for this volume.

Edited by
John W. Lucas
JET PROPULSION LABORATORY
PASADENA, CALIFORNIA

Copyright © 1972 by
The Massachusetts Institute of Technology

This book was printed by Alpine Press, Inc.
and bound by Colonial Press, Inc.
in the United States of America.

All rights reserved. No part of this book may be reproduced in any form or by any
means, electronic or mechanical, including photocopying, recording, or by any
information storage and retrieval system, without permission in writing
from the publisher.

Library of Congress Cataloging in Publication Data
Main entry under title:

Fundamentals of spacecraft thermal design.

   (Progress in astronautics and aeronautics, v. 29)
   "Technical papers selected from the AIAA 9th Aerospace Sciences Meeting,
January 1971, and the AIAA 6th Thermophysics Conference, April 1971, subsequently
revised for this volume."
   Includes bibliographies.
   1. Space vehicles—Thermodynamics—Congresses.
I. Lucas, John W., 1923-    ed. II. AIAA Aerospace Sciences Meeting, 9th,
New York, 1971. III. AIAA Thermophysics Conference, 6th, Tullahoma, Tenn.,
1971. IV. Series.
TL507.P75 vol. 29 [TL900]         629.1'08s         [629.47'01'5367]
ISBN 0-262-12059-3                                      72-5605

| | |
|---|---|
| Preface | xiii |
| Editorial Committee for Volume 29 | xvii |

## I  Surface Radiation Properties — 1

### I.1  Synthesis and Measurement — 1

Space Radiation Environmental Effects on Reactively
Encapsulated Zinc Orthotitanates and Their Paints — 3
G. A. ZERLAUT, J E. GILLIGAN, AND N. A. ASHFORD

Coloration of MgO Single Crystals by Low Energy Protons—
A Function of Selected Defects — 33
H. LEVIN, C. C. BERGGREN, AND W. M. PEFFLEY

Reflectance Measurements of Dielectric Coatings on a
Conductor Substrate — 53
GHASSANE M. CHAABANE, CHESLEY PIEROWAY, AND JOHN FRANCIS

Bidirectional Reflectance of a Randomly Rough Surface — 69
T. F. SMITH AND R. G. HERING

Bidirectional Reflectance Characteristics of Integrating
Sphere Coatings — 87
R. C. ZENTNER, R. K. MacGREGOR, AND J. T. POGSON

Particulate Radiation Effects on the Solar Absorptance of
Thermal Control Surfaces — 107
R. J. ANDRES, PAUL M. BLAIR JR., AND E. C. SMITH

Long-Duration Exposure of Spacecraft Thermal Coatings to
Simulated Near-Earth Orbital Conditions — 123
K. E. STEUBE AND R. M. F. LINFORD

### I.2  Contamination Effects — 135

Sensitivity of Thermal Surface Solar Absorptance to
Particulate Contamination — 137
O. HAMBERG AND F. D. TOMLINSON

Measurement and Application of Contaminant
Optical Constants — 153
R. C. LINTON

Restoration of Degraded Spacecraft Surfaces Using
Reactive Gas Plasmas   167
R. B. GILLETTE AND W. D. BEVERLY

## I.3 Space Flight Effects   187

Report on the Flight Performance of the Z-93 White Paint
Used in the SERT II Thermal Control System   189
N. JOHN STEVENS AND GEORGE R. SMOLAK

Radiation Degradation Analysis of Surveyor III Material   205
D. L. ANDERSON, B. E. CUNNINGHAM, AND R. G. DAHMS

Study of Thermal Control Surfaces Returned from
Surveyor III   221
PAUL M. BLAIR JR., W. F. CARROLL, S. JACOBS, AND L. J. LEGER

## II Thermal Analysis   241

Directional Property Effects on Radiant Heat Transfer
and Equilibrium Temperature   243
A. F. HOUCHENS AND R. G. HERING

Radiant Heat Transfer between Nongray Surfaces   269
R. G. HERING AND W. D. FISCHER

Thermal Contact Conductance of Turned Surfaces   289
M. MICHAEL YOVANOVICH

Thermal Conductance of a Row of Cylinders Contacting
Two Planes   307
M. MICHAEL YOVANOVICH

Re-entry Thermal Analysis of Variable Thickness
Spherical Vehicles   319
J. C. DUNN AND R. E. NICKELL

Coupling of Shape Change, Heating Distribution and
Internal Conduction for Ablating Bodies   333
JIN H. CHIN

Temperature Uncertainties Associated with Spacecraft
Thermal Analyses   349
R. G. GOBLE

Spacecraft Thermal Design Verification through Modeling  361
R. K. MacGREGOR

## III Heat Pipes  381

Effects of Friction on the Sonic Velocity Limit in Sodium
Heat Pipes  383
E. K. LEVY

Possible Application of Electro-Osmotic Flow Pumping in
Heat Pipes  401
M. M. ABU-ROMIA

Experimental Performance of Grooved Heat Pipes at
Moderate Temperatures  417
N. KOSOWSKI AND R. KOSSON

Operating Characteristics and Long Life Capabilities of
Organic Fluid Heat Pipes  431
A. BASIULIS AND M. FILLER

Design and Performance of Noncondensible Gas
Controlled Heat Pipes  445
J. D. HINDERMAN, E. D. WATERS, AND R. V. KASER

Feedback Controlled Variable Conductance Heat Pipes  463
W. B. BIENERT, P. J. BRENNAN, AND J. P. KIRKPATRICK

Design, Fabrication, and Testing of a Variable Conductance
Heat Pipe for Equipment Thermal Control  487
F. EDELSTEIN AND R. J. HEMBACH

A Variable Conductance Heat-Pipe Flight Experiment  505
J. P. KIRKPATRICK AND B. D. MARCUS

## IV Thermal Design  529

Apollo Telescope Mount/Thermal Systems Unit: Correlation
of Predicted Data and Test Results  531
UWE HUETER, J. MICHEAL CONNOLLY, AND PAUL A. CHRISTENSEN

Radiative, Ablative, and Active Cooling Thermal Protection
Studies for the Leading Edge of a Fixed-Straight Wing Space
Shuttle  547
A. V. GOMEZ, C. G. JOHNSTON, AND D. M. CURRY

Space Station Environmental Thermal Control System
Definition 579
JOSEPH C. CODY, R. M. BYKE, AND A. T. STELL

Index to Contributors to Volume 29 601

# PREFACE

This volume represents one of a sequence of publications under different formats covering the field of thermophysics for over a decade. The total content of subject matter, dealing with many physical processes of heat exchange and with techniques for heat flow control, ranges all the way from the basic concepts and simple design principles incorporated in the earliest spacecraft of a decade ago to the highly developed concepts employed in the designs of manned spacecraft and in multiply launched unmanned spacecraft of today. The first volume of the thermophysics sequence in the format of this Series, No.18 (1966), was preceded by two NASA Special Publications and by a number of other documents (see bibliography at the end).

Each volume represents the status of the field of thermophysics at the time of its publication. The sequence of volumes permits analysis of the evolution of the entire field of thermophysics technology and of the separate scientific disciplines which comprise the field. In the past, the development of the field of thermophysics, and more particularly the development of the constituent disciplines, has taken place in response to specific technical requirements set by advanced missions, to prediction of new requirements (either real or imaginary), to unknowns in the new environment of space, or to concepts which offered the promise of improved mission performance or reliability. By reviewing in this way how thermophysics and its constituent disciplines have developed in relation to the space program, it is possible to envision the directions for future development in response to still more ambitious goals. Thermophysics and its component disciplines will continue in the future, as in the past, to move in response to the demands placed upon them by new goals and in response to the initiatives of the workers in the field.

The vast storehouse of technical ideas and information which has been built up in response to the space program is available, of course, for application in many other fields. For example, the understanding of radiative properties of materials may find important use in solar energy utilization and control, minimization of thermal pollution, enhancement of agricultural productivity, more suitable architecture, and in improvement and refinement of the more conventional industrial techniques of heat transfer control.

We have learned to utilize the sun's energy in space to generate onboard electrical power and to control the temperature of space vehicles, but on earth we have not progressed significantly beyond the indirect utilization of solar energy through hydroelectric power from dams since the early days of the industrial revolution. In the not very distant future, however, mankind's demands for energy will have to be met through direct utilization of abundant solar energy or through the conversion of some form of atomic energy; the results of space-generated thermophysics technology may very well play a vital role in either or both of these major energy sources.

Waste heat from man's power systems is taxing the local heat sinks of the earth, the atmosphere, and the various small bodies of water. Therefore, this heat must eventually be converted to usable energy or dumped into the much more extensive sink of the ocean or the nearly infinite sink of outer space. Utilization or dumping will require a technology built upon the foundations of space-generated thermophysics.

In the field of agriculture, we may ask, have horticulturists examined the growth rate, vitamin content, water consumption rate, susceptibility to spoilage, and other parameters of agricultural products, in response to the spectral content of the solar energy incident on the plants? The solar spectrum has been measured with a high degree of precision, the spectral response of solar cells extensively studied, and spectrally selective interference filters developed to provide the maximum power output for any condition of use. Can agriculture benefit from similar kinds of filters? Does the near infrared solar spectrum contribute anything other than mere heat to agricultural productivity, or does it even work in the opposite direction by increasing the water consumption through evaporation, thus effectively reducing the productive acreage in arid localities?

We have learned to tailor the characteristics of very large space vehicles so that the thermal fluctuations are at a minimum, even in the absence of the "damping effect" of a convective atmosphere. In so doing, we have utilized surface coating systems originally designed for terrestrial applications as well as those developed specifically for space. The thermal control of habitable buildings, however, continues to be accomplished by the brute method of production of thermal energy or rejection of thermal energy to the local environment, depending on the demand of the temperature control system at the moment, a process that is inherently more wasteful than that employed in spacecraft. Although there are obvious economic considerations in the choice of thermal control systems for housing, there continues to be a disturbing absence of the application of space thermophysics technology to this problem.

Whether the technology of thermophysics born of the space age will eventually contribute to any of these fields --

energy production, agriculture, architecture--is a challenge to be resolved in the future. The technology which made it possible for man to walk on the surface of the moon and to obtain close-up pictures of the planet Mars cannot be allowed to end there. Mankind will surely put it to use in other fields, including those directly related to living conditions on earth.

The papers in this Volume were selected from the thermophysics sessions at the AIAA 9th Aerospace Sciences Meeting (New York, January 1971) and the AIAA 6th Thermophysics Conference (Tullahoma, Tennessee, April 1971).

The editor wishes to express deep thanks to several persons who contributed invaluable effort toward the preparation of this Volume. R. P. Caren organized the fine thermophysics sessions in New York. W. C. Snoddy, General Chairman, and E. Fried, Technical Program Chairman, were responsible for the excellent Thermophysics Conference. The Session Chairmen prepared reviews of their respective papers. Members of the 1971 Thermophysics Committee prepared additional reviews of the papers. W. F. Carroll helpfully provided a number of suggestions which were used in the development of this Preface. Dr. Martin Summerfield, Editor-in-Chief, provided guidance, and Miss Ruth Bryans provided editorial assistance throughout the preparation of this Volume. My secretary, Miss Nancy Parmelee, maintained correspondence and files on all the papers.

John W. Lucas
June 1972

Bibliography

Clauss, F. J., "Surface Effects on Spacecraft Materials," John Wiley & Sons, New York, 1960.

Blau, H. and Fischer, H., "Radiation Transfer from Solid Materials," The Macmillan Co., New York, 1962.

Richmond, J. C., "Measurement of Thermal Radiation Properties of Solids," NASA SP-31, 1963.

Katzoff, S., "Symposium on Thermal Radiation of Solids," NASA SP-55, 1965.

Heller, G. (editor), "Thermophysics and Temperature Control of Spacecraft and Entry Vehicles," Vol. 18, AIAA Progress in Astronautics and Aeronautics, Academic Press, New York, 1966.

Heller, G. (editor), "Thermophysics of Spacecraft and Planetary Bodies," Vol. 20, AIAA Progress in Astronautics and Aeronautics, Academic Press, New York, 1968.

Bevans, J. T. (editor), "Thermal Design Principles of Spacecraft and Entry Bodies," Vol. 21, AIAA Progress in Astronautics and Aeronautics, Academic Press, New York, 1969.

Bevans, J. T. (editor), "Thermophysics: Applications to Thermal Design of Spacecraft," Vol. 23, AIAA Progress in Astronautics and Aeronautics, Academic Press, New York, 1970.

Lucas, J. W. (editor), "Heat Transfer and Spacecraft Thermal Control," AIAA Progress in Astronautics and Aeronautics, The MIT Press, Cambridge, Mass., 1971.

Lucas, J. W. (editor), "Thermal Characteristics of the Moon," Vol. 28, AIAA Progress in Astronautics and Aeronautics, The MIT Press, Cambridge, Mass., 1972.

Tien, C. L. (editor), "Thermal Control and Radiation," Vol. 31, AIAA Progress in Astronautics and Aeronautics, The MIT Press, Cambridge, Mass. (to appear in 1973).

**Editorial Committee
for Volume 29**

John W. Lucas, Volume Editor
JET PROPULSION LABORATORY

Gary Arnett
NASA MARSHALL SPACE FLIGHT CENTER

Walter Bienert
DYNATHERM CORPORATION

Carl Boebel
AIR FORCE MATERIALS LABORATORY

Edward T. Chimenti
NASA MANNED SPACECRAFT CENTER

George Cunnington
LOCKHEED MISSILES & SPACE COMPANY

Robert G. Hering
UNIVERSITY OF IOWA

Billy P. Jones
NASA MARSHALL SPACE FLIGHT CENTER

Robert Kidwell
NASA GODDARD SPACE FLIGHT CENTER

Vernon G. Klockzien
THE BOEING COMPANY

Tom J. Love
UNIVERSITY OF OKLAHOMA

Bruce D. Marcus
TRW SYSTEMS, INC.

Herman Mark
NASA LEWIS RESEARCH CENTER

Robert E. Rolling
LOCKHEED MISSILES & SPACE COMPANY

Firouz Shahrokhi
UNIVERSITY OF TENNESSEE SPACE INSTITUTE

Seymour Siegel
THE AEROSPACE CORPORATION

Chang L. Tien
UNIVERSITY OF CALIFORNIA AT BERKELEY

Alfred E. Wechsler
ARTHUR D. LITTLE, INC.

Charles C. Wood
NASA MARSHALL SPACE FLIGHT CENTER

# I Surface Radiation Properties
## I.1 Synthesis and Measurement

# SPACE RADIATION ENVIRONMENTAL EFFECTS ON REACTIVELY ENCAPSULATED ZINC ORTHOTITANATES AND THEIR PAINTS

G.A. Zerlaut,* J.E. Gilligan,+ and N.A. Ashford†

IIT Research Institute, Technology Center, Chicago, Illinois

## Abstract

Diffuse hemispherical reflectance measurements were made in situ of reactively encapsulated pigments and their paints before and after exposure to ultraviolet and proton radiation, singly and combined. The reflectance spectra from these tests are analyzed to deduce degradation mechanisms, from which protective mechanisms are postulated. Most important of the results obtained have been the development of 1) a viable scheme for stabilizing pigments by reactive encapsulation, 2) a $K_2SiF_6$-treated zinc orthotitanate-pigmented potassium silicate paint that possesses a solar absorptance, $\alpha_s$, of 0.12 and exceptional stability to ultraviolet irradiation ($\Delta\alpha_s$ = 0.002 after 1000 ESH of ultraviolet), and 3) an Owens-Illinois 650 silicone-resin based paint pigmented with plasma-calcined, silicate-treated zinc orthotitanate, that exhibits a $\Delta\alpha_s$ of zero after 2500 ESH of ultraviolet radiation in vacuum.

## Introduction

The efficacy of reactive encapsulation as a method of stabilization of pigments against the degrading influence of space ultraviolet radiation was first established with Z93 and then, analogously, with S-13G. However, it was not until S-13,

---

Presented as Paper 71-449 at the AIAA 6th Thermophysics Conference, Tullahoma, Tenn., April 26-28, 1971. The work reported here was performed under NASA Contract NAS8-5379, "Development of Space-Stable Thermal Control Coatings," for NASA George C. Marshall Space Flight Center, with D.W. Gates acting as the project engineer. The efforts of F.O. Rogers and J. Brzuskiewicz in preparing pigments and paints and of Robert F. Boutin in performing the irradiation tests are gratefully acknowledged.

*Manager, Polymer Chemistry Research.
+Group Leader-Thermal Control, Polymer Chemistry Research.
†Research Chemist, Physical Chemistry Research.

an elastomeric silicone paint based on untreated ZnO, was discovered to be unstable that the protective mechanisms associated with Z93's exceptional stability were deduced. These early experiments showed that zinc oxide was the unstable component in S-13 and that, therefore, the potassium silicate employed in the stable Z93 provided a stabilizing influence on the otherwise unstable zinc oxide.[1]

The idea of utilizing potassium silicate as a treatment to stabilize ZnO prior to its incorporation into the elastomeric silicone binder was thus developed[2] and the concept of reactive encapsulation was subsequently postulated. The now widely employed S-13G thermal-control paint utilizes this thesis in its formulation, in which the zinc oxide is first reacted in slurry with potassium silicate, is extracted and is then dried prior to being milled into the silicone elastomer binder. These investigations were the subject of a previous communication to the American Institute of Astronautics and Aeronautics (3rd Thermophysics Conference).[2]

The emphasis in this paper is on the synthesis and characterization of zinc orthotitanate as a high index pigment with special attention devoted to the concept of reactive encapsulation: A number of reactive encapsulants are discussed. Although EPR was utilized in drawing the conclusions given, and especially in planning the research reported, no spin resonance spectra are presented in this paper. These data, representing EPR/optical spectroscopy correlations, are the subject of separate communications presently in preparation.

## Selection of $Zn_2TiO_4$ as the Pigment

The solar reflectance of a white pigmented coating depends upon the coating's ability to scatter light of a very broad wavelength distribution. From light scattering considerations, it can be shown that in a system designed for maximum reflection over the very broad wavelength region represented by the solar spectrum, it is necessary to maximize the pigment volume concentration (PVC) with pigment whose size does not exceed that defined by the Jaenicke[3] expression for the solar maximum (500 nm), and to employ thick coatings in order to compensate for the decreased infrared scattering power of such a system (compared to its scattering efficiency in the region of the solar maximum).

Implicit in the concept of maximization of solar reflectance by pigmented coatings is the stringent requirement for complete transparency, or lack of absorption, of the vehicle, whether polymeric or inorganic in nature. That is, the very process of light scattering by multiple refraction greatly increases the path length in the film of light interacting with the surface. Thus, absorption centers, or color, in the film tend to frustrate the randomness of internal reflection and refraction, and the statistical probability of photon absorption is greatly enhanced, and hence damage to the binder is increasingly probable.

Light scattering, or, more cogently, pigment optics, relate most importantly to the concept of ultraviolet stability in the following manner. In ultraviolet-absorbing semiconductor pigments such as rutile $TiO_2$ and $ZnO$, the high extinction for ultraviolet at wavelengths below their edge, (which is 376 nm in ZnO), serves to effectively screen the ultraviolet from the binder. On the other hand, in dielectric pigments such as $Al_2O_3$, $MgO$, $SiO_2$, etc., which are transparent in most of the solar ultraviolet (down to approximately 200 nm), and which therefore are effective scatterers of ultraviolet, the effective ultraviolet pathlength in the film is manifoldly increased and the probability of creating serious, observable spectral damage in the binder is increased by orders of magnitude. This would be unimportant only if completely stable binders become available.

Since approximately 8% of solar energy lies below the absorption edge of zinc oxide, one can readily see why it is very difficult to obtain coatings based on it with solar absorptances of less than 0.16. Also, since none of the three binders currently used (Sylvania's PS7 potassium silicate, GE's RTV-602 silicone elastomer, and Owens-Illinois 650 silicone resin) absorb in the 300- to 400-nm wavelength region, where 80% of the solar ultraviolet energy lies, employment of a stable semiconductor pigment with an edge at about 300 nm would serve two purposes. It would still effectively screen the harmful 200- to 300-nm ultraviolet from the deep layers of the paint system and at the same time would offer an excellent chance of diminishing the solar absorptance by the solar absorptance factor 0.06 (= 0.8 x 0.08), providing its refractive index is at least as high as ZnO (n = 2.00).

Another aspect of the stability problem that involves the preference of semiconductor over dielectric pigments concerns the location of damage sites (or color centers). The ultraviolet transparency of dielectric pigments such as $Al_2O_3$ or MgO means that damage is a bulk phenomenon, with its optical manifestations relating, among other things, to bulk diffusion properties in the crystal. Semiconductor pigments such as ZnO and $Zn_2TiO_4$ possess high extinction for damaging ultraviolet and

Table 1 Criteria for selection of pigment and properties of zinc orthotitanate

| Criteria | $Zn_2TiO_4$ |
|---|---|
| 1. High refractive index; $n \geq 2.0$ | $n = 2.4$ |
| 2. No absorption from edge to 2700 nm | no absorption from 325 to 2700 nm |
| 3. Stability to ultraviolet | can be stabilized by surface treatment |
| 4. Stability to charged particle radiation | largely undetermined; appears promising |
| 5. Edge between 290 and 300 nm | edge at 325 nm |
| 6. Synthesizable in laboratory, pilot and production | yes (not produced in quantity yet, i.e., $>5$ lbs ) |
| 7. Acceptable impurities and unreacted materials | yes (unreacted precursors subject of much study) |

the damage therefore is confined predominantly to the surface where it can be precluded more effectively, i.e., heat treatment, surface doping, reactive encapsulation, etc.

Of all pigments examined in the course of our investigations during the past several years, only zinc orthotitanate appears to satisfy a maximum of criteria established above. These are presented in tabular form in Table 1.

### Synthesis of Zinc Orthotitanate

Literature

Zinc orthotitanate is a spinel that is formed from 2 moles of ZnO and 1 mole of $TiO_2$. A complete discussion of the literature pertaining to zinc titanates will not be attempted here. However, the most pertinent literature results will be given.

In 1960, Dulin and Rase[4] of the State University of New York's College of Ceramics published a study of "Phase Equilibria in the System $ZnO-TiO_2$." They confirmed the existence and structure of the orthotitanate but also reported the definite existence of metatitanate as a compound ($ZnTiO_3$) having the hexagonal structure of ilmenite and stable up to a temperature of 925°C. Theirs was the first thorough study of zinc titanates in nearly 30 years and, along with Bartram and Slepetys[5] in 1961, to some extent cleared up the discrepancies in previous studies. Bartram and Slepetys listed the orthotitanate as most easily prepared from sulfate type anatase and zinc oxide; a reaction time of 3 hr at 800° to 1000°C is required. The metatitanate, they found, required chloride process rutile and an optimum temperature of 850°C. The solid solution phenomenon claimed by earlier writers appeared to be explained by the claim of Bartram and Slepetys to a third zinc titanate ($Zn_2Ti_3O_8$), the sesquititanate. This is a defect spinel structure made from anatase and zinc oxide in ratios of 2 moles ZnO to 3 moles $TiO_2$ reacted at a temperature of 700°C for at least 100 hr.

In 1962, Loshkarev in three papers found only orthotitanate as a compound using only rutile and zinc oxide and temperatures up to 1400°C.[6-8] The reaction between rutile and zinc oxide did not begin below 740°C. The existence of unreacted zinc oxide in the final product, regardless of composition, temperature or time, was observed by Loshkarev and has been confirmed by these studies. They report "very intense shrinkage" (from 15 to 18%) in forming the orthotitanate at temperatures above 1000°C. They therefore recommend slow heating when reaching this range. (We follow this advice in our studies; the high shrinkage is quite apparent.) The Russian papers do not concede the existence of the metatitanate, $ZnO \cdot TiO_2$, nor the sesquititanate listed by Bartram and Slepetys.

A more recent publication on the subject is a paper by Kubo et al in which they acknowledge the existence of the three titanates and report success in making the metatitanate of exceptional purity.[9]

Summarizing the literature, all workers agreed on the composition, crystal structure and characteristics of the orthotitanate. A few agreed upon the existence and structure of the metatitanate, and only one claimed the existence and structure of the defect spinel, $Zn_2Ti_3O_8$, which we call the sesquititanate. It was considered best therefore to first attempt to form an orthotitanate using the method of Bartram and Slepetys.

Synthesis
---

The synthetic schedule for the three zinc orthotitanates that were prepared is given in Table 2. The metatitanate was the yellowest of the three stoichiometries prepared and possessed an absorption edge similar, but considerably more gentle, in

Table 2  Synthesis schedule of several zinc titanates of different stoichiometry

| Batch No. | Ratios of Reactants ZnO | TiO$_2$ | Temp., °C | Time, hr | Structure |
|---|---|---|---|---|---|
| B-129 | 1 mol | 1 mol (rutile) | 850 | 17 | meta |
| B-130 | 1 mol | 1 mol (anatase) | 700 | 64 | sesqui |
| B-131 | 2 mol | 1 mol (anatase) | 700 | 64 | sesqui |
| B-132 | 2 mol | 1 mol (anatase) | 1050 | 17 | ortho |
| B-133 | 3 mol | 2 mol (anatase) | 1050 | 17 | ortho |

slope than the rutile from which it was prepared. Like the orthotitanates discussed later, the metatitanate possesses unreacted ZnO, which can be extracted easily with acetic acid. The spectra of metatitanate ($ZnTiO_3$) and sesquititanate ($Zn_2Ti_3O_8$) are presented in Fig. 1.

The sesquititanate is whiter than the metatitanate and its absorption edge at ~365 nm is intermediate between that of the metatitanate (~400 nm) and the orthotitanate (~325 nm).

The orthotitanate is a very white pigment, brighter to the eye and more reflective than either of the pigments from which it is prepared. The reflectance spectra of zinc orthotitanate are presented in Fig. 2. The "step" in the reflectance spectra of all four heats is interpreted as being due to unreacted ZnO

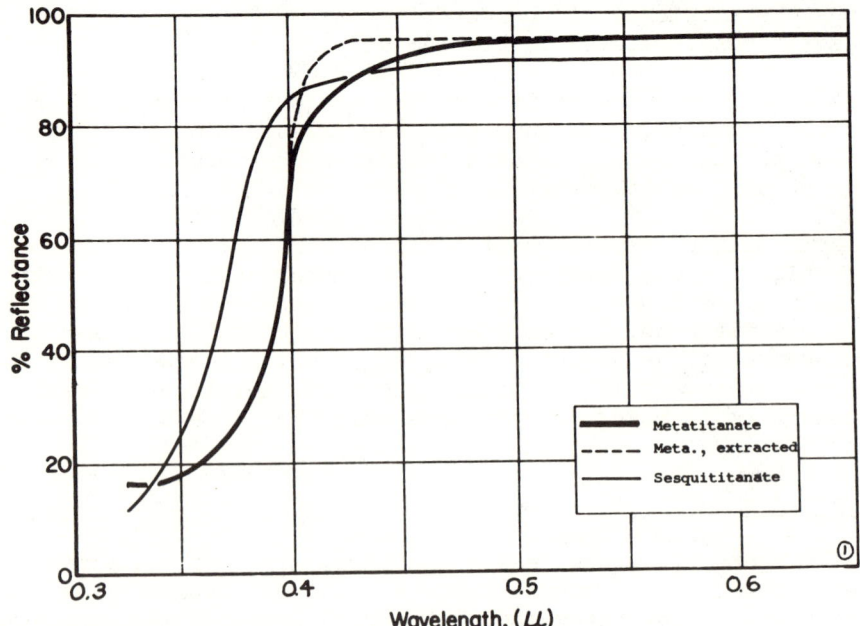

Fig. 1  Spectra of meta- and sesquititanates.

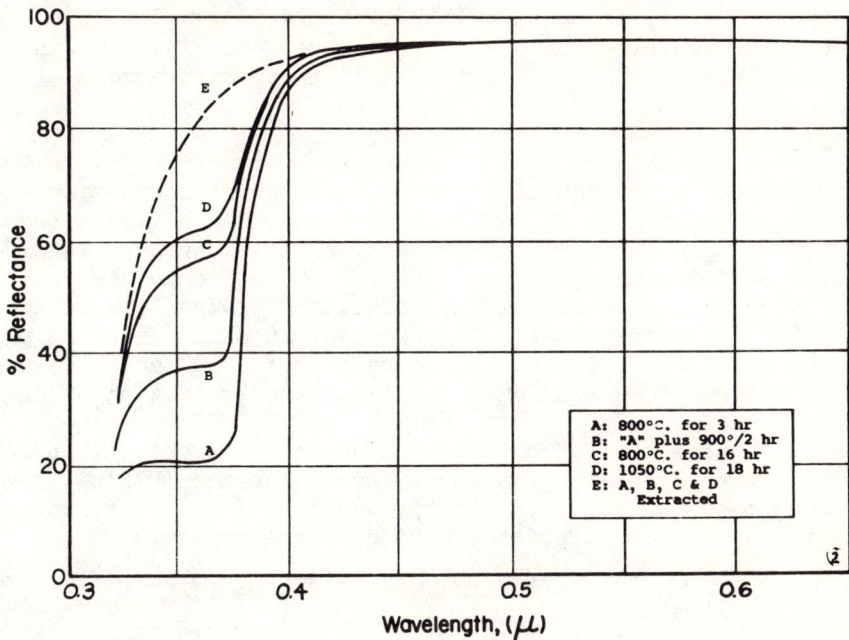

Fig. 2  Spectra of zinc orthotitanate.

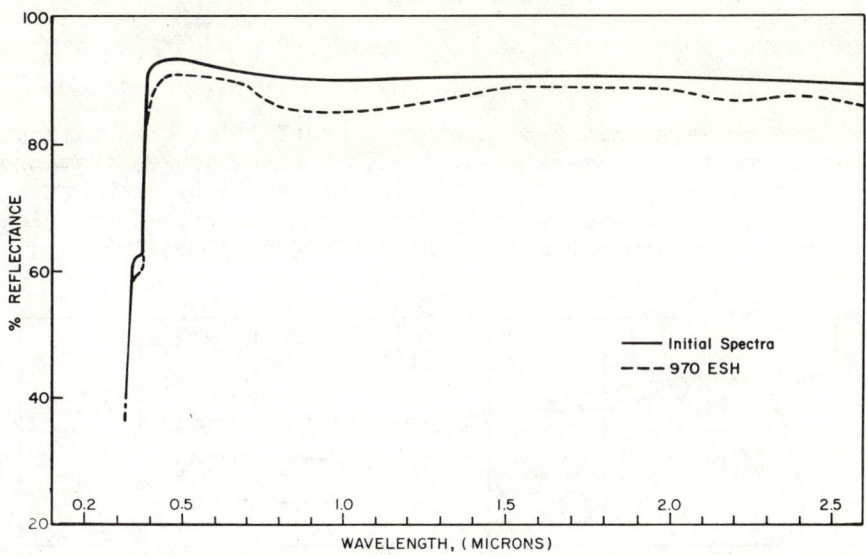

Fig. 3  Effect of Ultraviolet on B-229 control $Zn_2TiO_4$.

and in all cases, extraction with acetic acid results in a powder that exhibits the spectra of curve E in Fig. 2. Zinc oxide's presence has been confirmed by x-ray diffraction.

Irradiation

The effect of 970 ESH of ultraviolet irradiation on zinc orthotitanate prepared at 925°C is presented in Fig. 3. The absorption band at about 950-nm wavelength is characteristic of zinc orthotitanate and has been attributed to an electron trap associated with the $Ti^{+3}$ species.[10] Studies have shown that 0.5% excess zinc oxide is essential in minimizing the production of $Ti^{+3}$. This absorption band is fast-bleachable with oxygen and has been studied by Gilligan and Zerlaut.[11]

On the basis of electron spin resonance spectroscopy, in conjunction with optical spectroscopy employing the IRIF's (see Experimental), we attribute the production of $Ti^{+3}$ on irradiation to the photodesorption reaction of the general type

$$O_x^{-n} + Ti^{+4} \xrightarrow{h\nu} xO^{-(n-1)} + Ti^{+3} \qquad (1)$$

The importance of reaction temperature on the stability of zinc orthotitanate is seen from the effects of 1000 ESH of ultraviolet irradiation on pigment prepared at 1050°C (Fig. 4). The high temperature product exhibited a $\Delta\alpha_s$ of less than 0.01 but was of such hardness and agglomeration that the material could not be ground into paint. The 1050°C product has a Mho hardness of ~6.

Because we were unable to prepare a stable zinc orthotitanate at lower reaction temperatures, i.e., at temperatures that produce soft, small particle-sized material, the employment of surface treatments was necessitated. Both reactive encapsulation and plasma annealing were initiated (see Appendix).

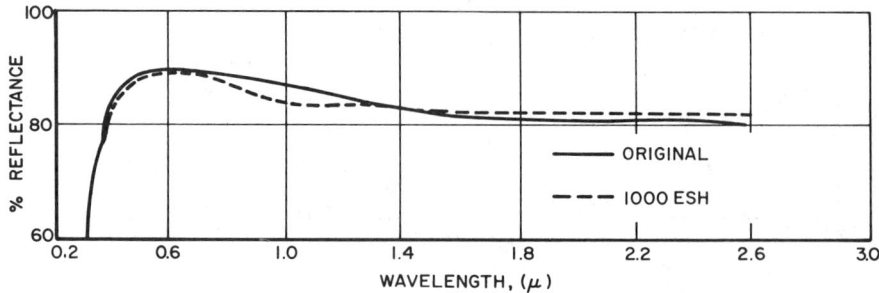

Fig. 4  Effect of ultraviolet on A-132 $Zn_2TiO_4$ (1050°C).

## Reactive Encapsulation

The complete results of plasma annealing of <u>untreated</u> zinc orthotitanate, although generally beneficial, will not be discussed in this paper. The plasma heat treatment of <u>silicated</u> zinc orthotitanate is of considerable importance and is included.

## Basic Considerations

We previously postulated that the effectiveness of silicate treatment of ZnO can be explained by the barrier mechanism concept in which the silicate coating on the pigment surface forms a barrier to charge and/or excitation transfer.[2] We are now of the opinion that protection may be afforded by other mechanisms, chief among which is the hole ($p^+$) withdrawing effect of the polynegative anions employed in reactive encapsulation.

Ultraviolet radiation produces electron-hole pairs with the electron entering the conduction band, leaving a hole ($p^+$) in the valence band. The holes combine with the chemisorbed oxygen (and other chemisorbed gases), now thought to be present on ZnO as $O^-$, releasing the oxygen at the surface

$$O^-_{ads} + p^+ \longrightarrow O \qquad (2)$$

Some lattice oxygen can be expected to be discharged by the holes, although this is believed to be unimportant in ZnO. The nonrecombined electrons in the conduction band either accumulate there, or fall into traps. Oxygen vacancies, i.e., from the discharge of lattice oxygen, will attract electrons because of their double positive charge.

On the basis of our studies of ultraviolet-irradiated zinc oxide[11, 12] involving optical spectroscopy and gas-adsorbate bleaching experiments, we have concluded that the predominant behavior in ZnO, that which manifests itself in the well-known broad infrared absorption that commences at approximately 1000 nm, is attributed to conduction band electrons. Shallow traps have been considered by others as being responsible for the observed infrared damage in ZnO. Nevertheless, we do not believe that a shallow trap with a transition energy of only about 0.03 eV, as necessitated by the infrared band observed, would permit the electron population required to be spectroscopically observable (i.e., $10^{14}$/cc). This is not to say though that shallow traps, lying 0.03 eV below the conduction band, are not present.

The earlier paper postulated that when the surface is treated with potassium silicate, the $O^-$ is displaced by the more highly charged, and thus more tightly bound, silicate anion. We now believe that the surface chemistry of protection is much more complicated and that it is only possible to say with any certainty that reactive encapsulation interferes with the oxygen chemistry at the pigment/binder interface. However, the most probable explanation involves the interaction of the "encapsulating" anion with the highly prevalent hydroxyl ions that are chemisorbed on metal oxide surfaces. Whether this interaction involves hydrogen bonding with the hydroxyls, or the displacement of the hydroxyls, remains to be determined. Also, whether high hole affinity of the polynegative anion precludes hole/$O^-_{ads}$ combination [as shown in Eq. (2)], or whether the $O^-_{ads}$ is displaced by the polynegative anion that acts as a physical barrier to hole/$O^-_{ads}$ combination, still remains to be determined.

Regardless of whether the barrier mechanism or the hole affinity concept prevails, it is obvious that oxygen is not photodesorbed when the electron-hole pair is produced in the pigment surface (where photon absorption takes place), and the kinetics of conduction electron formation are radically altered. Hence, as the electrons go into the conduction band, any holes left at the pigment/anion interface are captured preferentially by the polynegative anion, which is not desorbed. In the case of ZnO, the polynegative anion is silicate. We believe that a charge is then built up in the polynegative anion, attracting electrons from the conduction band, and reducing their concentration to the point that they are not spectroscopically observable, or thermodynamically important from the standpoint of spacecraft heat transfer.

Pigment Treatment Scheme

The complete history of all reactive encapsulation studies performed are presented schematically in Figs. 5 and 6. The open circles and squares in Figs. 5 and 6 represent surface-treated (reactively encapsulated) powders that were irradiated in the IRIF and/or employed as pigment material for silicate and silicone paints. The first row of "filled" circles represents PS-7 potassium silicate-based inorganic paints and the second row of filled circles (the last row) represents silicone paints prepared from Owens-Illinois 650 Glass Resin.

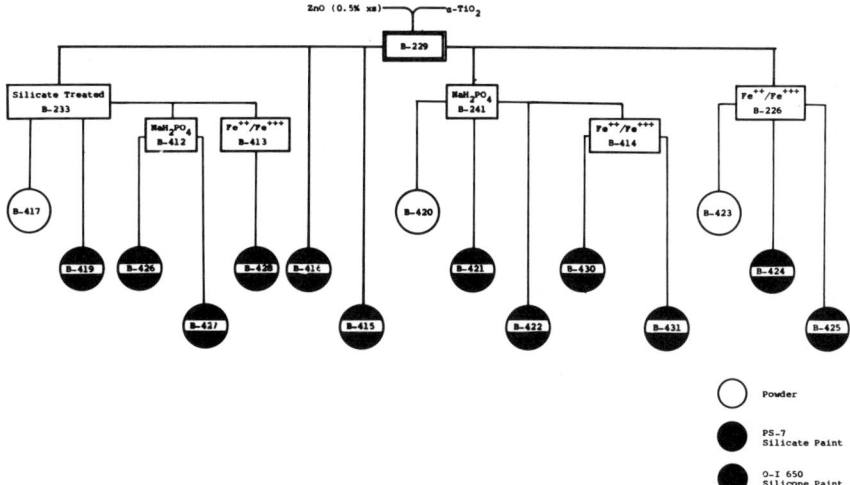

Fig. 5  Scheme for reactive encapsulation-I.

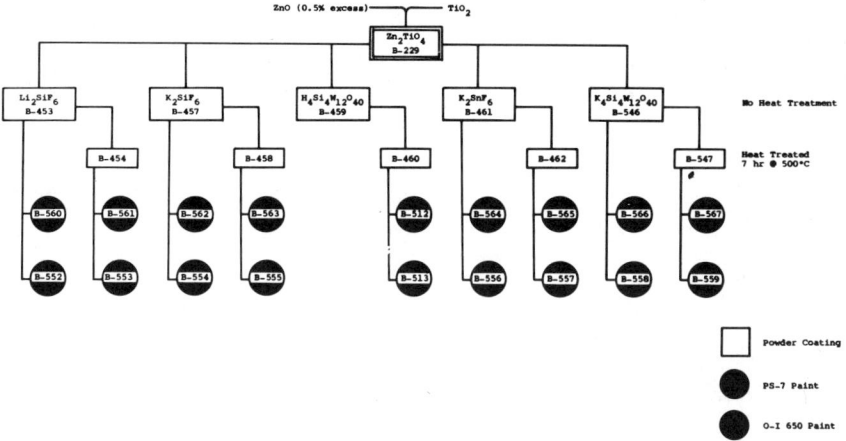

Fig. 6  Scheme for reactive encapsulation-II.

Results

Ultraviolet Irradiation

Pigments - Reactively Encapsulated: The effect of ultraviolet irradiation in vacuum on a series of zinc orthotitanate pigments that were reactively encapsulated with soluble alkali salts of the polyvalent anions of interest are shown in Table 3. This table presents the spectral reflectance changes for five (5) pertinent wavelengths.

Table 3  Effect of irradiation on the spectral reflectance of chemically treated $Zn_2TiO_4$ powders

| Figure no. | Batch no. | Chemical | Hr. @ temp., °C | Exposure (ESH) | Reflectance decrease % at wavelengths λ(nm): | | | | |
|---|---|---|---|---|---|---|---|---|---|
| | | | | | 362 | 400 | 700 | 950 | 2400 |
| 3 | B-229 | None | (Control) | | 2.0 | 4.0 | 4.0 | 4.5 | 2.8 |
| 7 | B-417 | PS7 | ... | 970 | 4.5 | 8.0 | 1.5 | 1.0 | 1.5 |
| | B-420 | $NaH_2PO_4$ | ... | 970 | 7.0 | 7.5 | 6.0 | 5.0 | 1.5 |
| | B-423 | Fe-CN | ... | 970 | 2.6 | 7.0 | 5.4 | 6.2 | 3.3 |
| 8 | B-453 | $Li_2SiF_6$ | ... | 1010 | 8.0 | 7.2 | 3.0 | 2.5 | 1.0 |
| 9 | B-454 | $Li_2SiF_6$ | 7 @ 500 | 1010 | 1.0 | 2.8 | 1.0 | 1.7 | 1.7 |
| | B-457 | $K_2SiF_6$ | ... | 1010 | 5.0 | 5.3 | 3.1 | 3.0 | 1.8 |
| 10 | B-458 | $K_2SiF_6$ | 7 @ 500 | 1010 | 2.7 | 3.0 | 1.6 | 2.2 | 0.3 |
| | B-461 | $K_2SnF_6$ | ... | 1010 | 8.0 | 6.5 | 1.0 | 1.5 | 1.5 |
| | B-462 | $K_2SnF_6$ | 7 @ 500 | 1010 | 9.0 | 6.0 | 1.3 | 2.5 | 1.7 |
| | B-546 | $K_4Si_4W_{12}O_{40}$ | ... | 1010 | 0.8 | 4.2 | 5.0 | 4.8 | 1.5 |
| | B-547 | $K_4Si_4W_{12}O_{40}$ | 7 @ 500 | 1010 | 0.5 | 3.3 | 2.1 | 2.1 | 1.3 |

The silicating of zinc orthotitanate clearly stabilizes the surface against formation of the broad damage spectra in the region 1000- to 2600-nm wavelength (Fig. 7). However, as with the silicated zinc oxide employed in IITRI's S-13G, the stability of the <u>powder</u> (as opposed to a silicate, or silicone, <u>paint</u>) is decreased in the 400-nm wavelength region by silicating. No explanation is available for this phenomenon except to note that, in the case of zinc oxide, the stability of the pigment in this region of the spectrum is greater than that of the potassium silicate.

The phosphate treatment (Batch B-420) improved the reflectance in the ultraviolet by essentially removing the shoulder at ~362-nm that has been attributed to unreacted zinc oxide.[10] The damage exhibited by the phosphated powder is greater than that sustained by an earlier specimen and cannot be explained except that it is conceivable that, in the present case, not all unreacted $NaH_2PO_4$ was removed from the surface by washing after phosphating.

Treatment with ferro/ferricyanide (Batch B-423) also has a deleterious effect on the spectral stability of zinc orthotitanate powders irradiated by ultraviolet.

Of the reactive encapsulants $Li_2SiF_6$, $K_2SiF_6$, $K_2SnF_6$ and $K_4Si_4W_{12}O_{40}$, all three fluorinated materials substantially reduced the infrared damage in the 700- to 2400-nm wavelength

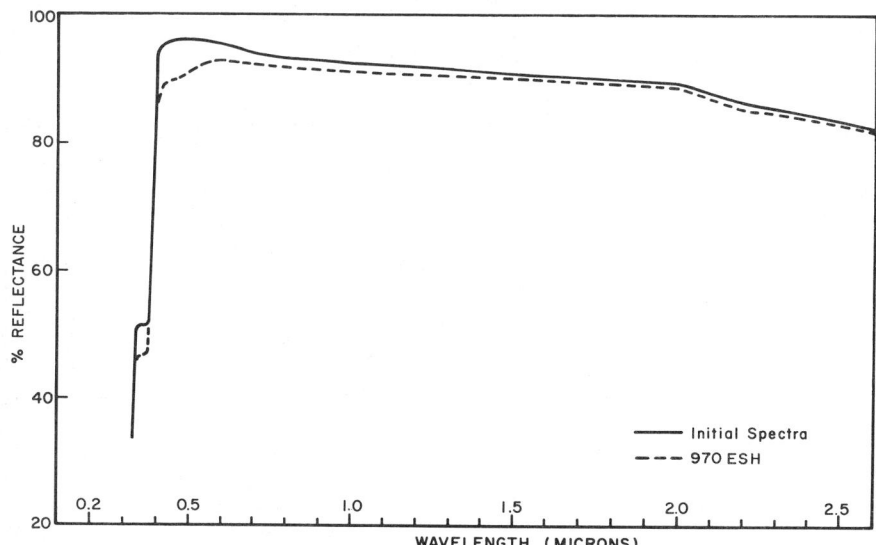

Fig. 7 Effect of uv on B-417 $Zn_2TiO_4$.

region that is characteristic of untreated zinc orthotitanate. Spectral reflectance data are given in Figs. 8-10 for the lithium and potassium silicofluoride-treated pigment. The silicotungstate and fluorostannate treatments had a highly deleterious effect on the stability of zinc orthotitanate in the 350- to 700-nm wavelength region (Batches B-461 and B-546).

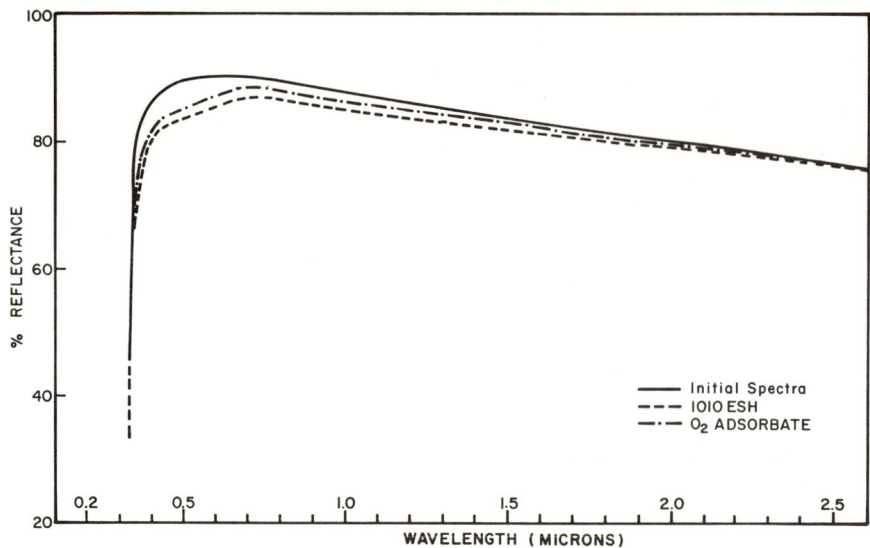

Fig. 8 Effect of uv on B-453 $Zn_2TiO_4$.

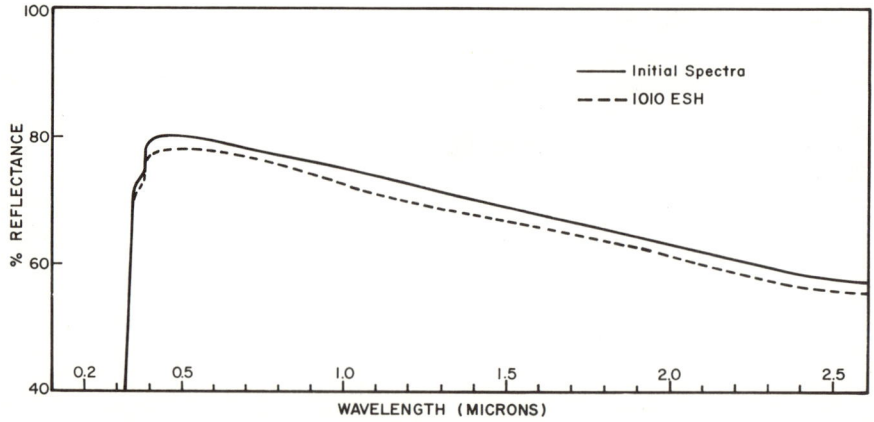

Fig. 9 Effect of uv on B-454 $Zn_2TiO_4$.

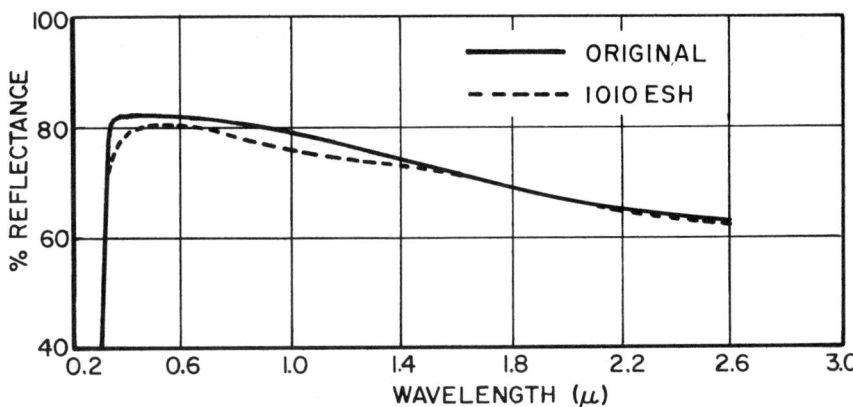

Fig. 10  Effect of uv on B-458 $Zn_2TiO_4$.

Heat treatment of all four chemically-treated zinc orthotitanate powders resulted in pigment having greater stability than the untreated control (B-229) in the infrared region 700- to 2400-nm wavelength. The lithium and potassium silicofluoride (hexafluorosilicate) treated pigment that was heat treated for 7 hr at 500°C exhibited improved stability in the near-ultraviolet and visible wavelength regions, as well (Figs. 8-10).

Although the heat-treated product prepared from the potassium silicotungstate-treated powder (B-547) was nearly as stable as the silicofluoride-treated products, this chemical treatment was not effective in removing the 362-nm shoulder absorption that is present in the $Zn_2TiO_4$ control--an absorption band that contributes significantly to the solar absorptance of coatings prepared from zinc orthotitanate. Treatment with the three fluorinated salts resulted in considerably decreased, if not completely eliminated, absorption at 362-nm wavelength (B-458); this change is presumably the result of the extraction of unreacted zinc oxide during the reflux operation.

<u>Pigments - Plasma Heat Treated</u>:  The data for a series of plasma-annealed zinc orthotitanate powders, both untreated and silicated, are presented in Table 4.  The data for the B-229 control pigment represent a successful attempt to establish the optimum effective surface temperature gradient, $\Delta T$, for zinc orthotitanate.

The subtleties in the damage spectra of plasma-annealed zinc orthotitanate have been discussed by two of the present authors in a previous communication.[11]  The infrared damage sustained

Table 4 Effect of uv irradiation on plasma-treated $Zn_2TiO_4$ (control and silicated)

| IITRI pigment | Figure no. | Plasma temp. ΔT (°C) | Exposure (ESH) | Reflectance decrease, % at wavelengths λ(nm): | | | | | Solar absorptance | |
|---|---|---|---|---|---|---|---|---|---|---|
| | | | | 362 | 425 | 700 | 950 | 2400 | $α_s$ | $Δα_s$ |
| B-229 | | 2000 | 1010 | 0.4 | 2.4 | 2.0 | 2.6 | 8.2 | 0.160 | 0.028 |
| B-229 | | 1400 | 1010 | 1.0 | 1.8 | 3.5 | 5.8 | -1.2 | 0.130 | 0.029 |
| B-229 | | 2450 | 1010 | -2.6 | 0 | 2.5 | 6.0 | 3.5 | 0.143 | 0.026 |
| B-229 | 11 | 1670 | 2500 | -4.0 | 0 | 0 | 0 | 10.0 | 0.145 | 0.010 |
| B-412[a] | | 1450 | 2500 | 1.4 | 3.3 | 0.3 | 0 | 0 | 0.140 | 0.003 |
| B-412[a] | | 1890 | 2500 | -2.7 | 0 | 0 | 0 | 2.5 | 0.150 | 0.008 |
| B-412[a] | 12 | 1670 | 2500 | 0 | 0 | 0 | 0 | 0.6 | 0.135 | 0 |

[a]Silicated and phosphated prior to plasma annealing.

Table 5 Effect of uv irradiation on $Zn_2TiO_4$-potassium silicate paints

| Batch no. | Figure no. | Pigment treatment | | Exposure (ESH) | Reflectance decrease, % at wavelengths λ(nm): | | | | | Solar absorptance | |
|---|---|---|---|---|---|---|---|---|---|---|---|
| | | Chemical | Heat hr / °C | | 362 | 425 | 700 | 925 | 2400 | $α_s$ | $Δα_s$ |
| B-419 | | PS7 | ... | 2400 | 4.0 | 7.0 | 1.0 | 1.0 | -1.0 | 0.136 | 0.020 |
| B-421 | | $HPO_4$= | ... | 2000 | 7.0 | 5.5 | 0.2 | -0.4 | -2.2 | 0.122 | 0.010 |
| B-424 | | Fe-CN | ... | 2000 | 4.2 | 5.0 | 1.8 | 1.0 | -2.2 | 0.154 | 0.011 |
| B-563 | 13 | $K_2SiF_6$ | 7 / 500 | 1200 | 2.2 | 2.8 | 0 | 0 | 0.5 | 0.120 | 0.002 |

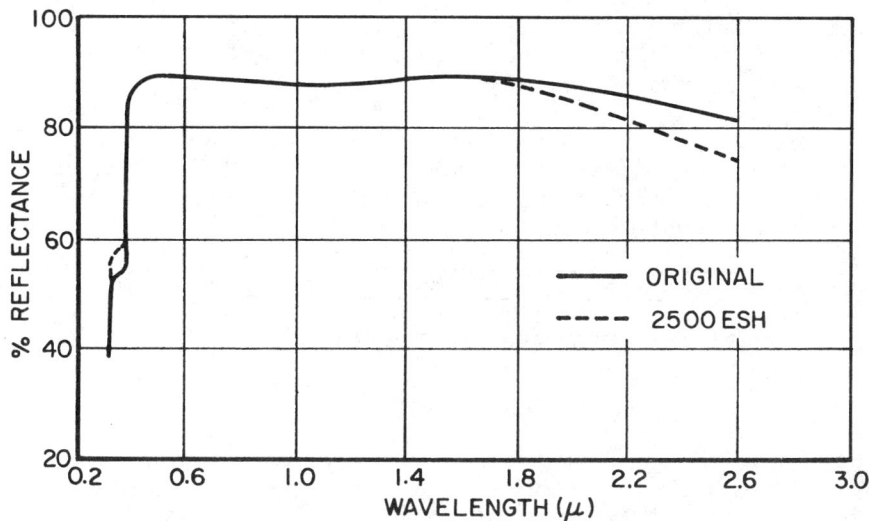

Fig. 11  Effect of uv on 1670°C ΔT $Zn_2TiO_4$.

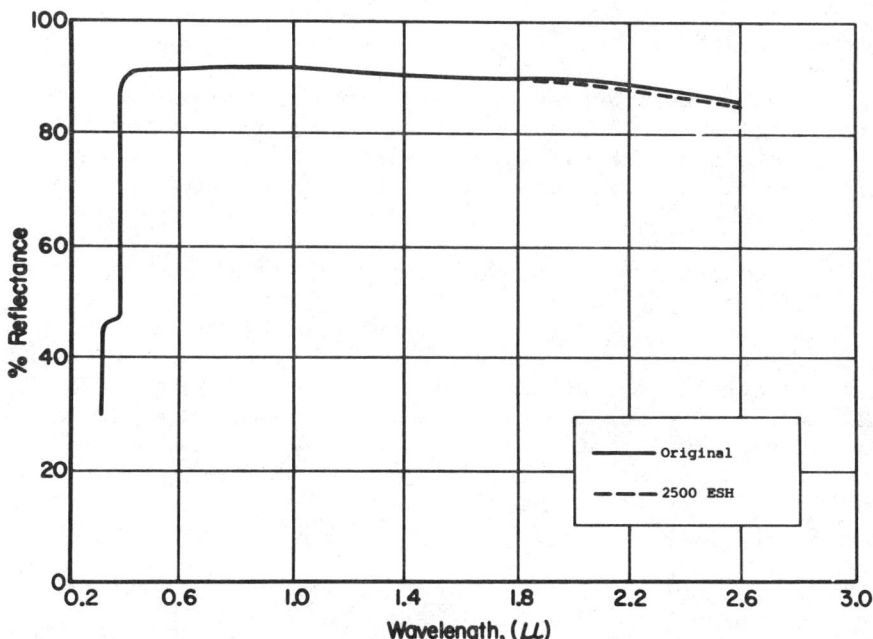

Fig. 12  Effect of uv on 1670°C ΔT silicated $Zn_2TiO_4$.

by the specimen annealed at a $\Delta T$ of 1670°C (Fig. 11) is attributed to conduction electrons as a result of free zinc oxide that is presumably condensed on the surface of the aerosol powder in the plasma.

Examination of the data in Table 4 clearly shows that a $\Delta T$ of 1670°C is essentially optimum for $Zn_2TiO_4$. This temperature is clearly shown to be superior for the silicated/phosphated zinc orthotitanate (Batch B-412) annealed at 1670°C (see Fig. 12). The data for the 1670°C product represents the greatest ultraviolet stability we have observed in a pure pigment powder; this is especially significant when we consider that it was irradiated for 2500 ESH of ultraviolet radiation in vacuum.

Paints

Silicate Paints: Examination of the data presented in Table 5 confirms the results obtained previously--namely that potassium silicate paints pigmented with zinc orthotitanate are, even in the absence of reactive encapsulation, quite stable to ultraviolet irradiation in vacuum.

Both specimens B-421 and B-563 possess excellent solar absorptances by virtue of the extraction of the zinc oxide-related shoulder in $Zn_2TiO_4$ at 368 nm by the acid phosphate and potassium hexafluorosilicate. Although the paint based on ferro/ferricyanide-treated pigment exhibited stability that was superior to the control paint (B-419), its higher solar absorptance of 0.15 is characteristic of iron cyanide-treated pigments.

The excellent stability exhibited by the silicate paint prepared from phosphated pigment should be noted (Batch B-421). In this case, unlike the phosphated-pigment irradiated as a powder, the phosphate treatment preceded pigmentation in the silicate vehicle and neutralization of any excess acid phosphate was assured by the highly alkaline silicate solution. (A similar formulation has been furnished the British RAE at Farnborough, England for a flight experiment to be flown on the Black Arrow* and to NASA-Goddard Space Flight Center for a flight experiment to be flown on OSO 8.+)

The potassium silicate paint B-563 based on the heat-treated, potassium hexafluorosilicate-treated $Zn_2TiO_4$, is equally stable to Z93 and possesses the lowest solar absorptance of any of the more stable coatings investigated to date (Fig. 13).

---

*J. Porter (RAE).
+J. Triolo (NASA-GSFC).

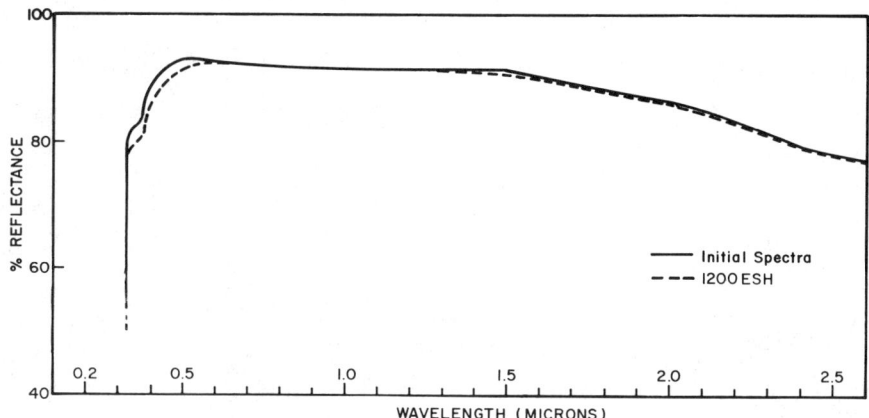

Fig. 13  Effect of uv on B-563 silicate paint.

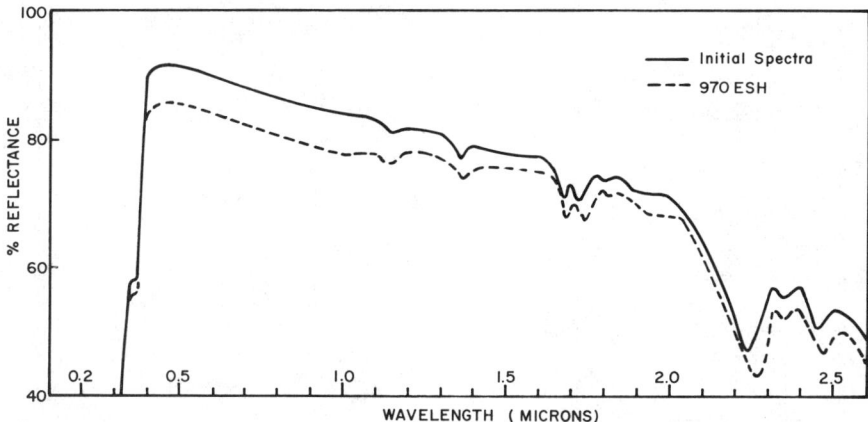

Fig. 14  Effect of uv on B-427 silicone paint.

Silicone Paints: The silicone paint prepared from untreated zinc orthotitanate was less stable in the near infrared (700- to 2600-nm wavelength) than the pigment powder itself (see Batch B-415, Table 6). Indeed, an Owens-Illinois 650 resin paint prepared from silicated/phosphated pigment (B-427, Fig. 14) was even less stable than the control paint. Even the paint prepared from pigment that had been treated with potassium hexafluorosilicate exhibited only slight improvement over the control paint. These disappointing results confirmed earlier studies that an adverse synergistic effect exists between the Owens-Illinois 650 resin and the surface of zinc orthotitanate (see Discussion for an explanation).

Table 6  Effect of uv irradiation on $Zn_2TiO_4$-silicone paints

| Batch no. | Figure no. | Binder | Pigment batch | Pigment treatment | Exposure (ESH) | Reflectance decrease at wavelengths $\lambda$(nm), %: 362 | 425 | 700 | 950 | 2400 | Solar absorptance $\alpha_s$ | $\Delta\alpha_s$ |
|---|---|---|---|---|---|---|---|---|---|---|---|---|
| B-415 | | OI-650 | B-229 | ... | 970 | 1.0 | 3.5 | 6.0 | 7.8 | 2.8 | ... | ... |
| B-427 | 14 | OI-650 | B-412 | PS7/HPO$_4$= | 970 | 2.2 | 5.2 | 6.0 | 6.9 | 3.9 | ... | ... |
| B-555 | | OI-650 | B-458 | K$_2$SiF$_6$ | 1200 | 2.0 | 4.5 | 4.5 | 5.7 | 3.2 | 0.212 | 0.041 |
| B-704[a] | | OI-650 | B-229 | ... | 2000 | -7.5 | -0.5 | 1.2 | 1.0 | 3.0 | 0.250 | 0.012 |
| B-705[a] | 15 | OI-650 | B-412 | PS7/HPO$_4$= | 2000 | -1.0 | 0 | 0 | 0 | 2.4 | 0.241 | 0.002 |
| B-574 | | RTV-602 | B-229 | ... | 1000 | 15.5 | 40.5 | 12.0 | 6.5 | 0.8 | ... | ... |

[a]Owens-Illinois 650 silicone resin paints prepared from Batch B-412 pigment plasma calcined for 1.1 sec at 1670°C.

Table 7  Summary of 1.2 kev proton-irradiation damage

| Figure no. | Material | Binder | Pigment oxide | Pigment treatment | Proton exposure $10^{15}$ p/cm$^2$ | $10^9$ p/cm$^2$-sec | Solar absorptance $\alpha_s$ | $\Delta\alpha_s$ |
|---|---|---|---|---|---|---|---|---|
| 16 | Z-93 | PS7 | ZnO | ... | 8.4 | 5.4 | 0.167 | 0.027 |
| | S-13G | RTV-602 | ZnO | PS7 | 2.5 | 4.9 | 0.212 | 0.022 |
| | Paint | PS7 | Zn$_2$TiO$_4$ | NaH$_2$PO$_4$ | 8.4 | 5.4 | 0.149 | 0.038 |
| 17 | Paint | OI-650 | Zn$_2$TiO$_4$ | Fe-CN | 2.5 | 4.9 | 0.269 | 0.005 |
| | Powder | H$_2$O | Zn$_2$TiO$_4$ | 1900°C Plasma | 8.4 | 5.4 | 0.197 | 0.041 |

The data in Table 6 show clearly that this adverse synergism between zinc orthotitanate and Owens-Illinois 650 resin can be completely overcome by plasma heat treatment of a silicate/phosphated $Zn_2TiO_4$. Indeed, irradiation of this paint (Fig. 15) resulted in an increase in solar absorptance of essentially zero after 2000 ESH of ultraviolet radiation.

The $Zn_2TiO_4$ paint (Batch B-574) prepared from RTV-602 silicone elastomer and B-229 control pigment was very badly damaged in the near ultraviolet and visible spectrum in only 1000 ESH (Table 6). Although this coating did not exhibit a noticeable "belly damage" at 950-nm wavelength as did the Owens-Illinois 650-resin analog, the $O_2$-bleaching in the 950-nm wavelength region is indicative of the $Ti^{+3} \rightleftarrows Ti^{+4}$ transition. The severe visible damage is attributed to two synergistic factors: The employment of the near-ultraviolet-scattering $Zn_2TiO_4$ pigment (compared to the absorber ZnO) in a binder that is less stable than Owens-Illinois 650 resin. (RTV-602 is less stable only by virtue of the requirement for amine curing. Owens-Illinois 650 resin heat cures by residual hydroxyl functionality.)

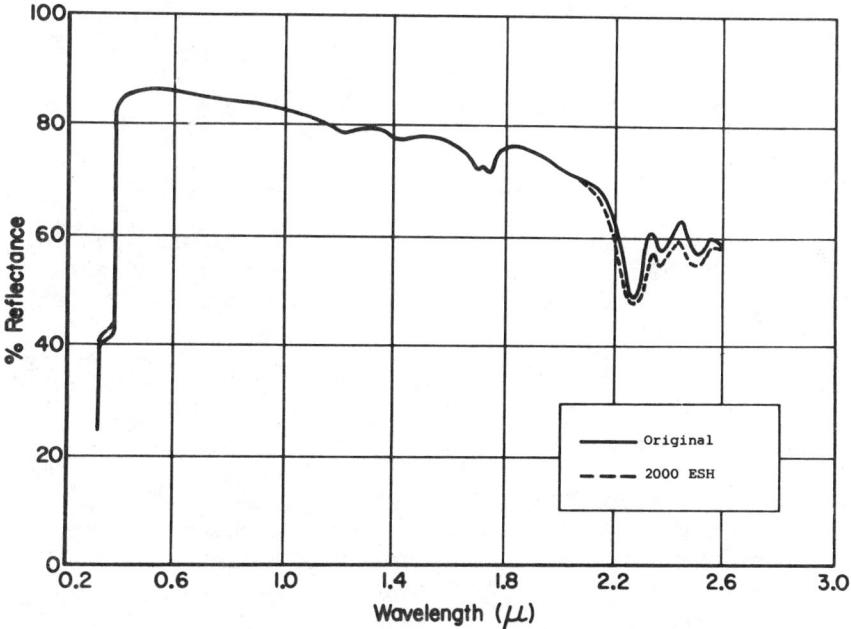

Fig. 15  Effect of uv on B-705 silicone paint.

## Proton Irradiation

The effects of irradiation of surface-treated pigments with 1.2 keV protons is shown in Table 7. The $\Delta\alpha_s$'s of 0.04 for the plasma-treated $Zn_2TiO_4$ and the phosphated $Zn_2TiO_4$-pigmented silicate paint are encouraging, since the fluence of 8.4 x $10^{15}$ p/cm$^2$ represents greater than one year in the solar wind stream. This compares favorably with Z93 (Fig. 16).

The most important observation is the stability exhibited by the Owens-Illinois 650 resin paint that was pigmented with cyanated $Zn_2TiO_4$ (Fig. 17). The ferro/ferricyanide treatment, first utilized by Morrison and Sancier[13] in studies of the defect state of ZnO, has been employed at IITRI in treating $Zn_2TiO_4$. We have been generally unsuccessful in using iron cyanide to effectively stabilize either ZnO or $Zn_2TiO_4$ to ultraviolet irradiation under ambient irradiation conditions. However, the ferro/ferricyanide treatment did indeed militate against the severe damage seen in wet-sprayed powders and silicate paints irradiated at high temperature.[14]

Although preliminary combined ultraviolet-plus-proton radiation experiments have been performed on untreated $Zn_2TiO_4$-pigmented potassium silicate and Owens-Illinois 650 paints, these data are the subject of another communication[15] and are not included here. The efficacy of reactive encapsulation, of ferro/ferricyanide treatment, or of plasma annealing has yet to be established in terms of either proton irradiation, or combined ultraviolet-plus-proton irradiation.

### Discussion, Summary, and Conclusions

It is necessary that an excess of ZnO (in our studies 0.5%) be present in the preparative stoichiometry to minimize the "surface" $Ti^{+4}$ available in $Zn_2TiO_4$ for the photodesorption reaction

$$O_x^{-n} + Ti^{+4} \xrightarrow{h\nu} xO^{-(n-1)} + Ti^{+3} \qquad (3)$$

from occurring, yielding an EPR-observable center and the concomittant broad 950-nm absorption band observable by <u>in situ</u> optical spectroscopy. The 950-nm absorption band is <u>fast</u> oxygen bleachable. On the basis of gas adsorbate kinetic studies reported earlier,[11] we have shown that the oxygen bleaching process associated with the 950-nm absorption peak depends upon the one-fourth power of oxygen pressure. Since the pressure dependencies of $O_2^-$, $O^-$ and $O^=$ adsorption are n = 1, 0.5 and 0.25, respectively, we previously concluded that

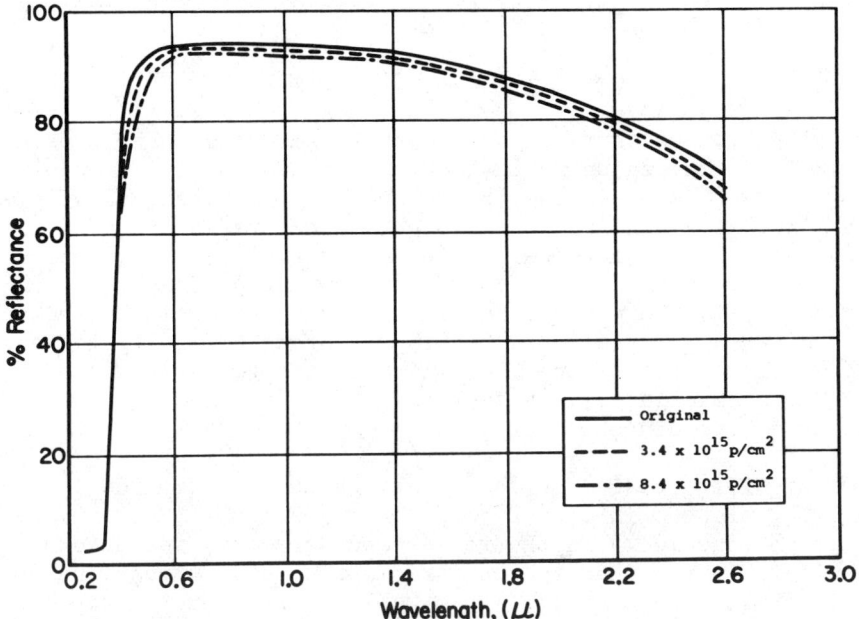

Fig. 16  Effect of protons on Z-93.

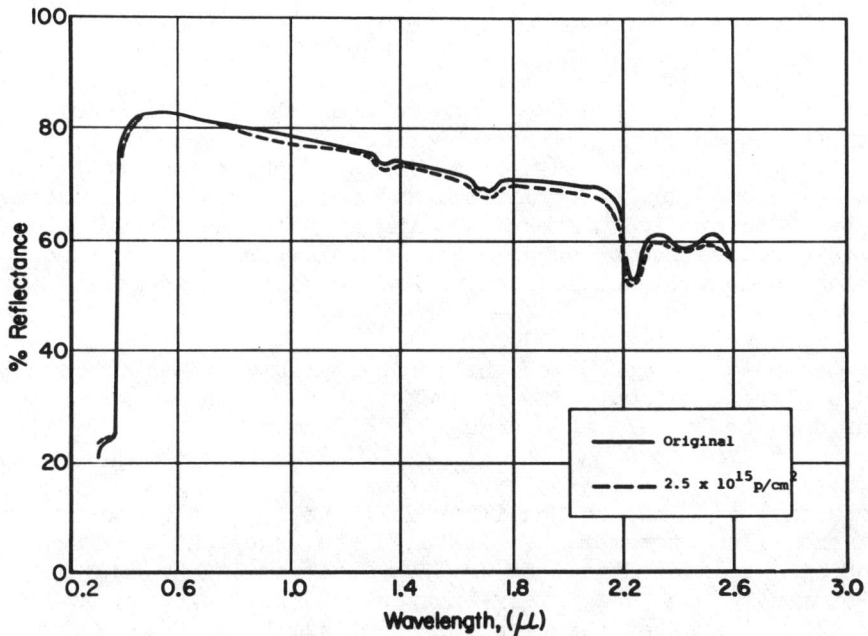

Fig. 17  Effect of protons on B-252 silicone paint.

Eq. (3) can be rewritten to represent a 2-step bleaching reaction in which O is subsequently reduced to $O^=$.

$$O + 2Ti^{+3} \longrightarrow O^= + 2Ti^{+4} \qquad (4)$$

Similarly, the infrared damage that commences (when present) in the 1500-nm region in $Zn_2TiO_4$ is associated with ZnO that is thermally produced--either by severe heating in air or by plasma annealing at too high a temperature. The presence of ZnO under both thermal circumstances has been observed by EPR spectroscopy and has been confirmed by x-ray powder diffraction. Furthermore, like conduction electron absorption in ZnO powder, the infrared absorption in $Zn_2TiO_4$ is broad and still increasing at the limits of the near-infrared spectrometer employed,** and, more importantly, is fast bleachable with an oxygen pressure dependence of n = 0.5.[11]

Of significance is the fact that residual zinc oxide in $Zn_2TiO_4$, present in the 925°C product, causes the 350-nm absorption (shoulder) in the reflectance spectra, yet does not result in infrared damage on irradiation, even though the conduction electron center is observable by EPR spectroscopy. However, it is not yet settled whether the EPR center(s) observed at g = 1.96 is in fact due to conduction electrons or is due to electrons in shallow traps.[16] (We stated earlier, however, that the observable infrared optical damage can be due only to conduction electrons.)

Also, whenever the zinc oxide-related ultraviolet shoulder is present at 350 nm, it can be removed by extraction with acetic acid or reaction with sodium acid phosphate, potassium and lithium hexafluorosilicate, and potassium hexafluorostannate. Conversely, plasma annealing at too high a temperature always causes the bleachable ZnO-like infrared damage to occur, which is usually accompanied by an increase in absorption at 350 nm.

It should be emphasized that proper plasma annealing results in a pigment without either conduction-electron damage in the infrared or $Ti^{+3}$-related damage at 950-nm.

Reactive encapsulation with potassium silicate, and with potassium and lithium hexafluorosilicate, potassium hexafluorostannate, and potassium silicotungstate, substantially reduces the $Ti^{+3}$-related damage in $Zn_2TiO_4$ observed at 950-nm. Heat treatment of the reactively-encapsulated $Zn_2TiO_4$ powders totally eliminates the possibility of damage at 950-nm for all

---

**Beckman DK-1 and DK-2A spectrometers.

treatments. Of these, surface treatment with lithium and, especially, potassium hexafluorosilicate stabilizes zinc orthotitanate in the entire solar spectrum. Potassium silicate paints prepared from the reactively encapsulated pigment were all quite stable with the potassium hexafluorosilicate-treated $Zn_2TiO_4$ producing a paint of exceptional stability ( $\Delta\alpha_s$ of 0.002 in 1000 ESH) and solar reflectance ( $\alpha_s < 0.12$).

Reactive encapsulation of zinc orthotitanate did not, by itself, result in stable silicone paints based on Owens-Illinois 650 resin. Indeed, these paints exhibit greater damage to ultraviolet irradiation at 950-nm than the pigment powders alone. However, plasma annealing ( $\Delta T = 1670°C$ ) of reactively-encapsulated $Zn_2TiO_4$ (silicated and phosphated) not only resulted in a pigment that exhibited no damage, but the Owens-Illinois 650 silicone paints produced therefrom exhibit a $\Delta\alpha_s$ of zero in 2500 ESH of ultraviolet radiation in vacuum.

We believe, on the basis of the optical spectroscopy reported here, and EPR studies to be reported later,[17] that the treatments with potassium silicate and hexafluorosilicate provide a barrier to photodesorption of oxygen, as with zinc oxide, wherein the holes are captured by the polynegative anion, thus preventing the photodesorption of $O^-$ and $O^=$ (or $O_2^-$) by reaction (2).

We attribute the failure of the reactive encapsulants $K_2O \cdot SiO_2$ and $K_2SiF_6$ to protect the $Zn_2TiO_4$ (in silicone paints) from 950-nm damage as being due to the interaction with the pigment surface of the free radicals formed in the Owens-Illinois 650 silicone resin (a triethoxymethylsilane condensation product) on ultraviolet-irradiation. A possible mechanism might be written as a two step process

$$1. \quad \begin{array}{c} CH_3 \\ -Si- \\ O \\ -Si- \\ CH_3 \end{array} \longrightarrow \begin{array}{c} CH_2 \cdot \\ -Si- \\ O \\ -Si- \\ CH_3 \end{array} + \cdot H \quad (5)$$

$$2. \quad \cdot H + Ti^{+4} \longrightarrow Ti^{+3} + H^+ \quad (6)$$

--a process that is precluded by plasma annealing of $Zn_2TiO_4$--especially if the pigment is first silicated. We believe that, although oxygen species hydrogen-bonded to the surface hydroxyl groups can reduce the surface before plasma annealing, the removal of the hydroxyl groups by plasma annealing and subsequent alteration of the surface, combined with the reactive encapsulant, results in a very different kind of reduction chemistry, certainly involving the free radicals produced in the silicone matrix. Ultimately, the oxygen species, though capable of reducing the surface by itself, most importantly acts as a shuttle for electrons provided by the matrix. Hence, removal or alteration of oxygen species from the surface is the removal of the conduit rather than the predominant source of damaging electrons.

In summary, both water and carbon dioxide, among other species, are highly absorbed on the surface of zinc orthotitanate.

$$\text{OH} \quad \overset{\text{O}\diagdown_{\text{C}}\diagup\text{O}}{\text{Ti/Zn}} \quad \overset{}{\text{Ti/Zn-O-Ti/Zn}} \quad \overset{\overset{\text{O}}{\underset{\text{C}}{\|}}}{\underset{\text{O}\diagup\quad\diagdown\text{O}}{\text{Ti/Zn-O-Ti/Zn}}}, \text{ etc.}$$

We believe that these species promote the association of $O^-$ and $O^=$, as well as $O_2^-$, on the surface, the photodesorption of which allows an electron to be injected into the surface where, in the case of zinc oxide, it is available for conduction absorption (i.e., broad based, free-carrier absorption) and, in the case of zinc orthotitanate, where it is available for reduction of $Ti^{+4}$ to $Ti^{+3}$. This concept is wholly consistent with the considerable improvements obtained by plasma annealing and reactive encapsulation of zinc orthotitanate. Plasma annealing dehydroxalates the surface (and oxidizes adsorbed carbonate, as well), thus reducing the sites for absorbed $O^-$ and $O^=$. Similarly, reactive encapsulation is thought to tie up the surface, and again, the availability of sites for absorption of $O_x^{-n}$ is greatly diminished. Thus, photodesorption of oxygen and the injection of an electron into the lattice surface is precluded and ultraviolet-induced damage does not occur.

Finally, the role of ferro/ferricyanide in militating against ultraviolet-induced damage in $Zn_2TiO_4$ powder and $Zn_2TiO_4$-pigmented silicate (but not in silicone) paints, and in proton-induced damage in $Zn_2TiO_4$-pigmented silicone, is of interest. This seeming disparity is rationalized in the following manner: The iron cyanide treatment described by Morrison et al,[13] and later by Sancier,[16] as providing a high capture cross section for electrons, has not in our studies, as he predicts from his studies of the defect state, provided an effective means of stabilizing ZnO (or $Zn_2TiO_4$) to ultraviolet-induced reflectance degradation in vacuum. (It should be pointed out, however, that Sancier[††] has suggested that we might obtain greater effectiveness at an order of magnitude greater concentration of $Fe^{++}/Fe^{+++}$ in the treatment solution. We point out, however, that this

---

[††]Private communications.

treatment has a deleterious effect on solar reflectance at the concentrations we employ and an increased level of $Fe^{++}/Fe^{+++}$ would seriously impair the reflectance of these coatings.)

The effectiveness of the iron cyanide treatment in stabilizing $Zn_2TiO_4$ to ultraviolet irradiation at high temperature and proton irradiation as a silicone paint is, however, noteworthy, and is attributed to the electron capture cross section of the $Fe^{++}/Fe^{+++}$ couple. Ultraviolet irradiation of the silicate paint at high temperature ($\sim 165°C$) changes the kinetics of silicate damage, and, most probably the character, with the result that ionization (as opposed to excitation) of the silicate occurs, injecting electrons into the surface of the $Zn_2TiO_4$ where the following reaction is enhanced.

$$e + Ti^{+4} \longrightarrow Ti^{+3} \tag{7}$$

Likewise, proton irradiation, which results in the production of an ionization track in the methyl silicone, along which secondary electrons are produced (a cascade phenomenon), also furnishes electrons for the reduction of $Ti^{+4}$. In both cases, the presence of the ferro/ferricyanide provides a barrier to $Ti^{+4}$ reduction by the capture of the electrons that are produced in the binder (silicate in the case of high temperature ultraviolet irradiation and silicone in the case of proton irradiation).

Although it is obvious from this discussion that these studies have posed more questions than have been answered, the efficacy of surface treatment by reactive encapsulants and/or plasma annealing, has been firmly established. It is also obvious that a total-system (pigment plus binder) approach is required. We have shown that these techniques offer a very promising and practical approach to the ultraviolet stabilization of the semiconductor pigment zinc orthotitanate.

Finally, we believe that the potential for stabilizing zinc orthotitanate paints to ionizing charged-particle radiation has at least been suggested and that a different surface treatment, employed in addition to the "hole-capturing" reactive encapsulants, will be required ultimately to do so.

Appendix: Experimental

Pigment Preparation

The basic $Zn_2TiO_4$ pigment, the reactive encapsulation procedures, and the potassium silicate- and silicone-paint preparation procedures are presented in a NASA contractor report.[18]

The plasma heat treatments were performed by E.P. Farley of Stanford Research Institute utilizing SRI's rf-excited, induction plasma facility: These experiments were performed under separate contract to the NASA-Marshall Space Flight Center.[19] The procedure employed with the specimens presented in Tables 4 and 6 consisted of passing an $Ar/O_2$ aerosol of the pigment being treated through the plasma reactor at reactor retention times of 1.1 sec, a mean boundary layer temperature gradient, $\Delta T$, of 1670°C (except for one test at 1450°C and another at 1890°C), and a chamber pressure of 13 Torr (except for one test at 32 Torr). The parameters are discussed fully in a recent SRI report by Farley.[20]

## Space Radiation Testing

Two facilities for space irradiation were employed in the tests described in this communication. IRIF-I is the subject of a communication to the 1967 Thermophysics Specialists Conference[21] and will not be described here. A 1-Kw General Electric A-H6 mercury-argon ultraviolet source was employed in the ultraviolet-simulation tests.

The Combined Radiation Environment Facility (CREF) system was employed in the proton-irradiation experiments and has been described in a NASA contractor report.[22] The performance characteristics of this facility was the subject of a communication to the 17th Annual Meeting of the Institute of Environmental Sciences.[15]

## References

[1] Zerlaut, G.A. and Rubin, G.A., "Development of Space-Stable Thermal-Control Coatings," IIT Research Rept. IITRI-U6002-36 (Triannual Rept.), NASA Marshall Space Flight Center, Huntsville, Ala., Contract NAS8-5379, Feb. 21, 1966.

[2] Zerlaut, G.A., Rogers, F.O., and Noble, G., "The Development of S-13G-Type Thermal Control Coatings," AIAA Progress in Astronautics and Aeronautics: (Thermal Design Principles of Spacecraft and Entry Bodies) edited by J.T. Bevans, Vol. 21, Academic Press, New York 1969, pp. 741-766.

[3] Jaenicke, W., Zeit. für Elektrochemie, Vol. 60, No. 2, 1956, p. 163.

[4] Dulin, F.H. and Rase, D.E., Journal of the American Ceramic Society, Vol. 43, No. 3, 1960, pp. 125-131.

[5] Bartram, S.F. and Slepetys, R.A., *Journal of the American Ceramic Society*, Vol. 44, No. 10, 1961, pp. 493-499.

[6] Loshkarev, B.A., *Steklo i Keram*, Vol. 19, No. 3, 1962, pp. 22-66.

[7] Loshkarev, B.A., *Steklo i Keram*, Vol. 19, No. 10, 1962, pp. 21-24.

[8] Loshkarev, B.A., *Trans. Uralsk Polyt. Inst. Symp.*, 1962, p. 117.

[9] Kubo, T. and Kato, M., et al., Kogyo Kagaku Zasshi, Vol. 66, No. 4, 1963, pp. 403-407.

[10] Ashford, N.A. and Zerlaut, G.A., "Development of Space Stable Thermal-Control Coatings," IIT Research Institute Repts. U6002-77 and 83 (Triannual Repts.) July 11, 1969 and Nov. 17, 1969, respectively.

[11] Gilligan, J.E. and Zerlaut, G.A., "The Role of Gaseous Adsorption in Thermal Coatings Degradation," AIAA Paper 69-1025, Los Angeles, California.

[12] Gilligan, J.E., "The Optical Properties Inducible in Zinc Oxide," *AIAA Progress in Astronautics and Aeronautics*: (Thermophysics of Spacecraft and Planetary Bodies) edited by G. Heller, Vol. 20, Academic Press, New York, 1967, pp. 329-347.

[13] Morrison, S.R. and Sancier, K.M., "Effect of Environment on Thermal Control Coatings," SRI Project PAD-6146 Final Rept., Oct. 1969, Jet Propulsion Lab., Pasadena, Calif.

[14] Zerlaut, G.A. and Ashford, N.A., "Development of Space-Stable Thermal-Control Coatings," IIT Research Institute Rept. IITRI-U6002-85 (Triannual Rept.), Feb. 20, 1970, IIT Research Institute, Chicago, Illinois.

[15] Gilligan, J.E. and Zerlaut, G.A., "The Space Environment Stability Problem in White Pigmented Coatings," 17th Annual Meeting, Institute of Environmental Sciences, April 1971, Los Angeles, Calif.

[16] Sancier, K.M., *Surface Science*, Vol. 21, 1970, p. 1.

[17] Ashford, N.A., Jarke, F.H., Gilligan, J.E. and Zerlaut, G.A., "EPR Investigations of Ultraviolet Irradiated $Zn_2TiO_4$," 162nd National ACS Meeting, Washington, D.C., Sept. 1971.

[18] Zerlaut, G.A., "Development of Space-Stable Thermal-Control Coatings," IIT Research Institute Rept. IITRI-U6002-94 (Triannual Rept.), Nov. 1970, NASA Contract NAS8-5379, IIT Research Institute, Chicago, Illinois.

[19] Bartlett, R.W., "Induction Plasma Calcining of Pigment Particles for Thermal Control Coatings," SRI Rept. 1, PMU-7083, NASA Contract NAS8-21270, Aug. 1968.

[20] Farley, E.P. and Bartlett, R.W., "Induction Plasma Calcining of Pigment Particles for Thermal Control Coatings," SRI Rept. 4, PMU-7083, NASA Contract NAS8-21270, Feb. 5, 1971.

[21] Zerlaut, G.A. and Courtney, W.J., "Space-Simulation Facility for in situ Reflectance Measurements," *AIAA Progress in Astronautics and Aeronautics*: (Thermophysics of Planetary Bodies), Edited by G. Heller, Vol. 20, Academic Press, New York, 1967, pp. 349-368.

[22] Gilligan, J.E. and Zerlaut, G.A., "Development of Space-Stable Thermal-Control Coatings," NASA Contract NAS8-5379, Rept. IITRI-U6002-90 (Triannual Rept.), July 1, 1970.

# COLORATION OF MgO SINGLE CRYSTALS BY LOW ENERGY PROTONS - A FUNCTION OF SELECTED DEFECTS

H. Levin,[*] C. C. Berggren,[+] and W. M. Peffley[+]

Hughes Aircraft Company, Culver City, Calif.

## Abstract

Radiative coloration of MgO single crystal by 1.0 kev protons was studied. Manipulation of defect densities important to solar absorptance was demonstrated by selective impurity control and chemical processing. It was also demonstrated that proper defect regulation can yield radiation stability. Despite restriction of the primary radiation to a depth of approximately 500 Å, i.e., a "surface" dose, changes in bulk defect densities were found predominant in the single crystal. Comparison of proton irradiated MgO powder with single crystal MgO revealed that changes in optical absorption are similar. This result is attributed to similarity in absorption spectra of prominent surface and bulk defect analogs. Bulk defect studies in single crystals thus provide a convenient tool for preliminary characterization and suppression studies of those defects (in candidate dielectric solids) which dominate radiative coloration.

---

Presented as Paper 71-450 at the AIAA 6th Thermophysics Conference, Tullahoma, Tenn., April 26-28, 1971. This research was supported by the NASA Ames Research Center, Contract NAS 2-5034. The authors thank Dr. F. A. Kröger, University of Southern California, Dr. J. L. Kolopus, Oak Ridge National Laboratory, and Drs. A. L. Gentile and O. M. Stafsudd of the Hughes Research Laboratories for their valuable assistance.

[*]Senior Technical Staff Assistant, Components and Materials Laboratory, Aerospace Group.

[+]Members of Technical Staff, Components and Materials Laboratory, Aerospace Group.

## I. Introduction

The continuing search for a white pigment, stable to particulate radiation and whose optical properties qualify its use in a thermal control coating for spacecraft, has led in recent years to the investigation of dielectric or wide band gap solids. These solids typically possess band gap energies $\geq 6.5$ ev. As such they possess inherent resistance to coloration resulting from photoionization by photons in the solar uv spectrum. Exploitation of such inherent radiation stability depends largely on the ability to identify pertinent defect states, i.e., those which absorb solar photons, and to limit their growth during irradiation. These localized defect states, often termed color centers, occupy energy levels within the energy gap between the valence band and the conduction band. They are thus able to trap (or to localize, as in the case of recombination centers such as $Fe^{3+}$ or $Fe^{2+}$) charge carriers generated by ionization events. If the trap containing a carrier (or the recombination center with a localized carrier) demonstrates optical resonance in the solar spectrum, its characterization and the control of its density become important to efforts designed to improve resistance to radiative coloration. Both native and foreign disorder may be involved. The former includes charge carriers, ion vacancies, displaced lattice ions, etc.; the latter includes ionic impurities substituted in the host lattice.

MgO is considered typical of the dielectric solids represented by such groups as the alkaline earth oxides and refractory oxides, at least in the larger sense. Among such dielectrics, MgO has enjoyed the most color center research interest.[1]

A number of bulk defects involving charge carrier trapping at native (intrinsic) defects have been identified as a result of single crystal MgO investigations. These include trapped electron centers, trapped hole centers, and aggregrate centers involving one or the other with other foreign (nonlattice) species. The best defined at present include the $F^+$, $F$, $V_1$, and $V_{OH}$ centers. In terms of the $F^+$ center (one electron trapped at an oxygen ion vacancy), the efforts range from the early work of Wertz,[2,3] and associates to more recent efforts by Henderson and associates.[4,5] F center (two electrons trapped at an oxygen ion vacancy) studies have recently been reported by Kroes[6] and Kappers, Kroes, and Hensley.[7] Trapped hole investigations include $V_1$ center (one hole bound at a magnesium ion vacancy) studies by Chen and Sibley[8] and $V_{OH}$ center (one hole trapped at a magnesium ion vacancy adjacent to an $OH^-$) studies by Kirklin, Auzins and Wertz[9] and by Glass and Searle.[10]

Major foreign (impurity) defect states include the ions of the transition elements, especially $Fe^{3+}/Fe^{2+}$, $Cr^{3+}/Cr^{2+}$, and $Mn^{2+}$. These have been investigated in the bulk by Haxby,[11] Soshea, Deckker and Sturtz,[12] Wertz and associates,[13] Schall,[14] Hansler and Segelken,[15] and Low.[16,17] Studies of the $OH^-$, another important impurity, have been reported by Glass and Searle. Such reported defect identification (modeling) and associated detection techniques are fundamental to an investigation of the radiative coloration of a dielectric solid such as MgO.

Initial anticipation in this work that defect generation by low energy protons, e.g., 1.0 kev, would be restricted to the projected range of the proton (an order of 500 Å) proved unfounded. Based on proton-induced defect changes which were measured, it is proposed that in the bulk the dominant coloration mechanism consists of electron-hole pair diffusion into the bulk (from a surface layer in which they are generated, i.e., the projected range) followed by trapping and recombination in the bulk. This diffusion dependence in the bulk greatly increases the influence of initial defects on the resultant optical damage.

## II. Color Center Generation and Detection

Discoloration effects were introduced by bombarding specimen targets with 1.0 kev protons (neutral beam) delivered at $\sim 3.0 \times 10^{11}$ $p^+/cm^2$ - sec in vacuum at room temperature. In situ (vacuum) measurement of optical and EPR spectra were obtained and analyzed to demonstrate proton-induced defect species and generation rates. These kinetic data were then correlated in terms of individual color centers important to solar energy absorption.

The selection of a charged particle irradiation source of low energy was based primarily on relevance to the solar wind, the continuous radiation environment in interplanetary space. Early measurements by the Mariner II satellite[18] and more recent confirmation by the Vela series of satellites[19,20] define this radiation as a plasma consisting of a proton flux of the order of $2. \times 10^8$ protons/$cm^2$ - sec (at low solar activity) with a mean energy of approximately 1.0 kev and a neutralizing electron component with a mean energy in the range 20 - 40 ev. A small $He^{2+}$ flux is also present.

The use of such a low energy incident particle also restricts primary damage to a "surface" layer. At least in the "surface layer" of a pigment array (whether compacted or in a binder continuum), such a radiation source introduces a degree of equivalence between radiation damage effects in a single crystal and a powder.

## Proton Targets

Target specimens were mostly single crystal wafers (approximately 0.1 cm thick) cleaved from large specimens. These specimens were obtained from two sources. High and moderate purity material was provided through the courtesy of C. T. Butler from the Research Materials Program of the Solid State Division of the Oak Ridge National Laboratory. Low purity material was obtained from the Norton Company (optical quality "Magnorite"). In terms of possible substitutional replacement of lattice ions, the impurities selected in Table I are likely candidates. In this manner they can generate intrinsic defect structure initially and during irradiation quench-in radiation-induced defect structure as trapping sites, individually or in association with intrinsic defects. With the exception of Li and Na (determined by flame photometry and atomic absorption analysis respectively) and N and F (estimated from typical analyses on this material previously made), the remaining elements were obtained from arc emission analyses.

Limited studies were also carried out with a relatively impure powder (Mallinkrodt Chemical, reagent grade power). This power was hydraulically compressed in a polished steel die (50 - 70 kpsi) at room temperature. Easy delamination of the pressed discs permitted random chip selection from within the pressed body. A weighed amount of these irregular chips ($\sim$ 0.4 cm x 0.4 cm x 0.13 cm) constituted the proton target. They were not annealed or given other thermal treatment prior to irradiation.

## Proton Irradiation Facility

This facility consists of an 8" o.d., horizontal vacuum station in which the specimen is mounted on a push-pull, rotary transfer arm. Pumping is provided by a 400 l/s ion pump, with a sorption pump and dry nitrogen gas aspirator providing initial pump-down.

The proton beam is generated by a trapped electron ionization source which was constructed by H. J. King and D. E. Schnelker of Hughes Research Laboratories. This source is described in detail by King and Zuccaro.[21] Mass separation is accomplished by means of a homogeneous magnetic field to yield a $H^+$ beam (> 99% purity) of uniform intensity over a diameter of 2.5 cm at the target specimen. Flux is variable over several orders of magnitude (up to $\sim 4.0 \times 10^{11}$ $p^+/cm^2$ - sec) and proton energy is variable up to 3.0 kev.

Table 1  Concentration of selected impurities in various MgO specimens (parts per million atomic)

| Element | Detection limit, ppma | Compressed powder- very low purity ppma | Single crystal, ppma | | | |
|---|---|---|---|---|---|---|
| | | | Low purity- reduced | High- purity, as received | High purity- reduced | High purity: Fe-doped, oxidized |
| Li | 1.0 | c | 4.9 | 6.0 | 6.0 | 6.0 |
| B  | 19.0 | nd[a] | nd | nd | nd | 18.6 |
| N  | ---- | c | c | $7.2^b$ | $7.2^b$ | $7.2^b$ |
| F  | ---- | c | c | $<1.0^b$ | $<1.0^b$ | $<1.0^b$ |
| Na | 0.1 | 830.0 | 28.7 | 9.3 | 12.2 | nd |
| Al | 3.7 | 12.7 | 16.4 | 11.9 | nd | 10.4 |
| Si | 14.0 | 108.0 | 20.8 | 23.6 | 38.7 | 36.5 |
| P  | ---- | c | c | $26.0^b$ | $26.0^b$ | $26.0^b$ |
| Ti | 4.2 | nd | $1.3^b$ | nd | nd | nd |
| V  | 4.0 | nd | nd | nd | nd | nd |
| Cr | 0.4 | nd | 1.3 | nd | nd | nd |
| Mn | 1.5 | nd | 20.6 | nd | 21.6 | nd |
| Fe | 1.8 | 7.6 | 39.7 | 2.0 | 5.0 | 541.0 |
| Co | 1.7 | nd | nd | nd | nd | nd |
| Ni | 1.4 | nd | 1.6 | nd | nd | nd |
| Cu | 0.2 | 1.2 | 0.3 | 0.3 | 0.4 | 0.4 |

[a] nd - not detected (below detection limit).

[b] Estimate.

[c] Not determined.

   Proton beam neutralization is accomplished by a separate thermal electron source (~1.0 ev electrons) mounted within the pumping station above and near the specimen target. Intermittent measurement of beam fluxes ($p^+$ or $e'$) is performed by a Faraday cup mounted to a transfer arm in the station.

   Most irradiations involve a flux of ~$3.0 \times 10^{11} p^+/cm^2$-sec. The hydrogen leak rate required for this flux establishes a station pressure of the order of $10^{-5}$ torr.

## Detection Gear

Proton-induced optical changes are obtained in situ (vacuum) by means of a modified dual beam, recording spectrophotometer (Cary Model 14). Both optical beams are folded into an optical integrating sphere (smoked MgO) mounted in the pumping station. A transfer arm permits target transfer from the irradiation position to the center of this sphere for optical measurements.

Optical changes were measured at room temperature ($\sim 20°C$) over the spectral range 230 - 1600 nm (0.23 - 1.6 micron), with the 230 - 800 nm region proving to be the most relevant. In the case of the compressed powder spectral measurement of diffuse reflectance was made. Since backscatter of even the longest wavelength of interest is estimated as complete within the specimen thickness ($\sim 0.13$ cm), the measured spectral reflectance $R_\lambda$ is simply related to the spectral absorptance $A_\lambda$ by the expression: $R_\lambda = 1 - A_\lambda$. Differences in these $A_\lambda$ values induced by radiation represent absolute change in absorptance and are so presented later.

Single crystals were mounted on a Suprasil substrate which was aluminized on the second surface. Reflectance measurements off this mirror were made, noting that they involve 2-pass transmission through the crystal bulk. It can be readily shown[22] that the total measured reflectance of a given incident wavelength $R_\lambda$ is closely approximated by the expression

$$R_\lambda \sim r + K_\lambda \exp(-2\alpha_\lambda x) \qquad (1)$$

where
    $r$ = Fresnel reflectance
    $K_\lambda$ = factor containing terms in $r$ and $R_m$
    $\alpha_\lambda$ = spectral absorption coefficient, $cm^{-1}$
    $x$ = crystal thickness, cm
    $R_m$ = spectral mirror reflectance

This expression yields values of $\alpha_\lambda$ and thus permits changes in absorption coefficient $\Delta \alpha_\lambda$ to be obtained. This optical parameter is of use in obtaining defect densities by means of the classical Smakula equation. Alternatively, direct optical characterization in terms of spectral absorptance $A_\lambda$ follows from the expression

$$A_\lambda = 1 - R_\lambda' \qquad (2)$$

where $R_\lambda$ = measured total spectral reflectance normalized to mirror reflectance.

Electron paramagnetic resonance (EPR) spectra are obtained after irradiation. Immediately upon termination of irradiation, the target specimen is dropped in the dark into a high-purity, fused silica recovery tube, connected to the station through a soft copper tubulation. This tubulation is quickly pinched off and the recovery tube immersed in liquid nitrogen. This recovery procedure permits approximate retention of terminal defect structure until the EPR spectrum (at $\sim 77°K$) can be obtained. This spectrum is measured with a modified Varian spectrometer (model V-4500; operated in the X-band near 9.2 GHz) while the recovery tube containing the specimen is mounted in a tuned microwave cavity (Varian V-4535). An appropriate cold nitrogen gas-dewar system permits specimen temperature control.

Individual defect densities discussed later are obtained by resolution of both optical and EPR spectra into individual, identifiable components. Integration of individual defect spectra is then carried out by use of appropriate spectral parameters and spin density standards.

### III. Defects Important to Solar Absorptance

This work is primarily concerned with those defect states, either initial or radiation-generated, which dominate the solar absorptance of the solid. Thus those defects displaying optical resonance in the near-uv, visible, or near-ir spectral regions are emphasized.

Materials processing control represents the ultimate tool for regulating these pertinent defects to yield resistance to proton-induced coloration. Data presented later verify this conclusion in the case of the lattice bulk, i.e., as observed with single crystals. Where surface effects may dominate, e.g., in the case of dielectric powders (consider an array of submicron single crystals), evidence for the existence of surface analogs of the $F^+$ center,[23] $F_c^+$ center,[24] and V centers[25] points to a similar conclusion.

The rationale for studying single crystals initially, rather than the polycrystalline state, as a means toward developing stable white pigments is based on a number of experimental factors. Many defect states are anisotropic, and EPR detection of their presence and subsequent radiation-induced transitions is possible only by proper orientation of the single crystal. Selective crystal orientation can be used as an aid in EPR spectral resolution, important to the determination of individual defect densities. In determining individual defect densities in the bulk from optical absorption spectra, definition of the optical path is necessary. In contrast with a transmitting single crystal, path determination in a backscattering particulate array of significant density is poorly estimated.

In view of these experimental factors favoring single crystal studies for initial evaluation of radiation stability, it is pertinent to examine the relevance to the ultimate pigment problem. Without special provisions, single crystal studies involve bulk defects. Powders, which possess high surface to bulk ratios, should be characterized primarily by surface defects. In spite of this difference, comparison of color center data for powders and crystals reveals the existence of many surface and bulk defects which are close analogs. In the alkaline earth oxides, for example, the S center has been identified[23] as an electron trapped at a surface anion vacancy, analogous to the $F^+$ center in the bulk. Similarly the S' center represents an electron trapped at a surface anion vacancy adjacent to a surface cation vacancy, analogous to the $F_c^+$ center in the bulk. Similar analogy has been found for the case of trapped hole defects, Lunsford[25] reporting a $V_1$-type center in the surface with properties (EPR, bleaching, etc.) very similar to those of the $V_1$ center in the bulk. Chemisorption of gas molecules at surface defects can lead to unique paramagnetic species in the surface. For example $O_2$ adsorption at an S center yields the $O_2^-$ species in the surface with discharge of the EPR line of the S center.

Powders have certain unique features, such as the asymmetry of surface forces, and a large surface to bulk ratio which is conducive to gas adsorption. The effect of these differences between powder and single crystal state does not seem to introduce major differences in proton-induced spectral changes. Typical results for low-purity powder (Fig. 1) and crystal (Fig. 2) may be compared. Resolution into component spectra is based on an assumed Gaussian distribution of each component and a least squares fit to the measured data. The close similarity in individual defect spectra between powder and crystal would seem to bear out reported EPR results indicating close similarity of defect analogs in the bulk and in the surface. It seems reasonable to conclude that radiation effects studies in single crystals measured in situ present no serious simulation problems while offering real experimental advantages.

Since 99% of the solar power flux occurs at a wavelength $\lambda > 0.3\mu$, the important absorbing color centers are seen to be those located at 4.3, 3.6, 3.1, 2.7, and 2.3 ev. Little

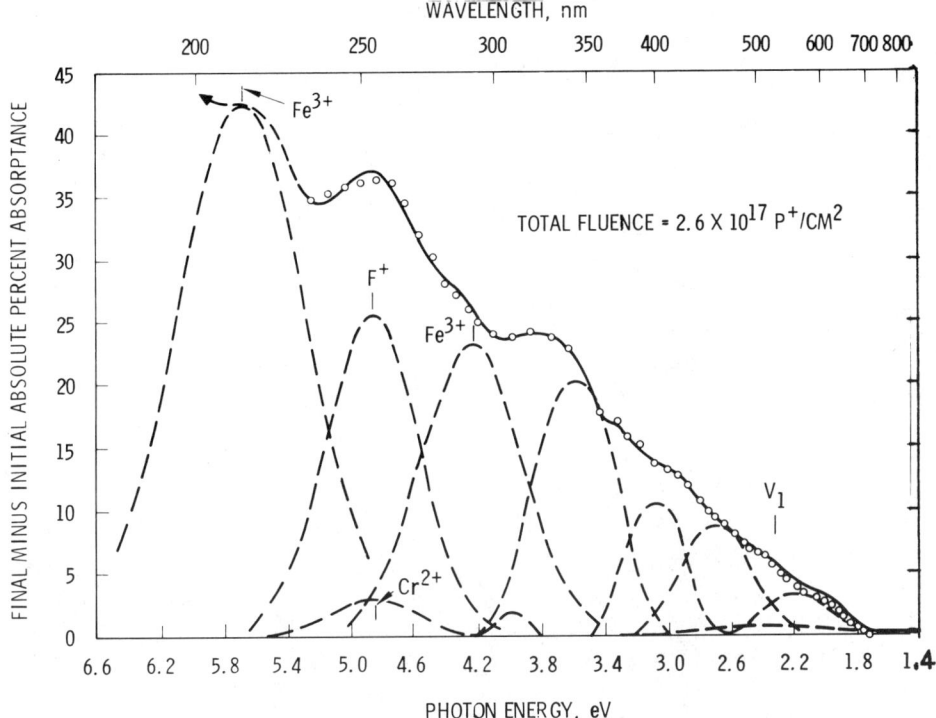

Fig. 1 Proton-induced absorption change in low-purity MgO powder (compressed).

optically resonant structure of pertinence here is observed at $h\nu < \sim 1.6$ ev ($\lambda > \sim 0.78\mu$). The 4.3 ev band represents $Fe^{3+}$ excitation and is important because strong tailing into the visible region occurs at high $Fe^{3+}$ densities. It is not photo-ionized at room temperature. The 2.3 ev band represents absorption into a center (termed $V_1$) containing a photoionizable hole trapped at a magnesium ion vacancy in the lattice, $V''_{Mg}$. The three other centers (displayed by the 3.6, 3.1, and 2.7 ev bands) have not yet been modeled. The 3.6 ev band is believed associated with an electron trapped at an intrinsic lattice defect. Three different types of identifiable defects are thus observed to exert important influence on proton-induced coloration.

## $Fe^{3+}$ Band (4.3 ev)

This band displays the presence of a major impurity defect. Its proton-induced growth reflects net $h\cdot$-trapping at $Fe^{2+}$, a trapping and recombination center, as follows:

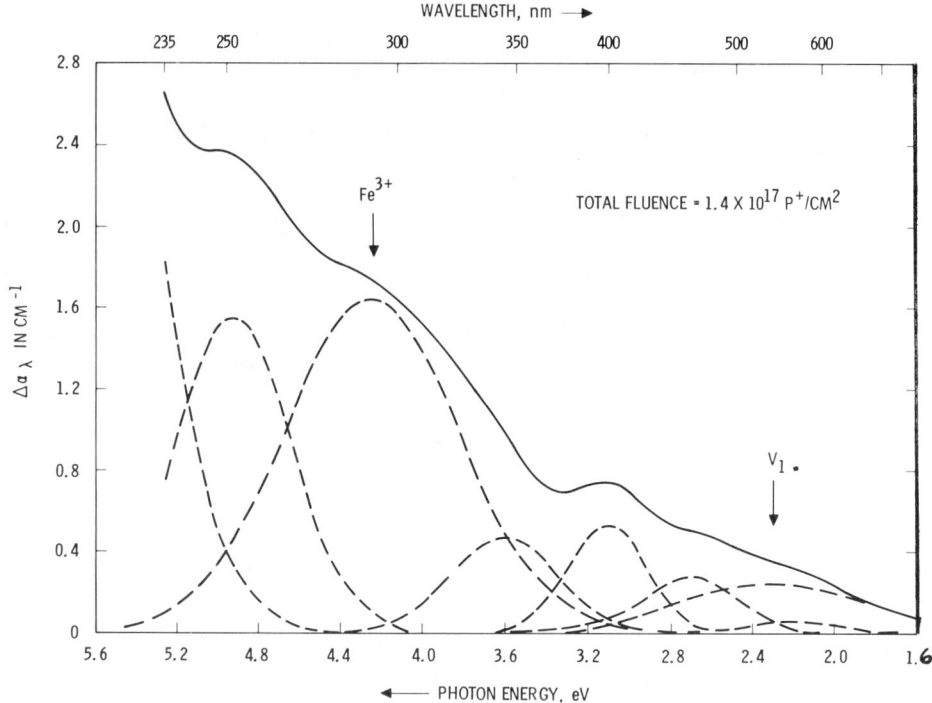

Fig. 2 Proton-induced change in spectral absorption coefficient in low-purity MgO single crystal (reduced).

$$\left(Fe^{2+}\right)^x + h^{\cdot} \longrightarrow \left(Fe^{3+}\right)^{\cdot} \qquad (3)$$

where the symbol x designates a neutral substituent in the lattice (after Kröger).[26]

Upon heating in a reducing or inert atmosphere to 1200 - 1500°C, it is well established[13] that some loss of lattice oxygen occurs from the surface. A first effect is that the surface contains excess electrons. The extra electrons can be localized by $Fe^{3+}$, thus reducing its density.

For every two $Fe^{3+}$ thus reduced, charge compensation requires that one magnesium ion vacancy, $V''_{Mg}$, be lost, either by migration to the surface or to a dislocation. The net effect is a reduction in initial densities of both $Fe^{3+}$ and $V''_{Mg}$ in the bulk. These high temperature induced electron-transfer processes may be depicted by the following equations, where

$$\left(O^{2-}\right)^x \xrightarrow{\Delta} \tfrac{1}{2} O_2\uparrow + 2e' + \left(V_O^{\cdot\cdot}\right)_s \qquad (4)$$

$$2\left(Fe^{3+}\right)^{\bullet} + 2e' + V_{Mg}'' \rightarrow 2\left(Fe^{2+}\right)^x + \left(V_{Mg}''\right)_s \qquad (5)$$

$$\left(O^{2-}\right)^x + 2\left(Fe^{3+}\right)^{\bullet} + V_{Mg}'' \rightarrow 2\left(Fe^{2+}\right)^x \qquad (6)$$

The effect of such chemical reduction of a high-purity MgO single crystal at 1450 - 1500°F for 4 hr in dry nitrogen gas is given in Table 2. Subsequent proton irradiation effects on the $Fe^{3+}/Fe^{2+}$ ratio are also given. The $Fe^{3+}$ densities were obtained from room temperature optical absorption spectra and confirmed by EPR spectra measured near 77°K. Since $Fe^{1+}$ is only stable at $\leq 20°K$, the $Fe^{2+}$ density was obtained by subracting the $Fe^{3+}$ density from the total iron content as determined by arc emission spectrographic analysis.

Note that the total iron content ($Fe^{2+} + Fe^{3+}$) varies by a factor of two even though both specimens were cleaved from the same ingot. Such variation is not uncommon and does not impair the relevance of the data. The decrease in initial $Fe^{3+}$ from $2.0 \times 10^{17}$ to $0.4 \times 10^{17}$ ions/cm$^3$ is as anticipated, with the $Fe^{3+}/Fe^{2+}$ partition favoring the $Fe^{2+}$. What isn't anticipated is a decrease in growth of the $Fe^{3+}$ density by a factor of three at a doubled proton fluence. Comparison of the changes in corresponding spectral absorption coefficients induced by irradiation (Figs. 3 and 4) reveals a parallel, net decrease in $h^{\bullet}$-trapping (see 2.3 ev band) and $e'$-trapping (see 3.6 ev band). It would appear that such chemical reduction (in conjunction with controlled cooling) effects sufficient decrease in densities of both $h^{\bullet}$- and $e'$-traps so as to result in an increased probability for $e'$-$h^{\bullet}$ recombination. The net result is a slower growth of the $Fe^{3+}$ band during proton irradiation.

An alternate technique of enhancing the recombination rate of proton-generated charge carriers is by introducing a dominant

Table 2  Effect of reduction on $p^+$ - generated growth of $Fe^{3+}$ density

| Specimen | Fluence, $10^{17}$ $p^+/cm^2$ | Density, $10^{17}$ ions/cm$^3$ | |
|---|---|---|---|
| | | $Fe^{2+}$ | $Fe^{3+}$ |
| High purity (as received) | 0 | 0.7 | 2.0 |
| | 1.1 | 0.1 | 2.6 |
| High purity (reduced) | 0 | 5.4 | 0.4 |
| | 2.3 | 5.2 | 0.6 |

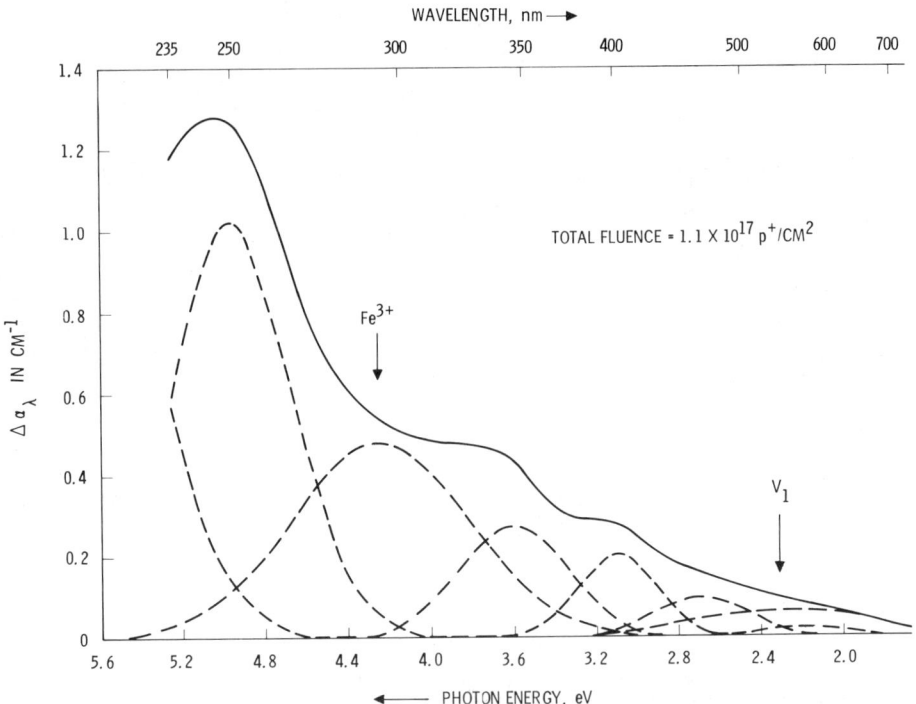

Fig. 3 Proton-induced change in spectral absorption coefficient in high-purity MgO single crystal (as-received).

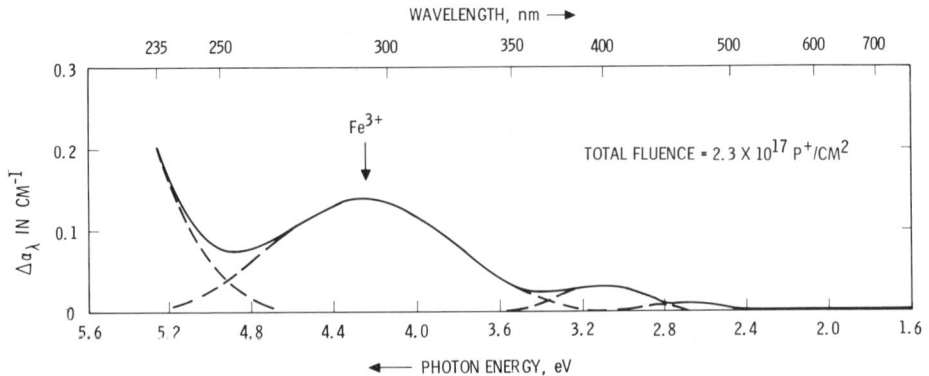

Fig. 4 Proton-induced change in spectral absorption coefficient in high-purity MgO single crystal (reduced).

density of recombination centers. Thus when the high-purity (as-received) crystal shown in Table 2 was doped with iron to raise the density by about two orders of magnitude (i.e., $Fe^{2+}$ from $0.7 \times 10^{17}$ to $4.3 \times 10^{19}$ ions/cm3; $Fe^{3+}$ from $2.0 \times 10^{17}$ to $1.5 \times 10^{19}$ ions/cm3), little optical change was noted during irradiation to a total fluence of $1.8 \times 10^{17}$ $p^+/cm^2$. Iron-doping has no practical relevance to thermal control because of the unacceptable $Fe^{3+}$ optical absorption at densities in which recombination at $Fe^{3+}/Fe^{2+}$ is dominant. However selective doping with valence pairs such as $Ce^{4+}/Ce^{3+}$ or interacting bound native defect states such as $F^+/F$, which are spectrally located in the uv further removed from the visible region, does have practical relevance.

In addition to e' - h˙ pair generation in the stopping region (projected range), the lattice ion displacement probability is an order of one ion/proton.[22] Such ion vacancy generation is in addition to the initial ion vacancy densities created by processing or impurities, e.g., charge compensation requiring one $V''_{Mg}$ for each two substitutional $Fe^{3+}$. Thus this region approaches an amorphous condition after proton fluences of the order of $10^{17}$ $p^+/cm^2$. One result of this condition would be that proton channeling would be eliminated and the projected range would fall by about an order of magnitude. The effect of amorphous condition on e' - h˙ pair generation, recombination, and diffusion was not studied. In this regard comparison of color centers generated by irradiation (ionizing- and displacement type) in silica glass and crystalline quartz is instructive. Such comparison[27] reveals considerable similarity. A tentative conclusion is that spectral changes should not be markedly different; however defect growth rates may become slower.

## $V_1$ Center

This center consists of a h˙ trapped at a $V''_{Mg}$ or more accurately a h˙ localized on one of the oxygen ions nearest neighbor to a $V''_{Mg}$. Any materials processing technique which a) minimizes the density of $V''_{Mg}$ or b) increases competition for the capture of holes will reduce proton-induced growth of this center. In terms of the former, minimum densities of soluble trivalent ions in the lattice (e.g., $Fe^{3+}$, $Cr^{3+}$, etc.) and/or suppression of trivalency by chemical reduction is effective. In terms of the latter, competitive capture at recombination centers has been discussed earlier. Capture at other hole traps is also possible; and where the bound state displays optical absorption outside the solar spectral region, improved resistance to coloration results.

In the case of $V_1$ center generation (as was true also for p⁺-induced changes in iron state densities) most of the p⁺-induced effect seemed to be accounted for in the bulk. As a check, deliberate ionization of the bulk was performed by an ~ 1 Mev gamma ray source. Selected data on two differently processed specimens, γ-irradiated to $V_1$ center saturation, are presented in Table 3.

Table 3  Gamma-irradiation of bulk crystal

| Specimen | Condition | Density, $10^{17}/cm^3$ | | | |
|---|---|---|---|---|---|
| | | $(Fe^{3+})$ | $(Fe^{2+})$ | $(Fe)_{tot}$ | $(V_1)_{Sat}$ |
| A(low iron, oxidized) | Before | 4.6 | 4.1 | 8.7 | 0.79 |
| | After | 7.7 | 1.0 | | |
| B(high iron, reduced) | Before | 0.2 | 42.0 | 42 | 1 |
| | After | 30.0 | 12.0 | | |

On the basis of such bulk ionizing irradiation, the low initial $(Fe^{3+})$ in specimen B would indicate a reduced initial $(V_{Mg}'')$ in the bulk and consequently reduced ability to form $V_1$ centers. In contrast, $V_1$ center formation in specimen A should be enhanced because of its higher initial $(Fe^{3+})$. Despite variation in total iron content by a factor of 5 between the two specimens, it does not seem unreasonable therefore to anticipate similar $V_1$ center growth and saturation. This similarity in $V_1$ center growth was in fact observed as a result of p⁺-irradiation (Fig. 5).

3.6 ev Center

Henderson and Wertz[1] consider the 3.6 ev band (as well as a 1.2 ev band not shown) "as due to intrinsic lattice defects" which Hansler and Segelken[15] associate with a trapped electron. Peria[28] observed no photoionizability at room temperature, indicative of a relatively deep excited state. Comparison of Figs. 3 and 4 reveals that high temperature preprocessing in inert gas yields a lower proton-generated growth rate for this center. Such preprocessing was terminated by an intermediate cool-down rate (~25°C/min.). This is believed slow enough to minimize the e'-trap densitites originated by dislocations and yet fast enough to prevent aggregation of other uv-resonant

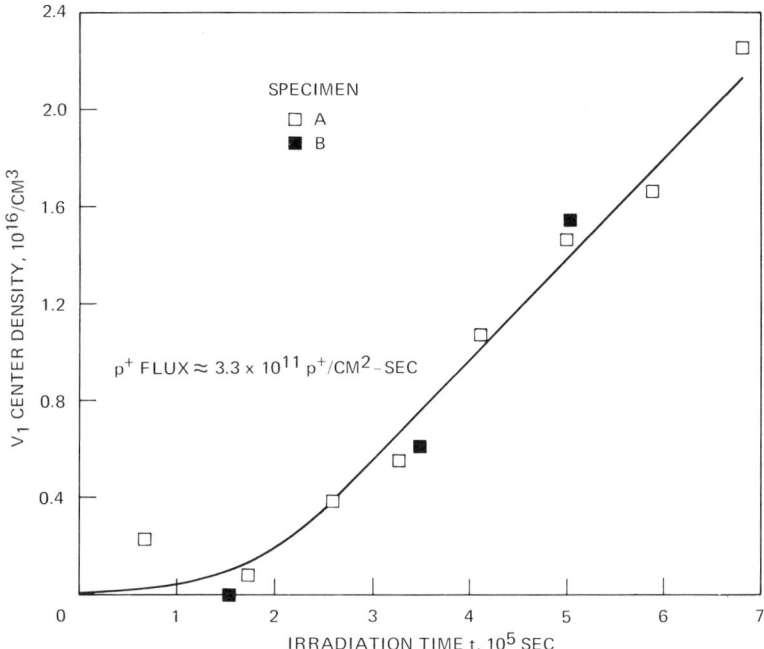

Fig. 5   Proton-induced growth of $V_1$ center in single crystal.

centers into defect aggregates exhibiting visible and near-ir optical absorption.

Thermal Design

The spacecraft thermal designer is more directly concerned with how such proton-generated defect growth affects spectral absorptance in the solar region. Such practical influence of defects is displayed in Fig. 6 (low purity -- see column 4 in Table 1) and Fig. 7 (high purity -- see column 6 in Table 1). Comparison reveals one major variation, namely the total iron content. Both crystal specimens were processed simultaneously, after which $(\text{total iron})_{\text{Fig. 6}} \approx 8 \,(\text{total iron})_{\text{Fig. 7}}$. Processing was designed to minimize the initial density of the 3+ state of iron and anneal some of the intrinsic lattice defect structure. Both uniradiated specimens exhibit similar $\alpha_S$ values of $\sim 0.03$. However, after comparable radiation (equivalent to $>10$ yr exposure to the solar wind), the $\alpha_S$ (low purity) has tripled while the $\alpha_S$ (high purity) is largely unchanged.

References to $\Delta\alpha_S$ values for single crystal dielectrics are believed indicative of irradiation behavior in powder state, provided defect structure introduced by grinding or other size

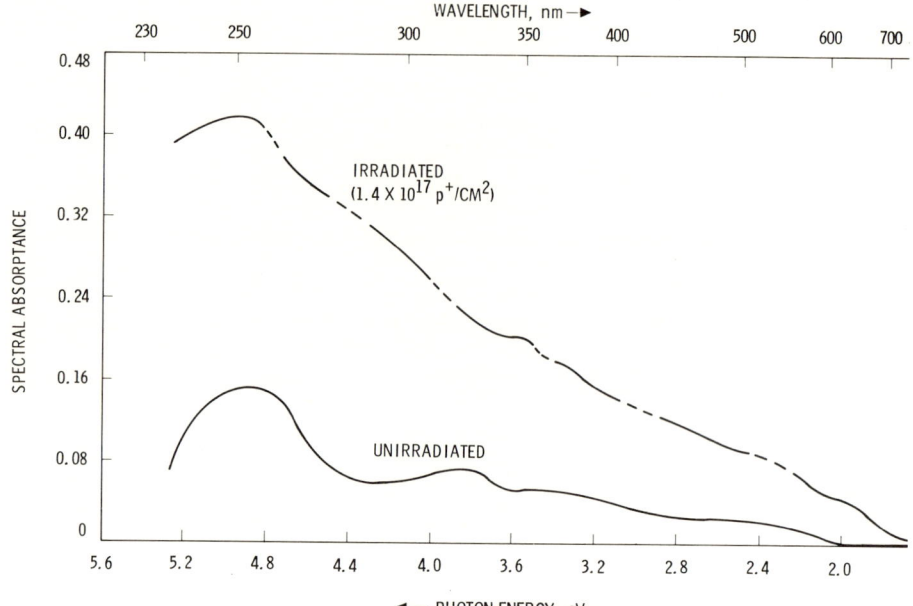

Fig. 6 Spectral absorptance of low-purity MgO single crystal (reduced) before and after proton irradiation.

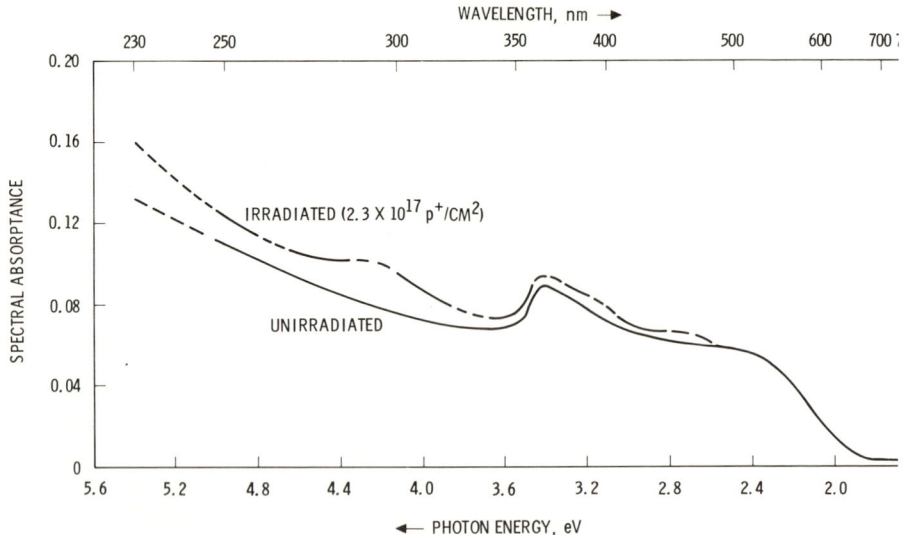

Fig. 7 Spectral absorptance of high-purity MgO single crystal (reduced) before and after proton irradiation.

reduction is removed by appropriate annealing. Such similar behavior would not be anticipated for low band gap energy materials, e.g., ZnO and $TiO_2$, which typically are subject to photodesorption of lattice oxygen. Large increases in the surface to bulk ratio, as in going from the crystal to the powder, in such cases would permit photodesorption at the surface to dominate the radiative process. Thus, while dominant processes in the surface and bulk are believed similar for the dielectric solid, this is not true for a semiconductor solid.

IV. Proposed Proton Coloration Mechanism

Previously cited work of others has involved radiation damage introduced uniformly in the bulk. This results from the use of particles or photons (electrons, neutrons, alpha-particles, γ-rays, or X-rays) sufficiently energetic to penetrate the crystal thickness. In contrast, low-energy protons (1 - 3 kev, such as associated with the solar wind) are stopped in a surface layer. Collison theory[22] when modified by an approximation for chanelling predicts a projected range of the order of 500 Å. However, consistent evidence[29] demonstrates both h· - and e'-trapping by defects deep in the bulk (several orders of magnitude beyond the proton-range). For example, crystal specimen A (see Fig. 5) was $p^+$-irradiated to a fluence of $2.27 \times 10^{17}$ p+/$cm^2$ ($6.8 \times 10^5$ sec). Bulk measurements (no distinction between surface and bulk) showed that this resulted in a decrease in $[Fe^{2+}]$ of $\sim 1.8 \times 10^{17}/cm^3$, i.e., from $4.1 \times 10^{17}/cm^3$ to $2.3 \times 10^{17}/cm^3$, averaged over a bulk thickness $\approx 0.1$ cm. On the basis of an incident surface of 1.0 $cm^2$, the bulk crystal volume $\approx 0.1$ $cm^3$, a volume containing $4.1 \times 10^{16}$ $Fe^{2+}$ initially. These ions are initially uniformly distributed in the bulk. With the $p^+$-range in the crystal estimated as $\sim 5 \times 10^{-6}$ cm, the volume associated with this range $\sim 5 \times 10^{-6}$ $cm^3$ and can account for only $2.1 \times 10^{12}$ $Fe^{2+}$ initially. Even if all of these $Fe^{2+}$ in the "surface" layer changed valence state during $p^+$-bombardment, they could not account for the measured decrease in $[Fe^{2+}]$.

It appears that net radiative coloration is dominated by defect generation in the bulk. Thus despite trapping at defects (e.g., ion vacancies) generated directly by the protons in a surface layer, the over-all rate of coloration appears regulated by the rate of diffusion of $p^+$-generated e' - h· pairs into the bulk from a steady state density in the surface. Based on irradiation conditions employed in this work, this constant surface density is estimated to be $2 \times 10^{13}$ pairs/$cm^3$. This diffusion-dependent process is considered to include trapping (which yields various color centers) and recombination in the bulk.

## References

[1] Henderson, B., and Wertz, J. E., "Defects in the Alkaline Earth Oxides," *Advances in Physics*, Vol. 17, No. 70, Nov. 1968, pp. 749-855.

[2] Wertz, J. E., Auzins, P., Weeks, R. A., and Silsbee, R. H., "Electron Spin Resonance of F Centers in Magnesium Oxide; Confirmation of the Spin of Magnesium-25," *The Physical Review*, Vol. 107, No. 6, Sept. 1957, pp. 1535-1537.

[3] Wertz, J. E., Orton, J. W., and Auzins, P., "Electron Spin Resonance Studies of Radiation Effects in Inorganic Solids," *Discussions of the Faraday Society*, Vol. 31, April 1961, pp. 140-150.

[4] Henderson, B., and King, R. D., "Dose Dependence of F Center Production by Fast Neutrons in Magnesium Oxide," *The Philosophical Magazine*, Vol. 13, No. 126, June 1966, pp. 1149-1156.

[5] Henderson, B., King, R. D., and Stoneham, R. D., "The Temperature Dependence of the F Band in Magnesium Oxide," *Journal of Physics C*, Series 2, Vol. 1, No. 3, June 1968, pp. 586-593.

[6] Kroes, R. L., "Color Centers in Magnesium Oxide," Ph.D. thesis, 1968, Univ. of Missouri, pp. 1-69.

[7] Kappers, L. A. Kroes, R. L., and Hensley, E. B., "$F^+$ and $F'$ Centers in Magnesium Oxide," *Physical Review B*, Vol. 1, No. 10, May 1970, pp. 4151-4157.

[8] Chen, Y., and Sibley, W. A., "Study of Ionization-Induced Radiation Damage in MgO," *The Physical Review*, Vol. 154, No. 3, Feb. 1967, pp. 842-850.

[9] Kirklin, P. W., Auzins, P., and Wertz, J. E., "A Hydrogen$^-$ Containing Trapped Hole Center in Magnesium Oxide," *The Journal of Physics and Chemistry of Solids*, Vol. 26, No. 6, June 1965, pp. 1067 - 1074.

[10] Glass, A. M., and Searle, T. M., "Reactions Between Vacancies and Impurities in Magnesium Oxide. II $Mn^{4+}$- Ion and $OH^-$ - Ion Impurities," *The Journal of Chemical Physics*, Vol. 46, No. 6, March 1967, pp. 2092-2101.

[11] Haxby, B. V., "A Study of Color Centers in Magnesium Oxide," Ph.D. thesis, 1957, Univ. of Minnesota, pp. 1-70.

[12] Soshea, R. W., Dekker, A. J., and Sturtz, J. P., "X-Ray-Induced Color Centers in MgO," The Journal of Physics and Chemistry of Solids, Vol. 5, No.'s 1/2, 1958, pp. 23-33.

[13] Wertz, J. E., Orton, J. W., and Auzins, P., "Spin Resonance of Point Defects in Magnesium Oxide," Journal of Applied Physics, Vol. 33 supplement, No. 1, Jan. 1962, pp. 322-328.

[14] Schall, P., "Effect of Atomic Displacements on Color Center Production in MgO," The Journal of Physics and Chemistry of Solids, Vol. 28, No. 7, July 1967, pp. 1211 - 1223.

[15] Hansler, R. L., and Segelken, W. G., "Correlation of Thermoluminescence in MgO and Valence Changes of Iron and Chromium Impurities Detected by EPR," The Journal of Physics and Chemistry of Solids, Vol. 13, No.'s 1/2, 1960, pp. 124-131.

[16] Low, W., "The Paramagnetic Resonance Spectrum of $Fe^{3+}$ in the Cubic Field of MgO," Proceedings of the Physical Society, Vol. B69, No. 11, Nov. 1960, pp. 1169-1170.

[17] Low, W., and Weger, M., "Paramagnetic Resonance and Optical Spectra of Divalent Iron in Cubic Fields. II Experimental Results," The Physical Review, Vol. 118, No. 5, June 1960, pp. 1130-1136.

[18] Neugebauer, M., and Snyder, C. W., "Mariner 2 Observations of the Solar Wind," Journal of Geophysical Research, Vol. 71, No. 19, Oct. 1966, pp. 4469-4484.

[19] Hundhausen, A. J., Asbridge, J. R., Bame, S. J., and Strong, I. B., "Vela Satellite Observations of Solar Wind Ions," Journal of Geophysical Research, Vol. 72, No. 7, April 1967, pp. 1979-1987.

[20] Montgomery, M. D., Bame, S. J., and Hundhausen, A. J., "Solar Wind Electrons: Vela 4 Measurements," Journal of Geophysical Research, Vol. 73, No. 15, Aug. 1968, pp. 4999 - 5003.

[21] King, H. J., and Zuccaro, D. E., "Solar Wind Simulation Techniques," CR-73443, April 1970, NASA Ames Research Center.

[22] Levin, H., Honnold, V.R., and Berggren, C. C., "Study of Color Center Formation in White Powder Compounds," CR-73337, July 1969, NASA Ames Research Center.

[23] Nelson, R. L., Tench, A. J., and Harmsworth, B. J., "Chemisorption on Some Alkaline Earth Oxides," *Transactions of the Faraday Society*, Vol. 63, No. 6, June 1967, pp. 1427-1446.

[24] Lunsford, J. H., and Jayne, J. P., "Formation of $CO_2^-$ Radical Ions when $CO_2$ Is Adsorbed on Irradiated Magnesium Oxide," *The Journal of Physical Chemistry*, Vol. 69, No. 7, July 1965, pp. 2182 - 2184.

[25] Lunsford, J. H., "A Study of Irradiation-Induced Active Sites on Magnesium Oxide Using Electron Paramagnetic Resonance," *The Journal of Physical Chemistry*, Vol. 68, No. 8, Aug. 1964, pp. 2312-2316.

[26] Kröger, F. A., *The Chemistry of Imperfect Crystals*, Wiley and Sons, New York, 1964, Chap. 7, pp. 192-214.

[27] Schulman, J. H., and Compton, W. Dale, *Color Centers in Solids*, The Macmillan Company, New York, 1962, Chap. 10, pp. 295-299.

[28] Peria, W. T., "Optical Absorption and Photoconductivity in Magnesium Oxide Crystals," *The Physical Review*, Vol. 112, No. 2, Oct. 1958, pp. 423-433.

[29] Levin, H., Berggren, C. C., and Peffley, W. L., "Study of Color Center Formation in White Powder Compounds," Contract NAS 2-5034, Dec. 1970, NASA Ames Research Center.

# REFLECTANCE MEASUREMENTS OF DIELECTRIC COATINGS ON A CONDUCTOR SUBSTRATE

Ghassane M. Chaabane,[*] Chesley Pieroway,[†] and John Francis[‡]

University of Oklahoma, Norman, Okla.

## Abstract

The prime objective of this work was to measure the directional reflectance of a dielectric coating on a metallic substrate and compare the results with an analytical model. The samples studied were made from commercially available microscope and cover glass slides. The slides were vapor-deposited with gold to form the metallic substrate. For studies of a diffuse substrate with dielectric coating, the glass slides were roughened prior to vapor-depositing the gold. The directional reflectance of the samples was measured using a goniometer to position the sample. Comparisons of the measured experimental reflectance and the values predicted by the analytical model are presented.

## Nomenclature

| | | |
|---|---|---|
| $n_d$ | = | refractive index of the dielectric coating |
| $n_m$ | = | refractive index of the metallic substrate |
| $n_a$ | = | refractive index of air |
| $\tau$ | = | optical distance |
| $\tau_o$ | = | optical thickness |
| $\theta$ | = | polar angle |

---

Presented as Paper 71-448 at the AIAA 6th Thermophysics Conference, Tullahoma, Tennessee, April 26-28, 1971. This work was supported by NSF Grant GK 3649.

[*]Graduate Student.
[†]Graduate Student; presently Captain, U. S. Air Force.
[‡]Associate Professor of Aerospace, Mechanical, and Nuclear Engineering.

| | | |
|---|---|---|
| $\mu$ | = | cosine of the polar angle |
| $\mu_L$ | = | cosine of leaving polar angle |
| $\mu_i$ | = | cosine of incident polar angle |
| $I_\nu(\mu)$ | = | monochromatic directional intensity |
| $\rho_s(\mu)$ | = | Fresnel's reflectance for the conductor substrate dielectric interface |
| $\rho_d(\mu)$ | = | Fresnel's reflectance for the dielectric-air interface |
| $I_o(\mu_i')$ | = | incident intensity |
| $\Delta\mu_i', \Delta\Phi_i'$ | = | incident solid angle |
| $\rho_D$ | = | diffuse reflectance of the roughened metallic substrate |
| $\rho_{ss}(\mu)$ | = | specular reflectance of the roughened metallic substrate |
| $R_s(\mu_i', \mu_i')$ | = | $R_s(\mu_L')$ |
| $R_s(\mu_L')$ | = | intensity ratio in the specular direction |
| $R_T(\mu_L', \mu_i)$ | = | intensity ratio in the directions $(\mu_L', \mu_i')$ |
| $\sigma$ | = | rms surface roughness |
| $\lambda$ | = | wavelength |
| $a'$ | = | correlation length |

Superscripts

| | | |
|---|---|---|
| + | = | positive direction |
| - | = | negative direction |
| ' | = | outside the coating |

## Introduction

The use of coatings as heat shields as well as heat absorbers has caused a great deal of activity in the development of analytical models for the determination of the directional reflectance of coatings. To date, little work has been done to evaluate these models. In this paper, we present experimental measurements of the directional reflectance of coatings on a metallic substrate and compare these results with developed theories. In order to measure the directional reflectance, a goniometer was constructed by modifying a surveying transit head as shown in Fig. 1. The goniometer was constructed to allow mapping as much of the hemisphere above the sample as possible while holding the incident direction fixed. This was accomplished by mounting the globar and incident optics on an arm that extended from the axis of rotation of the sample. The incident optics rotated with the sample, keeping the incident direction fixed as the leaving direction was changed. Front surface spherical mirrors were used for focusing the radiation from a globar onto the sample and for collecting the reflected radiation. The goniometer was mounted on a specially designed stand to allow as much freedom of motion as possible. The collecting optics were mounted on an optical bench located on a separate concrete table. All angles could be measured accurately to within 30 min. A schematic diagram of the external optics is shown in Fig. 2. Measurements were made with a Perkin-Elmer model 112 monochromator. All samples studied in this paper were axially symmetric so that only variations in the plane of incidence were recorded.

Fig. 1  Experimental apparatus.

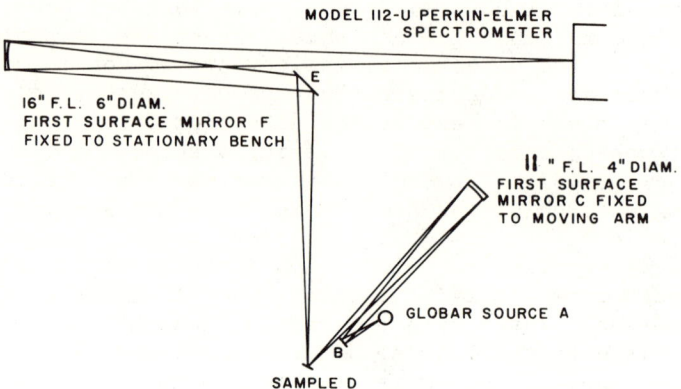

Fig. 2  External optics schematic diagram.

## Analysis

The analytical model employed assumes monodirectional radiation incident on a smooth dielectric coating. The coating and substrate are assumed to be infinite in extent with parallel surfaces. The problem is limited to the isothermal case, with the coating thickness much greater than one wavelength. The coating is considered to be a homogeneous absorbing medium, and emission is neglected. Figure 3 is a diagram of the conductor-coating combination model employed. The transport equation for the axially symmetric case is

$$\mu \frac{dI_\nu(\tau,\mu)}{d\tau} = - I_\nu(\tau,\mu) \tag{1}$$

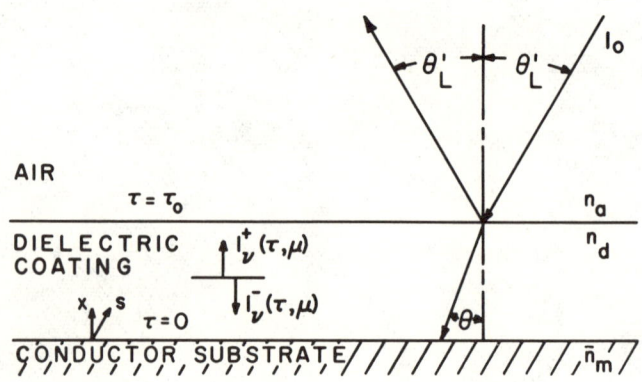

Fig. 3  Coating-conductor model.

Writing the equation in terms of the intensities directed along the positive and negative directions separately, Eq. (1) becomes

$$\mu(dI_\nu^+/d\tau) = - I_\nu^+(\tau,\mu) \tag{2}$$

and

$$-\mu(dI_\nu^-/d\tau) = - I_\nu^-(\tau,\mu) \tag{3}$$

with the solutions

$$I_\nu^+(\tau,\mu) = c_1 e^{-\tau/\mu} \tag{4}$$

and

$$I_\nu^-(\tau,\mu) = c_2 e^{(\tau-\tau_o)/\mu} \tag{5}$$

Two separate sets of boundary conditions will be utilized to model the specimen studies experimentally. For perfectly smooth interfaces, the reflections are specular and follow Fresnel's equations. For this case, the boundary conditions are expressed as

$$\text{at } \tau = 0 \qquad I_\nu^+(0,\mu) = \rho_s(\mu) I_\nu^-(0,\mu) \tag{6}$$

and

$$\text{at } \tau = \tau_o \qquad I_\nu^-(\tau_o,\mu) = \rho_d(\mu) I_\nu^+(\tau_o,\mu)$$

$$+ [1 - \rho_d(\mu_i')] n_d^2 I_o(\mu_i') \tag{7}$$

where $\rho_s(\mu)$ is Fresnel's reflectance for the conductor substrate and $\rho_d(\mu)$ is Fresnel's reflectance for the dielectric coating-air interface.

The ratio of the intensity leaving the direction $\mu_L'$ to the intensity incident on the surface is denoted as $R_s(\mu_L')$ and is given by

$$R_s(\theta_L') = R_s(\mu_L') = I_\nu(\mu_L')/I_o(\mu_i') = \rho_d(\mu_i')$$

$$+ \frac{\rho_s(\mu_i') e^{-2\tau_o/\mu_i}[1 - \rho_d(\mu_i)][1 - \rho_d(\mu_i')]}{[1 - \rho_s(\mu_i)\rho_d(\mu_i) e^{-2\tau_o/\mu_i}]} \tag{8}$$

Equation (8) represents the reflectance of a dielectric coating on a smooth conductor substrate. It is important to note that the results are only for the specular substrate case and only in the specular direction.

The second set of boundary conditions utilized was for a roughened substrate with a smooth dielectric coating. The approach used is the same as given by Love and Francis[1], that is, the portion of the incident radiant flux refracted into the coating is attenuated by the coating and accounted for in the boundary condition as it is reflected from the roughened substrate.

The boundary conditions thus become

$$\text{at } \tau = \tau_o \qquad I_\nu^-(\tau_o,\mu) = \rho_d(\mu) I_\nu^+(\tau,\mu) \qquad (9)$$

and in the specular direction

$$\text{at } \tau = 0 \qquad I_\nu^+(0,\mu) = \rho_{ss}(\mu_i) I_\nu^-(0,\mu) +$$

$$2 \int_o^1 \rho_D I_\nu^-(0,\mu) \mu\, d\mu +$$

$$\frac{\rho_D}{\pi} \mu_i' \Delta\mu_i' \Delta\Phi_i' I_o(\mu_i') [1 - \rho_d(\mu_i')] e^{-\tau_o/\mu_i} +$$

$$n_d^2 \rho_{ss}(\mu_i) I_o(\mu_i')[1 - \rho_d(\mu_i')] e^{-\tau_o/\mu_i} \qquad (10)$$

where $\rho_D$ is the diffuse reflectance of the metallic substrate, and $\rho_{ss}$ is the specular reflectance of the metallic substrate.

Substitution of Eqs. (2) and (3) into Eqs. (9) and (10) permits determination of $c_1$ and $c_2$.

If we express $R(\mu_L',\mu_i')$ as the ratio of the intensity of radiation leaving in the $\mu_L'$ direction to the intensity incident in the $\mu_i'$ direction, the solution in the specular direction becomes

$$R_s(\mu_L',\mu_i) = \frac{I_L(\mu_L')}{I_o(\mu_i')} = \rho_d(\mu_i') + \frac{1}{(n_d)^2} [1 - \rho_d(\mu_i')] e^{-\tau_o/\mu_i} c_1^* \qquad (11)$$

with $\mu_i = \mu_L$, and $\overset{*}{c}_1$ is given by

$$\overset{*}{c}_1 = \{\rho_{ss}(\mu_L')n_d^2[1 - \rho_d(\mu_L')]$$

$$+ \rho_D \mu_i \Delta\mu_i \Delta\Phi_i[1 - \rho_d(\mu)]e^{-\tau_o/\mu_i}\}/$$

$$\{[1 - \rho_{ss}(\mu_i) \rho_d(\mu_i')e^{-2\tau_o/\mu_i}] \cdot$$

$$[1 - 2\int_0^1 \frac{\rho_D \rho_d(\mu)e^{-2\tau_o/\mu}}{1 - \rho_{ss}(\mu) \rho_d(\mu)e^{-2\tau_o/\mu}} \mu d\mu]\} \quad (12)$$

For the non specular direction, the equation simplifies somewhat. The last term in Eq. (10) is not present, and the intensity ratio is given by

$$R_T(\mu_L', \mu_i') = [1/(n_d)^2][1 - \rho_d(\mu_i')]e^{-\tau_o/\mu_i} \overset{*}{c}_1 \quad (13)$$

where $\overset{*}{c}_1$ is given by Eq. (12), with $\rho_{ss}(\mu_L')$ set equal to zero.

In all cases, the critical angle was taken into account in the integration.

Following the suggestion of Houchens and Hering,[2] Beckmann's model was used for the reflectance of the diffuse substrate.

For large $\sigma/\lambda$, Beckmann[3] obtained the following relationship:

$$\rho_D(\theta_i', \Phi_i', \theta_L', \Phi_L') = \frac{\overline{FC}}{\pi \cos\theta_L' \cos\theta_i'} e^{-\overline{FD}} \quad (14)$$

$\overline{C} = 4\pi C/A^2 \quad C = 2\pi[1 + \cos\theta_L' \cos\theta_i' +$

$\qquad\qquad\qquad\qquad \sin\theta_L' \sin\theta_i' \cos(\pi - \Phi)]/A$

$\overline{D} = 4\pi D/A^2 \quad A = 2\pi(\cos\theta_i' + \cos\theta_L')$

$\overline{F} = (a'/2\sigma)^2 \quad D = \sin^2\theta_i' + \sin^2\theta_L' +$

$\qquad\qquad\qquad\qquad 2\sin\theta_L' \cdot \sin\theta_i' \cos(\pi - \Phi)$

and for the specular direction

$$\rho_{ss}(\mu) = \exp\left[-(4\pi \cos\theta_i' \, \sigma/\lambda)^2\right] \qquad (15)$$

For purposes of comparison, $R_T(\mu_L', \mu_i')/R_s(\mu_L', \mu_i')$, which is the ratio of the reflected intensity to the intensity in the specular direction, was calculated on the computer and compared with experimental data.

## Experimental Procedure

The coating substrate combination was formed by taking commercially available microscope and cover glass slides and vapor-depositing the substrate on one surface.

The optical thickness of the slides was determined by measuring the transmittance and utilizing the manufacturers refractive index data in the analytical expression from Francis and Love[4]:

$$\text{transmittance} = \frac{[1 - \rho_d(\mu)]^2 \, e^{-\tau_o/\mu}}{1 - \rho_d^2(\mu) e^{-2\tau_o/\mu}} \qquad (16)$$

For the specular samples, the slides were then placed in a vacuum chamber, and a 2000 - 2500 Å thick gold coating was vapor-deposited on one surface.

For the diffuse samples, one face of the slides was either sandblasted or ground, using carborundum powder. The surface roughness was measured using a Bendix model 4 profilometer, and the samples were then placed in the vacuum chamber, and a 2000 - 2500 Å gold coating was deposited on the roughened surface. Before vapor deposition of the gold coating, the samples were cleansed with distilled water and ethyl-alcohol to remove any contamination on the surface.

The samples were placed in the goniometer and data taken using a globar source and a Perkin-Elmer model 112 with thermopile detector to measure the reflected radiation. For the samples with a specular substrate, intensity ratios are presented. This ratio was determined by dividing the reading obtained at various polar angles by the reading obtained with the globar focused directly at the monochromator entrance slit. For the roughened substrate samples, the ratio of the radiation reflected in the direction of the polar angle $\theta$ to that reflected in the specular direction is presented. For these samples, only an incident polar angle of $30°$ is presented. For all of the experimental studies, only variations in the plane of incidence are presented.

## Discussion of Results

The results for the specular substrate are presented in Figs. 4 - 7. The analytical prediction for reflectance given by Eq. (8) is presented as the solid line, and two runs of experimental data are represented by the symbols ○ and □. As seen from these figures, the theory compares quite favorably with experimental results. Data were taken at 1, 2, 4, and 5 µ, but only 2 and 4 µ data are presented in this paper. At 5 µ, the transmittances of all slides with the exception of the Corning 0211 were zero, and, therefore, the results for these slides would simply be Fresnel's expression.

For the diffuse substrate, it was necessary to modify the analytical model to account for the effect of the finite collecting solid angle. The procedure followed was similar to that utilized by Francis and Love.[5] The specular reflectance was modified by multiplying the specular reflectance terms in the analytical expression by the ratio of the portion of the collecting solid angle which is contained in the specular collecting solid angle to the total solid angle. The analytical solution is shown as the solid line in Figs. 8 - 13. The circles indicate experimental data in the same figures.

As mentioned by Houchens and Hering,[2] the ratio $(a'/\sigma)^2$ was found to be very important. Varying $(a'/\sigma)^2$ alters the value of the wings of the curves. An increase in $(a'/\sigma)^2$ would increase the values of $R_T/R_S$ in the wings. The correlation length a' was obtained by trial and error. Equations (12) and (13) were modified to account for the finite solid angle and programed on the computer. A trial-and-error procedure was followed to obtain the value of a' which best satisfied the experimental results. The value of $(a'/\sigma)^2 = 18$ was found to be the best for the data here presented. The general shape of the analytical curve and the experimental data were in good agreement. However, departure from theory did occur in the wings of the curves. This was probably due to inaccurate determination of $\rho_D$ in the Beckmann model. Bechmann's model was multiplied by the hemispherical reflectances of the metallic substrate to account for the finite conductivity of the metal. The hemispherical reflectance was obtained from Fresnel's expression. Data in the wings of the curves were difficult to obtain and, because of the low magnitudes involved, subject to greater experimental error.

In addition, the surface roughness was determined from profilometer measurements. It is felt by the authors that these readings may not be an accurate expression for the value σ in Beckmann's model.

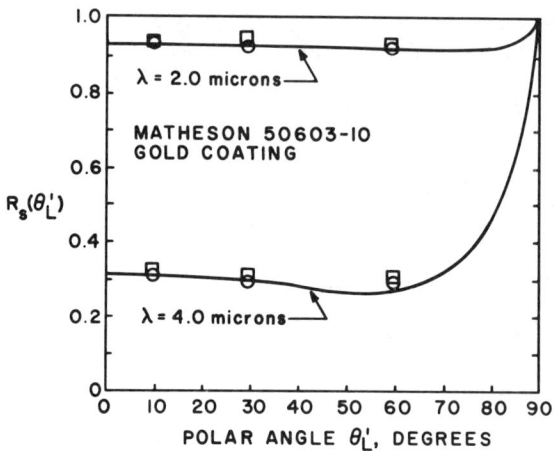

Fig. 4 Specular reflectance for a Matheson 50603-10 coating.

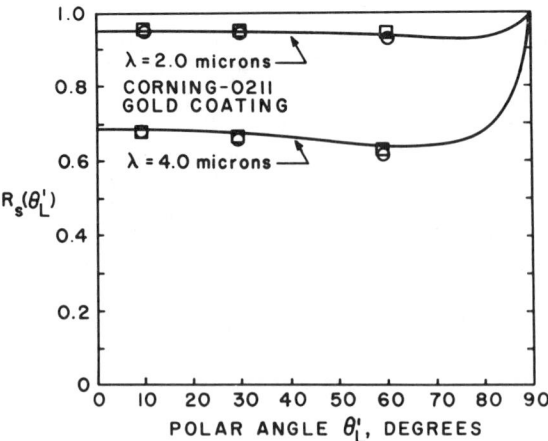

Fig. 5 Specular reflectance for a Corning 0211 coating.

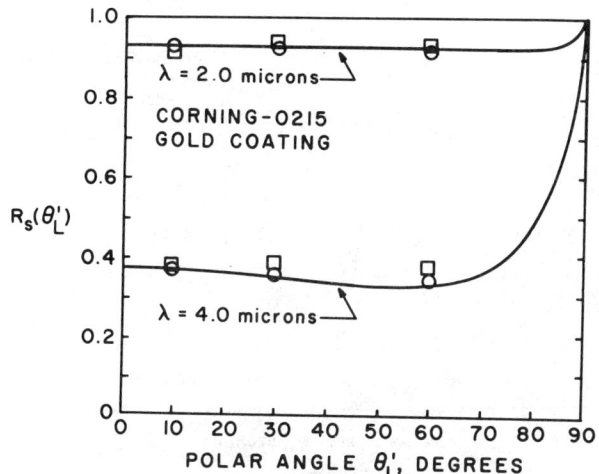

Fig. 6  Specular reflectance for a Corning 0215 coating.

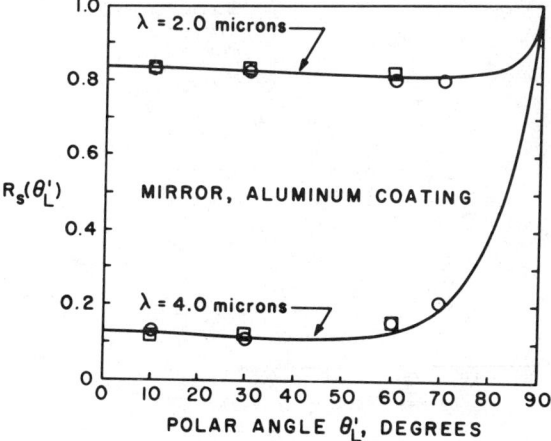

Fig. 7  Specular reflectance for an aluminum mirror.

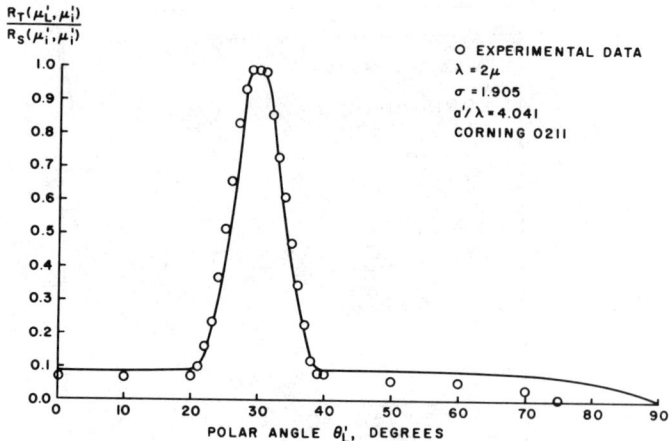

Fig. 8   Reflectance ratio of 2μ for a roughened Corning 0211 slide with a gold substrate (σ = 1.905μ).

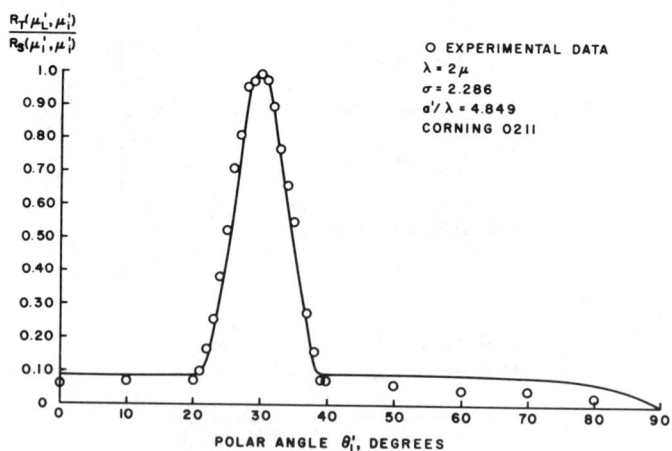

Fig. 9   Reflectance ratio at 2μ for a roughened Corning 0211 slide with a gold substrate (σ = 2.286μ).

# REFLECTANCE OF COATINGS ON A CONDUCTOR SUBSTRATE 65

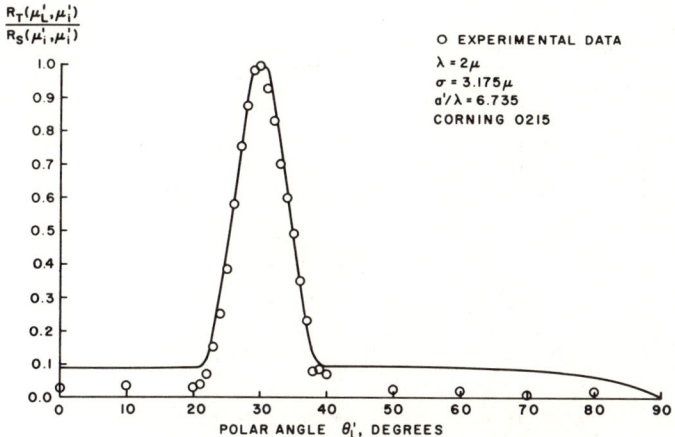

Fig. 10  Reflectance Ratio at 2μ for a roughened Corning 0215 slide with a gold substrate ($\sigma = 3.175\mu$).

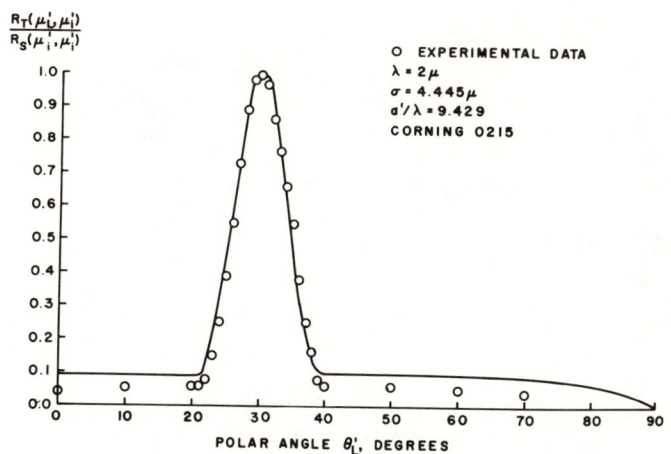

Fig. 11  Reflectance Ratio at 2μ for a roughened Corning 0215 slide with a gold substrate ($\sigma = 4.445\mu$).

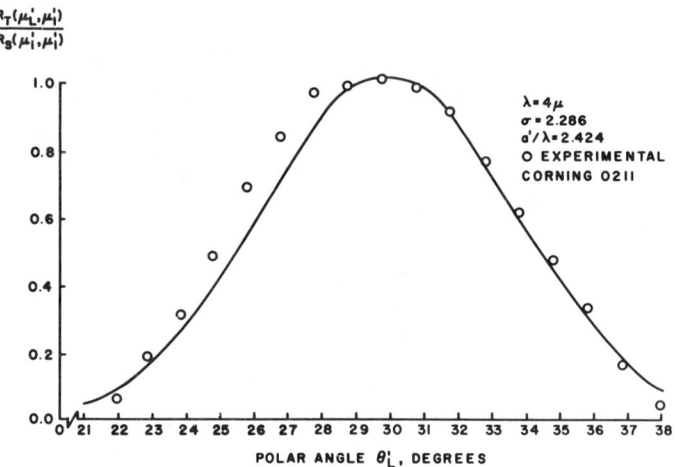

Fig. 12  Reflectance ratio at 4µ for a roughened Corning 0211 slide with a gold substrate ($\sigma = 2.286\mu$).

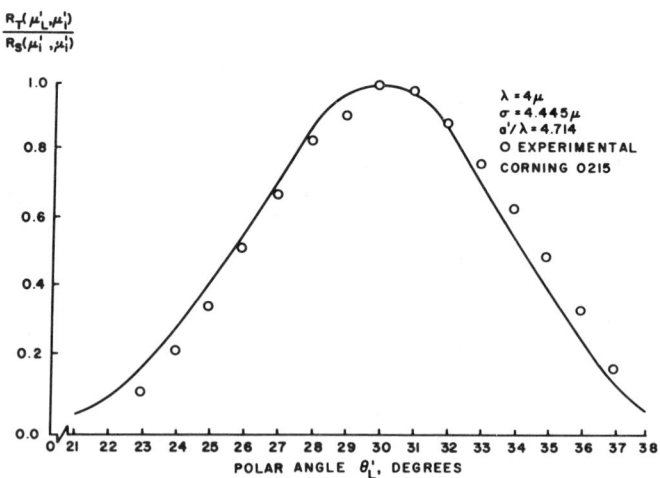

Fig. 13  Reflectance ratio at 4µ for a roughened Corning 0215 slide with a gold substrate ($\sigma = 4.445\mu$).

## References

[1] Love, T. J. and Francis, J. E., "Reflection of Monodirectional Flux by a Coating on a Substrate," *International Journal of Heat and Mass Transfer*, Vol. 11, 1967, pp. 369-374.

[2] Houchens, A. F. and Hering, R. G., "Bidirectional Reflectance from Rough Metal Surfaces," *AIAA Progress in Astronautics and Aeronautics: Thermophysics of Spacecraft and Planetary Bodies*, edited by G. B. Heller, Vol. 20, Academic Press, New York, 1967, pp. 65-89.

[3] Beckmann, P. and Spizzichino, A., *The Scattering of Electromagnetic Waves from Rough Surface*, Macmillan Company, New York, 1963.

[4] Francis, J. E. and Love, T. J., Jr., "Effect of Optical Thickness on Directional Transmittance and Emittance of Dielectric," *Journal of the Optical Society of America*, Vol. 56, No. 6, 1966, pp. 779-782.

[5] Love, T. J. and Francis, R. E., "Experimental Determination of Reflectance Function for Type 302 Stainless Steel," *AIAA Progress in Astronautics and Aeronautics: Thermophysics of Spacecraft and Planetary Bodies*, edited by G. B. Heller, Vol. 20, Academic Press, New York, 1967, pp. 115-135.

# BIDIRECTIONAL REFLECTANCE OF A RANDOMLY ROUGH SURFACE

T. F. Smith[*] and R. G. Hering[†]

University of Illinois at Urbana-Champaign, Urbana, Ill.

## Abstract

A bidirectional reflectance model is developed for a one-dimensionally rough surface consisting of V-shaped roughness elements with randomly distributed slopes. Multiple reflections within and shadowing by adjacent roughness elements are accounted for in the model. A distribution function in terms of rms slope is utilized to specify the probability that a macroscopic surface area element contains microscopic roughness elements of specific slope. For rms slope less than 0.01, the surface reflects specularly, and energy experiencing multiple reflections is unimportant. Multiple reflections become increasingly significant for larger rms slopes. This importance, however, diminishes as the direction of incident energy approaches grazing incidence. The model exhibits characteristics similar to those observed in recent bidirectional reflectance data.

## Nomenclature

A   = total reflecting area

I   = radiant intensity

$\ell$   = length of roughness element surface

---

Presented as Paper 71-465 at the AIAA 6th Thermophysics Conference, Tullahoma, Tenn., April 26-28, 1971. This paper presents results of research supported in part by Jet Propulsion Laboratory, California Institute of Technology, Contract No. 951661.
  [*]Research Assistant, Department of Mechanical Engineering; presently Assistant Professor, Department of Mechanical Engineering, The University of Iowa, Iowa City, Iowa.
  [†]Professor, Department of Mechanical Engineering; presently Professor and Chairman, Department of Mechanical Engineering, The University of Iowa, Iowa City, Iowa.

$m$ = rms slope

$n$ = number of reflections

$P$ = distribution function

$\alpha$ = roughness element slope angle

$\theta',\theta$ = directions of incident and reflected energy

$\xi^*,\eta^*$ = dimensionless coordinates

$\rho$ = roughness element surface reflectance

$\rho_{bd}$ = bidirectional reflectance

$\rho_d$ = directional reflectance

$\rho_{ij,n}$ = dimensionless reflectance factors

$\chi$ = roughness element included angle

## Introduction

Experiments confirm that bidirectional reflectance is strongly dependent upon direction and wavelength of incident energy as well as the surface characteristics. In view of the ranges for directions of incidence and reflection as well as for wavelength of incident energy, the bidirectional reflectance measurements required for a single surface are enormous. Consequently, it is important to develop bidirectional reflectance models that describe the characteristics of the measurements. In developing such models, however, the surface characteristics must be specified. Of the various surface characteristics, roughness is recognized as one that influences both the magnitude and spatial distribution of reflected radiant energy. Surface roughness effects are generally categorized in terms of optical roughness, that is, the ratio of a characteristic roughness height of the surface to a characteristic wavelength of incident energy.

For optical roughness values less than unity, the magnitude of reflected energy is essentially that of a smooth surface of the same material. The spatial distribution of reflected energy, however, is markedly influenced by diffraction effects. The bidirectional reflectance model developed by Beckmann[1] and examined by Houchens and Hering[2] is characteristic of models developed for this optical roughness range. This model accounts for incident energy that experiences only one contact within surface asperities while neglecting shadowing by adjacent roughness elements. For optical roughness values greater than unity, both the magnitude and spatial distribution of reflected energy are influenced by roughness. The methods of geometrical optics may be employed to develop bidirectional reflectance models[3-7] for this optical roughness range. Except for the model developed by Hering and Smith,[7] all reported models[3-6] only partially account for multiple reflections within and shadowing by adjacent roughness elements. These effects are expected to be important in this optical roughness range. The model presented by Hering and Smith,[7] however, is limited to a surface composed of identical roughness elements. It is the purpose of this study to develop a bidirectional reflectance model for a randomly rough surface of optical roughness value greater than unity.

A schematic of the rough surface chosen for study is illustrated in Fig. 1. It consists of a one-dimensionally randomly rough surface of parallel and symmetric V-shaped roughness elements. All roughness elements have identical surface property values, equal-length surfaces, and upper edges lying in a common plane. Roughness element surfaces are taken as specularly reflecting, with direction-independent reflectance. Roughness element depth is assumed large relative to wavelength of incident energy so that the methods of geometrical optics are applicable. Multiple reflections within and shadowing by adjacent roughness elements are accounted for in the development. A distribution function in terms of rms slope is utilized to specify the probability that a macroscopic surface area element contains microscopic roughness elements of specific slope. Since previously reported bidirectional reflectance models[2,3,5] for large optical roughness values are expressed in terms of only roughness element slope, a distribution function of the

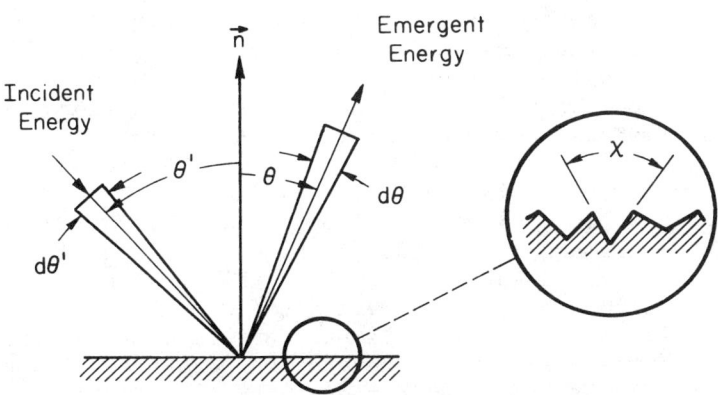

Fig. 1 Schematic diagram of system.

form employed here is appropriate. Emphasis is placed on development of a bidirectional reflectance model and examination of the characteristics of this model in view of recently reported measurements.

## Analysis

A typical macroscopic surface area element of a rough surface is illustrated in Fig. 1. Radiant energy incident within an infinite cylindrical sector subtending an angular increment $d\theta'$ about the $\theta'$ direction is reflected into a similar sector subtending an angular increment $d\theta$ about the $\theta$ direction. Bidirectional reflectance is defined as[7]

$$\rho_{bd}(\theta',\theta) = \frac{2}{\pi} \frac{dI_r^+(\theta)}{I^-(\theta') \cos\theta' \, d\theta'} \qquad (1)$$

where $I^-(\theta')$ and $dI_r^+(\theta)$ denote intensities of incident and reflected energy, respectively. To develop an expression for reflected intensity, the reflection phenomenon that incident energy undergoes within microscopic roughness elements must be considered. Since the upper edges of the roughness elements lie in a common plane, it is necessary to analyze only a single roughness element. The total energy reflected is then obtained by properly summing contributions from all roughness elements.

A typical microscopic surface roughness element (Fig. 2) has equal-length surfaces $\ell$, surface reflectance $\rho$, and included angle $\chi$. Roughness element slope angle $\alpha$ is related to included angle by $\alpha = (\pi - \chi)/2$. Since roughness element surfaces are specularly reflecting, the reflectance factors developed by Hering and Smith[7] are applicable for each element. Development and further discussion of the reflectance factors are available elsewhere.[7] It is sufficient for the present study to recognize that the reflectance factors denote the fraction of energy incident on the roughness element per unit time from the $\theta'$ direction which after n specular reflections within the element emerges in the $\theta$ direction. Analysis of the reflectance factor results establishes that there may exist two elements of different included angle values which reflect incident energy into the $\theta$ direction after n interreflections within the roughness elements. Consider first energy incident on the roughness element surface receiving the greater amount of directly incident energy and reflected into the $\theta$ direction after n reflections. Roughness element surface of this type is designated surface 1. Let $\chi_1$ denote the roughness element included angle for this reflection path. The energy incident on the roughness element is

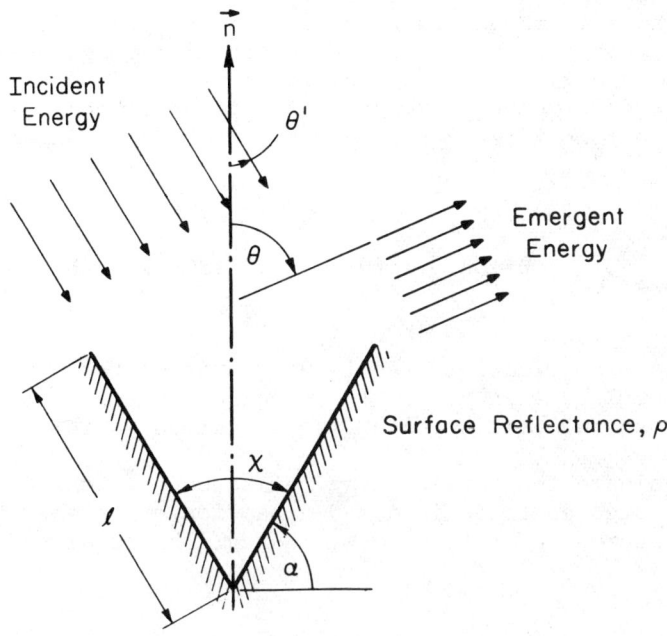

Fig. 2  Typical surface roughness element.

$$(\pi/2) I^-(\theta') \cos \theta' \, d\theta' \, 2\ell \sin (\chi_1/2)$$

where $2\ell \sin (\chi_1/2)$ is the roughness element opening area. The reflected energy is

$$\sum_{j=1}^{2} \rho_{1j,n}(\theta',\theta) \frac{\pi}{2} I^-(\theta') \cos \theta' \, d\theta' \, 2\ell \sin\left(\frac{\chi_1}{2}\right)$$

where $\rho_{1j,n}(\theta',\theta)$ are the reflectance factors. Specific expressions for the reflectance factors are given later. For the element with the other value for included angle, energy incident on the roughness element surface receiving the lesser amount of directly incident energy is also reflected into the $\theta$ direction after n reflections. Roughness element surfaces of this type are designated as surface 2. Reflected energy from elements with this included angle $\chi_2$ is evaluated by replacing subscript 1 with 2 in the preceding expression. However, the probability of finding elements with included angle $\chi_2$ within the macroscopic area may be quite different from that of finding elements with included angle $\chi_1$. To include this aspect in the analysis, the contribution to reflected energy from each roughness element must be weighed with the probability, $P(\chi)\,d\chi$, that the macroscopic area contains roughness elements of included angle $\chi$. An expression for $P(\chi)$ is given later. Thus, reflected energy emergent from the roughness elements in the $\theta$ direction after n reflections within the elements follows as

$$\sum_{i=1}^{2}\sum_{j=1}^{2}\rho_{ij,n}(\theta',\theta)\,\frac{\pi}{2}\,I^{-}(\theta')\,\cos\theta'\,d\theta'\,2\ell\sin\left(\frac{\chi_i}{2}\right)P(\chi_i)\,d\chi_i$$

where the summation over the i index extends to 2 only for $\theta'$ directions less than $\chi_2/2$. The roughness element with included angle $\chi_2$ is fully illuminated for $\theta'$ values in this range.

To evaluate the total energy reflected into the $\theta$ direction, the reflected energies in the $\theta$ direction are summed for all possible number of reflections. Hence, the total reflected energy is

$$\frac{\pi}{2}\,dI_r^{+}(\theta)\cos\theta\,d\theta\,A = \sum_{n=1}^{\infty}\left[\sum_{i=1}^{2}\sum_{j=1}^{2}\rho_{ij,n}(\theta',\theta)\right.$$
$$\left.\cdot\frac{\pi}{2}\,I^{-}(\theta')\cos\theta'\,d\theta'\,2\ell\sin\left(\frac{\chi_i}{2}\right)P(\chi_i)\,d\chi_i\right] \quad (2)$$

where A is the total reflecting area. Since the macroscopic area contains roughness elements of different opening areas, the total reflecting area is related to the probability that the macroscopic area contains roughness elements with specific included angles. The total reflecting area is then given by

$$A = 2\ell\int_{0}^{\pi}\sin\left(\frac{\chi}{2}\right)P(\chi)\,d\chi \quad (3)$$

Combining Eqs. (2) and (3) with the definition of Eq. (1), bidirectional reflectance is expressed as

# BIDIRECTIONAL REFLECTANCE OF A ROUGH SURFACE

$$\rho_{bd}(\theta',\theta) = \frac{2}{\pi \cos\theta \, d\theta \int_0^{\pi} \sin\left(\frac{\chi}{2}\right) P(\chi) \, d\chi} \sum_{n=1}^{\infty}$$

$$\cdot \left[ \sum_{i=1}^{2} \sum_{j=1}^{2} \rho_{ij,n}(\theta',\theta) \sin\left(\frac{\chi_i}{2}\right) P(\chi_i) \, d\chi_i \right] \quad (4)$$

The relationship between $d\theta$ and $d\chi_i$ is developed after the expressions for the reflectance factors are presented.

It is convenient for the present study to rephrase the reflectance factor expressions[7] in terms of directions of incidence and reflection as well as the number of reflections incident energy experiences before emerging from the roughness element. The reflectance factors follow as

$$\left. \begin{array}{l} \rho_{11,n}(\theta',\theta) = \dfrac{\xi_n^* \sin(\chi_1/2 + \theta')}{2 \sin(\chi_1/2) \cos\theta'} \rho^n \\[2ex] \rho_{12,n}(\theta',\theta) = \dfrac{(1 - \xi_{n+1}^*) \sin(\chi_1/2 + \theta')}{2 \sin(\chi_1/2) \cos\theta'} \rho^n \\[2ex] 0 \leq \theta' < \chi_1/2 \end{array} \right\} \quad (5)$$

where $\rho$ is the roughness element surface reflectance and $\xi^*$ is a dimensionless distance measured from the apex along surface 1. $\chi_1$ is the roughness element included angle for which incident energy undergoing n reflections emerges from the element in the $\theta$ direction. Expressions for $\xi^*$ and $\chi_1$ are

$$\left. \begin{array}{l} \xi_k^* = \dfrac{\sin[(k - 1/2)\chi_1 + \theta']}{\sin(\chi_1/2 + \theta')} \\[2ex] \chi_1 = \dfrac{1}{n}\left[\pi - \theta' - \dfrac{\theta}{(-1)^n}\right] \\[2ex] n_1 = \left\{ \dfrac{\pi - \theta'}{\chi_1} + \dfrac{1}{2} \right\} \\[2ex] 0 \leq \theta' < \chi_1/2 \end{array} \right\} \quad (6)$$

where $n_1$ is utilized to distinguish which reflectance factor of Eq. (5) is zero. (Here and throughout the analysis section, the notation $\{\chi\}$ denotes the operation of taking the integer part of $\chi$. If $\chi$ is precisely an integer, then $\{\chi\}$ is that integer minus 1.) If $n_1 = n$, then $\rho_{12,n}(\theta',\theta)$ is zero, but, if $n_1 = n + 1$, then $\rho_{11,n}(\theta',\theta)$ is zero. For values of $n_1$ other than those cited, both reflectance factors are zero.

The reflectance factors in Eq. (2) account for incident energy that is initially reflected from surface 1. To account for incident energy that is initially reflected from surface 2, the reflectance factors $\rho_{21,n}(\theta',\theta)$ and $\rho_{22,n}(\theta',\theta)$ are employed. The expressions for $\rho_{11,n}(\theta',\theta)$ and $\rho_{12,n}(\theta',\theta)$ may be utilized to evaluate $\rho_{21,n}(\theta',\theta)$ and $\rho_{22,n}(\theta',\theta)$, respectively, by replacing $\theta'$ with $-\theta'$, $\theta$ with $-\theta$, $\xi_k^*$ with $\eta_k^*$, $\chi_1$ with $\chi_2$, and $n_1$ with $n_2$ in Eqs. (5) and (6). For roughness element included angles such that $\chi_1/2 \leq \theta' < \pi/2$, shadowing is present and $\xi_k^*$ is replaced with $(\xi_k^* - \xi_o^*)$, where

$$\left. \begin{array}{l} \xi_o^* = \dfrac{\sin(\theta' - \chi_1/2)}{\sin(\theta' + \chi_1/2)} \\[1em] n_1 = \left\{ \dfrac{\pi - 2\theta'}{\chi_1} + 1 \right\} \end{array} \right\} \chi_1/2 \leq \theta' < \pi/2 \qquad (7)$$

Should $\xi_k^*$ (or $\eta_k^*$) exceed unity in the foregoing expressions, it is replaced with unity. Furthermore, should $\chi_1$ (or $\chi_2$) exceed $\pi$, the corresponding reflectance factors are zero.

The relationship between $d\chi_i$ and $d\theta$ is obtained by reference to the expression for $\chi_i$ in Eq. (6). Since only the magnitude of this variation is important here, the following expression results:

$$d\chi_i = d\theta/n \qquad (8)$$

Taking the angular increment $d\theta$ as fixed, Eq. (4) for bidirectional reflectance can be written as

$$\rho_{bd}(\theta',\theta) = \dfrac{2}{\pi \cos\theta \int_0^\pi \sin\left(\dfrac{\chi}{2}\right) P(\chi)\, d\chi} \sum_{n=1}^\infty$$

$$\cdot \left[ \sum_{i=1}^2 \sum_{j=1}^2 \rho_{ij,n}(\theta',\theta) \sin\left(\dfrac{\chi_i}{2}\right) \dfrac{p(\chi_i)}{n} \right] \qquad (9)$$

For roughness elements with included angle $\pi$ which occurs only when $n = 1$ and $\theta = \theta'$, the summation within the brackets of Eq. (9) is divided by a factor of two, since $d\chi_i$ is allowed to vary only one-half of $d\theta$. Otherwise, values of $\chi_i$ greater than $\pi$ result. By use of Eq. (9), the contribution of multiply reflected energy ($n > 1$) to bidirectional reflectance can be evaluated.

Incident energy that makes only one contact with the rough surface may be reflected into any of the possible $\theta$ directions. Multiply reflected incident energy, however, emerges in $\theta$ directions confined within the following limiting values:

$$\frac{(n-1)\theta' - \pi}{n+1} \le \theta \le \frac{\pi + (n-1)\theta'}{n+1} \qquad \theta' < \frac{\pi}{2n} \qquad (10a)$$

$$\frac{(n+1)\theta' - \pi}{n-1} \le \theta \le \frac{\pi + (n-1)\theta'}{n+1} \qquad \theta' \ge \frac{\pi}{2n} \qquad (10b)$$

for n even. For n odd, the limiting values are multiplied by minus one. Examination of these limits reveals that the specular direction ($\theta = \theta'$) lies within the ranges for n even though, for n odd, the limits encompass the backscattering direction ($\theta = -\theta'$). The limiting values approach the specular and backscattering directions as n increases. Thus, incident energy that is multiply reflected emerges in $\theta$ directions at or near the specular and backscattering directions.

It can be demonstrated that bidirectional reflectance expressed by Eq. (9) satisfies the Helmholtz reciprocity requirement. Hence, no distinction need be made between directional-hemispherical and hemispherical-directional reflectance. Directional reflectance $\rho_d(\theta')$ is given as

$$\rho_d(\theta') = \frac{\pi}{2} \int_{-\pi/2}^{\pi/2} \rho_{bd}(\theta',\theta) \cos\theta \, d\theta \qquad (11)$$

with $\rho_{bd}(\theta',\theta)$ evaluated from Eq. (9).

To obtain a distribution function in terms of roughness element included angle, roughness element slopes ($\tan \alpha$) are taken to have a normal distribution with zero mean value and rms slope m. Thus,

$$P(\tan \alpha) = \frac{1}{m}\sqrt{\frac{2}{\pi}} \exp\left[-\frac{\tan^2 \alpha}{2m^2}\right] \qquad (12)$$

Similar distribution functions[1,5] have been utilized to describe the surface contour. The distribution function in terms of roughness element included angle follows as

$$P(\chi) = \frac{\csc^2(\chi/2)}{m\sqrt{2\pi}} \exp\left[-\frac{\cot(\chi/2)}{2m^2}\right] \tag{13}$$

For small rms slopes, the macroscopic surface is smooth and becomes rougher as rms slope increases.

## Results and Discussion

Before presenting bidirectional reflectance results, it is of interest to examine rms slope values for rough surfaces prepared using common techniques. Only a limited amount of information is available for rms slopes observed on engineering surfaces. Rms slopes calculated from profilometer measurements[8,9] for grit-blasted aluminum and ground glass surfaces were within the ranges of 0.1 to 0.15 and 0.06 to 0.12, respectively. Various correlations of bidirectional reflectance models with measurements have employed rms slopes either calculated or estimated from reflectance data. Comparisons[2,10] with measurements for rough glass and for rough aluminum alloy surfaces coated with pure aluminum employed rms slopes within the range of 0.02 to 0.2. An rms slope value[5] of 0.247 correlated a model and measurements for rough glass surfaces. Comparisons[4] of a model with previously reported measurements were performed with rms slope values of 0.5 to 11.0. For the comparisons reported by Treat and Wildin,[6] an rms slope value of approximately 0.2 yields a probability distribution similar to that used in the cited study. Taking into account the preceding observations, it is estimated that the rms slope range of interest is 0.01 to 1.0.

Bidirectional reflectance distributions are presented in Figs. 3 and 4 for directions of incident energy of $0°, 30°, 60°,$ and $80°$ with high surface reflectance (0.9) and low surface reflectance (0.1) values. Results shown in Fig. 3 are for rms slopes of 0.1 and 0.25, and those in Fig. 4 are for rms slopes of 0.5 and 1.0. Some general characteristics of the bidirectional reflectance model are evident in these figures. First, greater amounts of reflected energy are found in directions of reflection other than the specular direction for fixed direction of incidence as rms slope increases. For a fixed rms slope, however, the surface becomes more specular as direction of incidence approaches grazing incidence. Bidirectional reflectance measurements[11] exhibit characteristics similar to those observed here.

# BIDIRECTIONAL REFLECTANCE OF A ROUGH SURFACE

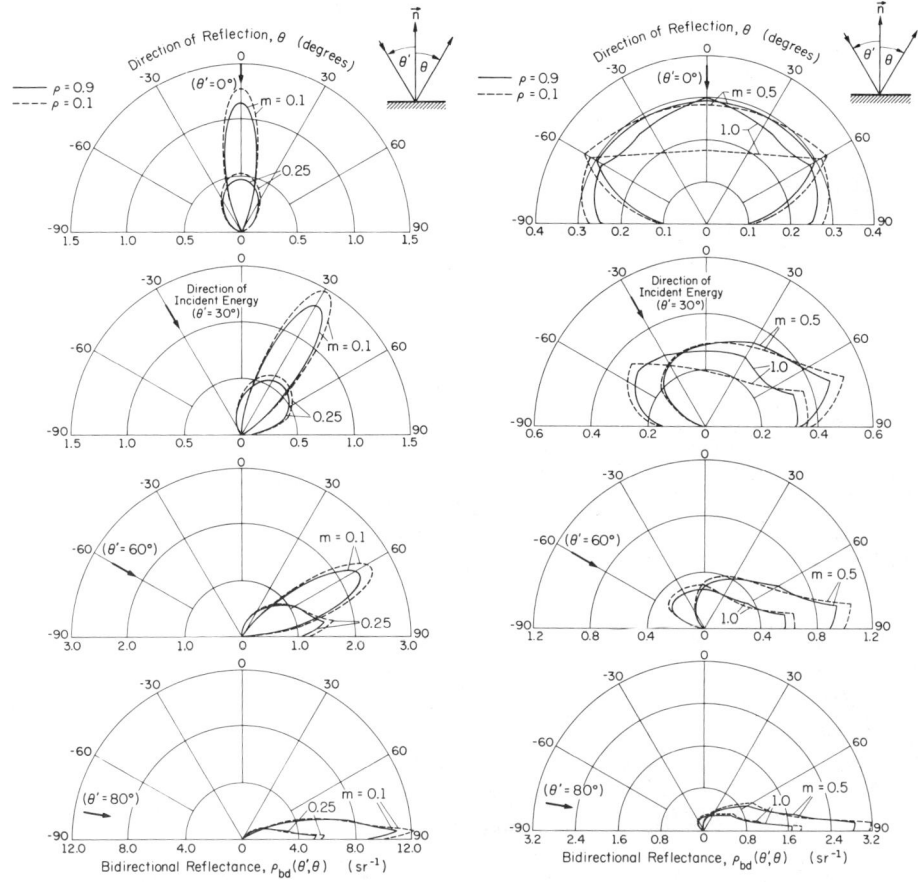

Fig. 3 Bidirectional reflectance distributions (divide $\rho_{bd}$ values read from figure by 10 for $\rho = 0.1$).

Fig. 4 Bidirectional reflectance distributions (divide $\rho_{bd}$ values read from figure by 10 for $\rho = 0.1$).

Distributions are not presented for rms slope of 0.01 but are limited to a spread of ±4° about and maxima in the specular direction for all directions of incidence. Except for near-normal incidence, distributions in Figs. 3 and 4 have maxima at directions of reflection greater than the specular direction. Off-specular peaks are also characteristic of measurements[11] for rough surfaces. Throughout this discussion, the direction of reflection where the off-specular peak occurs is referred to as the off-specular direction. For the rough surfaces in Fig. 3, the difference between the off-specular direction and the specular direction increases as rms slope or direction of incidence increases. However, as direction of incidence approaches grazing incidence, this difference decreases, and the off-specular directions are identical for both rough surfaces. Previously reported measurements[5,11] show similar trends for this difference. The off-specular direction also depends weakly on surface reflectance. Distributions for direction of incidence of 30° with rms slope of 0.25 exhibit off-specular directions of 38° and 37° for high and low surface reflectance, respectively. The off-specular direction for low reflectance is attributed to reduced importance of multiply reflected energy.

Except for the rougher surface with low reflectance, distributions illustrated in Fig. 4 for normal incidence are nearly diffuse. A diffusely reflecting surface would be represented in the polar plots by a semicircle with radius $\rho_d(\theta')/\pi$ and would have a $\rho_{bd}$ value of approximately 0.3 for the parameters given in Fig. 4. Specific values of $\rho_d(\theta')$ are presented later. Measurements[3,11] also show that diffuse distribution of reflected energy is attained only for near-normal incidence.

For direction of incidence of 30°, the rougher surface in Fig. 4 exhibits a peak at a direction of reflection of -50°. Distributions for surfaces with rms slopes larger than those presented here exhibit large amounts of backscattering and have maxima that can even exceed those in the off-specular direction. Backscattering peaks have also been observed in measurements[3] for aluminum alloy surfaces.

In order to examine the importance of multiply reflected energy, bidirectional reflectance distributions are presented in Fig. 5 for rms slope of 1.0. Results are shown for directions of incidence of 0°, 30°, 60°, and 80° with high and low reflectance values. Where distinguishable, distributions are displayed for once- and twice-reflected energy as well as the necessary number of reflections to account for all reflected energy. Distributions for near-normal incidence exhibit a strong dependency on multiply reflected energy. Up to six reflections are necessary to account for all reflected energy for high reflectance, but only four reflections are required for low reflectance. The directions of reflection which receive contribution from multiply reflected energy can be easily observed from this polar plot. All directions of reflection receive once-reflected energy, but only directions of reflection within the range of ±60° receive twice-reflected energy. The corresponding directions of reflection for six reflections lie within the range of ±25.71°. As direction of incidence increases, the importance of multiply reflected energy decreases.

BIDIRECTIONAL REFLECTANCE OF A ROUGH SURFACE 81

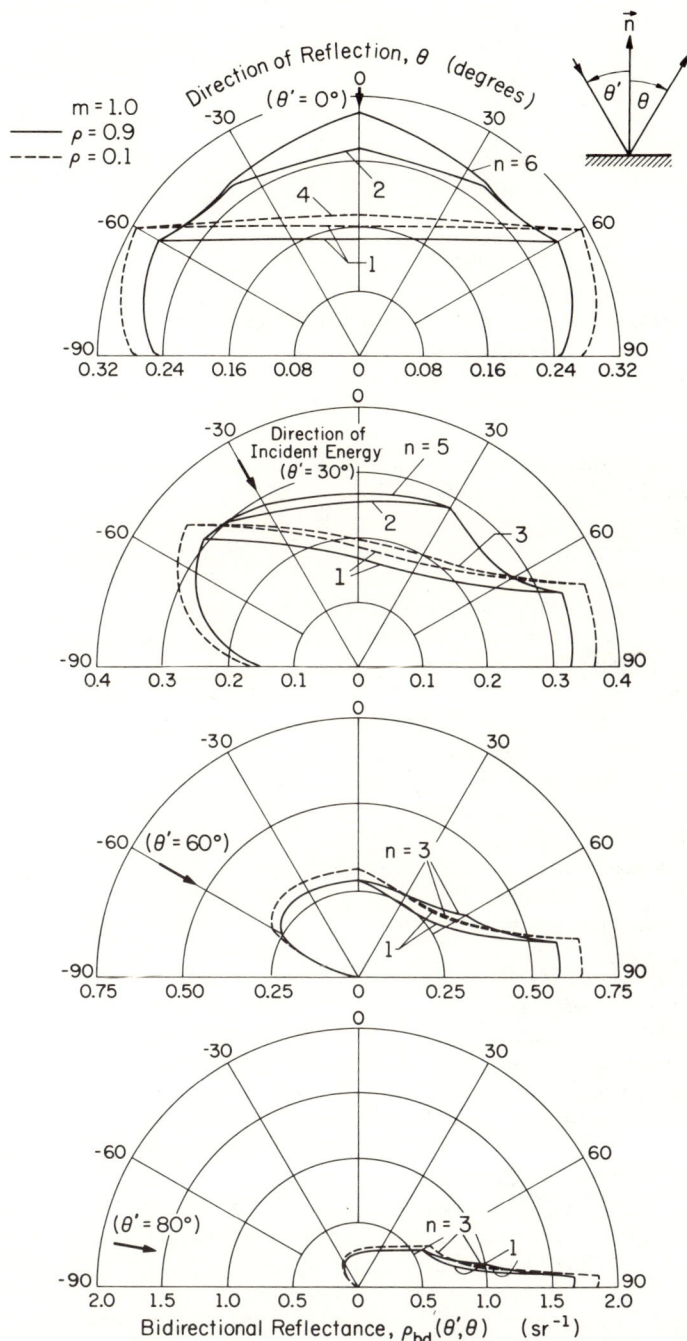

Fig. 5 Bidirectional reflectance distributions (divide $\rho_{bd}$ values read from figure by 10 for $\rho = 0.1$).

At near-grazing incidence, only three reflections must be included. Distributions with high reflectance for once- and twice-reflected energy differ, but those for low reflectance are essentially identical. In view of the dependence on multiply reflected energy demonstrated here, reported bidirectional models[3-5] may not adequately account for all reflected energy.

Shadowing effects are present when incident energy partially illuminates any roughness element. Except for normal incidence when all roughness elements are fully illuminated, shadowing effects become increasingly important as direction of incidence or rms slope increases. Shadowing effects may be examined by dividing directions of incidence into two ranges. For directions of incidence less than 45°, incident energy that experiences one contact within roughness elements is reflected from fully illuminated elements. The importance of shadowing effects increases for multiply reflected energy as direction of incidence approaches 45°. Shadowing effects can be neglected in analyses accounting for incident energy that has experienced up to eight reflections for direction of incidence of 10°. However, at a direction of incidence of 30°, shadowing effects must be included for multiply reflected energy. For directions of incidence greater than 45°, incident energy undergoing one contact with the rough surface is reflected from partially illuminated roughness elements into directions of reflection within the range of -90° to (3θ'- 180°). Shadowing effects must be included in analyses of multiply reflected energy for directions of incidence greater than 45°. It should be noted that the importance of incident energy reflected from partially or completely illuminated roughness elements is related to the rms slope of the rough surface.

Directional reflectance results are presented in Fig. 6 for high and low reflectance values with rms slopes of 0.1, 0.25, 0.5, and 1.0. Results for rms slopes less than 0.01 as well as for rms slope of 0.1 with high reflectance are indistinguishable from those for a smooth surface of the same material. Directional reflectances for rms slopes of 0.1 and 0.25 decrease with increasing direction of incidence until grazing incidence is attained. This reduction at large directions of incidence is a result of multiple reflections within roughness elements of small slopes. As rms slope increases, multiple reflections become increasingly important for near-normal incidence. For rms slope of 1.0, directional reflectances at near-normal incidence are lower than those for other directions of incidence. A ninefold reduction in roughness element reflectance results in approximately a factor of 10 decrease in directional reflectances.

## BIDIRECTIONAL REFLECTANCE OF A ROUGH SURFACE

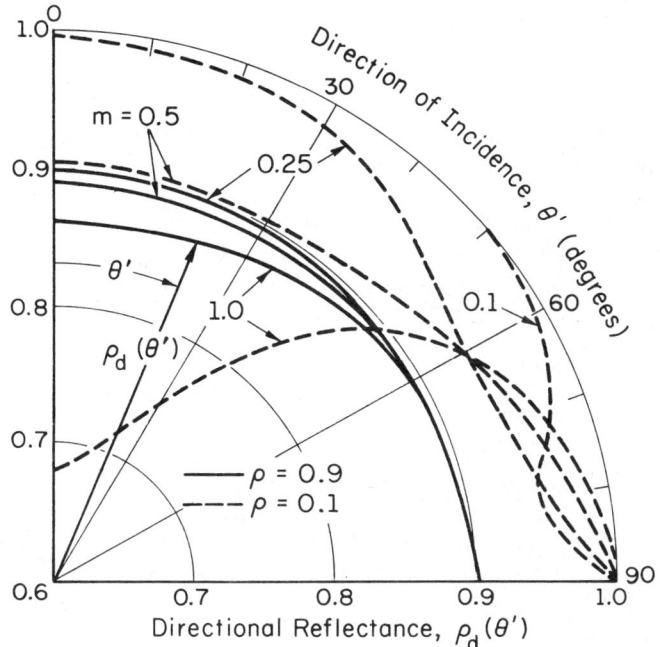

Fig. 6  Directional reflectance (divide $\rho_d$ values read from figure by 10 for $\rho = 0.1$).

## Conclusions

A bidirectional reflectance model was developed for a one-dimensionally randomly rough surface consisting of parallel and symmetric V-shaped roughness elements. Roughness element surfaces are taken as specularly reflecting with direction-independent reflectance. Multiple reflections within and shadowing by adjacent roughness elements are accounted for in the model. A distribution function in terms of rms slope is utilized to specify the probability that a macroscopic surface area element contains macroscopic roughness elements of specific slope. For rms slopes less than 0.01, the rough surface reflects as a specular reflector. As rms slope increases, greater amounts of reflected energy are found in directions of reflection other than the specular direction. Bidirectional reflectance distributions for other than near-normal incidence exhibited maxima in directions of reflection greater than the specular direction. A nearly diffuse distribution of reflected energy is attained only for rms slope of 1.0 and near-normal incidence. For rms slope of 1.0 and direction of incidence of 30°, the distributions showed maxima near the backscattering direction. Multiple reflections and shadowing effects become increasingly important as rms slope increases. The former effects, however, diminish as direction of incidence approaches grazing incidence. Directional reflectance showed a strong dependency on rms slope for low roughness element surface reflectance values. The model exhibited characteristics similar to those observed in recent bidirectional reflectance data.

## References

[1] Beckmann, P. and Spizzichino, A., The Scattering of Electromagnetic Waves from Rough Surfaces, Macillan Co., New York, 1963.

[2] Houchens, A. F. and Hering, R. G., "Bidirectional Reflectance of Rough Metal Surfaces," AIAA Progress in Astronautics and Aeronautics: Thermophysics of Spacecraft and Planetary Bodies, Vol. 20, edited by G. B. Heller, Academic Press, New York, 1967, pp. 65-89.

[3] Torrance, K. E. and Sparrow, E. M., "Theory for Off-Specular Reflection from Roughened Surfaces," Journal of the Optical Society of America, Vol. 57, No. 9, 1967, pp. 1105-114.

[4] Look, D. C., Jr. and Love, T. J., "Investigation of the Effects of Surface Roughness upon Reflectance," AIAA Progress in Astronautics and Aeronautics: Heat Transfer and Spacecraft Thermal Control, Vol. 24, edited by J. W. Lucas, The MIT Press, Cambridge, Mass., 1971, pp. 123-142.

[5] Smith, A. M., Muller, P. R., Frost, W. and Hsia, H. M., "Super- and Subspecular Maxima in the Angular Distribution of Polarized Radiation Reflected from Roughened Dielectric Surfaces," AIAA Progress in Astronautics and Aeronautics: Heat Transfer and Spacecraft Thermal Control, Vol. 24, edited by J. W. Lucas, The MIT Press, Cambridge, Mass., 1971, pp. 249-269.

[6] Treat, C. H. and Wildin, M. W., "Investigation of a Model for Bidirectional Reflectance of Rough Surfaces," AIAA Progress in Astronautics and Aeronautics: Thermophysics: Application to Thermal Design of Spacecraft, Vol. 23, edited by J. T. Bevans, Academic Press, New York, 1970, pp. 77-92.

[7] Hering, R. G. and Smith, T. F., "Apparent Radiation Properties of a Rough Surface," AIAA Progress in Astronautics and Aeronautics: Thermophysics: Application to Thermal Design of

Spacecraft, Vol. 23, edited by J. T. Bevans, Academic Press, New York, 1970, pp. 337-361.

[8] Renau, J. and Collinson, J. A., "Measurements of Electromagnetic Backscattering from Known Rough Surfaces," The Bell System Technical Journal, Vol. 44, No. 10, 1965, pp. 2203-2226.

[9] Latta, M. R., "The Scattering of 10.6 Micron Radiation from Ground Glass Surfaces," Journal of the Optical Society of America, Vol. 59, No. 4, 1969, p. 493.

[10] Smith, T. F. and Hering, R. G., "Comparison of Bidirectional Reflectance Measurements and Model for Rough Metallic Surfaces," ASME Proceedings of the Fifth Symposium on Thermophysical Properties, 1970, pp. 429-435.

[11] Torrance, K. E. and Sparrow, E. M., "Off-Specular Peaks in the Directional Distribution of Reflected Thermal Radiation," Journal of Heat Transfer, Vol. 88, Ser. C, 1966, pp. 223-230.

# BIDIRECTIONAL REFLECTANCE CHARACTERISTICS OF INTEGRATING SPHERE COATINGS

R. C. Zentner[*], R. K. MacGregor[†], and J. T. Pogson[†]

The Boeing Company, Seattle, Wash.

## Abstract

Increasingly complex characterizations of the reflectance properties of real materials are required for further advances in the accuracy of radiative heat-transfer computations. A bidirectional reflectometer with the capability for measurements over the wavelength range from 0.35 to 15 μ has been developed to fulfill this requirement. The reflectometer is described, and data are presented for samples of smoked magnesium oxide and gold-coated sandpaper. The magnesium oxide data are compared with previously published bidirectional reflectance data. The sandpaper with a vapor-deposited gold coating exhibited total hemispherical reflectance in excess of 80%. It is evaluated for application as an infrared integrating sphere coating.

## Introduction

There were three major reasons for the development of the bidirectional reflectometer. First, bidirectional reflectance is required for the accurate prediction of heat transfer with engineering materials such as solar cells, rocket nozzles, and heat shields. These are components exposed to solar heating or high-intensity directional radiant heating. The second requirement for bidirectional reflectance data is in the evaluation of classical thermal analysis techniques by more recent analytic methods employing complex directional property characterizations. And finally, most of the existing bidirectional property-measuring facilities are biased toward measurements in the visible wavelength range with a high degree of angular resolution. These data are important for visibility studies and solar reflection analysis for optical and limited thermal

---

Presented as Paper 71-77 at the 9th Aerospace Sciences Meeting, New York, N. Y., January 25 - 27, 1971.
  [*]Engineer, Engineering Division.
  [†]Specialist Engineer, Engineering Division.

design purposes. However, the equipment designed here is specifically directed toward operation in the infrared wavelength range. The most significant wavelength range limitation was imposed by the limited availability and relatively high cost of detection equipment and large-aperture first surface mirrors.

Integrating sphere coatings have been chosen as candidates for these initial bidirectional reflectance measurements because of the great general interest in their properties. The magnesium oxide properties provide a check with existing data, and the gold-coated sandpaper is a coating recently suggested by Sherrell and Shahrokhi[1] for use in an infrared hemispherical-hemispherical reflectometer. The sandpaper exhibits a high reflectance in the infrared wavelength range and is much more durable than magnesium oxide coatings.

## Definition of Bidirectional Reflectance

Figure 1 shows the geometry associated with the definition of bidirectional reflectivity. Reflection terms are designated by primed symbols. In this notation, the bidirectional reflectivity is defined

$$\rho(\theta, \phi, \theta', \phi', \lambda) = \frac{I'(\theta', \phi', \lambda) \pi}{I(\theta, \phi, \lambda) \cos \theta \, dw}$$

where $\lambda$ = wavelength, I = radiant flux intensity (monochromatic power per unit projected area per unit solid angle), dw = solid angle of incident flux. This may be stated,

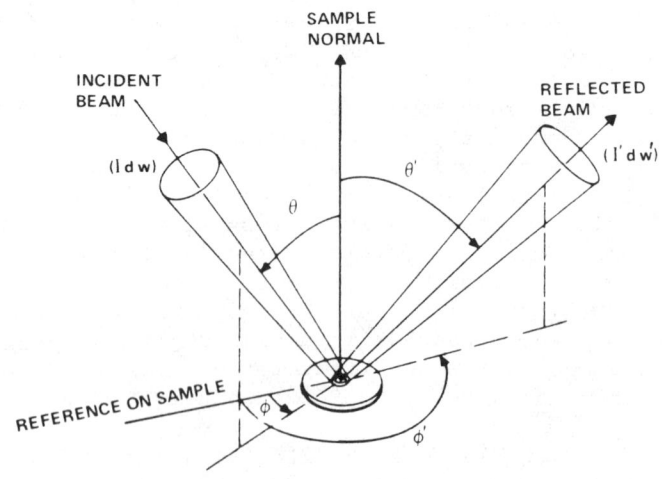

Fig. 1 Bidirectional reflectance geometry.

"bidirectional reflectivity is the ratio of reflected monochromatic intensity at angles $\theta'$, $\phi'$ to the intensity which would result by perfectly diffuse total (100%) reflection of the incident flux."

The desired bidirectional test data consist of an "intensity map" of the reflection hemisphere above the sample for conditions of constant illumination. The values of relative intensity and bidirectional reflectance are proportional to detector response as long as the viewed sample area projected in the direction of the detector is constant over all of the detector positions utilized and the detector system response is linear over the range of energy experienced.

Only "relative bidirectional reflectance" data are discussed here. These data characterize the spatial distribution of reflections and are suitable by themselves for evaluation of integrating sphere coatings. Unless otherwise specified, the data presented here are adjusted to unity in the direction of the surface normal.

### Equipment Description

The experimental system for measurement of bidirectional reflectance is shown on Fig. 2. Major components are source

Fig. 2 Bidirectional reflectometer facility.

head assembly with source chopper, source optics, and sample holder; detector assembly with transfer optics, monochromator, and detector; and the related electronics for supplying source and detector power and signal amplification and recording. The optical system is shown schematically on Fig. 3. Operation is described in the following paragraphs.

Source Head Assembly

Three sources are available, depending on wavelength range of interest. A universal mount allows each source to be located at the object focal point of mirror M-1. Polychromatic emission from the source passes through the incident beam chopper and through stops S-1 and S-2 to toroidal mirror M-1. The expanding beam is focused from the front surface reflection of M-1 upon the focal stop S-4. In the process, the beam is reflected from front surface plane mirrors M-2 and M-3 and passes between these along a path symmetrical around and concentric with axis y-y. Polychromatic source emission, chopped and passing through stop S-4, reflects from the toroidal surface of M-4 and images on the center of the sample. The globar spot is about 0.44 in. in diameter. The tungsten source (quartz iodine lamp), with its smaller filament area, produces uniform intensity of illumination over a spot about 0.25 in. in diameter.

The source mirror M-4 and source diagonal M-3 are mounted together on the source polar yoke assembly. These may be rotated about axis y-y to provide illumination at angles $\theta$. The source polar yoke is held in position by a locking shaft. Variations of the illumination azimuth angle $\phi$ are obtained by rotating the sample in the source head assembly.

Detector motion is accomplished by motions of the source head assembly. The assembly rotates through 360° about axis x-x (Fig. 3), sweeping detector azimuth angles with the aperture S-5. This detector azimuth motion $\phi'$ is provided by a variable-speed motor and gear train on the vertical support which rotates the entire source head. Continuous 360° rotation is achieved by passing all electrical and cooling water connections to the source head through a slip ring and rotating fluid coupling on the back of the assembly.

Detector polar motion $\theta'$ is provided by rotating the source head assembly at its attachment point on the table. A 10-in. diam. ball-bearing unit preloaded in thrust stabilizes the unit where it mounts to the table. Motion is controlled by a gear drive unit beneath the table top which is operated by a hand crank.

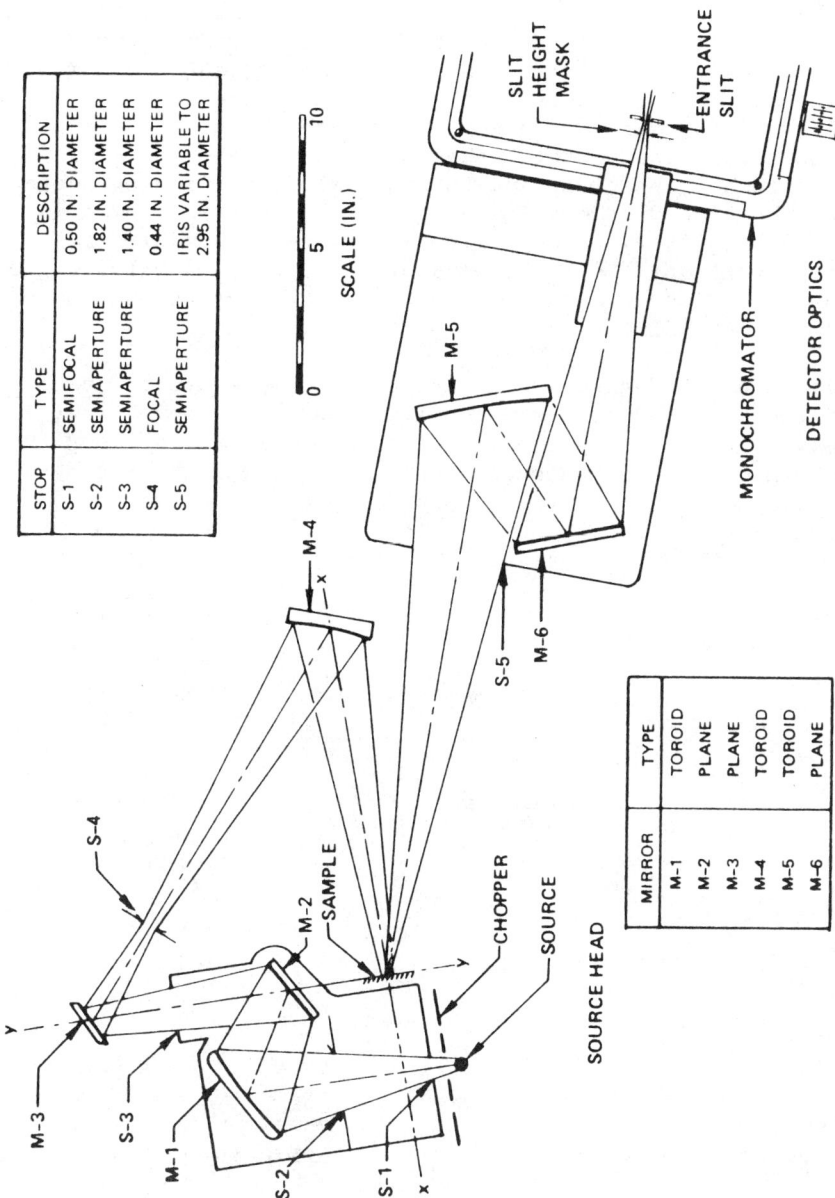

Fig. 3 Schematic of bidirectional reflectometer.

Detector Assembly

The detector assembly is shown schematically on Fig. 3. It is positioned so that a region on the center of the sample is imaged by the transfer optics onto the monochromator entrance slit. At the 1:1 magnification of the optical transfer system, the energy reflected by a region on the sample equal in size to the open slits which enters the detector aperture stop S-5 is focused by mirror M-5 and enters the monochromator. The size of the main detector mirror M-5 is such that the f/4.5 and f/3.5 monochromator aperture limits are filled, maximizing energy inputs. The size of the area on the sample that is viewed is controlled by slit width and slit height masking and is inversely proportional to the cosine of the detection polar angle $\theta'$. The detection area exhibits constant cross section when projected in the direction of the detector.

The illumination/detection areas on the sample are arranged so that the detected area always lies within the region that is uniformly illuminated by the incident beam. Figure 4 shows the source beam overlap of detection area that is achieved at detector viewing angles up to 75° for fully open (2.0-mm wide) entrance slits masked to 3/8-in. high. This represents a practical limit of viewing angle for the apparatus.

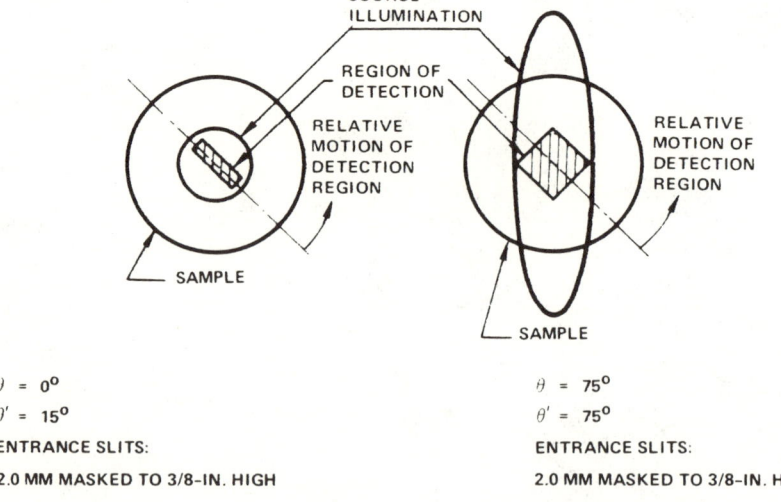

Fig. 4  Characteristics of the over-illumination optical design.

The monochromator is a Perkin Elmer model 98 with interchangeable prisms and detectors. Energy incident on the entrance slit is spectrally dispersed by transmission through the prism and focused on the exit slit. Energy passing through the exit slit is concentrated onto thermocouple, lead sulfide cell, or photomultiplier tube detectors. The energy in a narrow-wavelength band which reaches the detector consists of sample emission, chopped source energy, and background radiant flux reflected from the sample. The detector output consists of an alternating signal superimposed on a d.c. signal. The total signal is relayed to the preamplifier and fed into the proper channels of the a.c. amplifier shown in Fig. 2.

As the chopper blade rotates, it also interrupts a small light beam in the trigger assembly on the source head. The trigger assembly provides a square wave input to the amplifier at the same frequency as the source beam chopping. The amplifier filters the detector signal and amplifies only the part of the signal which has an a.c. frequency and phase orientation exactly like the trigger signal. The amplifier successfully rejects any unchopped background radiation that strikes the detector.

Data Collection

A sample of interest is mounted at the focal point of source and detector optics. Interchangeable components are selected, and the monochromator is adjusted to the wavelength of interest. The source incidence angle $\theta$ and azimuth $\phi$ are set up. Source, chopper, and detector systems are activated. The reflection hemisphere is scanned for the location of maximum detector response. The system is adjusted to provide maximum values on the strip chart recorder at this position.

Data are collected by continuously rotating the source head at a low rate of speed while continuously recording on the strip chart. The reflection hemisphere is scanned at constant detector polar angle $\theta'$ and steadily changing detector azimuth angle $\phi'$. As each 360° detector azimuth angle rotation is completed, the detector polar angle is changed incrementally.

A continuous strip chart record of detector response over the reflection hemisphere is collected in this way. All the data for a single set of incident angles and wavelength are collected on this continuous record.

## Experimental Results

The experimental program conducted with the bidirectional reflectometer consisted of a detailed examination of two engineering materials: smoked magnesium oxide and vapor-deposited gold on sandpaper. The magnesium oxide was utilized as a standard for verification of the reflectometer calibration. The goldized sandpaper was suggested[1] as a coating for integrating spheres in the infrared.

The magnesium oxide sample was examined over wavelengths varying from 0.5 - 2.0 µ. Source incidence angles ranged from normal incidence to 75° near grazing angles. Samples of the resultant data for 0.5 - 0.2-µ wavelengths are shown in Figs. 5 and 6, respectively. For near normal incidence angles, the reflective character of magnesium oxide is very diffuse. Note that in Figs. 5 and 6 an ideally diffuse reflector would have a relative reflected flux value always equal to unity.

The 0.5-µ data in Fig. 5 compare well with the data reported by Brandenberg and Neu[2] at 0.507 µ. Brandenberg and Neu also presented variations in the bidirectional reflectance at 75° from normal as compared to the value at 15° from normal. With the 15° incidence, the bidirectional reflectance at 75° from normal was reported by Brandenberg and Neu to be 77% of its value at 15°. With 75° incidence angle, the variation was reported to be 250% higher than at the 15° reflection angle. The present study includes measurements at 15° incidence, although the data are not shown on Fig. 5. The percentage variation between 15° and 75° reflection directions are 70 and 240%, respectively, which compare well with the previously reported data. In general, the Brandenberg and Neu data always show higher reflectance at the larger angles from normal. These differences may be due to the different thicknesses of the MgO samples and differences in their preparation. Brandenberg and Neu used a 1.0-mm-thick coating, whereas the coating used in this study was nearly 2.5-mm thick.

A comparison of Figs. 5 and 6 indicates that an increase in forward scattering was obtained at the 2-µ illumination for incidence angles greater than 45° in comparison to the 0.5-µ illumination. For incidence angle less than 45°, the reflected flux remained more nearly diffuse at the 2-µ illumination as compared to the 0.5-µ results.

It should be noted that degradation of the MgO was observed in both the directional-hemispherical data and in the bidirectional reflectance data over the 24-hr time period required for

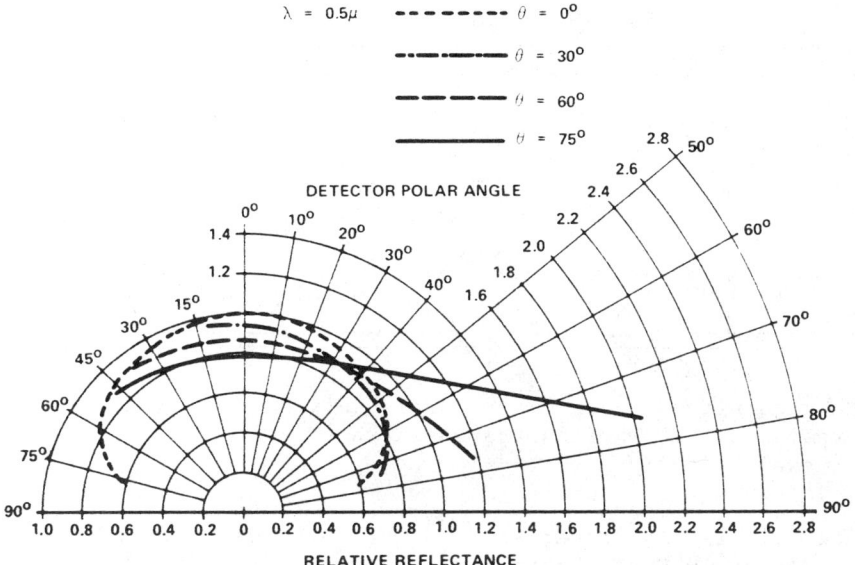

Fig. 5  Directional reflectance characteristics of magnesium oxide ($\lambda = 0.5$ μ).

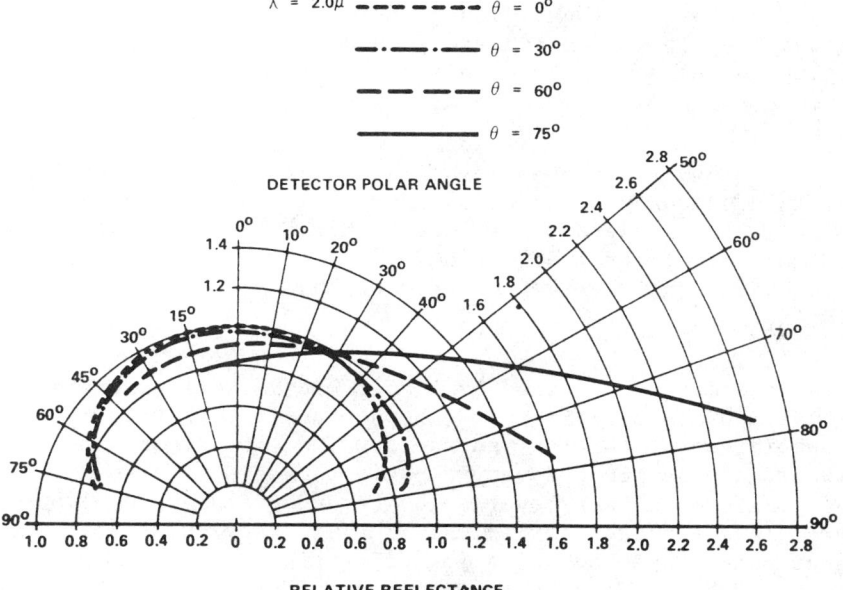

Fig. 6  Directional reflectance characteristics of magnesium oxide ($\lambda = 2.0$ μ).

Table 1   Gold Sandpaper Samples

| Grit | Trademark |
|------|-----------|
| 150  | Carborundum J255E |
| 180  | Armour |
| 220  | Carborundum A115F |
| 400  | Carborundum C935R |

data acquisition. The directional-hemispherical reflectance decreased by 1%, whereas the strength of the specular component increased slightly.

The goldized sandpaper samples were examined over wavelengths from 0.5 to 10.0-$\mu$. The source incidence (polar) angles ranged from near normal to 75° grazing angles. The source azimuth positions were in approximately the same plane of incidence as the directions from which gold was plated onto the samples. Table 1 contains a description of the substrate materials that were vapor deposited with gold. The vapor deposition was accomplished by plating from a gold-coated filament several inches away located alternately on opposing 45° incidence angles in the same azimuth plane. As noted, this is the source azimuth plane in all bidirectional data.

Table 2 shows directional-hemispherical reflectances for the goldized sandpaper samples and for a smooth gold surface. The major trend in the data is toward increasing reflectance for the more finely grained surfaces. There is no significant wavelength sensitivity present over that of the flat gold coating.

The samples of goldized sandpaper had markedly different visual appearance. Figure 7 shows scanning electron microscope images of the four surfaces. The 150-grit cloth-backed carborundum sample had very closely packed particles and the most uniform surface coverage. The 400-grit sample also had very closely packed particles. The 180 and 220 samples were paper backed and had voids in the particle coverage amounting to 10 - 20% of the surface area. As indicated by Table 2 data, the 150-grit sample was slightly darker in appearance.

Table 2  Hemispherical reflectance of gold sandpaper samples

| Sample Wavelength, μ | Gold sandpaper samples, grit | | | | Flat gold specimen[3] |
|---|---|---|---|---|---|
| | 150 | 180 | 220 | 400 | |
| 0.5  | 0.318 | 0.447 | 0.428 | 0.433 | 0.477 |
| 2.0  | 0.637 | 0.713 | 0.698 | 0.806 | 0.991 |
| 5.0  | 0.644 | 0.790 | 0.843 | 0.808 | 0.994 |
| 10.0 | 0.635 | 0.784 | 0.846 | 0.854 | 0.994 |
| 15.0 | 0.682 | 0.737 | 0.834 | 0.847 | 0.994 |

The specularities of the four samples are compared on Fig. 8. At 5 μ, the 180 and 220-grit samples show a significant specularity with the maximum reflection several times greater than the average. The 400-grit sample was surprisingly specular,

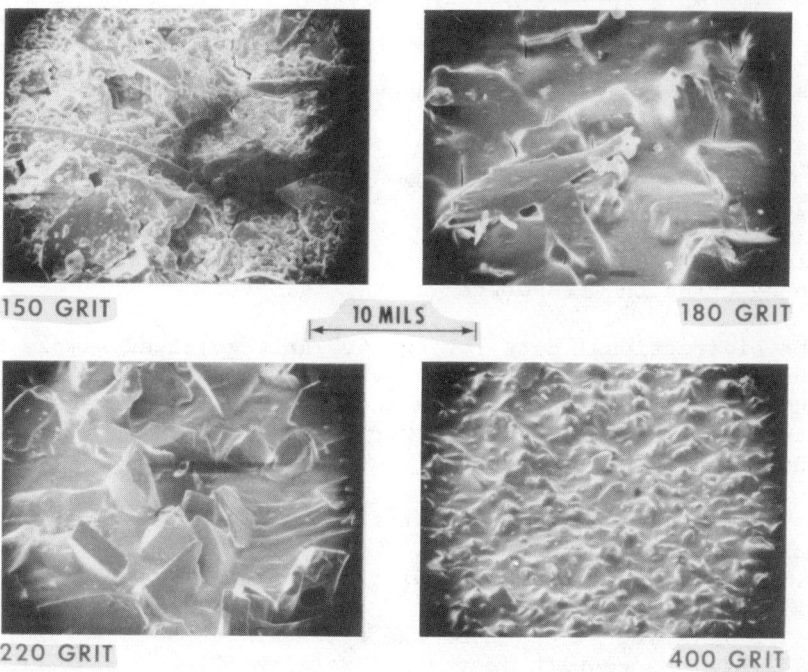

Fig. 7  Microscope photographs of gold sandpaper samples.

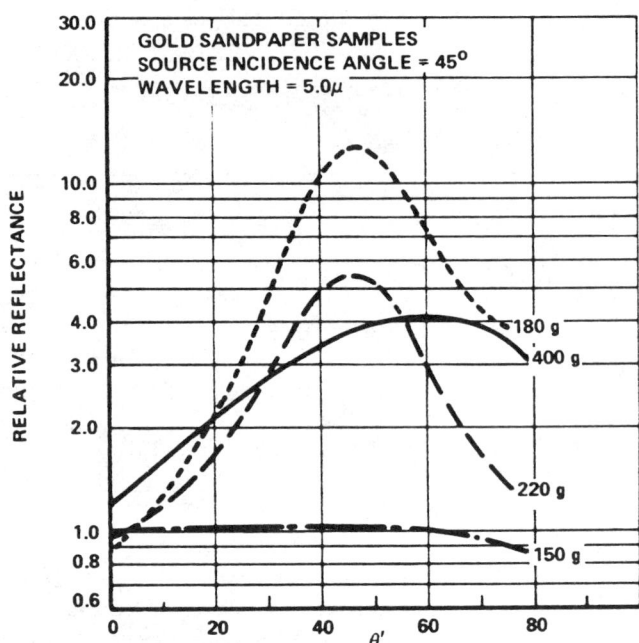

Fig. 8   Directional reflectance characteristics of gold sandpaper samples ($\lambda = 5.0~\mu$).

considering its very diffuse appearance at visible wavelengths. The 150-grit sample was the only one that could be termed diffuse at 5 μ.

The 400-grit sample was selected for further evaluation because it represented a compromise between maximization of total reflectance and achievement of minimum specularity. Figures 9-16 describe the properties of this sample.

The bidirectional data for the 400-grit goldized sample are illustrated in the polar plots of Figs. 9-12 for four incident illumination angles. The data in each plot are adjusted to unity in the direction of the surface normal. Several observations may be made from these figures:

1) The specular component of reflectance increases with increasing source incidence angle.
2) The direction of the specular component of reflectance is significantly off-specular.
3) The relative strength of the specular component exhibits a negligible change with increasing wavelength.
4) As the wavelength increases, the direction of the specular component of reflection approaches the theoretical specular direction (as defined by geometric optics).

BIDIRECTIONAL REFLECTANCE                99

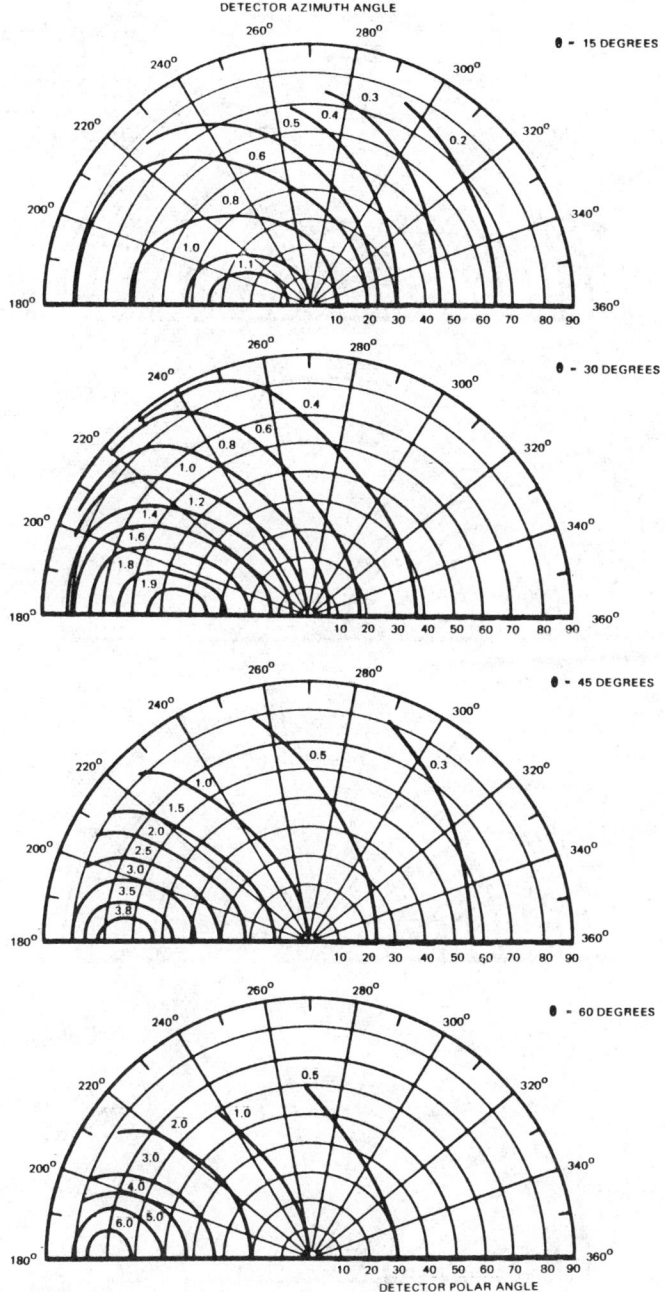

Fig. 9  Bidirectional reflectance characteristics of 400-grit gold sandpaper sample ($\lambda = 0.5$ μ).

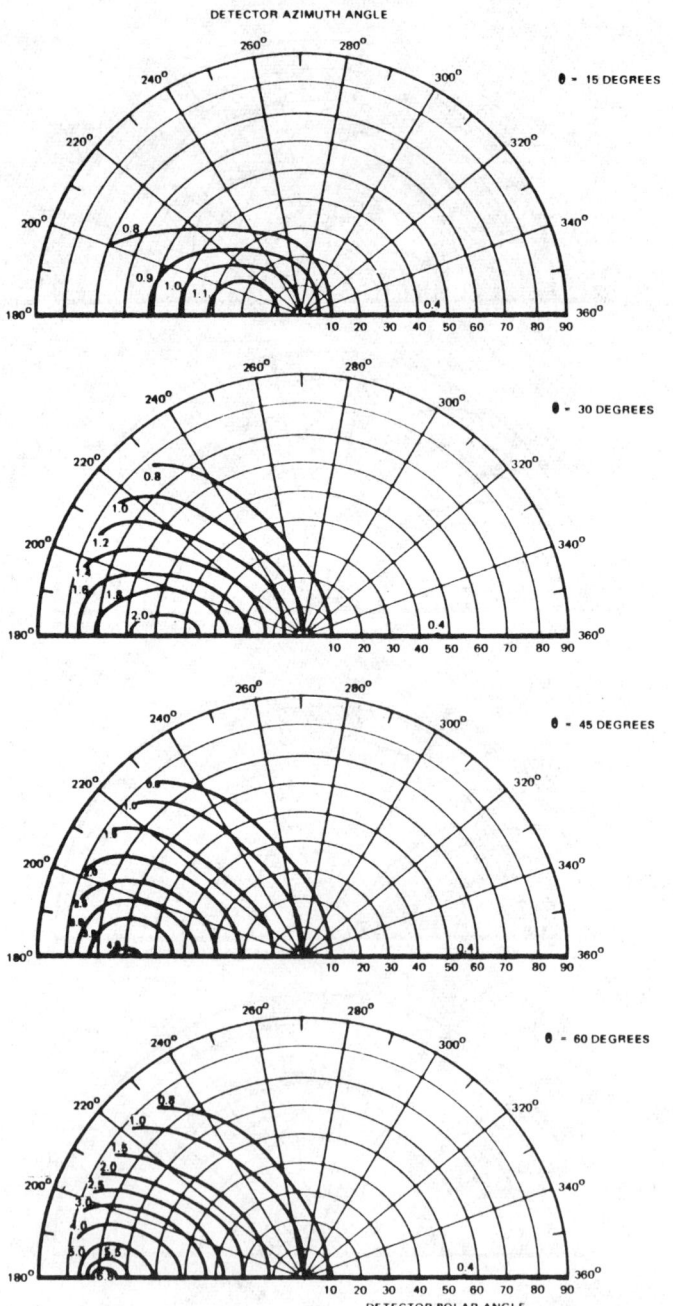

Fig. 10  Bidirectional reflectance characteristics of 400-grit gold sandpaper sample ($\lambda = 2.0\ \mu$).

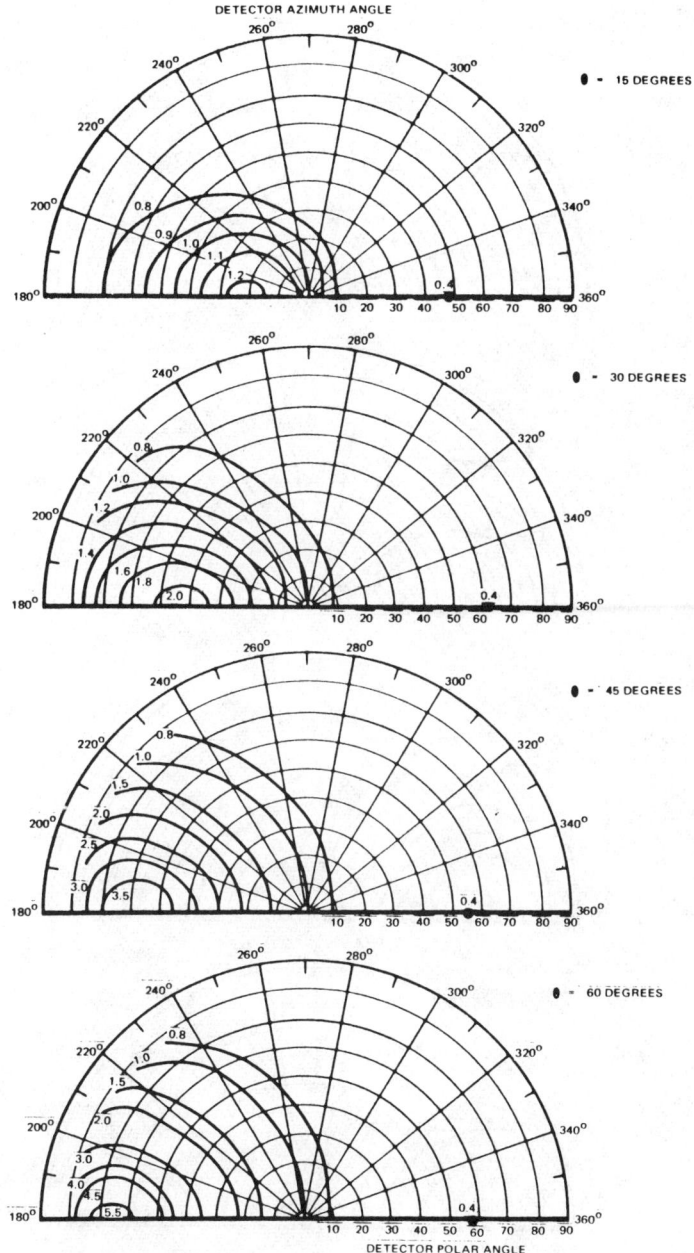

Fig. 11 Bidirectional reflectance characteristics of 400-grit gold sandpaper sample ($\lambda = 5.0\ \mu$).

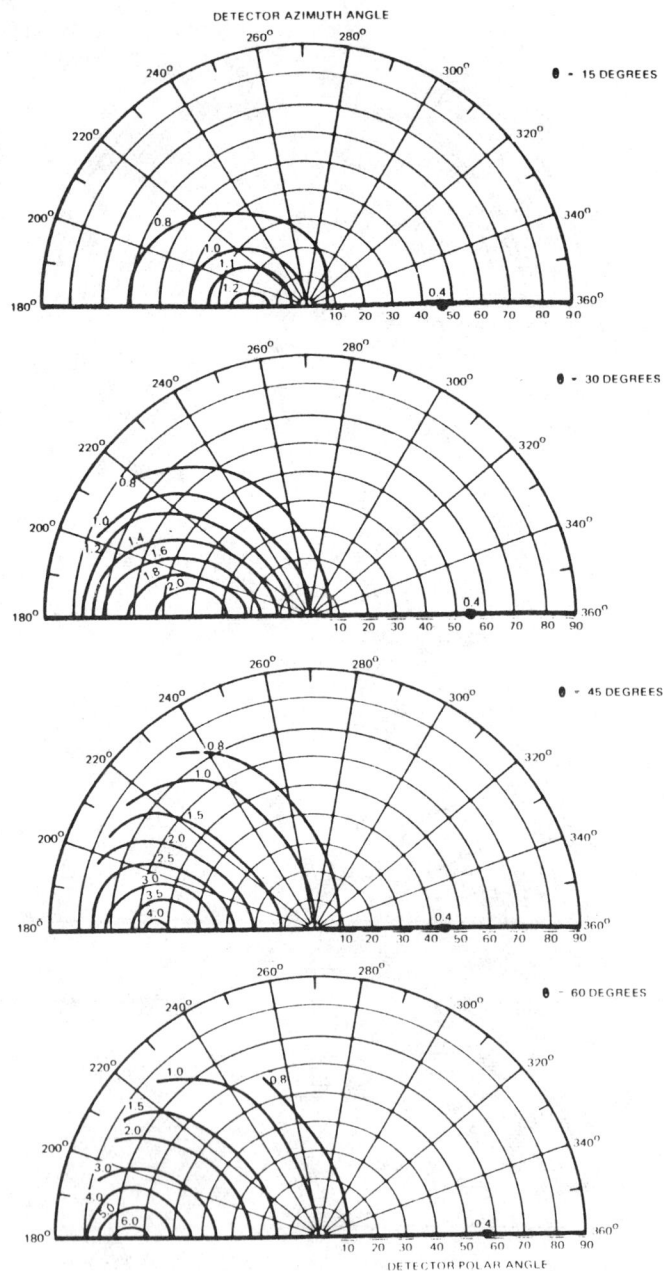

Fig. 12  Bidirectional reflectance characteristics of 400-grit gold sandpaper sample ($\lambda = 10.0$ μ).

Although the relative strength of the specular component of reflectance of the "goldized" sandpaper is three times that shown by the MgO, it should be noted that the "goldized" sandpaper is quite diffuse when compared with other engineering materials. Specular materials such as polished aluminum show specular component strengths 3 to 4 orders of magnitude greater than the normal components of reflection.[4] Even white paints show specular component strengths 2 to 3 orders of magnitude greater than the normal components of reflection.[4]

Figures 13-16 illustrate the relative reflectance data in the plane of incidence as collected from Figs. 9-12, respectively. The figures describe the sensitivity of reflection distributions to the direction of incidence. They compare the plane of incidence data from corresponding polar plots at constant wavelength. That is, the data in the polar plots are presented relative to the normal reflectance for each incidence angle. Comparisons between incidence angles for a given wavelength in each polar plot may be accomplished by noting the corresponding relative reflectances at $\theta' = 0$ in Figs. 13-16. Also as may be noted in Figs. 13-16, the relative reflectance curves for the various source polar angles cross at approximately a 25°-30° detector polar angle for all four wavelengths.

## Conclusions

The operation of the bidirectional reflectometer and its ease of data collection and reduction has been demonstrated. The accuracy of the instrument has been verified by comparison of collected results with the published literature. New data have been presented to characterize the reflectance of magnesium oxide further.

Data have been generated to characterize the reflectance of "goldized" sandpaper. As suggested by Sherrell and Shahrokhi[1], the increase in the specular component of reflection for the "goldized" sandpaper as compared to MgO is small and will introduce negligible errors when utilized as an integrating sphere coating in the infrared. Although the hemispherical reflectance in the infrared is low for application to integrating spheres, Sherrell and Shahrokhi[1] have shown higher values for their goldized sandpaper samples.

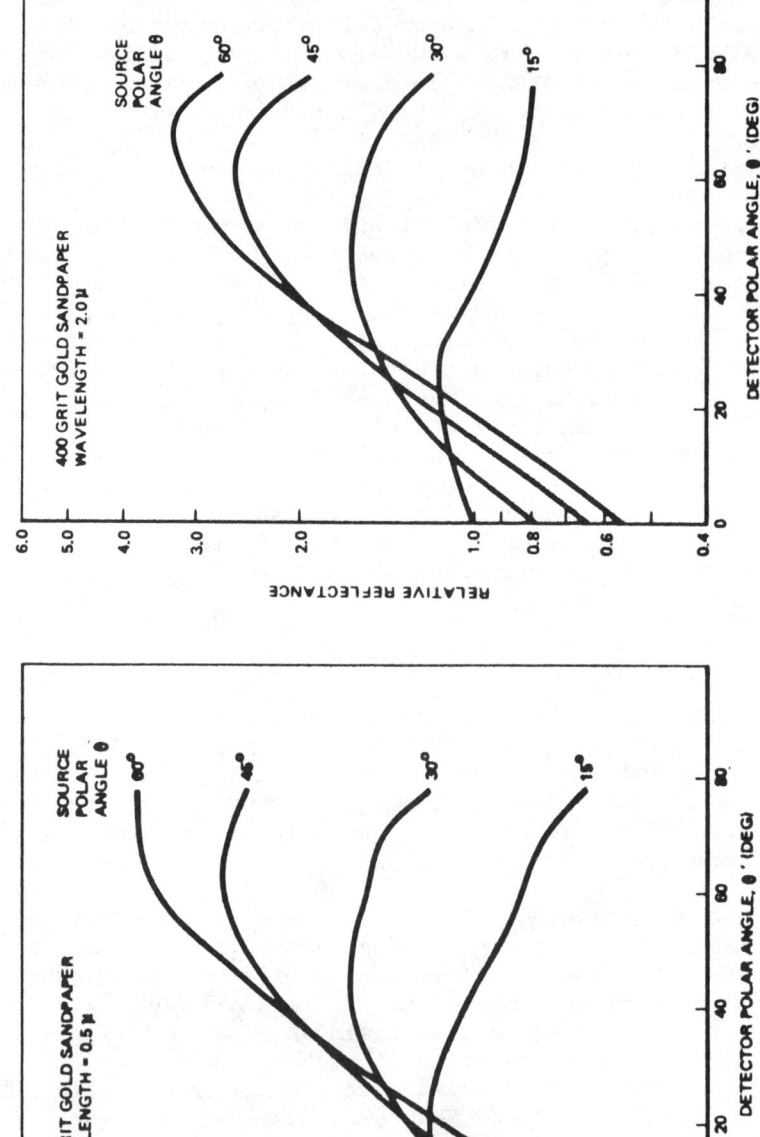

Fig. 13  Directional reflectance of 400-grit gold sandpaper sample ($\lambda = 0.5$ μ).

Fig. 14  Directional reflectance of 400-grit gold sandpaper sample ($\lambda = 2.0$ μ).

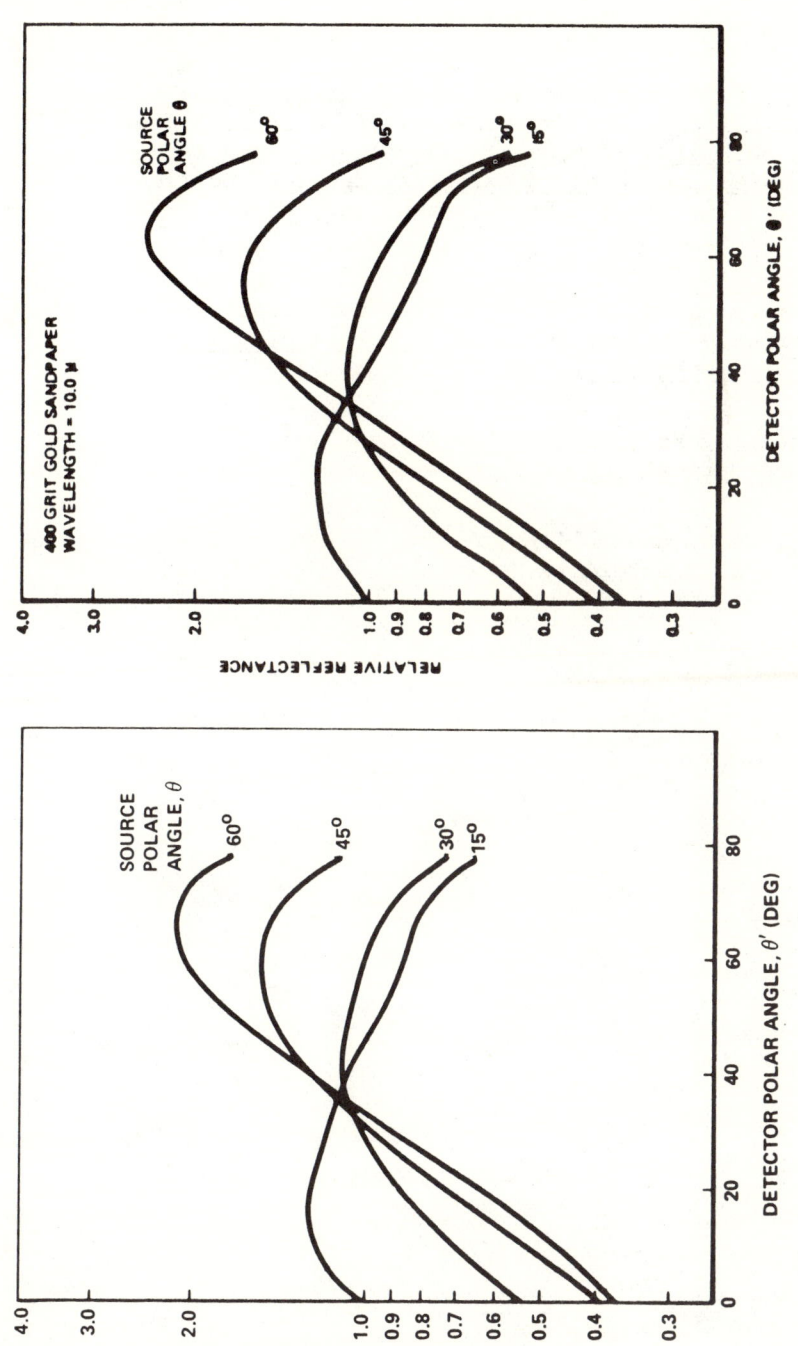

Fig. 15  Directional reflectance of 400-grit gold sandpaper sample ($\lambda = 5.0\ \mu$).

Fig. 16  Directional reflectance of 400-grit gold sandpaper sample ($\lambda = 10.0\ \mu$).

## References

[1] Sherrell, F. G. and Shahrokhi, F., "Determination of Hemispherical Emittance by Measurements of Infrared Bihemispherical Reflectance," *AIAA Progress in Astronautics and Aeronautics: Heat Transfer and Spacecraft Thermal Control*, edited by J. W. Lucas, Vol. 24, The MIT Press, Cambridge, Mass., 1971, pp. 169-183.

[2] Brandenberg, W. M. and Neu, J. T., "Unidirectional Reflectance of Imperfectly Diffuse Surfaces," *Journal of the Optical Society of America*, Vol. 56, No. 1, Jan. 1966, pp. 97-103.

[3] Kingslake, R. ed., *Optical Components*, Vol. III of *Applied Optics and Optical Engineering*, Academic Press, New York, 1965.

[4] Keating, G. M. and Mullins, J. A., "Vectorial Reflectance of the Explorer IX Satellite Material," TN D-2388, Aug. 1964, NASA.

PARTICULATE RADIATION EFFECTS
ON THE SOLAR ABSORPTANCE OF
THERMAL CONTROL SURFACES

R. J. Andres,[*] Paul M. Blair Jr.[†] and E. C. Smith[‡]

Hughes Aircraft Company, Culver City, Calif.

Abstract

A computerized model has been developed which allows the calculation of the absorbed energy profile in materials exposed to particulate radiation. Experiments were designed and conducted on spacecraft thermal control surfaces simulating an absorbed dose profile from synchronous altitude radiation. Several types of finishes were exposed to three and ten year missions and changes in reflectance were measured in vacuum and air. Materials tested include white paints, black paints, metalized second surface mirrors (Kapton, Teflon FEP and quartz), and metals. Relative influence of electrons and protons was determined.

Introduction

A number of studies have been conducted on the effect of electrons and protons on the solar absorptance of various spacecraft thermal control finishes.[1-3] In these studies spectral reflectance changes were measured after exposure to electrons or protons with no attempt made to simulate the space radiation environment. The study reported within is quite different.

A spacecraft operating at synchronous altitude is subjected to a wide energy and flux spectrum of both electrons and

---

Presented as Paper 71-453 at the AIAA Thermophysics Conference, Tullahoma, Tenn., April 26-28, 1971. The radiation tests and optical measurements were conducted by General Dynamics, Fort Worth, Texas, under the direction of John Romanko, Paul Cheever and J. K. Miles.
    *Member of Technical Staff.
    †Section Head.
    ‡Senior Staff Physicist.

protons. Optical damage in thermal control finishes may occur from both the electrons and protons, with the magnitude of spectral reflectance changes a function of the absorbed energy dose. Because the incident particles vary in energy and kind, the absorbed dose profile in a material is not uniform with thickness. The low energy protons, as an example, are deposited near the surface of the material whereas higher energy electrons deposit only a small fraction of their energy near the surface and penetrate quite deeply within the material. The spectral reflectance changes caused by these particles are a function of the absorbed energy dose-depth profile. Thus to predict the behavior of the spectral reflectance of a thermal control finish at synchronous altitude it is necessary to simulate the particulate radiation absorbed dose profile in the laboratory.

When this study was undertaken it was necessary to develop methods of calculating the absorbed dose profile in materials exposed to a given radiation environment. A model was developed and a computer program written for this calculation for both electrons and protons. To reduce machine time the thermal control finishes understudy were divided into two groups according to their stopping power. A single absorbed dose profile was calculated for each group. No material within a group varied greatly from this profile.

Experiments were designed based on the calculated dose profiles. It was found that one proton energy and two electron energies were sufficient to simulate the profile to a depth of about 0.003 in. The absorbed dose profile could not be extended to a greater depth because of the upper energy limit of the accelerator (140 kev) used in these studies.

Irradiation and the spectral reflectance measurements before and after exposure were conducted in vacuum. Following some of the tests, which simulated 3- and 10-yr exposures to particle radiation at synchronous orbit, air was admitted to the test chamber. The reflectance of the samples was remeasured to determine if any spectral reflectance changes occurred due to exposure to air. Spectral reflectance changes that occurred when air was admitted are discussed.

## Space Dose Profile Calculation

Prior to conducting the irradiation tests two studies were undertaken to establish the necessary basis from which the experiments were designed. First a method was developed for the calculation of energy depth-dose profiles for shallow penetrations of electrons and for low energy protons.

Secondly, a method was formulated for a rational approach to the simulation of the space radiation environment with particle accelerators. These studies are briefly reviewed.

In calculating the electron dose-depth profile proper account must be taken of the effects of electron back scattering. The moments-method calculations of Spencer show that errors greater than factor of two will result if backscattering is ignored. If Monte Carlo methods are used to calculate the dose, realistic substrate thicknesses must be used. For this study electron energy dose-depth profiles were developed using the work of Spencer.[4]

The calculation of the dose depth profile for an electron of energy $E_o$ incident normal to the surface is given by

$$D_n(Z) = \phi_{E_o} \; S_o \; J(Z/r_o) \qquad (1)$$

where $\phi_{E_o}$ = flux of electrons of energy $E_o$, $S_o$ = stopping power of material for electrons of energy $E_o$, $J(Z/r_o)$ = Spencers normalized dose-depth data, $Z$ = penetration range, and $r_o$ = range of electron of energy $E_o$.

The space fluence of electrons, $\phi$, at synchronous orbit can be obtained from Vette.[5]

Since the space fluence is omnidirectional it is necessary to convert the results from Eq. (1) to a dose-depth profile from this type of source. This is done, using the Gross transformation,

$$D_o(Z) = (Z) \int_Z^\infty D_N(R) \frac{dR}{R^2} \qquad (2)$$

where $D_N(R)$ = energy dose-depth from a unidirectional normally incident fluence of electrons and $D_o(Z)$ = energy dose-depth for an omnidirectional fluence of electrons.

The calculation of proton energy dose-depth profiles was based on the assumption that protons move in straight lines when passing through a material. Although not rigorously true at low energies, adequate dose-depth profiles can be calculated. As the proton moves along its path and it is assumed that the particle losses energy continuously, the residual energy of the particle is given by

$$R(E) = R(E_o) - t \qquad \text{if } t \leq R(E_o) \qquad (3)$$

where $R(E_o)$ = incident energy of the proton and t = thickness of material traversed by the proton. Calculation of the residual energy after the particle has moved some distance yields the energy deposited in that distance. This is done using Eq. (4):

$$D(t) = \int_0^\infty S_M(E) \phi(E) dE \qquad (4)$$

where $D(t)$ = is the remaining energy to be deposited in the material, $S_M(E)$ = stopping power of the material for proton of energy E, and $\phi(E)$ = differential energy flux spectrum of protons from King.[6] The omnidirectional dose is obtained from a Gross transformation applied to Eq. (4) as for electrons.

A number of materials were tested in this program. Energy dose-depth profiles were not calculated for each material. Instead, materials having similar stopping power were grouped into two classes represented by aluminum and epoxy. Table 1 gives the material classes and those materials falling into each class. Grouping of materials was done to reduce computer time and does not significantly alter the results. Dose-depth profiles for individual materials deviate less than an order of magnitude from the representative curve. Energy dose-depth profiles were calculated for each class of material. These resulting dose-depth profiles for the two classes of materials are given in Figs. 1 and 2 for a 3-year exposure to synchronous orbit.

The dose-depth profiles for a ten year exposure to synchronous orbit were determined by extrapolation. Because some

Table 1  Material classes grouped according to similar stopping power

| Material class | Materials assigned to this class |
|---|---|
| Aluminum | Aluminum<br>Quartz<br>Teflon<br>Inorganic paint (Z-93, H-2)<br>Fiberglass (no binder) |
| Epoxy | Epoxy<br>Kapton<br>Polyester<br>Silicone - alkyd paint (3 M black velvet) |

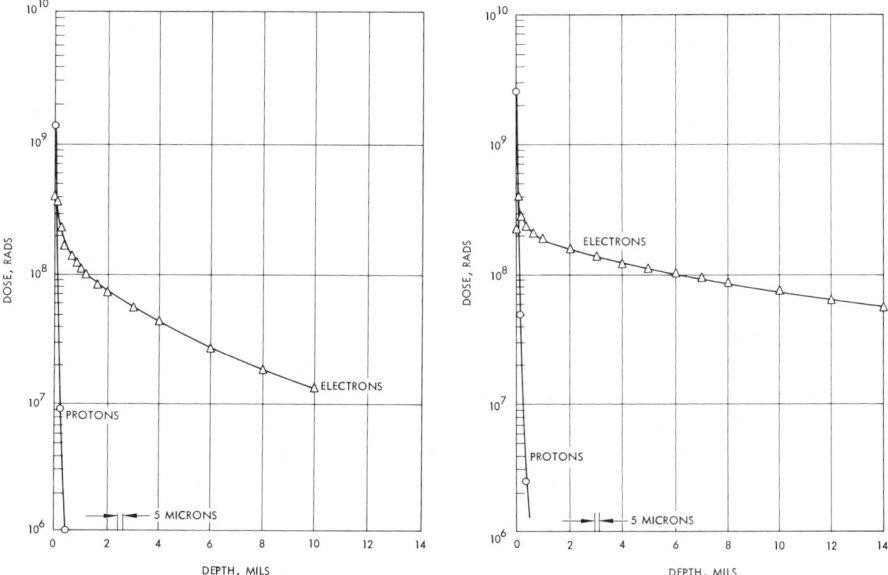

Fig. 1  Energy (Rads) dose-depth profile at synchronous orbit in aluminum after 3 yr of exposure.

Fig. 2  Energy (Rads) dose-depth profile at synchronous orbit in epoxy after 3 yr of exposure.

uncertainty exists concerning the low energy spectrum of the cited data source[5,6] an additional factor of 3 was imposed beyond the linear extrapolation of the 3 year data. This additional factor was felt to be a conservative estimate of the uncertainty in published data, due to limited measurement, and also to account for possible variations in the radiation environment due to solar activity. Thus the ten year data was experimentally established at a fluence approximately 10 times greater than that established for a simulated 3 year exposure.

The next step in the study was to define the experimental parameters of energy and fluence. It was desired to minimize the number of energies and by selecting the appropriate fluence for each energy to simulate as closely as possible the space energy dose-depth profile.

The space dose was converted to an integral dose-depth profile as is shown in Fig. 3. Accelerator particle energies and fluences were selected by starting at the highest available energy and varying fluence and energy until the integral dose curve from the accelerator matched the space dose. The resulting particle energies and fluences are shown in Table 2 and are those used in the test program. Figs. 4 and 5 show the resulting accelerator dose profiles compared to the space dose. Aluminum type materials were matched to depth of about 0.002 in. while epoxy type materials matched the dose profile of space to about 0.004-in. depth. This was the maximum depth possible to which the space dose could be simulated due to the upper limit of particle energy available (140 kev) from the accelerator chosen for this study.

In Fig. 6 aluminized Teflon FEP bonded to a substrate is shown. It was desired to match the dose not only in the Teflon but also in the polyester adhesive. This was accomplished to a depth of nearly 0.003 in. The electron particle fluence and energy was different than shown in Table 2 for this 10-yr exposure. The electron exposure data was, $7 \times 10^{15}$ e-/cm$^2$ of 50 kev and 140 kev electrons.

## Materials Tested

The materials chosen for testing are typical thermal control finishes used on several Hughes synchronous orbit satellites. Paints, metals, second surface mirrors, and a composite structural material were selected. Table 3 describes the surfaces and gives initial solar absorptance of each.

## Facilities

The experiments reported within were conducted at the Nuclear Aerospace Research Facility, Division of General Dynamics, Fort Worth, Texas. The facilities used have been fully described elsewhere.[1,2] A Texas Nuclear

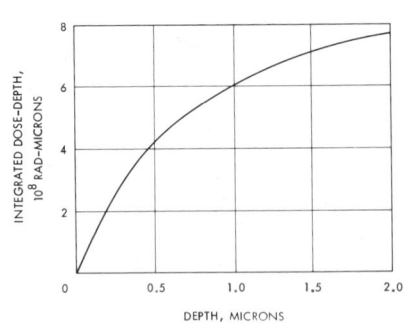

Fig. 3 Integral energy dose-depth curve for protons in aluminum at synchronous orbit after 3 yr exposure

Table 2  Accelerator fluence and rates used in exposing materials to simulated synchronous orbit electrons and protons

| Material class | Electron energy | | Proton energy |
|---|---|---|---|
| | 50 kev | 100 kev | 140 kev |
| Aluminum | | | |
| 3-yr exposure | | | |
| fluence (particles/cm$^2$) | $9.0 \times 10^{13}$ | $1.4 \times 10^{15}$ | $8.3 \times 10^{13}$ |
| rate (particles/cm$^2$/sec) | $1.0 \times 10^{12}$ | $1.0 \times 10^{12}$ | $2.0 \times 10^{11}$ |
| 10-yr exposure[a] | | | |
| fluence (particles/cm$^2$ | $1.1 \times 10^{15}$ | $1.2 \times 10^{16}$ | $4.8 \times 10^{14}$ |
| rate (particles/cm$^2$/sec) | $1.0 \times 10^{12}$ | $1.0 \times 10^{12}$ | $5.0 \times 10^{11}$ |
| Epoxy | | | |
| 3-yr exposure | | | |
| fluence (particles/cm$^2$) | $4.2 \times 10^{14}$ | $1.0 \times 10^{15}$ | $5.5 \times 10^{13}$ |
| rate (particles/cm$^2$/sec) | $1.0 \times 10^{12}$ | $1.0 \times 10^{12}$ | $2.0 \times 10^{11}$ |
| 10-yr exposure[a] | | | |
| fluence (particles/cm$^2$) | $5.1 \times 10^{15}$ | $1.2 \times 10^{16}$ | $3.3 \times 10^{14}$ |
| rate (particles/cm$^2$/sec) | $1.0 \times 10^{12}$ | $1.0 \times 10^{12}$ | $5.0 \times 10^{11}$ |

[a]Includes a factor of 3 to allow for variation in radiation environment due to solar activity.

Model 150-1H Neutron Generator was used as the source of electron and protons. The samples, mounted on a rotating drum, were individually exposed to the radiation beam. Upon completion of radiation of one sample the drum was indexed such that irradiation of the next sample could begin.

The spectral reflectance of each sample was measured using a Perkin Elmer-13U Spectrophotometer and a vacuum mounted, MgO coated integrating sphere. The drum, to which the samples were mounted, was indexed placing a selected sample at the wall of the sphere. The reflected sample beam was referenced to a beam reflected from a MgO coated reference block also at the

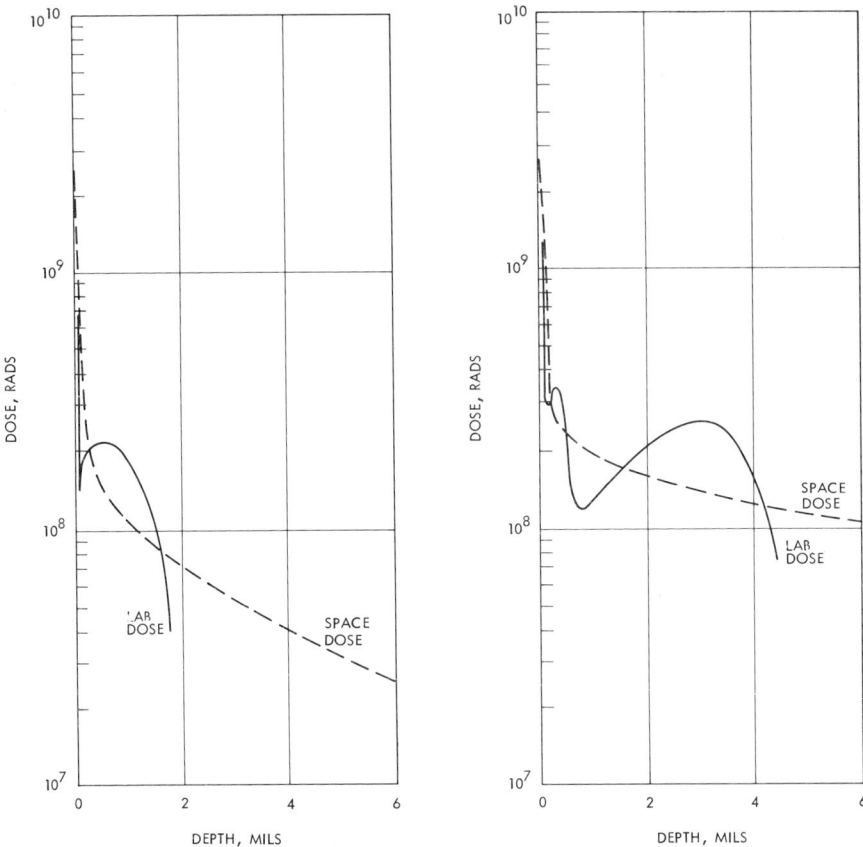

Fig. 4  Total energy dose-depth profile from space and accelerator in aluminum for a 3-yr synchronous orbit exposure

Fig. 5  Total energy dose-depth profile from space and accelerator in epoxy for a 3-yr synchronous orbit exposure

wall of the sphere. Spectral reflectance data was obtained from 0.25 μm to 2.6 μm.

The samples and integrating sphere were mounted in an ion and sublimation pumped vacuum chamber. The attached accelerator and beam tube are oil diffusion pumped (liquid nitrogen trapped) to reduce pressure in that part of the system to $10^{-6}$ torr. The diffusion pump is then valved off and vacuum maintained in the accelerator by the ion pump of the sample chamber.

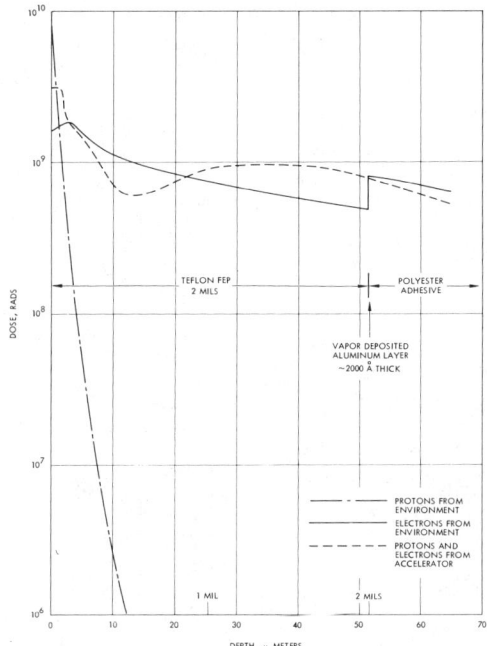

Fig. 6  Energy dose-depth profile in aluminized Teflon FEP (2-mils thick) bonded with polyester adhesive after 10-yr mission in synchronous orbit.

A qualification test was conducted in the facility prior to sample exposure. This was done to insure that the chamber-accelerator system was free of contamination which might influence the observed data. A 7940 quartz second surface mirror (silvered) was chosen as the qualification sample since this material has exhibited no optical changes in tests reported by other workers.[7] Samples were exposed to both protons and electrons simulating later tests. No spectral reflectance changes were found in the measured spectrum after exposure.

## Results

The materials listed in Table 3 were exposed to the fluences of electrons and protons shown in Table 2. All the samples were first exposed to protons and then followed by electron irradiation. The reflectance of some samples was measured following proton exposure only. The reflectance of all samples

Table 3  Thermal control surfaces and the effect of radiation on their solar absorptance

| Thermal control surface | Initial solar absorptance $\alpha_s$ | Change in solar absorptance, $\Delta\alpha_s$ | | | |
|---|---|---|---|---|---|
| | | 3 yr exposure | | 10 yr exposure | |
| | | Protons | Protons plus electrons | Protons | Protons plus electrons |
| **Second Surface Mirrors** | | | | | |
| Teflon FEP, Type A, 2 mils thick, aluminized: | | | | | |
|   as received | 0.14 | +0.00 | +0.00 | NS[a] | NS |
|   Laminated to aluminum foil with polyester adhesive | 0.16 | NS | NS | NM[a] | +0.03 |
| Kapton, 1/2 mil thick aluminized | 0.33 | +0.05 | +0.06 | NS | NS |
| Quartz, Corning 7940, 7 mils thick, silvered | 0.07 | NM | +0.00 | NM | +0.00 |
| **White Paints** | | | | | |
| Z-93 (ZnO in potassium silicate) | 0.18 | +0.04 | +0.07 | NS | NS |
| H-2 (China clay in potassium silicate) | 0.16 | +0.00 | +0.06 | NS | NS |
| **Metals** | | | | | |
| Polished aluminum, 6061-T4 | 0.17 | NM | -0.02 | NS | NS |
| Vapor deposited aluminum on Kapton | 0.10 | NM | -0.01 | NS | NS |
| **Black Paints** | | | | | |
| CTL-15 epoxy | 0.94 | NM | -0.01 | NS | NS |
| 3M Black Velvet | 0.96 | NS | NS | NM | -0.02 |
| **Composite** | | | | | |
| Epoxy fiberglass laminate, carbon black impregnated | 0.95 | NS | NS | NM | -0.04 |

[a] NS - no sample exposed; NM - solar absorptance not measured.

was measured following the electron exposure. The test sequence was completed without breaking vacuum.

The change in solar absorptance resulting from the irradiation of the test samples is shown in Table 3. Reflectance changes are shown in Figs. 7 - 10. Duplicate samples were exposed for all types of materials. Data reported in Table 3 is the average of the two samples, with no sample differing from that reported in Table 3 by more than ±0.01 solar absorptance units. The reflectance data is for selected individual samples.

The effect of air on the radiation damaged spectral reflectance of the test surfaces was determined using a second set of test samples. Three radiation damaged materials exhibited a significant change in reflectance after air was let into the vacuum chamber. Both white paints (Z-93 and H-2) behaved in a similar manner. Damage in the visible spectrum was reduced (bleached) as the reflectance increased when air was admitted.

The reflectance in the infrared (near 2.0 μm) decreased. The infrared reflectance had increased during irradiation due to loss of water from the silicate binder and a reduction in reflectance upon return to air is normal. As noted the bleaching test was conducted on a different sample of Z-93 than shown in Fig. 8. In this second test group the electron dose was one-half that of the previously reported sample. At this lower dose level no infrared spectral damage occurred. Selected spectral reflectance changes resulting from air exposure of irradiated samples are shown in Table 4.

The spectral reflectance of Kapton which had degraded quite severely during the irradiation environment, increased when air was admitted to the vacuum chamber. The reflectance at 0.6 μm increased about 10% and air bleaching was noted over the spectral region of 0.4 μm to 0.9 μm.

No mechanical property tests were conducted in conjunction with this study. The samples were, however, examined after all optical measurements were completed. In the Teflon 10-yr exposure samples, surface cracking, not extending through the sample, was observed. Cracking was not noted in the 3-yr exposure Teflon FEP sample. No other surfaces appeared to have suffered physical property changes in the radiation test.

## Conclusion

A series of tests were designed and performed to develop design information on thermal control surfaces. Data were obtained on the change in solar absorptance due to simulated synchronous orbit radiation. The optical measurements were conducted in vacuum.

Table 4  The spectral reflectance change observed when air was admitted to the vacuum chamber following proton and electron irradiation

| Wavelength, μ meters | Reflectance change, percent | | |
|---|---|---|---|
| | Z-93 White paint | H-2 White paint | Kapton film |
| 0.38 | ... | +4 | +1 |
| 0.45 | +3 | +4 | +1 |
| 0.60 | +6 | +5 | +10 |
| 1.1  | -3 | -2 | +1 |
| 2.1  | -4 | -5 | +1 |

The results indicate that white paints are not stable, low solar absorptance surfaces for the radiation environment of synchronous orbit. The thermal designer must use a second surface mirror material such as quartz or Teflon FEP if stability in solar absorptance is critical. If the mission is planned for long periods, > than 3 yr, and absolute stability is required then quartz is the only available choice.

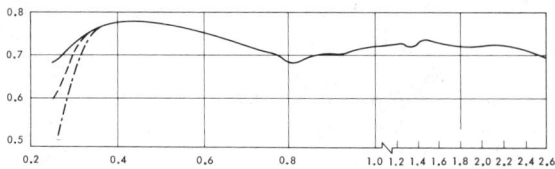

Teflon, FEP, aluminized, 3-yr exposure

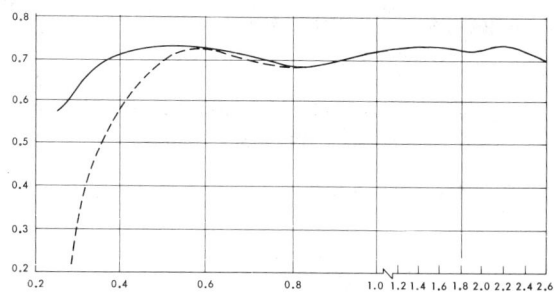

Teflon FEP, aluminized, laminated to aluminum with polyester adhesive, 10-yr exposure

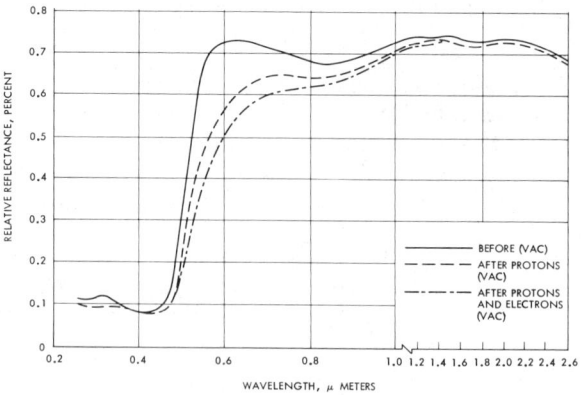

Kapton, aluminized, 3-yr exposure

Fig. 7  Effect of electron and proton irradiation on spectral reflectance.

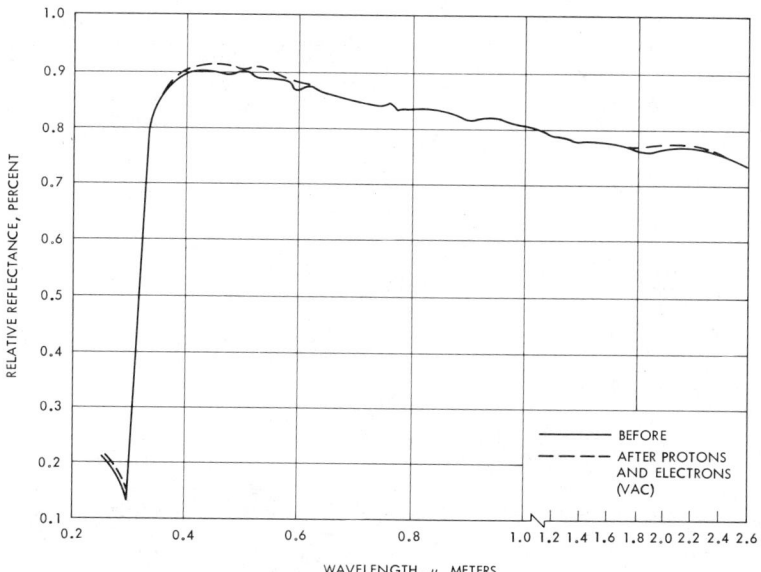

Quartz mirror, 10-yr exposure

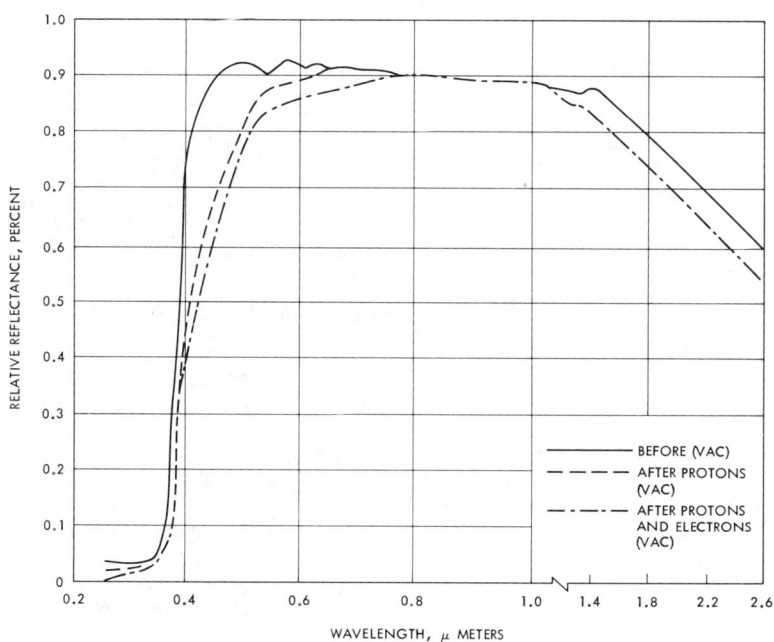

Fig. 8  Z-93 white inorganic paint 3-yr exposure; effect of proton and electron irradiation on spectral reflectance

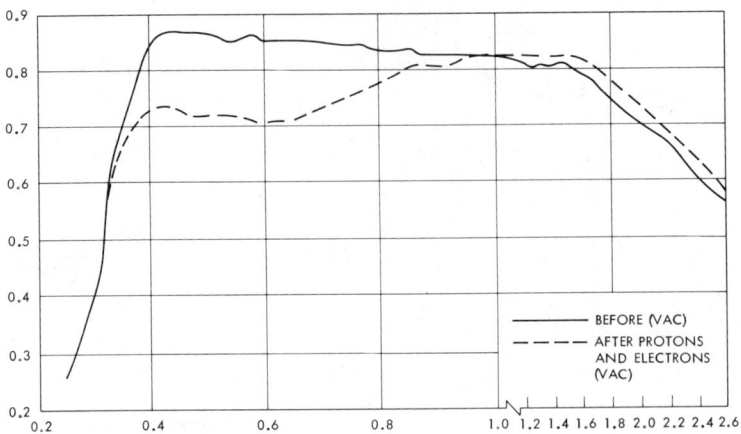
H-2 white inorganic paint 3-yr exposure

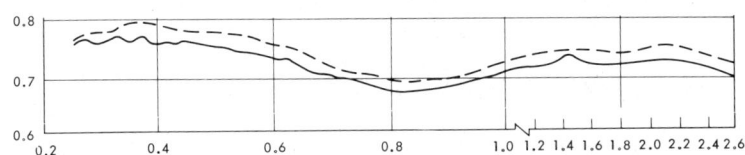
Polished aluminum 3-yr exposure

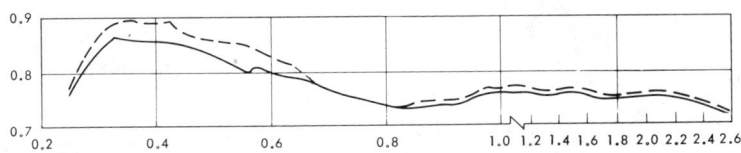
Vapor deposited aluminum, 3-yr exposure

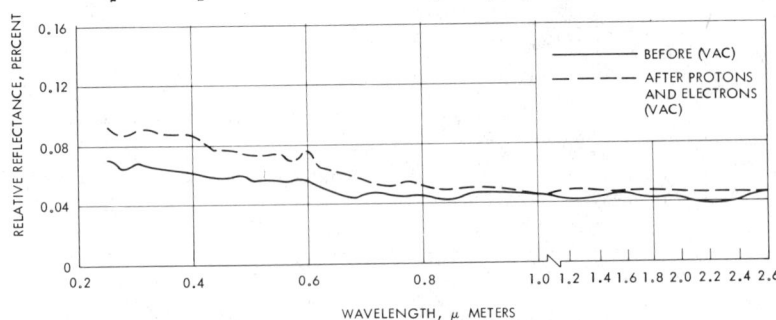
CTL-15 epoxy black paint, 3-yr exposure

Fig. 9  Effect of proton and electron irradiation on spectral reflectance.

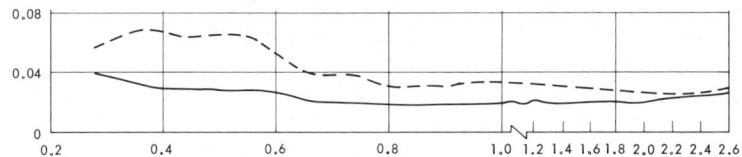

3M black velvet, 10-yr exposure

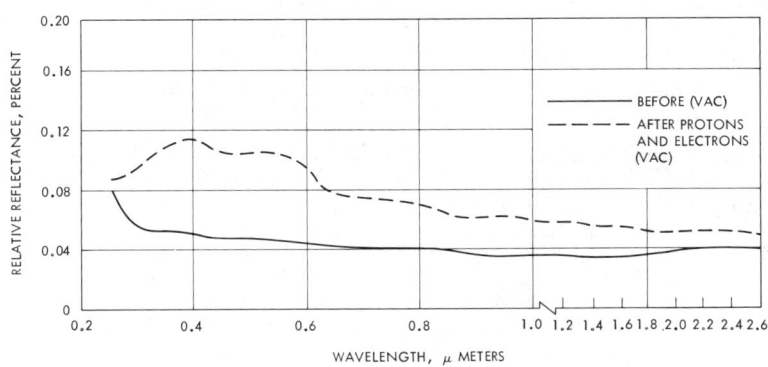

Epoxy fiberglass laminate, 10-yr exposure

Fig. 10 Effect of proton and electron irradiation on spectral reflectance

References

[1]Cheever, P. R., Miles, J. K., and Romanko, J., "In Situ Measurements of Spectral Reflectance of Thermal Control Coatings Irradiated in Vacuo," AIAA Progress in Astronautics and Aeronautics: Thermophysics of Spacecraft and Planetary Bodies, Vol. 20, edited by G. B. Heller, Academic Press, New York, 1967, pp. 281-296.

[2]Miles, J. K., Cheever, P. R., and Romanko, John, "Effects of Combined Electron-Ultraviolet Irradiation on Thermal Control Coatings in Vacuo at 77°K," AIAA Progress in Astronautics and Aeronautics: Thermal Design Principles of Spacecraft and Entry Bodies, Vol. 21, edited by Jerry T. Bevans, Academic Press, New York, 1969, pp. 725-740.

[3]Brown, R. R., Fogdall, L. P., and Cannady, S. S., "Electron-Ultraviolet Radiation Effects on Thermal Control Coatings," AIAA Progress in Astronautics and Aeronautics: Thermal Design Principles of Spacecraft and Entry Bodies, Vol. 21, edited by Jerry T. Bevans, Academic Press, New York, 1969, pp. 697-724.

[4]Spencer, L. V., "Energy Dissipation by Fast Electrons," Monograph 1, Sept. 10, 1959, National Bureau of Standards.

[5]Vette, J. I., et al, "Models of the Trapped Radiation Environment, Vol. III: Electrons at Synchronous Altitudes," NASA SP-3024, 1967.

[6]King, J. H., "Models of the Trapped Radiation Environment, Vol IV: Low Energy Protons," NASA SP-3024, 1967.

[7]Greenberg, S. A. and Vance, D. A., "Low Solar Absorptance Surfaces with Controlled Emittance: A Second Generation of Thermal Control Coatings," <u>AIAA Progress in Astronautics and Aeronautics: Thermophysics of Spacecraft and Planetary Bodies</u>, Vol. 20, edited by G. B. Heller, Academic Press, New York, 1967, pp. 297-314.

# LONG-DURATION EXPOSURE OF SPACECRAFT THERMAL COATINGS TO SIMULATED NEAR-EARTH ORBITAL CONDITIONS

K. E. Steube* and R. M. F. Linford[†]

McDonnell Aircraft Company, St. Louis, Mo.

## Abstract

Thermal control materials intended for use on the NASA Skylab have been exposed to conditions simulating 1 1/2 yr in near-earth orbit. Using two solar constants over the ultraviolet wavelengths, the samples were exposed for 4 1/2 mo; the degradation of their optical properties was determined in situ at frequent intervals. The solar absorptance data obtained during this experiment have provided a basis for the final selection of the Skylab coating materials.

## Introduction

A long-duration experiment has recently been completed at McDonnell Aircraft Company (MCAIR) for the purpose of selecting thermal control coatings for the components of the NASA Skylab program. This program involves a manned, long-duration, near-earth orbital mission. Degradation of the thermal control coatings must be minimal over a period of at least 1 1/2 yr.

The available long-term stability data for the materials of interest were somewhat limited. Some flight data were taken on missions such as Mariner but much of the exposure on these vehicles included the particulate fluxes of the

---

Presented as Paper 71-454 at the AIAA 6th Thermophysics Conference, Tullahoma, Tenn., April 26-28, 1971. The cooperation and advice of E. L. Rusert was most valuable throughout this experiment. The assistance of C. R. Johnson with the data acquisition is also gratefully acknowledged.

*Senior Engineer, Materials Laboratories.

†Senior Engineer, Materials Laboratories.

interplanetary solar wind; Skylab will experience only the limited fluxes of the South Atlantic Anomaly. Of the data generated from laboratory simulated exposures the majority were obtained using greatly accelerated solar intensities. Intensities in excess of 10 solar constants were common. Some rate-dependent effects have been reported[1,2]; therefore it was decided that this experiment would be conducted with a maximum of two solar constants below 400 nm. A xenon arc-lamp was used for the exposure as the simulation of the solar spectrum in this region was considerably better than that obtainable with a mercury lamp. Such attention to the simulation was considered vital; degradation of many coatings is known to be spectrally dependent[3] and the mercury-lamp spectrum is weighted to the 253.7 nm resonance line; this weighting could cause excessive degradation due to the artificially high number of more energetic protons in the exposure.

## Materials

The materials evaluated in this test were 1) aluminized teflon sheet, 0.005 in. thick, type A, FEP grade, supplied by G. T. Schjeldahl Co.; 2) fiberglass cloth with the sizing burned off at 1000°F for 30 min; 3) Tedlar (DuPont 200 XRB 108 WHPVG) film 0.002-in. thick bonded to the black side (Viton A) of DuPont 85-302 thermal curtain; 4) white radiative coating (Z-93) - zinc oxide pigment in potassium silicate binder, 10 mils thick, prepared in accordance with IIT Research Institute (IITRI) specifications; 5) IITRI S13G coating applied over a white epoxy undercoat; and 6) IITRI S13G coating applied over a black epoxy undercoat. (With the exception of the fiberglass material, all of the samples were applied to 7/8-in.-diam aluminum disks, 0.032-in. thick.)

The proposed applications of these materials are tabulated in Fig. 1.

## Equipment

The experiment was conducted in a space simulation facility designed especially for the in situ evaluation of the degradation of optical properties of spacecraft thermal control materials[4] (Fig. 2).

A major problem associated with long-duration exposures of optical surfaces in vacuo is that contamination from the vacuum pumping systems will invalidate the simulation. To reduce this risk, the MCAIR facility was equipped with ion

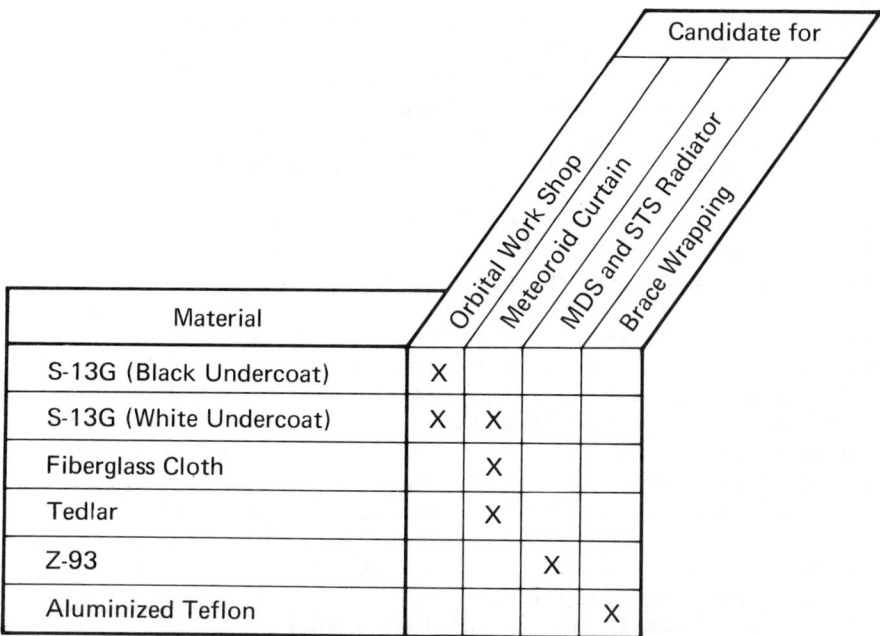

Fig. 1   Test materials.

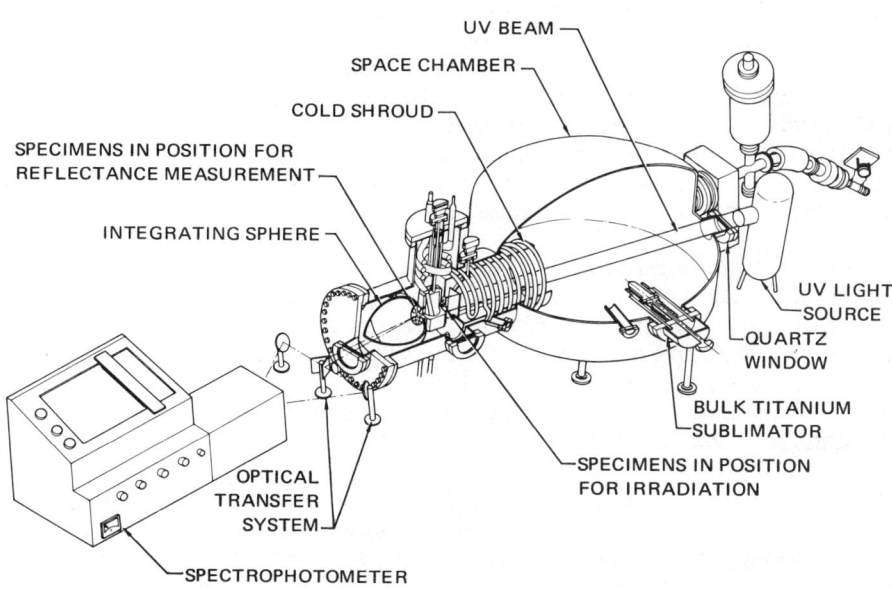

Fig. 2   Test facility.

and titanium sublimation pumps backed by an oil-free turbomolecular pump. An additional precaution, the installation of a system of shields and liquid-nitrogen-cooled shrouds, was taken to prevent the deposition of titanium on the samples during sublimation. As a check on these precautions, a second surface mirror was installed adjacent to the samples, and its optical properties were measured frequently. Throughout the long exposure period of this test, no trace of contamination was detected. During the solar exposures, the pressure in the system was maintained below $1 \times 10^{-7}$ t.

Solar simulation was achieved with a high-pressure, 1000-w xenon arc-lamp. An unfiltered beam was focused on the specimens through a 4-in.-diam quartz window in the chamber wall to provide uniform irradiance in the sample plane. To determine the irradiance at each sample position, a matching array of thermopiles, calibrated against an NBS irradiance standard, was installed. The spectral distribution of the radiation was measured with a filter-radiometer (Eppley model MK 3) and was found to be a good match to the solar spectrum. Approximately 10% of the total radiation incident on the samples was in the 200-400 nm wavelength band. The ultraviolet content of the radiation was taken as this percentage of the indicated total flux incident on each thermopile detector.

The array of thermopiles was mounted on the opposite side of the sample mount, which could be rotated to allow the radiation measurements to be made. The mount was also cooled with liquid nitrogen as part of the sample-temperature control system. Since the thermopile calibration was affected by this cold support as well as by the vacuum environment, the variation in sensitivity was noted during the evacuation and cooling phases of the experiment.

Each of the sample positions was provided with a small heater so that thermal balance could be maintained between the cold support and the thermal input from the xenon lamp. Sample temperatures were measured with a fine-wire thermocouple located on the back of the aluminum substrate. During the uv exposure, the sample temperature was maintained between 70° and 90°F.

Total normal reflectance measurements of the samples were made with a 10-in.-diam integrating sphere mounted inside the vacuum system. The inside of this sphere was coated with a barium sulphate paint,‡ and all measurements were

---

‡Eastman White Reflectance Paint.

referenced to the sphere wall. Both prior to and following the exposure, the reflectance of this reference surface was measured with respect to a freshly prepared magnesium oxide sample. Despite the long duration of the test, no change in the properties of the sphere was detected. The measuring and reference beams were transmitted through an optical transfer system from a Beckman DK2A Spectrophotometer, and the reflectance measurements were made at wavelengths between 350 and 2300 nm. Solar absorptance values were determined by digitizing the data and running it through a 96-point computer integration.

Emittance values were determined with a Gier-Dunkle DB100 Infrared Reflectometer. These measurements were made both prior to the installation of the samples and within 48 hr of the completion of the exposure.

## Experimental Procedure

Emittance measurements were first made on the as-prepared samples. Reflectance measurements were made on the samples as soon as they were installed in the vacuum chamber. A similar set of measurements was taken after the system had been under vacuum for approximately two days. This procedure was designed to detect any changes due to outgassing or contamination of the samples. Following this, the xenon lamp was ignited, the exposure initiated, and thermal balance established. Details of the operational parameters are summarized in Table 1. Sample temperatures were checked each day. Thermopiles were read at each reflectance-measuring session. These sessions were a few days apart at the beginning of the exposure but were stretched to a biweekly schedule as the test progressed.

Table 1  Test parameters

| | |
|---|---|
| Sample Temperature | $70° - 90°F$[a] |
| Pressure | $7 \times 10^{-8}$ torr |
| UV Exposure Intensity | 2 Suns (Approx) |
| Exposure Duration | 7000 ESH |
| In Situ Measurement Interval | 2 Weeks |

[a]Fiberglass sample reached 170°F due to poor contact with chill plate.

At the conclusion of the exposure period, the vacuum system was backfilled to atmospheric pressure, and, after 8 hr elapsed, reflectance measurements again were made on the samples to determine any recovery.

Emittance of all samples was again determined approximately 48 hr after the conclusion of the exposure test.

## Results

Degradation data for the paint samples (Fig. 3) and for the sheet-type materials (Fig. 4) show the following:

1) Z-93 was fairly stable throughout the exposure. The solar absorptance increase ($\Delta\alpha_s$) was only +0.02 after 7000 ESH. No bleaching was detected on returning the sample to ambient pressure.

2) S13G on black epoxy degraded rapidly during the early part of the exposure ($\Delta\alpha_s$ = +0.05 after 800 ESH and then deteriorated further at a slower rate ($\Delta\alpha_s$ = +0.07 after 6600 ESH). On returning to ambient pressure, the sample partially recovered (the net $\Delta\alpha_s$ = +0.04).

3) S13G on a white epoxy base was slightly less unstable than the black-based sample. The degradation was more gradual ($\Delta\alpha_s$ = +0.05 after 6950 ESH). Limited recovery occurred (the net $\Delta\alpha_s$ = +0.04).

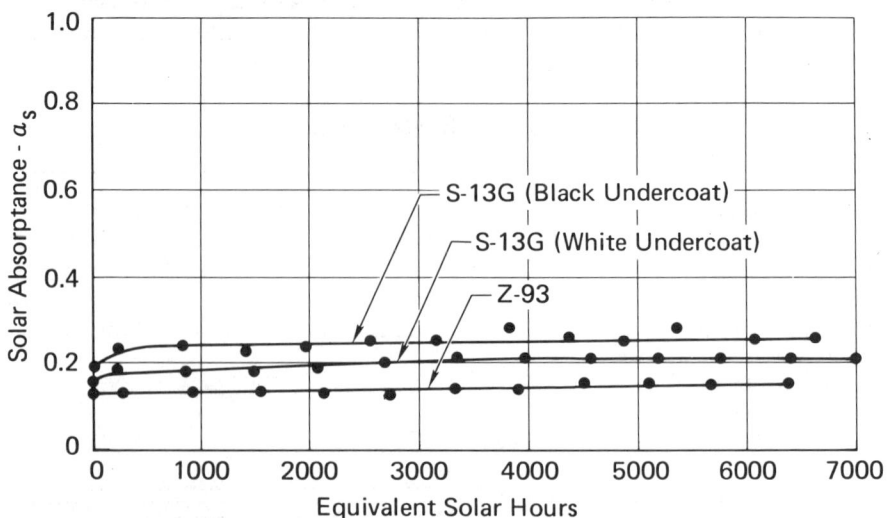

Fig. 3  Degradation data - paint samples.

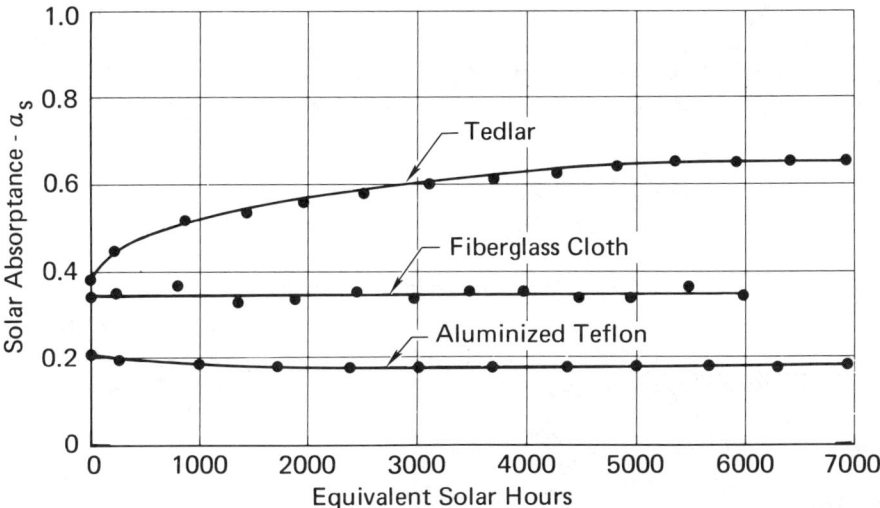

Fig. 4 Degradation data sheet material samples.

4) The aluminized teflon exhibited unusual behavior ($\Delta\alpha_s$ = -0.02 after 1000 ESH). Following this improvement, the sample was entirely stable. No recovery was observed.

5) Fiberglass cloth was completely stable.

6) Tedlar film on thermal curtain material degraded severely. After the first 850 hr $\Delta\alpha_s$ = +0.12, and by the end of the exposure $\Delta\alpha_s$ = +0.27. Minimal recovery was observed. The extent of the damage to the sample is shown in the computer-plotted reflectance curves drawn in Fig. 5.

Table 2 summarizes the solar absorptance data.

The emittance data for the samples are summarized in Table 3. No significant change was observed as a result of the exposure.

## Discussion

The major problems associated with long-duration exposures have been overcome. Continuous exposure of the samples was achieved, except for the short breaks for reflectance measurements and an occasional change of xenon lamp. Repeated checks of the reflectivity of the second surface mirror confirmed that no significant sample contamination occurred.

These checks also indicated that the optical properties of the integrating sphere had not changed. Before and after the test, measurements of a freshly prepared magnesium oxide sample confirmed the stability of the sphere coating as a reference.

Table 2  Total solar absorptance change for test duration

| Material | $\Delta\alpha_s$ | $\Delta\alpha_s$ after recovery | Total equivalent solar hours |
|---|---|---|---|
| Aluminized teflon | -0.03 | -0.03 | 7721 |
| Fiberglass cloth | 0.00 | 0.00 | 6039 |
| Tedlar | +0.27 | +0.26 | 6495 |
| Z-93 | +0.02 | +0.02 | 7067 |
| S-13G (white undercoat) | +0.05 | +0.04 | 6960 |
| S-13G (black undercoat) | +0.07 | +0.04 | 6582 |

Table 3  Total normal emittance before and after exposure

| Material | Infrared emittance Before exposure | After exposure |
|---|---|---|
| Aluminized teflon | 0.823 | 0.825 |
| Fiberglass cloth | 0.895 | 0.896 |
| Tedlar | 0.905 | 0.905 |
| Z-93 | 0.919 | 0.921 |
| S-13G (white undercoat) | 0.898 | 0.900 |
| S-13G (black undercoat) | 0.898 | 0.903 |

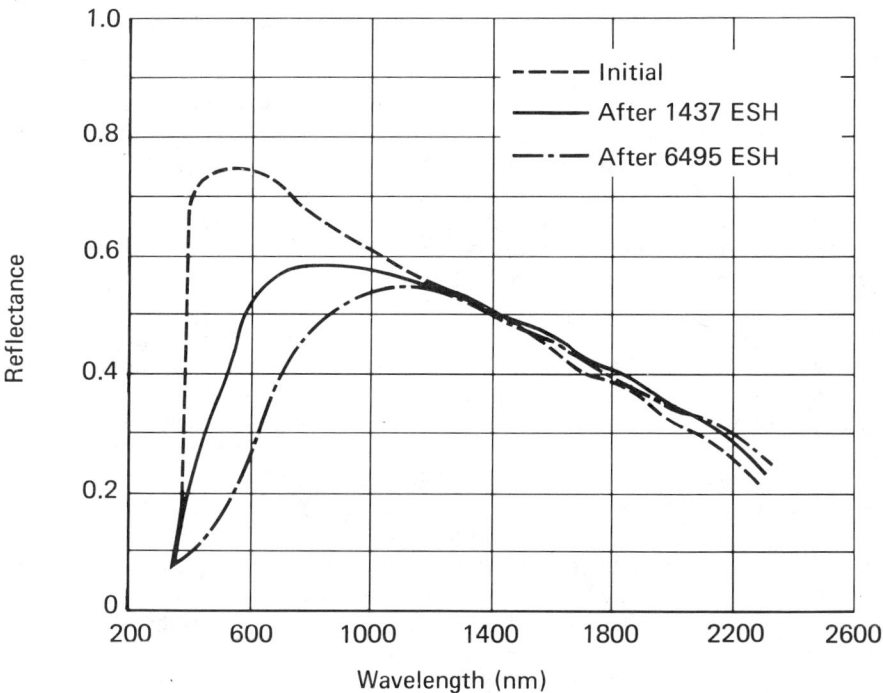

Fig. 5  Spectral reflectance of Tedlar.

From the digitized reflectance data, the computer programs could derive solar absorptance values, absolute spectral reflectance curves, and a plot of the percentage of degradation as a function of wavelength. Such information is valuable for monitoring the progress of the experiment.

Comparisons cannot be made to the results from the aluminized teflon, fiberglass cloth, and Tedlar samples (Fig. 4) inasmuch as little information is available in the technical literature. A decrease in the solar absorptance of the aluminized teflon in the early hours of the test is attributed to the outgassing of surface contamination, such as water vapor and solvent residue. Otherwise this material appears very stable under solar radiation. Some scatter appears in the solar absorptance data for the fiberglass cloth due to the irregularities in this open-weave sample. Tedlar film appears to be most unsuitable for any spacecraft surface exposed to solar radiation. It quickly turned brown in the early stages of the exposure, and the degradation was still worsening at the end of the test.

Both Z-93 and S13G paints have been studied frequently with regard to their stability to solar exposure. Zinc oxide/potassium-silicate paints, like Z-93, have been proved stable both in satellite[5,6] and laboratory[1,2] tests. The degradation observed in these experiments ($\Delta\alpha_s = +0.02$) is very similar to that measured for the Z-93 sample on OSO-III[6], Pegasus[5], and the laboratory data of Zerlaut[7]. However, much larger values of $\Delta\alpha_s$ have been recorded in experiments in which the sample temperature was allowed to rise[8] or greatly accelerated solar exposure were used.[1,2]

Both coating thickness and substrate preparation are expected to affect the solar absorptance of white paints such as S13G. In these tests, the sample undercoated with white epoxy paint had an initial value of $\alpha_s = 0.16$, whereas, for the black-backed sample, $\alpha_s = 0.19$. The nature of the undercoat also affected the degradation rates: $\Delta\alpha_s = +0.05$ for the white undercoat, and $\Delta\alpha_s = +0.07$ for the black undercoat. A possible explanation for this difference is the migration of undercoating constituents into the surface of the coating. The black-undercoated sample also had a greater recovery. Direct comparison of these results with published data is complicated by the substrate contribution, but some agreement can be observed. As with the Z-93 data, measurements made with greatly accelerated solar exposures must be regarded with some suspicion.

## References

[1] Zerlaut, G. A., Harada, Y., and Thompkins, E. H., "Ultraviolet Irradiation of White Spacecraft Coating in Vacuum," SP-55, p. 391, 1965, NASA.

[2] Streed, E. R. and Arvesen, J. C., "A Review of the Status of Spacecraft Thermal Control Materials," 7MX-60406, 1967 NASA.

[3] Arvesen, J. C., "Spectral Dependence of Ultraviolet-Induced Degradation of Coatings For Spacecraft Thermal Control," Progress in Astronautics and Aeronautics: Thermophysics of Spacecraft and Planetary Bodies, edited by G. B. Heller, Vol 20, 1967, p. 265.

[4] Clifford, D. W. and Schmitt, R. J., A Space Simulation Facility for In Situ Evaluation of Spacecraft Material Degradation, IES/AIAA/ASTM 3rd Space Simulation Conference, Proceedings, p. 157, 1968.

[5]Schafer, C. F., and Bannister, T. C., "Thermal Control Coating Degradation Data From the Pegasus Experiment Packages," AIAA Progress in Aeronautics and Astronautics: Thermophysics of Spacecraft and Planetary Bodies, edited by G. B. Heller, Vol. 20, 1967, pp. 457-473.

[6]Millard, J. P., "Results From the Thermal Control Coatings Experiment on OSO-III, "Progress in Aeronautics and Astronautics: Thermal Design Principles of Spacecraft and Entry Bodies, edited by J. T. Bevans, Vol. 21, 1969, p. 769.

[7]Zerlaut, G. A., Study of In Situ Degradation of Thermal Control Surfaces," IIT Research Institute Rept., IITRI-U6061-9, 1967.

[8]Greenberg, S. A., Interim Report, Contract NAS3-7630, NASA.

# I Surface Radiation Properties
## I.2 Contamination Effects

# SENSITIVITY OF THERMAL SURFACE SOLAR ABSORPTANCE TO PARTICULATE CONTAMINATION

O. Hamberg[*] and F. D. Tomlinson[†]

The Aerospace Corporation, El Segundo, Calif.

## Abstract

The sensitivity of thermal surface solar absorptance to particulate contamination was evaluated. Experiments were conducted using a mirror surface that was incrementally obscured with a limited number of contaminant types and sizes. Aluminum oxide and fly ash particles were selected to bound high and low contaminant reflectance extremes. Size sensitivity was demonstrated by using 3- and 20-$\mu$ aluminum oxide particles. Measurements were made of the obscured surface areas and the corresponding changes in reflectance. From these, the related solar absorptance changes were derived analytically. The changes in absorptance indicated high sensitivity to contaminant composition and quantity, low sensitivity to particle size, and linearity with obscured area.

## Nomenclature

$E$ = solar energy absorbed, w/cm$^2$

$H_\lambda$ = solar spectral irradiance, w/cm$^2$-$\mu$

$\alpha_\lambda$ = spectral absorptance

$\alpha_s$ = solar absorptance

$\lambda$ = wavelength of light, $\mu$

$\rho$ = spectral hemispherical reflectance

$x$ = fraction of total mirror surface contaminated

---

Presented as Paper 71-473 at the AIAA 6th Thermophysics Conference, Tullahoma, Tenn., April 26-28, 1971.

[*]Section Manager, Equipment Section.

[†]Member of the Staff.

## Introduction

The particulate contamination considerations of thermal surfaces require a knowledge of the particles that will settle and remain on the surface and the resulting changes in absorptance and emittance. The sensitivity of thermal control surfaces to contamination can be determined experimentally by artificially contaminating the surfaces with a variety of materials and measuring the resulting effect.

An experiment was designed to contaminate second surface mirrors artificially to various degrees and to measure the respective obscured area plus the reflectance change of the surfaces. The related solar absorptance changes were derived analytically from the spectral reflectance measurements.

The projected surface area obscured by the contaminants was selected as the parameter to measure contamination level. This was based on the hypothesis that the reflectance of a contaminated surface would change basically in proportion to the percentage of its surface area obscured. If the reflectance of a clean mirror equals $\rho_1$, the reflectance of the contaminant equals $\rho_2$, and the fractional area of the mirror obscured by the contaminant equals x, then the total reflectance of the clean mirror portion would be $\rho_1(1 - x)$, the total reflectance of the contaminant portion would be $\rho_2 x$, the total reflectance of the contaminated mirror would be $\rho_1(1 - x) + \rho_2(x)$, and the change in reflectance as a result of contamination would be

$$\Delta \rho = (\rho_1 - \rho_2) x \qquad (1)$$

The change in absorptance for an opaque surface would then be

$$\Delta \alpha = (\alpha_2 - \alpha_1) x \qquad (2)$$

## Experiment

The samples selected for the experiment were second-surface thermal mirrors with a specified solar absorptance not greater than 0.06 at wavelengths from 0.3 to 2.5 µ. They measured 1 × 1 in., were 9 mils thick, and were made of quartz with a metallic silver coating. Figure 1 shows the spectral reflectance of a typical mirror made to this specification.

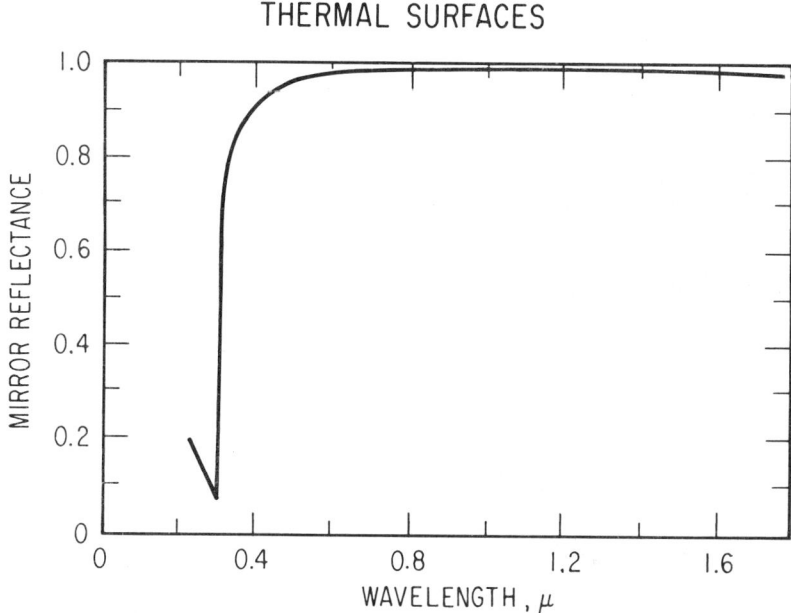

Fig. 1 Mirror spectral reflectance.

Aluminum oxide and fly ash were selected to present the extreme characteristics of high and low reflectance contaminants. It was also desired to determine if different changes in reflectance would occur for the same type and degree of contamination, but using different particle sizes. For this purpose 3- and 20-µ nominal-size aluminum oxide particles were selected. The fly ash mean size was approximately 4µ as determined by photomicrographic inspection during the course of the experiments.

The device used to contaminate the test specimens is shown in Fig. 2. It consisted of a test-tube container from which a dust cloud was sprayed into a chamber. The dust entered into a 3-1/2 in.-diam. by 11-in.-high chamber through a bottom opening. The mirror specimen was mounted on a special holder and held external to the chamber until the cloud materialized. At that time, the specimen was inserted through a slit in the chamber wall. Particles from the cloud were allowed to fall out and settle on the surface. After a suitable settling period, the mirror was withdrawn. Seven mirrors were used in the tests conducted with each contaminant. Each mirror was cleaned prior to the test and its reflectance was measured. Six of the mirrors were then incrementally contaminated, and their contaminated reflectances were compared with their clean reflectances to obtain changes. Following this, the amount of area obscured by the contaminant was measured. The seventh mirror was maintained clean for calibration and served to represent zero contamination. The final measurement was the 100% contamination of the mirror. This was simulated by packing a watchglass with the contaminant and obtaining its spectral reflectance.

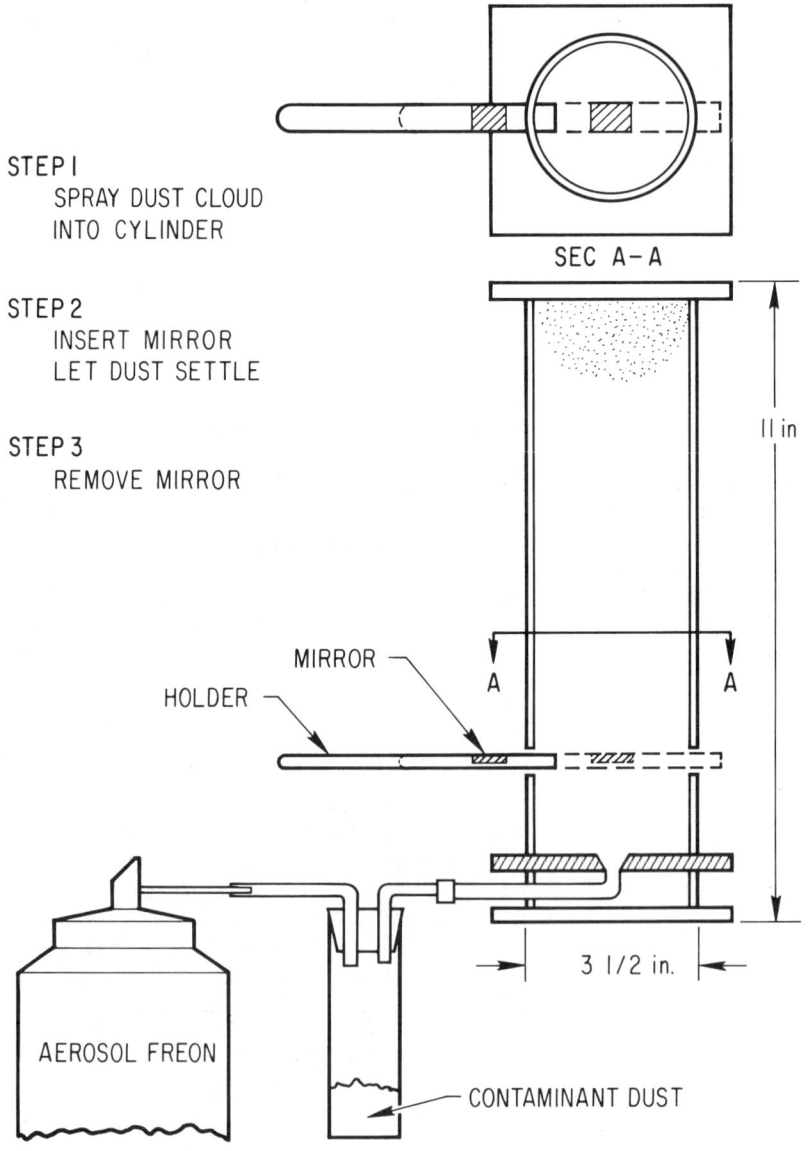

Fig. 2 Artificial contamination device.

Photomicrographs of 100X magnification were used to count and size particles. The projected surface area of the particles was measured manually on the photographs by projecting the photomicrographs on a grid. When magnified, a 1-µ particle had an apparent diameter of 1/32 in. and could be resolved readily. The total area obscured was obtained by estimating the percentage area of each grid which was obscured by particles. The relative accuracy of this type of measurement was evaluated by having several observers independently measure the area obscured in several projected photographs using this method. The maximum variation was found to be less than ± 12%. The actual measurements of obscured area used to obtain the experimental data were made by a single observer. The repeatability of the data obtained in this manner is estimated to vary by less than ± 5%. Figure 3 shows a typical photomicrograph of 20-µ aluminum oxide. For this sample, the obscured area measured 10.9%.

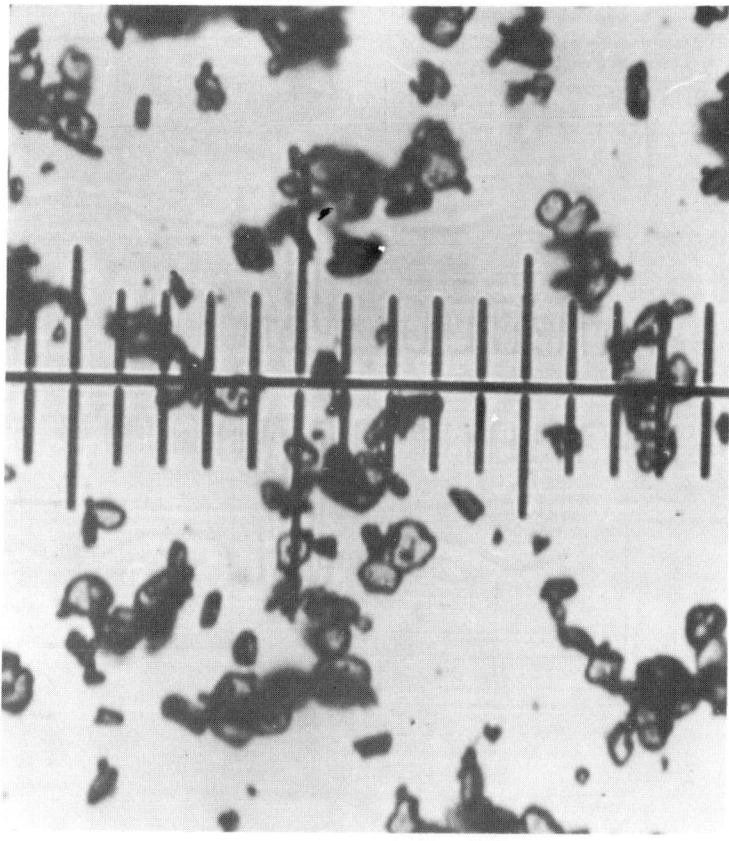

Fig. 3   Photomicrograph (20-µ aluminum oxide).

It can be observed that the distribution of particles within the area of approximately 0.08 mm$^2$ covered by the photograph was not uniform and that some agglomeration of particles had taken place. Since obscured area was the parameter of interest, the agglomeration was not considered a problem.

Figure 4 shows the locations on the specimen where photographs were taken. An average of six photographs spaced at equal intervals were taken at each contamination level. Since a relationship between change in reflectance and area obscured by the contaminants was desired, it was necessary only to measure the contamination level of the area seen by the reflectometer beam. The beam width is variable, depending on the wavelength of the monochromatic light beam, and measured from 0.2 to 1.8 mm over the range of wavelengths selected. The length is constant at 12 mm. The projected area of the beam therefore varied from 2.4 to 21.6 mm$^2$. It was estimated that the photographs fell within a tolerance band of ±1 mm relative to the projected beam area. The assumption was then made that the surface area obscured within this tolerance band was reasonably representative of the obscuration within the projected beam area.

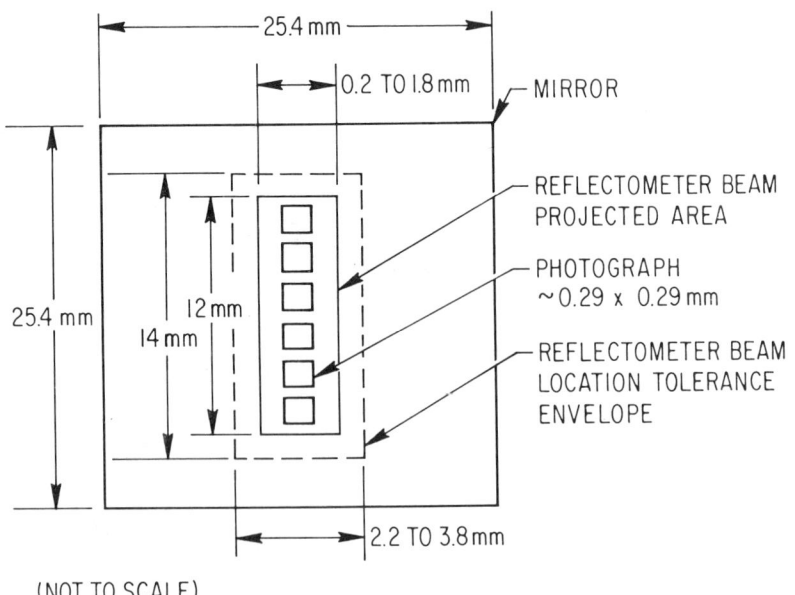

Fig. 4 Photograph locations.

The degree of sampling of the total area of the mirror and that portion enclosed by the beam and its tolerance band is shown in Table 1. The decision to take an average of six photographs within the beam envelope provided average measurements of 0.9 to 1.6% of the area potentially illuminated by the reflectometer beam.

Table 1  Area sampled

|  | Area, mm$^2$ | % mirror area | % envelope area |
|---|---|---|---|
| Mirror | 645 | 100 | ... |
| Beam envelope | 31 - 53 | 4.8 - 8.2 | 100 |
| Six photos | 0.49 | 0.07 | 0.9 - 1.6 |

The changes in reflectance as a result of contamination were measured with a Gier-Dunkle model no. 4A Integrating Sphere reflectometer, which is equipped with a Perkin Elmer model 98 Monochromator plus associated electronics and a data recording system. Six wavelengths ranging from 0.25 to 1.6µ were selected.

## Test Results and Discussion

The percent area obscured is the mean value from the measured projections. The average number of projections used to obtain this mean was six per contamination level. Table 2 shows typical variabilities of the area obscuration measurements as represented by data for the 3-µ aluminum oxide. The mean value of the obscuration measurements made at each contamination level was used to represent the total percent area obscured. It can be noted that the variability of obscuration as represented by the standard deviation of the individual measurements was relatively high at the low obscuration levels and became a lower percentage of the mean area obscured as the contamination increased. As discussed previously, the single observer's repeatability was estimated as ±5%. The variability of the measurements above 5% can therefore be attributed to nonuniformity of the contaminant distribution over the observed area. This nonuniformity contributed to some inaccuracy in the derived mean area; however, no specific value can be assigned to the error from the true mean. For future experiments, it is planned to increase the number of measurements or decrease the variability in the contamination distribution at the lower contamination levels.

Table 2  Typical variability of obscuration measurements (3-μ aluminum oxide).

| Measurement | Contamination levels | | | | | |
|---|---|---|---|---|---|---|
| | #1 | #2 | #3 | #4 | #5 | #6 |
| | Percent area obscured | | | | | |
| 1 | 1.28 | 4.96 | 13.89 | 25.76 | 31.01 | 36.42 |
| 2 | 2.15 | 4.89 | 9.83 | 24.10 | 30.40 | 32.15 |
| 3 | 2.56 | 10.51 | 9.50 | 24.21 | 29.50 | 31.99 |
| 4 | 1.67 | 8.14 | 14.74 | 21.69 | 27.50 | 29.65 |
| 5 | 1.96 | 9.49 | 10.17 | 26.79 | 29.08 | 33.29 |
| 6 | 3.18 | 6.44 | 11.46 | 23.36 | ..... | ..... |
| 7 | 2.06 | 5.26 | ..... | ..... | ..... | ..... |
| Mean percent area obscured | 2.12 | 7.09 | 11.60 | 24.32 | 29.50 | 32.70 |
| Standard deviation of percent area | 0.58 | 2.16 | 2.0 | 2.0 | 2.14 | 1.48 |
| Standard deviation as percent of mean area | 27 | 30 | 17 | 8 | 7 | 5 |

Figure 5 shows typical data points derived by measuring the changes in monochromatic reflectance and plotting them against the measured percent area obscured for the 3-μ aluminum oxide. A straight line fit of the data was obtained by means of a least-square fitting procedure forced through the origin.[1] Most of the data correlated reasonably well to the straight lines obtained, as judged by visual observation. A better fit of the data was obtained at the high wavelengths than at the low wavelengths. Figures 6 and 7 show the best straight line fits for the data obtained from 3-μ aluminum oxide, and fly ash, respectively.

Relative to the 3-μ aluminum oxide-contaminated mirror (Fig. 6), it can be noted that the reflectance increased at a wavelength of 0.25μ, increased even further at 0.30μ, and then decreased at 0.40μ compared to a clean mirror. At 0.65, 1.2, and 1.6μ the reflectance increased slightly compared to 0.4μ.

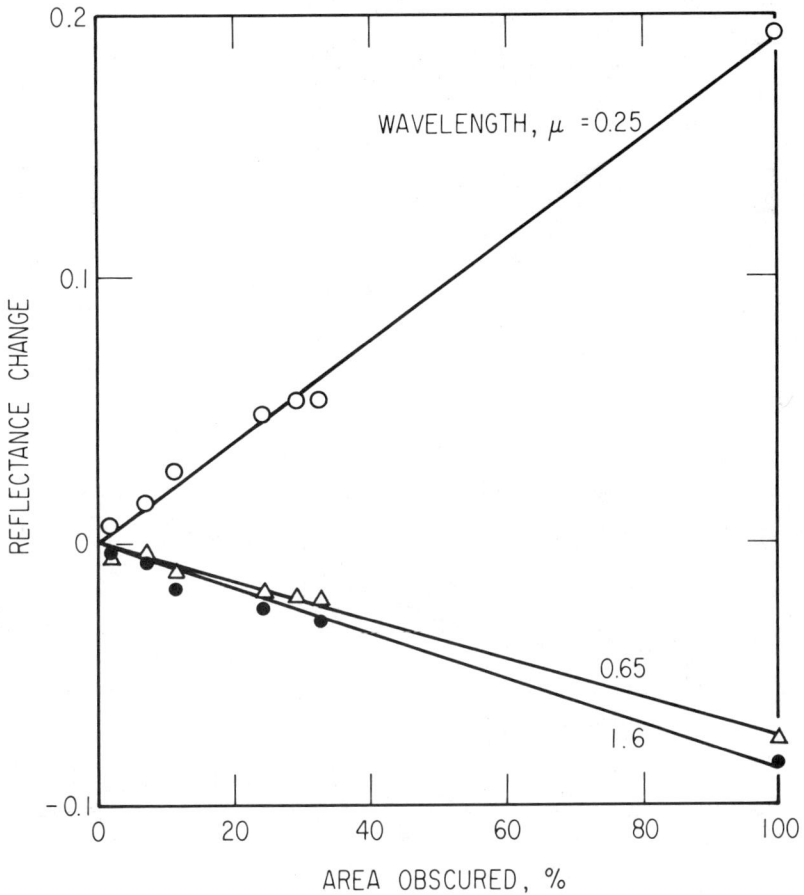

Fig. 5  3.0-μ aluminum oxide contaminant.

For the 20-μ aluminum oxide changes of a similar nature occurred, and good agreement with the 3-μ best fit lines was obtained. The fly-ash-contaminated mirrors (Fig. 7) showed a decrease in reflectance at all wavelengths measured. It can be noted that for a given percent obscured area the reflectance changes for fly ash are much greater than for the aluminum oxide at all wavelengths.

Figure 8 is shown to provide a rationale for the observed reflectance of contaminated mirrors. The bottom curve labeled "fly ash" was obtained by measuring the reflectance spectrum of fly ash packed on a watchglass. This curve would be equivalent to having the mirror surface covered 100% with fly ash. The aluminum oxide reflectance was derived in a similar manner and would be equivalent to 100%

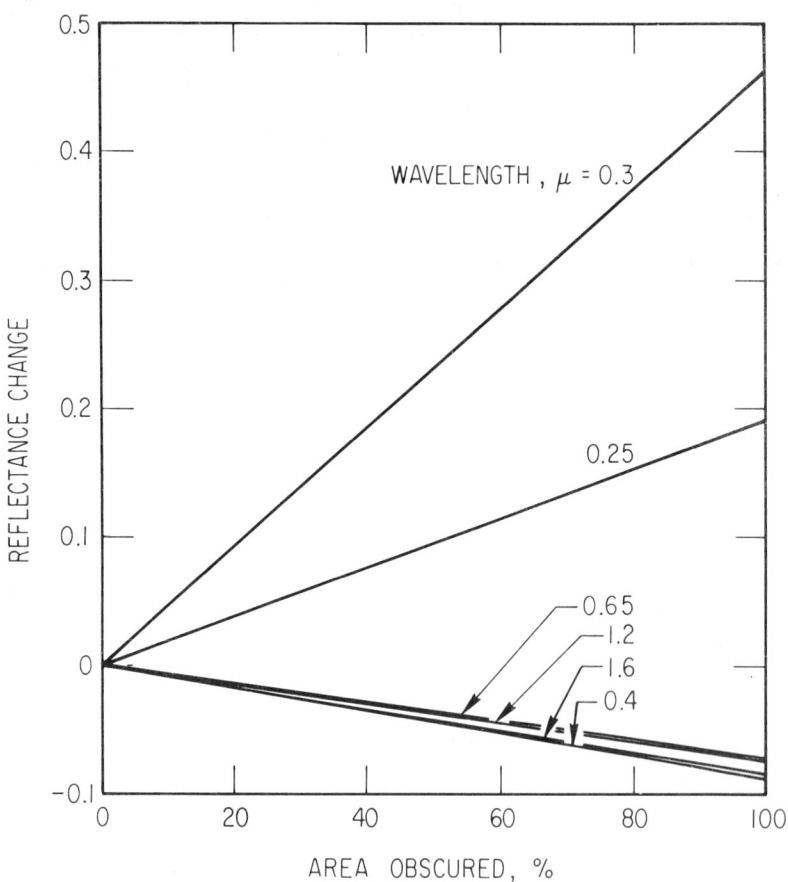

Fig. 6  3.0-μ aluminum oxide contaminant.

aluminum oxide obscuration. For reference, the reflectance curve of a clean mirror is shown.

When comparing the clean mirror reflectance with the aluminum oxide reflectance, it can be noted that the greatest difference in reflectance exists at the 0.32-μ wavelength and that the aluminum oxide has a higher reflectance than the clean mirror at wavelengths of 0.25 to approximately 0.33μ. This observation explains why the greatest increase in reflectance of contaminated mirrors was found at 0.3-μ wavelength for the 3- and the 20-μ aluminum oxide. Similar observations and corresponding explanations can be noted at the other wavelengths.

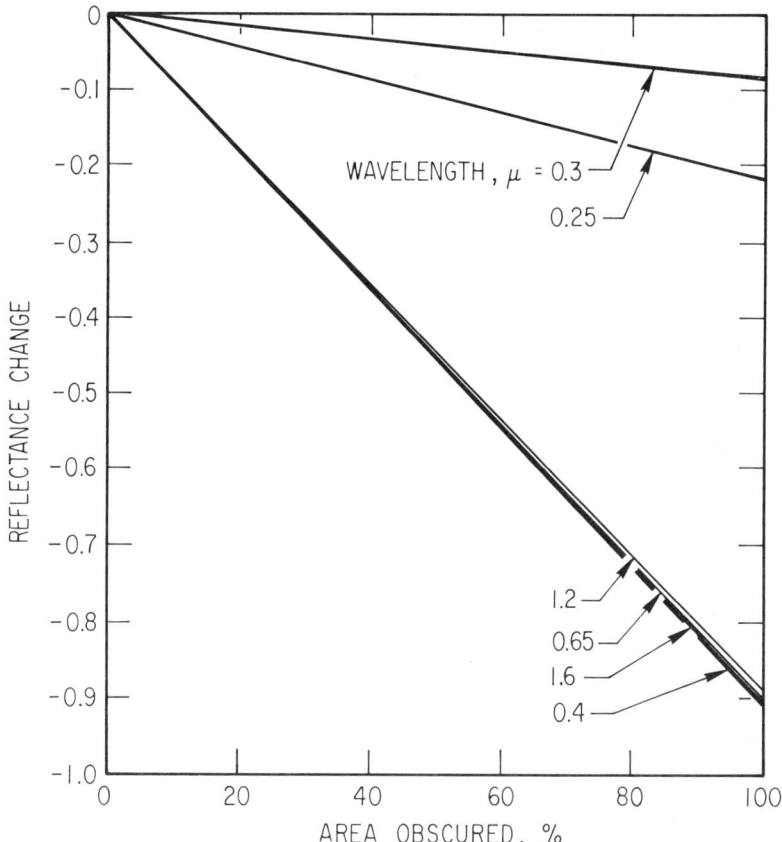

Fig. 7 Fly ash contaminant.

For the fly ash contaminant, Fig. 8 shows that the least difference in reflectance between the clean mirror and the fly ash also exists at a wavelength of $0.32\mu$ and that the clean mirror has a higher reflectance than the fly ash. This observation explains why the least change in reflectance of contaminated mirrors was found at $0.3\mu$ (Fig. 7) and why the change decreased the reflectance. As in the case of aluminum oxide, similar observations can be noted at the other wavelengths.

It is of interest that the spectral changes in reflectance measured were in good agreement with the theory postulated in Eq. (1).

Using the experimental data, the solar absorptance changes were calculated by assuming a Johnson solar spectral

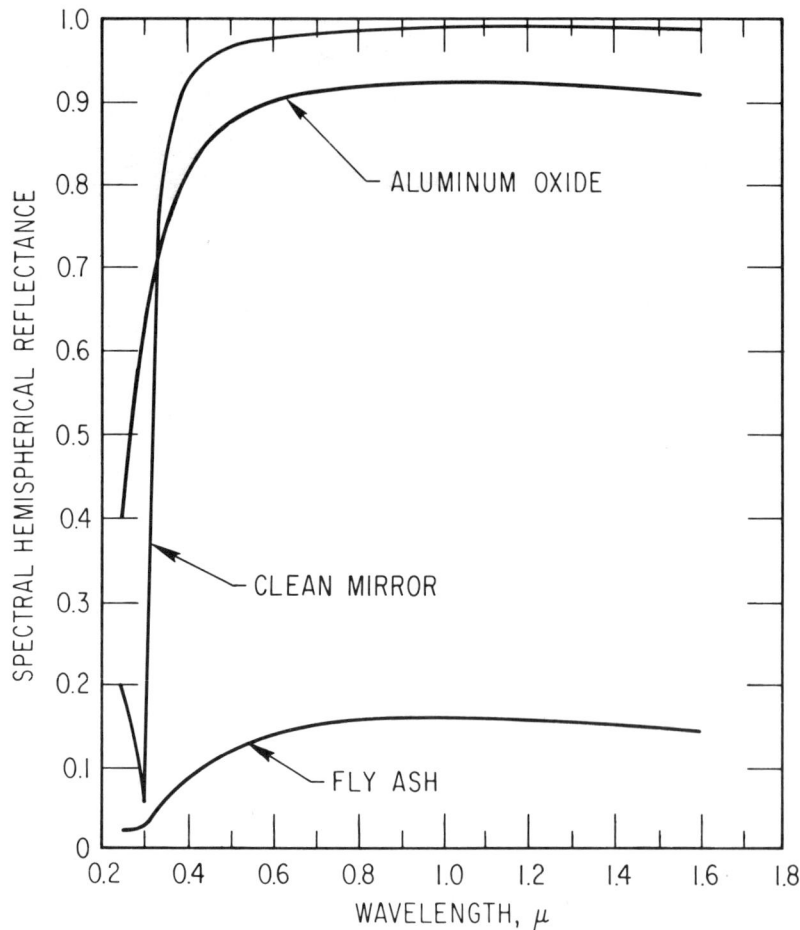

Fig. 8 Mirror and contaminant reflectances.

irradiance distribution.[2] The curves for the contaminated mirrors (Fig. 9) were constructed by fitting the general shape of the clean mirror curve to the experimental reflectance data. By integrating the resultant curves with respect to the solar energy spectrum shown on the Johnson curve, the solar absorptance was computed using the following relationship.[3]

Solar energy absorbed is

$$E = \int_{0.25}^{2.5} \alpha_\lambda H_\lambda \, d\lambda \qquad (3)$$

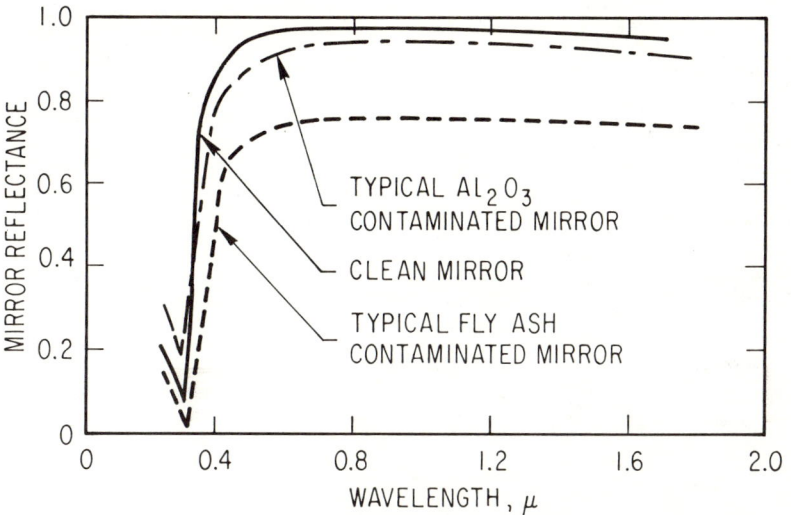

Fig. 9  Contaminated mirror reflectance.

Solar absorptance is

$$\alpha_s = \frac{\int_{0.25}^{2.5} \alpha_\lambda H_\lambda \, d\lambda}{\int_{0.25}^{2.5} H_\lambda \, d\lambda} \qquad (4)$$

Figure 10 shows the results of this analysis. It can be seen that, for the same obscured area, the 20- and 3-μ aluminum oxide particles produced approximately equal changes in solar absorptance, with the 3-μ particles producing slightly higher changes. A comparison of the solar absorptance changes produced by the fly ash with the absorptance changes produced by equal obscured areas of aluminum oxide shows that the fly ash produced changes that are higher by an order of magnitude. Changes in solar absorptance due to particulate contamination by aluminum oxide and by fly ash were found to be basically linear functions of the obscured area. The changes were found to be in approximate agreement with the hypothesis postulated in Eq. (2) that the absorptance of a contaminated surface changes in proportion to the amount of area obscured by the contaminants and in proportion to the difference of the absorptances of the clean surface and the contaminant.

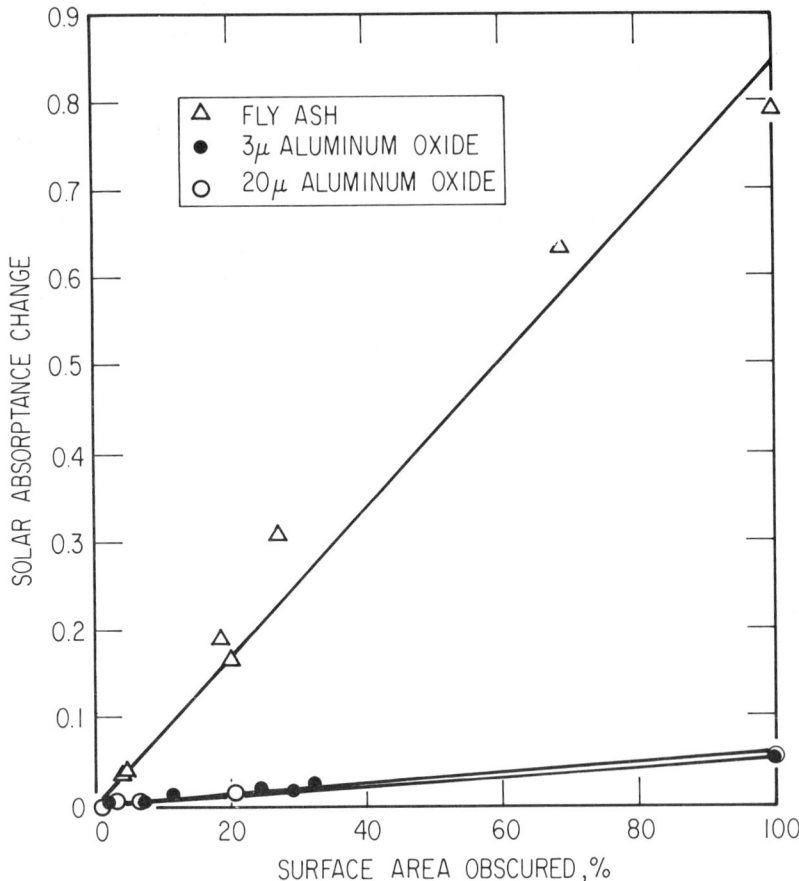

Fig. 10 Solar absorptance change vs surface area obscured.

## Conclusions

As a result of the study it was concluded that, for the type of surface examined, changes in solar absorptance are highly sensitive to contaminant compositon and quantity. The standards used to specify the cleanliness of this type of surface should therefore recognize not only the quantity of particles but also the solar absorptance of the contaminants. Since thermal performance of a surface is a function of emittance as well as absorptance, changes in emittance as a result of contamination will also have to be considered.

The significant linear correlation between changes in absorptance and obscured area, as expected from theory, and the insensitivity to size as demonstrated by the aluminum oxide, indicate that relatively simple techniques to predict changes in absorptance possibly can be developed. However, it should be noted that the data available at this time are too limited in the types and sizes of surfaces and contaminants considered and in the number of experimental samples measured to enable prediction of absorptance changes under untested conditions. It is considered necessary to conduct further experiments to validate a general theory.

## References

[1] Acton, F.S., <u>Analysis of Straight Line Data</u>, John Wiley & Sons, Inc., New York, 1959.

[2] Glasstone, S., <u>Sourcebook on the Space Sciences</u>, D. Van Nostrand Co., Inc., Princeton, N.J., 1965.

[3] Wiebelt, J.A., <u>Engineering Radiation Heat Transfer</u>, Holt, Rinehart, and Winston, New York, 1966.

# MEASUREMENT AND APPLICATION OF CONTAMINANT OPTICAL CONSTANTS

R. C. Linton*

NASA Marshall Space Flight Center, Huntsville, Ala.

## Abstract

The degrading effects of contamination on optical surfaces are predictable if the optical constants of the contaminant layers are known. Experimental and analytical procedures have been developed to enable calculation of the constants from data obtained by standard optical measurements of contaminated optical surfaces. The application of this approach to the analysis of data from real-time contamination monitors is discussed. Comparisons of theoretically calculated degradation to measured changes in optical properties indicate that the effects of contamination are dependent on the observed wavelength, contaminant layer thickness, and optical constants.

## I. Introduction

Recently, interest in real-time monitoring of accumulated contamination in space has emerged. The quartz-crystal microbalance (QCM) provides one effective way of indicating the quantity adhering to certain types of surface and, assuming homogeneity and the density, similarly can indicate the contaminant layer thickness.[1] As different types of contaminants of the same quantity may have quite different optical effects, evaluation of the effects of contamination in space for understanding the potential damage to optical surfaces requires representative optical measurements.

In selecting optical measurements to describe the effects of contamination in a certain spectral region, the optical properties such

---

Presented as Paper 71-460 at the AIAA 6th Thermophysics Conference, Tullahoma, Tenn., April 26-28, 1971.

*Physicist, AST Solar Studies Group, Space Thermophysics Division, Space Sciences Laboratory

as specular reflectance, transmittance, and scattering are generally monitored.[2] The index of refraction n and the extinction coefficient k of material are optical constants that, if known, can define the reflective and transmissive changes of a contaminated optic as a function of thickness, wavelength, and measurement angle of incidence. This paper presents several techniques for calculating the optical constants of materials from data obtained in the laboratory and in space by various experimental approaches. A discussion of the application of these type data to the analysis of contamination-monitoring measurements is included. Examples of the predicted degradation of gold-, platinum-, and magnesium-fluoride-overcoated aluminum mirrors are given using the optical constants of a contaminant film determined by measurements typically available in optical monitors.

## II. Analytical Approach

Reflection and absorption experiments are customarily discussed in terms of the complex refractive index N, which, by means of the relation

$$N = n - ik \tag{1}$$

contains the index of refraction n and the extinction coefficient k. These latter constants are spectrally dependent but, assuming an isotropic, homogeneous layer, are thickness independent. As an example of the relation of these constants to measurable optical properties, the normal incidence reflectance R of an opaque material is given by

$$R = \left[(n-1)^2 + k^2\right] / \left[(n+1)^2 + k^2\right] \tag{2}$$

The determination of the optical constants of a material is generally accomplished by numerical techniques developed for solutions of Maxwell's equations with appropriate boundary conditions between two or more media. The polarized components of reflected and transmitted light are given by the resulting Fresnel equations, which contain as variable parameters the thickness, optical constants, wavelength, polarization, and angle of incidence of the source light in measurements. For a single layer of an opaque film, measurement of reflected light intensities at two very different angles of incidence can provide sufficient data to determine the optical constants at a given wavelength.[3] To determine the optical constants of one component of a multilayered optic such as a contaminated mirror, the optical constants of the substrate mirror film(s) and the contaminant layer thickness must be known.

Two computer programs that function by numerical solution of these equation forms with inputs from two different types of optical measurements have been developed for determining optical constants of contaminant layers. The intent is to provide, with an optical monitor, a "quick-look" evaluation of optical damage which, at the same time, generates input for determination of the optical constants for predicting the degrading effects as a function of contaminant layer thickness.

The two-angle method requires the measurement of reflected light intensities at two very different angles of incidence, the contaminant layer thickness, and the wavelength and polarization of the measuring incident light beam. The two-thickness method requires the measurement of reflected light intensities only at near-normal incidence, but for two different thicknesses of the contaminant layer and the wavelength.

In each case the equations for the polarized components of reflected light are solved by iterative techniques. The parameters n and k are constrained to certain ranges of values, and the resulting values are selected by successive approximations to the applicable equations until the computer-calculated reflectance values agree, within certain defined errors, with the experimental optical properties data.

The relative advantages and disadvantages of the two methods are balanced by the experimental requirements for simplicity. Measurements of reflected light intensities at two or more angles of incidence are subject to certain errors because of nonparallelism and polarization of the incident light which, in some cases, can limit the usefulness of the method. It is useful to have an alternate method available which requires less complicated experimental facilities and removes some limiting constraints of the two-angle method. Thus, a method of measuring reflected light intensities at near-normal incidence for different contaminant layer thicknesses has been developed and applied.

### III. Experimental Facilities

To determine contaminant optical constants, it is necessary to know the corresponding values for the substrate layer(s) on which the contaminant layer is deposited. The available experimental and analytical approaches are similar, and the laboratory equipment required is the same. For vacuum ultraviolet studies of this nature in the spectral range 100 to 600 nm, the major equipment used includes a capillary discharge lamp, a scanning monochromator of the normal incidence type, and a reflectometer. The thin-film thicknesses were approximated by interpreting the changes in the beat frequency of a quartz crystal microbalance during in-situ tests. A triple reflection polarizer[4] has been built for determining the state of polarization.

The studies have been confined to the vacuum ultraviolet region of the spectrum to take advantage of the marked increase in absorption by most materials in this range and because, for many materials, the effects of contamination are thereby more pronounced.

The source of light used in the vacuum ultraviolet measurements is a Hinterregger-type discharge lamp operated in the windowless mode. The power supply for the lamp is a d.c.-regulated source operated in the range 1000-2000 v d.c., 50-500 mA.

The monochromotor used is a 1-m, normal-incidence type with a 600 lines/mm grating blazed for maximum efficiency in the vacuum ultraviolet. A 0.15-m (6-in.) oil diffusion pump, backed by a large mechanical roughing pump, evacuates the system to a vacuum of about $1.33 \cdot 10^{-4}$ newton/m$^2$; a liquid-nitrogen-cooled trap helps avoid the backstreaming contamination problem. An exit slit adjustable in width and height is kept open at 0.1 mm for optimum resolution with acceptable signal level. Polarization characteristics of this instrument have been studied in the literature,[5] and these data, combined with applicable experimental results, provide sufficient information to account for the effect in measurements.

The detector is a dual-beam reflectometer. An oscillating mirror inside the reflectometer alternately reflects the incident light to the reference and the sample measuring channels. Two photomultiplier tubes, connected optically to the respective channels by sodium-salicylate-coated light pipes, detect the monochromatic signals, and the output is ratioed electronically. From Fig. 1, a schematic of the dual beam reflectometer, it may be seen that the sample channel light pipe is rotated about a suspended sample; both are rotatable to give reflectance as a function of angle of incidence.

The general procedure used for determining the optical constants of contaminant films has been to expose "clean" mirrors, whose properties are previously known or determined, to vapor fluxes of the contaminating material in vacuum. If the contaminants are in the form of unirradiated films of materials such as DC-704 or DC-705 vacuum pump oils, experience has shown that exposure to air following the deposition modifies the physical properties of the film such that re-evacuation to a low vacuum causes immediate sublimation of the deposited film. This is not observed if the films are exposed to uv irradiation, because of the characteristic permatizing effects of such energetic radiant flux. The need for in-situ coating and measurement is thus imperative for extended studies in the vacuum uv.

## IV. Results and Applications

To illustrate the applicability of contaminant optical constants to the study of component degradation, a material such as vacuum-pump oil was considered sufficiently representative. The commercially available and widely used diffusion-pump fluids (DC-704 and DC-705) were obtained for this purpose from Dow Corning Corp., Midland, Mich. The optical constants of one of these potential contaminants were measured in the wavelength range 180-300 nm, and computer-calculated reflectance changes were obtained showing the effects on gold-, platinum-, aluminum-, and magnesium-flouride-overcoated aluminum. Optical constants in the vacuum ultraviolet from 115-200 nm were obtained from another source[6] and used for similar calculations in the spectral region.

The oil films were deposited by evaporation of the bulk fluid under a vacuum of $6.65 \cdot 10^{-3}$ newtons/m$^2$. The fluid was held in a small ceramic boat heated by a wound filament wire. Thin-film thickness was monitored during the coating process by a quartz crystal microbalance calibrated, assuming the oil density, for that purpose. The thicknesses were checked by measurements on an automated ellipsometer at the University of Alabama. Thin-film layers of DC-704 from 2-15 nm thick were deposited on opaque gold and platinum thin films supported by quartz substrates.

After deposition of the thin-film material, the samples were exposed to the ambient environment and placed in the dual-beam reflectometer discussed in Sec. III of this paper. To prevent the previously mentioned re-evaporation of the oil films, the measurements were restricted to the wavelengths available at atmospheric pressure (about 180 nm and above). The experimental equipment used is described in Sec. III.

All measurements of reflected light intensities were taken at 20° and 70° angles of incidence. The two-angle method computer program was then used to determine the optical constants of the deposited oil films. Basic to such a determination, however, is a set of values for the gold and platinum thin films themselves. These were obtained previously by similar means in the vacuum ultraviolet, and the results are shown in Figs. 2 and 3. These data are very similar, and one would expect similar effects on both because of deposition of the vacuum-pump oils.

The optical constants of DC-704 for the wavelengths available in the laboratory are shown in Fig. 4. Since the pump oils can be observed to be relatively transparent in the visible region, it is not surprising that the extinction coefficient k drops to zero at the higher wavelengths. This parameter k is related to the standard absorption coefficient $\alpha$ through the relation

$$\alpha = 4\pi k/\lambda$$

where $\lambda$ is the wavelength.

Now, assuming the optical constants of a thin-film contaminant and its substrate optic, the optical property changes for any thickness can be calculated straightforwardly using the inverse form of the two-angle method computer program. The reflectance or transmittance of a contaminated optic can then be predicted as a function of wavelength, thickness, angle of incidence, and polarization.

Some calculated results for DC-704 on platinum and gold mirrors are given in Figs. 5 and 6. As expected, the effects are very similar. These results are for specular reflectance at near-normal incidence, assuming unpolarized light. The periodic nature of the degradation as a function of DC-704 thickness is a result of interference effects, perhaps better understood by comparison with the data in Fig. 7, which are based on optical constants from the literature.[7] Examination of the curve in Fig. 7 for the positions of the minima and maxima leads to the effective quarter-wave, etc., thickness of magnesium-fluoride-overcoated aluminum. In a way, then, the effect of DC-704 contaminant films is shown to act as an interference coating. Since it is well known that magnesium fluoride acts as an excellent reflectance-increasing coating,[7] one might expect similar effects, assuming homogeneity, isotropy, and smoothness, for DC-704 on reflecting mirrors. The data in Figs. 8-10 show such effects for DC-704 on aluminum, platinum, and gold mirrors; these data were calculated using the optical constants for DC-704 obtained from E. T. Arakawa of Oak Ridge National Laboratories.[6] For the platinum and gold mirrors, large increases in reflectance of the contaminated optic at certain wavelengths, generally confined to the lower wavelengths of the vacuum ultra violet, are indicated.

The effects of DC-704 on magnesium-fluoride-overcoated aluminum mirrors are shown in Figs. 11-13 for the three most-used, effective thicknesses of magnesium fluoride. The effects of contamination are very strongly wavelength-dependent and are similarly linked to the thickness of magnesium fluoride. For instance, at the Lyman-Alpha wavelength (126 nm; Fig. 10), a half-wave thickness of magnesium fluoride and aluminum does not show appreciable degradation until approximately three monolayers of DC-704 are deposited, whereas the quarter-wave thickness curve is very sensitive to fractions of monolayers of DC-704.

These data show that the effects of contamination do not follow any general pattern. The degradation is a strong function of film thickness, wavelength, and optical element characteristics. Evaluation of the results of an optical contamination monitor shows that, in order to determine the contaminant optical constants, sufficient data need to be taken at either two angles of incidence or two thicknesses. Then, the degradation can be calculated for any thickness and any type of optical element.

## References

[1] Hass, G. ed., *Physics of Thin Films*, Vol. 3, Academic Press, New York, 1966, pp. 1-52.

[2] Arnett, G. M., ed., "Lunar Excursion Module RCS Engine Vacuum Chamber Contamination Study," NASA TMX-53859, Marshall Space Flight Center, Ala., July 8, 1969.

[3] Tousey, R., "On Calculating the Optical Constants from Reflection Coefficient," *Journal of the Optical Society of America*, Vol. 29, No. 6, June 1939, pp. 235-239.

[4] Arakawa, E. T., Horton, V. G., Hamm, R. N., and Williams, M. W., "A Triple Reflection Polarizer for Use in the Vacuum Ultraviolet," Applied Optics, Vol. 8, No. 3, March 1969, p. 667.

[5] Matsui, A. and Walker, W. C., "Polarization of Three Vacuum-Ultraviolet Monochromators Measured with a Biotite Polarizer," Journal of the Optical Society of America, Vol. 60, No. 1, Jan. 1970, pp. 64-65.

[6] Arakawa, E. T., personal communication, Jan. 1971, Oak Ridge National Labs., Tenn.

[7] Canfield, L. R., Hass, G., and Waylonis, J. E., "Further Studies on $MgF_2$ Overcoated Aluminum Mirrors," Applied Optics, Vol. 5, No. 1, Jan. 1966, pp. 45-49.

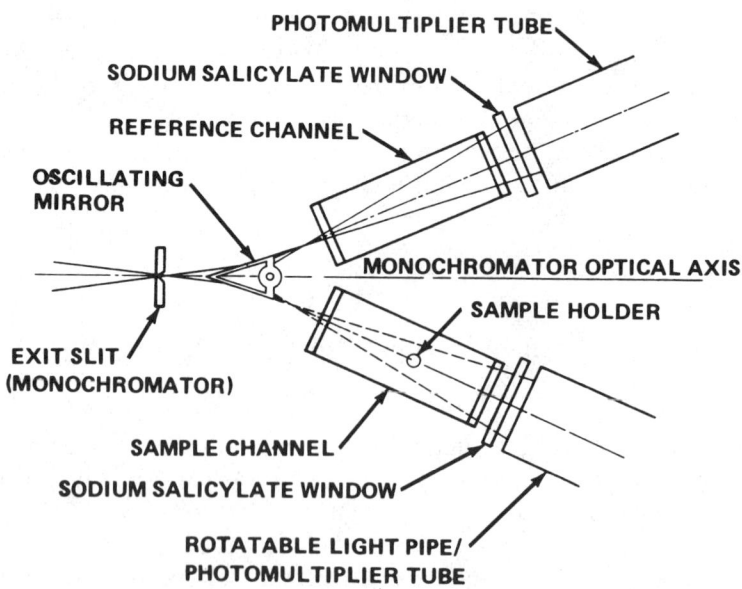

Fig. 1 Dual-beam reflectometer.

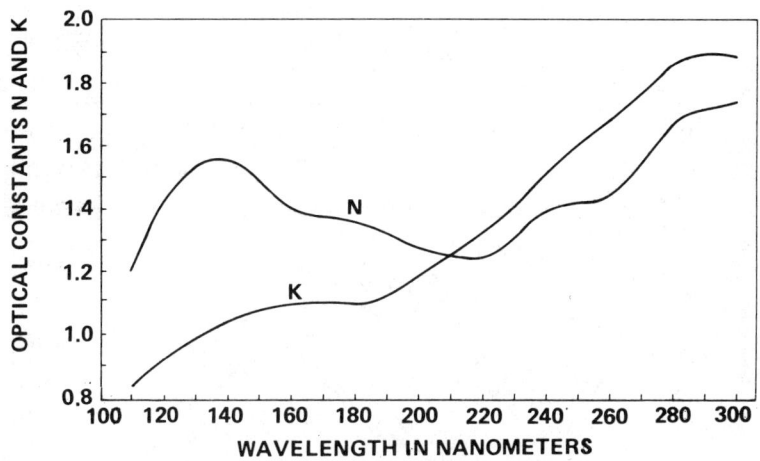
Fig. 2 Optical constants of gold.

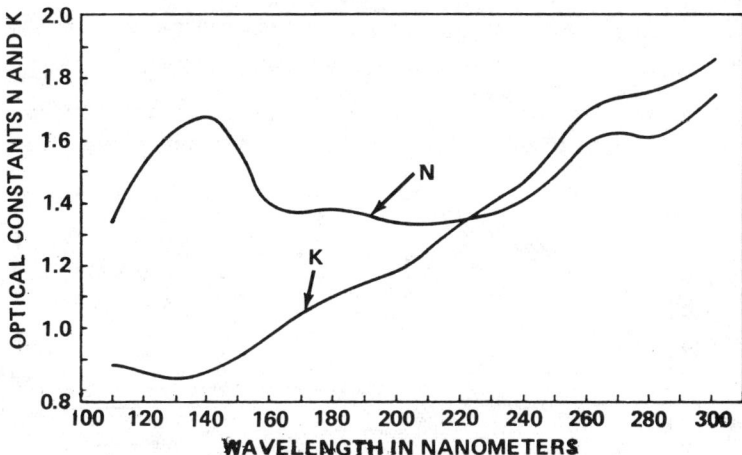
Fig. 3 Optical constants of platinum.

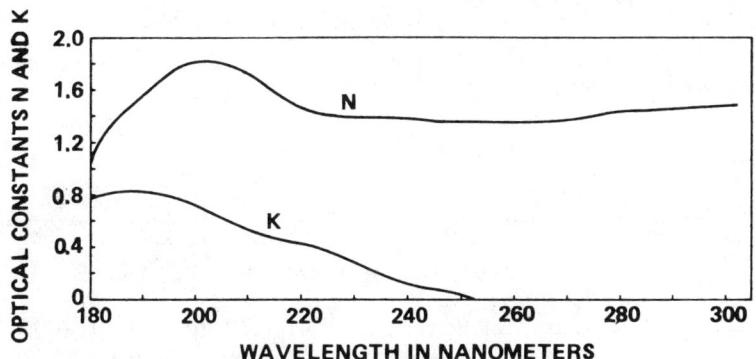
Fig. 4 Optical constants of DC-704.

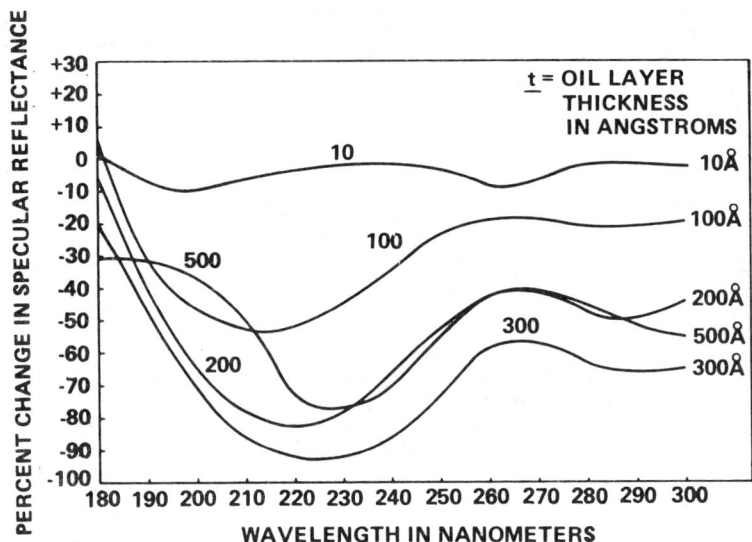

Fig. 5 Ultraviolet degradation of platinum mirror properties due to deposition of DC-704 vacuum pump oil.

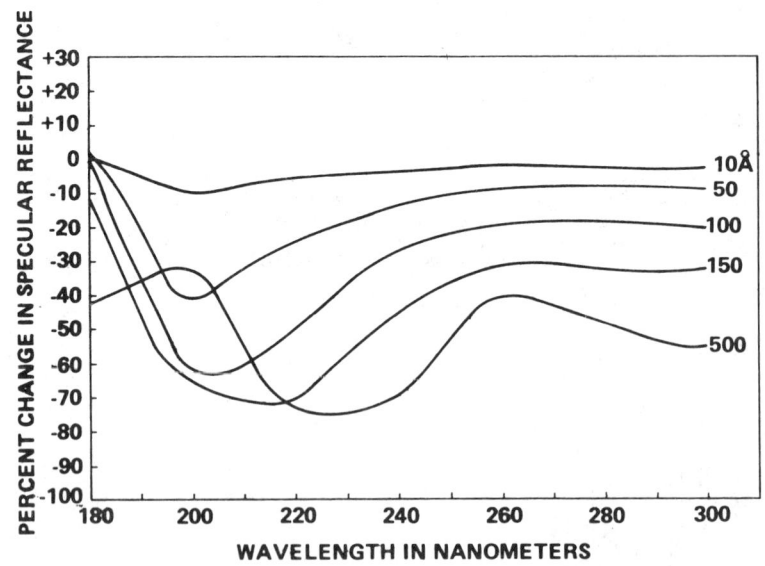

Fig. 6 Ultraviolet degradation of gold mirror properties due to deposition of DC-704 vacuum pump oil.

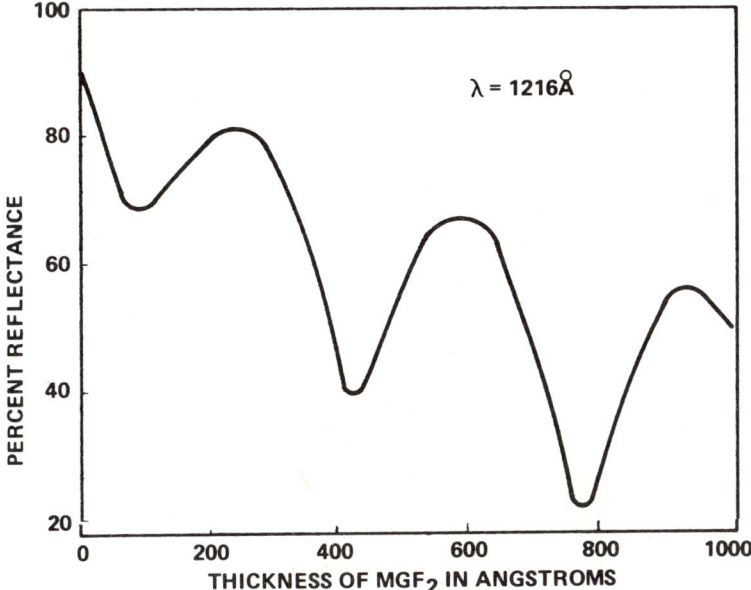

Fig. 7 Normal-incidence reflectance of magnesium fluoride on aluminum.

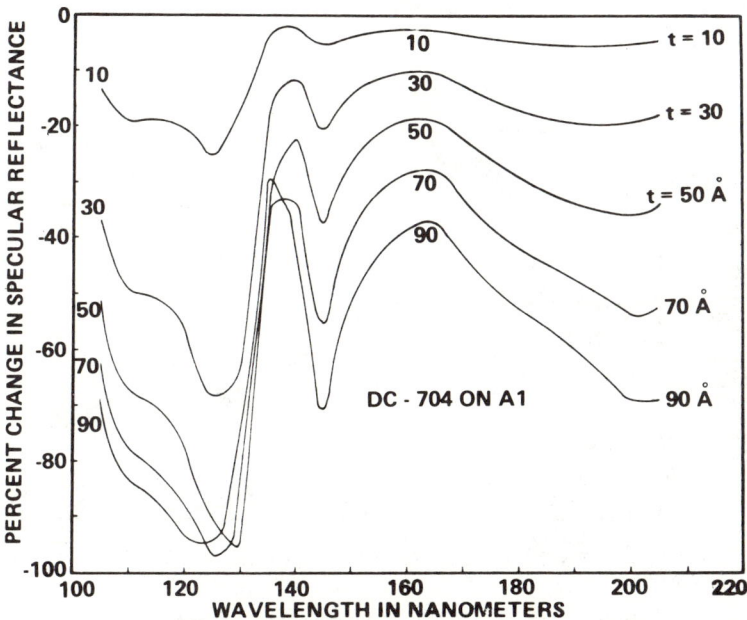

Fig. 8 Reflectance changes of an aluminum mirror due to deposition of DC-704.

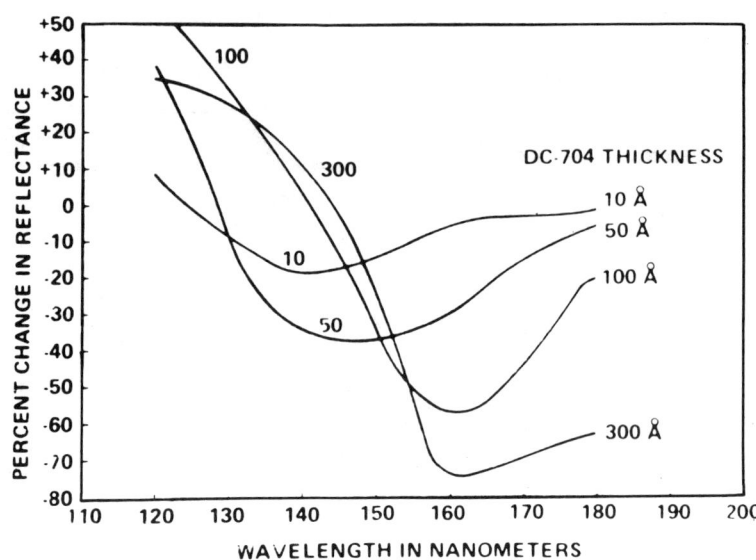

Fig. 9 Vacuum ultraviolet degradation of platinum mirror properties due to deposition of DC-704.

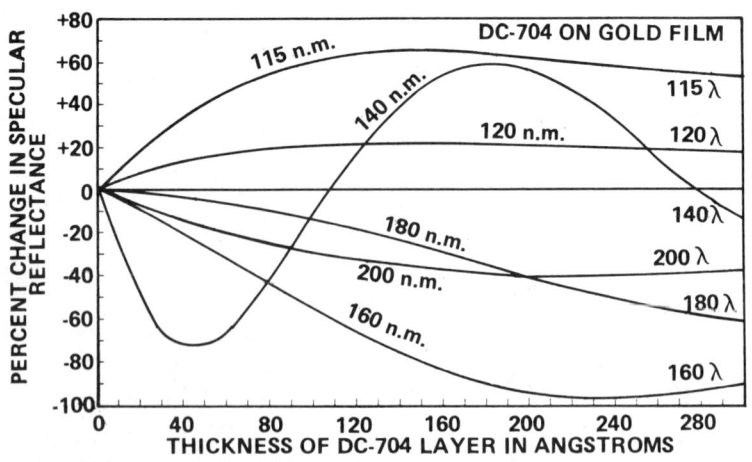

Fig. 10 Reflectance changes of a gold mirror due to deposition of DC-704.

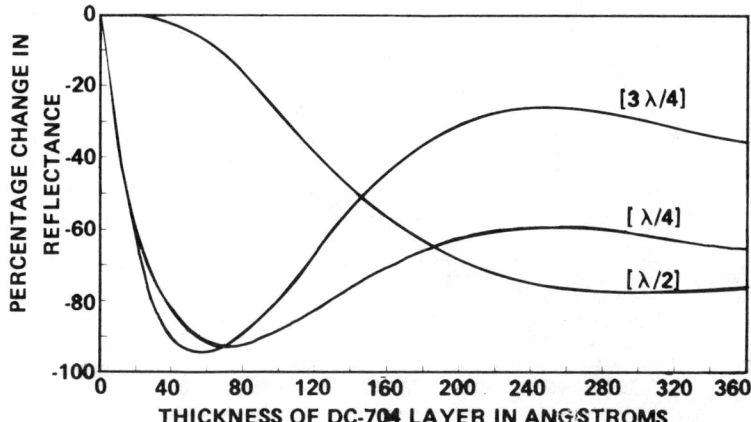

Fig. 11 Reflectance loss for magnesium-fluoride-overcoated aluminum contaminated with 0-300 Å of DC-704 for three effective magnesium fluoride thicknesses at wavelength 1200 Å.

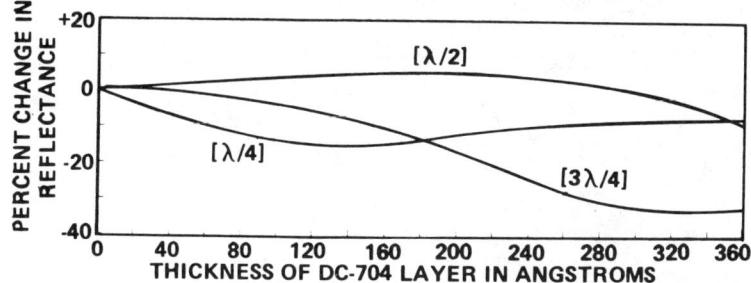

Fig. 12 Reflectance loss of magnesium-fluoride-overcoated aluminum contaminated with 0-300 Å of DC-704 for three effective thicknesses of magnesium fluoride at wavelength 1600 Å.

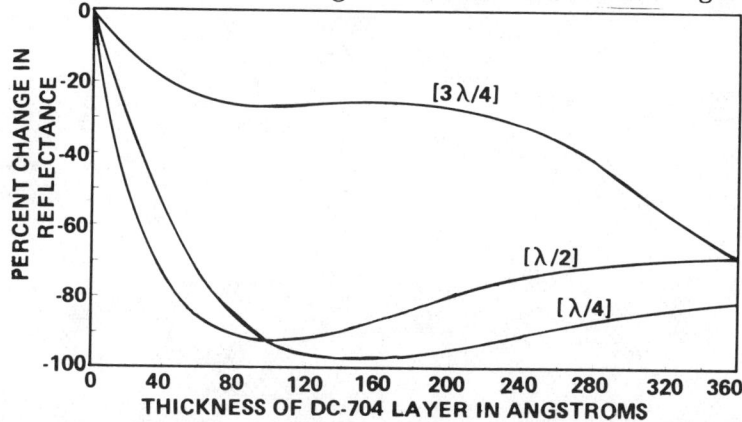

Fig. 13 Reflectance loss of magnesium-fluoride-overcoated aluminum contaminated with 0-300 Å of DC-704 for three effective thicknesses of magnesium fluoride at wavelength 2000 Å.

RESTORATION OF DEGRADED SPACECRAFT SURFACES
USING REACTIVE GAS PLASMAS

R. B. Gillette* and W. D. Beverly[†]

The Boeing Company, Seattle, Wash.

Abstract

Effects of a low-temperature oxygen plasma on typical spacecraft thermal-control coatings were studied. Coatings were degraded by irradiation in high vacuum prior to plasma exposure. Five types of metal-oxide pigmented coatings and a specimen of the Surveyor III spacecraft coating recovered from the moon by Apollo 12 astronauts were evaluated. The objective of this work was to evaluate the feasibility of restoring the reflectance of degraded coatings by oxygen plasma treatment. Ultimately such a plasma treatment process may be useful for prolonging coating lifetime in space. Results showed that both proton and ultraviolet radiation damage could be partially or completely eliminated on all specimens evaluated. Results also suggested that oxygen plasma treatment may be a means of increasing reflectance of freshly prepared coatings. The principle effect of plasma treatment was elimination of the short wavelength band (peaking in the range of 0.2 to 0.42µ) produced during proton and ultraviolet radiation.

Introduction

Thermal control and optical surfaces will pose a severe problem for long-lifetime space vehicles because of degradation resulting from the natural space radiation environment and deposition of contaminant films. The degrading components of the natural space radiation environment usually include protons,

---

Presented as Paper 71-463 at the AIAA 6th Thermophysics Conference, Tullahoma, Tenn., April 26-28, 1971. The assistance of Larry Fogdall, Sheridan Cannaday, and William Illi in performing radiation experiments and providing a portion of reflectance data is appreciated by the authors.
*Research Specialist, Fluid and Mechanical Systems Group.
[†]Research Engineer, Fluid and Mechanical Systems Group.

electrons, and electromagnetic radiation from the sun. Contaminants in space result from various sources including rocket plumes, outgassing organic materials, and life support system leakage and waste dumps. Degradation of coatings can also occur during functional or thermal/vacuum testing in vacuum chambers. In this case radiation is usually simulated sunlight, and contaminants result from outgassing organic materials and backstreaming of pump oils.

Three general approaches can be considered to overcome the effects of thermal or optical property changes in space: 1) the system can be overdesigned (if possible) to compensate for anticipated changes; 2) fresh coatings can be deposited in-situ; or 3) the thermal or optical properties of the original coatings may be restored. Compensation for degradation by system overdesign will probably require weight, volume and cost penalties and in some instances will not be possible. Applying fresh coatings in space would be cumbersome and difficult. Restoration of coating properties by either removing a contaminant film or eliminating radiation damage appears to be the most promising approach. Recent experiments[1,2] have shown that low-temperature oxygen plasmas can provide the dual function of removing certain types of contaminant films and eliminating radiation damage. In one earlier publication,[1] the use of an oxygen plasma for removing a contaminant film from $MgF_2$/Al and LiF/Al coated ultraviolet reflecting mirrors was discussed. The film had been deposited on the mirror during proton irradiation in a laboratory vacuum chamber. In the other publication,[2] details of the experimental plasma apparatus and preliminary data on plasma restoration of radiation-damaged white coatings were discussed.

The present paper discusses results of additional oxygen plasma restoration experiments on various types of white spacecraft coatings which were irradiated with protons, ultraviolet radiation, and the natural lunar environment.

### Apparatus

A schematic of the plasma generation apparatus used in these experiments is shown in Fig. 1. The over-all system includes gas supply apparatus, an rf generator, a glass-wall reaction chamber, and a vacuum pumping system. In operation, oxygen was bled into the reaction chamber through a suitable flowmeter and throttling valve. A pressure of one atmosphere was maintained in the flow meter, and flows in the order of 250 standard $cm^3$/min were used in tests. A pressure of about 0.5t was maintained in the reaction chamber by adjustment of the vacuum pump throttling valve. The oxygen was excited in the inlet

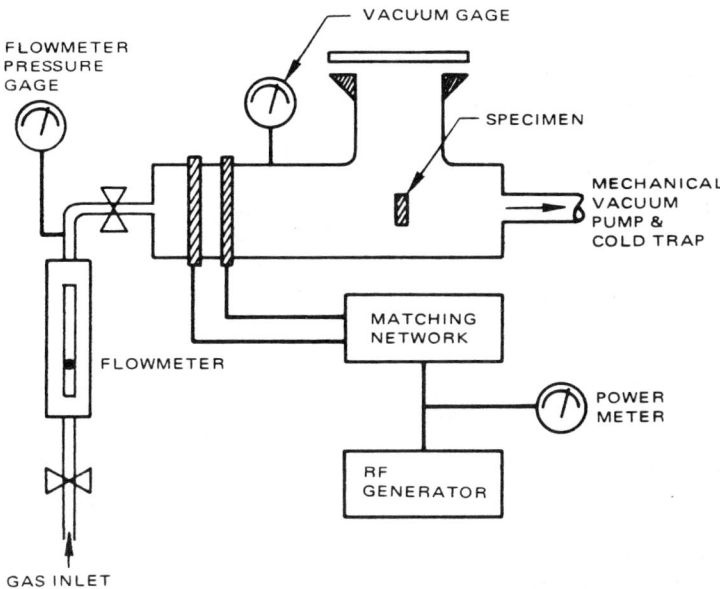

Figure 1  Schematic of plasma generator facility used for exposing thermal coatings.

end of the reaction chamber by passing it through an intense capacitive rf field. Approximately 300 w of power were transmitted to the rf antenna, an unknown portion of which was used for exciting the gas. The excited gas (plasma) then flowed through the glass reaction chamber (7.6-cm-diam pipe) at a linear axial velocity of about 8350 cm/min. The specimens were placed on the wall of the reaction chamber at a position about 58 cm downstream from the rf field. Since the chamber wall was glass, the specimens were isolated electrically.

Composition of a plasma created under these conditions has been discussed in the literature[3,4] and thus will not be reviewed here. It will be noted, however, that the primary excited constituents are neutral atomic oxygen $O(^3P)$ and excited molecular oxygen $O_2(^1\Delta g)$. Lesser quantities of both positive and negative oxygen ions and free electrons also exist in the plasma. No attempt was made in this research to ascertain either species or concentrations present, with the exception of atomic oxygen. It was estimated from titration experiments[5] with nitric oxide (NO) that the flux of atomic oxygen on the test specimen surface was in the order of 2 to $4 \times 10^{19}$ atoms/ $cm^2$-sec.

Hemispherical reflectance measurements performed immediately before and after plasma treatment were accomplished at atmospheric pressure using a Gier-Dunkle integrating sphere coupled to a Beckman DK-2A spectrophotometer. Reflectance data before and after irradiation were measured in evacuated integrating spheres operating in conjunction with Beckman DK-2A spectrophotometers.

## Experimental Results

A summary of the various types of specimens evaluated and qualitative results of plasma restoration experiments is given in Table 1. Eight different formulations of white coatings were evaluated including $TiO_2$ in methyl silicone, $Al_2O_3$ in potassium silicate ($K_2SiO_3$), ZnO treated with $K_2SiO_3$ in methyl silicone (two formulations), ZnO in methyl silicone (two formulations), ZnO in $K_2SiO_3$, and calcined china clay (primarily aluminum silicate) in $K_2SiO_3$ (recovered from the Surveyor III spacecraft on the moon). Three different environments were used to degrade the specimens prior to plasma treatment: 1) protons; 2) ultraviolet radiation; and 3) the natural lunar radiation environment. A detailed description of the experimental apparatus and procedures used to irradiate specimens nos. 1, 2, 3, 4, 7, 8, and 9 has been presented by Fogdall et al.[6,7] A description of the china clay specimen (no. 10) and the lunar environment have been discussed in a recent contractor report.[8] All specimens were exposed to ambient pressure conditions following irradiation, for periods of time varying from an hour up to 2 years. Specimen no. 4 was exposed to a nitrogen plasma prior to oxygen plasma treatment.

As can be noted in the table, oxygen plasma treatment was generally quite successful in restoring degraded specimens to pre-irradiation reflectance conditions. The degree of restoration, however, varied from a small amount to a reflectance higher than pre-irradiation values. Table 1 also shows, qualitatively, the rate of restoration experienced in the oxygen plasma. The rate of restoration, i.e., the length of exposure time required to saturate the recovery effect, varied from several minutes to several hours.

Reflectance data on the various types of thermal coatings are given in Figs. 2-11. The change in reflectance ($\Delta R$) is plotted vs wavelength. A positive value of $\Delta R$ represents an increase in reflectance. The after-irradiation and after-air-exposure

Table 1  Summary of thermal coating test results

| No. | Coating type | Radiation type | Radiation exposure | Restoration experienced in oxygen plasma | Rate of restoration |
|---|---|---|---|---|---|
| 1 | TiO$_2$/methyl silicone | Ultraviolet | 1130 eq. space sun hr (ESH) | Partial | Slow |
| 2 | Al$_2$O$_3$/potassium silicate | Ultraviolet | 1130 ESH | ~100% | Very rapid |
| 3 | K$_2$SiO$_3$ treated ZnO/methyl silicone (early S-13g) | Ultraviolet | 1130 ESH | Partial | Not known |
| 4 | ZnO/methyl silicone (S-13) | Ultraviolet | 1130 ESH | Partial[b] | Moderate |
| 5 | ZnO/methyl silicone (B-1060)[a] | Ultraviolet | 3600 ESH | Partial | Rapid |
| 6 | ZnO/potassium silicate (Z-93) | Ultraviolet | 2600 ESH | ~100% | Very rapid |
| 7 | Early S-13g | Protons (40 kev) | 10$^{16}$ p/cm$^2$ | Partial | Slow |
| 8 | Treated ZnO/methyl silicone (Goddard S.F.C. Type R) | Protons (40 kev) | 10$^{16}$ p/cm$^2$ | Partial | Moderate |
| 9 | Z-93 | Protons (40 kev) | 10$^{16}$ p/cm$^2$ | Small | Moderate |
| 10 | Surveyor III Plasmo Clay recovered from moon | Natural lunar environment | ~31 months | Partial | Rapid |

[a] A Boeing Company formulation of K$_2$SiO$_3$ treated ZnO in methyl silicone.
[b] Nitrogen plasma treatment priot to the oxygen plasma produced no reflectance change.

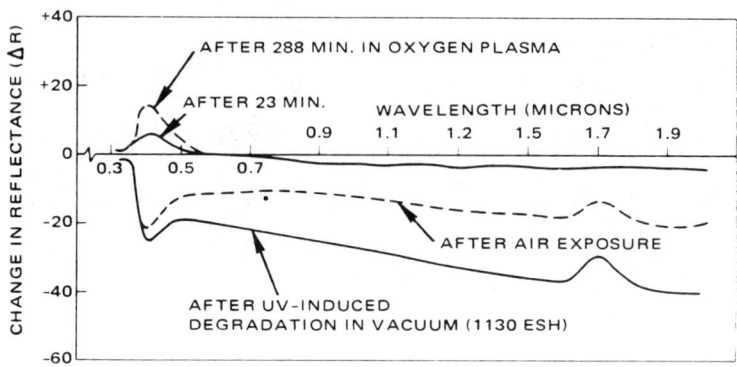

Figure 2  Effect of oxygen plasma on uv-degraded $TiO_2$/methyl silicone.

curves are both referenced to a virgin specimen. The after-plasma-exposure curves are referenced to the irradiated/air-exposed condition. Thus, air-exposure bleaching (restoration) is separated from oxygen-plasma bleaching wherever possible.

In the case of the uv-irradiated $TiO_2$/methyl silicone specimen (Fig. 2), substantial bleaching occurred upon exposure to air. The plasma treatment of 288 min eliminated a large portion of the residual damage in the wavelength region shorter than 0.5μ, however, and caused additional degradation at wavelengths beyond 0.7μ. The response of this specimen to plasma restoration was rather slow. The uv-irradiated $Al_2O_3$/potassium silicate specimen (Fig. 3) is a rather dramatic example of oxygen plasma effects. Negligible bleaching occurred when the specimen was exposed to air following irradiation. At wavelengths shorter than about 0.4μ, the oxygen plasma increased the reflectance to a value higher than the pre-irradiation condition. At longer wavelengths, however, some damage remained after plasma treatment. The rate of restoration of this specimen was very rapid, as noted by the large change in a 1-min exposure time. Results on this specimen suggest that it may be possible to increase the reflectance of freshly prepared coatings by oxygen plasma treatment.

Data on the uv-irradiated, early S-13g specimen (Fig. 4) showed that the oxygen-plasma eliminated a large portion of the absorption band centered at 0.42μ. The rate of plasma restoration was not determined because reflectance measurements were not performed after short plasma exposures.

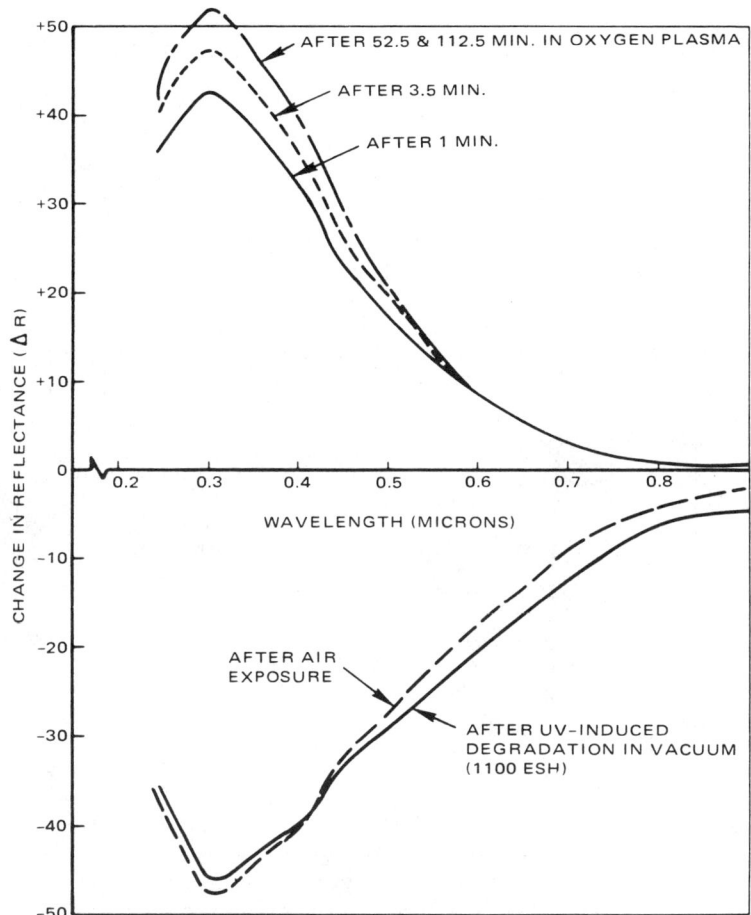

Figure 3 Effect of oxygen plasma on uv-degraded $Al_2O_3$/potassium silicate coating.

ts obtained with the S-13 specimen (Fig. 5) were similar
e S-13g specimen. Although intermediate plasma-exposure-
curves are not shown in Fig. 5, the rate of recovery was
nined to be moderate. After 14 min, the restoration was
half of the 42-min exposure value. Effects of oxygen
a treatment on the B1060 formulation of ZnO/methyl sili-
are shown in Fig. 6. A large portion of the short-wave-
1 absorption band was eliminated by plasma treatment.
restoration was experienced, resulting in saturation
a 25-min exposure. Results of plasma treating the uv-
iated Z-93 specimen (Fig. 7) indicated both extremely

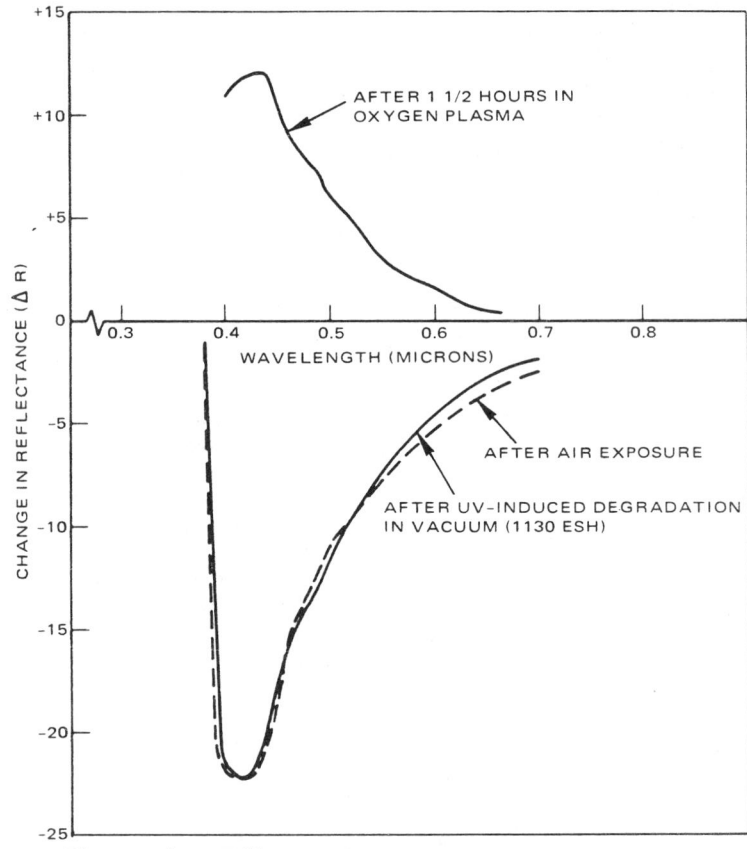

Figure 4  Effect of oxygen plasma on uv-degraded ZnO/methyl silicone coating (early S-13g).

rapid recovery (3-min saturation) and higher than pre-irradiation reflectance values in the wavelength region below about 0.475μ.

The proton-irradiated S-13g specimen (Fig. 8) exhibited a substantial increase in reflectance after plasma treatment. However, it is interesting to note that the peak of the restoration ΔR curve does not correlate with the degradation band peak. This is contrary to the general results obtained on the uv-irradiated specimens. The rate of restoration on this specimen was rather slow, as evidenced by the 99-min exposure saturation time. Reflectance data for the proton-irradiated, treated ZnO/methyl silicone coating (type R) is shown in Fig. 9. Moderately rapid restoration was experienced, with

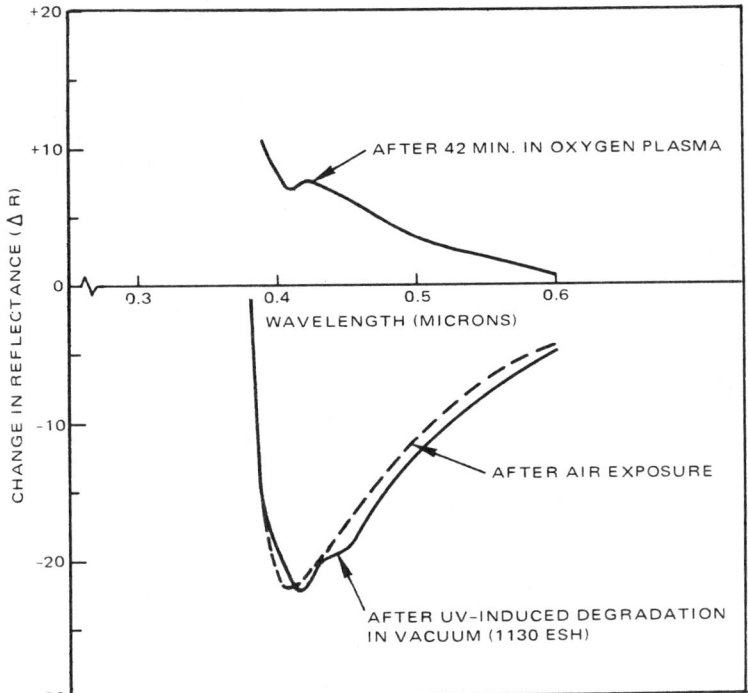

Figure 5  Effect of oxygen plasma on uv-degraded ZnO/methyl silicone coating (S-13).

saturation occurring after 61 min. Again, the peak of the restoration ΔR curve does not correlate with the peak of the degradation curve.

The reflectance of the proton-irradiated Z-93 coating was increased only by a small amount during plasma treatment, as noted in Fig. 10. The ΔR produced during irradiation was about -41% at 0.4μ, whereas only about a 5% increase occurred during plasma treatment at a wavelength of 0.5μ. No restoration was obtained in the infrared wavelength region; however, it is possible that air exposure could have eliminated radiation damage prior to plasma treatment. Reflectance data were not available to show effects of air exposure following irradiation.

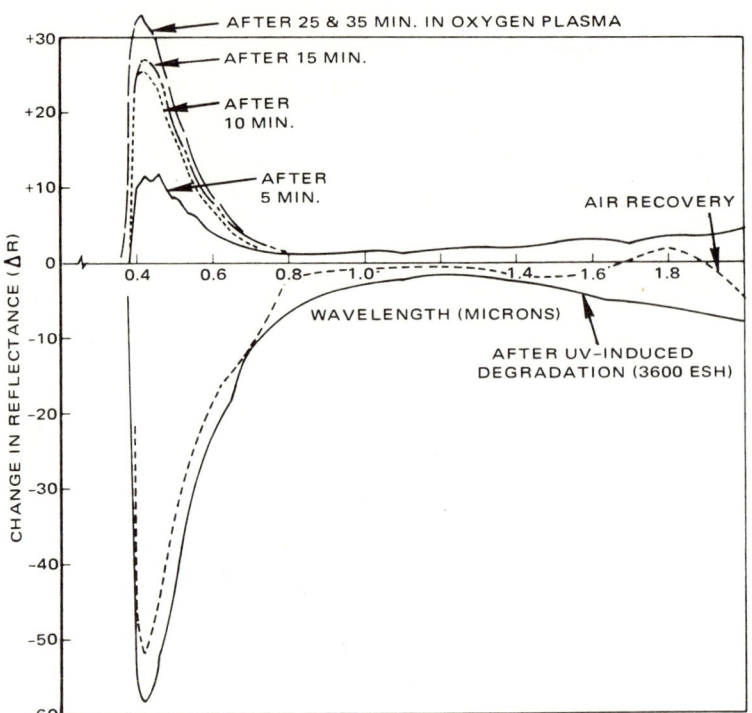

Figure 6  Effect of oxygen plasma on uv-degraded ZnO/methyl silicone coating (B1060).

Results of the plasma restoration experiment on the Surveyor III spacecraft china-clay coating are shown in Fig. 11. As noted in the figure, the lunar environment induced degradation in a band peaking at 0.4μ. The peak ΔR of the degraded specimen after about a 1-yr exposure in the laboratory was -58%. Oxygen plasma exposure caused rapid restoration in a band centered at the same wavelength as the degradation band. This observation suggests that uv was the principle cause of degradation in the lunar environment (based on preceding observations on proton and uv-degraded specimens). An increase of about 31% (at 0.4μ) occurred after a 12-min exposure. The plasma treatment also increased the reflectance in the infrared region to a value higher than an unirradiated specimen.

An experiment was conducted with a ZnO/$K_2SiO_3$ coating (specimen no. 6) to determine the effect of uv irradiation on reflectance after both air and oxygen-plasma exposures. The objective was to determine the rate of degradation after

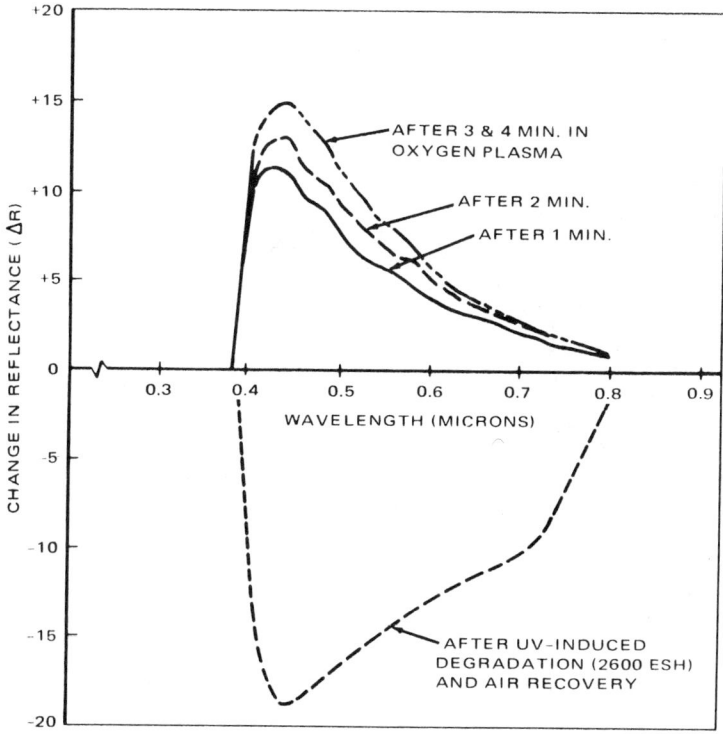

Figure 7   Effect of oxygen plasma on uv-degraded ZnO/potassium silicate.

bleaching or restoration had occurred.  Results are shown in Fig. 12, which is a plot of hemispherical reflectance (at 0.4μ) vs ultraviolet exposure time.  The specimen was first irradiated with 1400 ESH of ultraviolet radiation.  The reflectance change (ΔR) occurring upon air exposure was noted to be about +3.5%.  It was then irradiated with an additional 1200 ESH, and remeasured both in-vacuum and after a 4-min oxygen plasma exposure.  In-vacuum reflectance data indicated that the rate of degradation following air-exposure was comparable to that of the initial exposure.  The increase in reflectance during the 4-min plasma treatment was about 16%.  Following plasma treatment, the specimen was irradiated with an additional 480 ESH of ultraviolet, and reflectance was again measured in vacuum. Reflectance at 0.4μ only decreased from 85 to 84%, during the latter increment of uv exposure, indicating a rate of degradation no greater than that of an untreated specimen.  Similar data were not taken on other specimens.

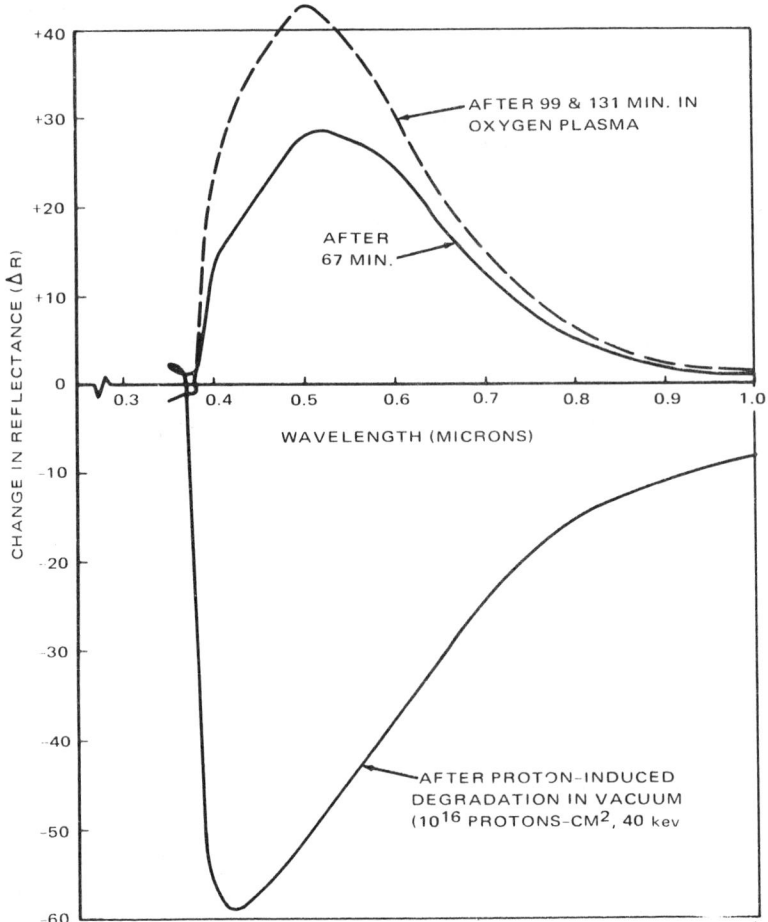

Figure 8  Effect of oxygen plasma on proton-degraded ZnO/methyl silicone coating (early S-13g).

## Discussion

The foregoing experimental results show that exposure of typical white spacecraft paints to plasmas containing atomic oxygen can induce a significant increase in reflectance on most specimens. The predominant effect of plasma treatment is elimination of the short-wavelength absorption band that is present in both proton and ultraviolet irradiated specimens. It is anticipated that plasma treatment also will induce a reflectance

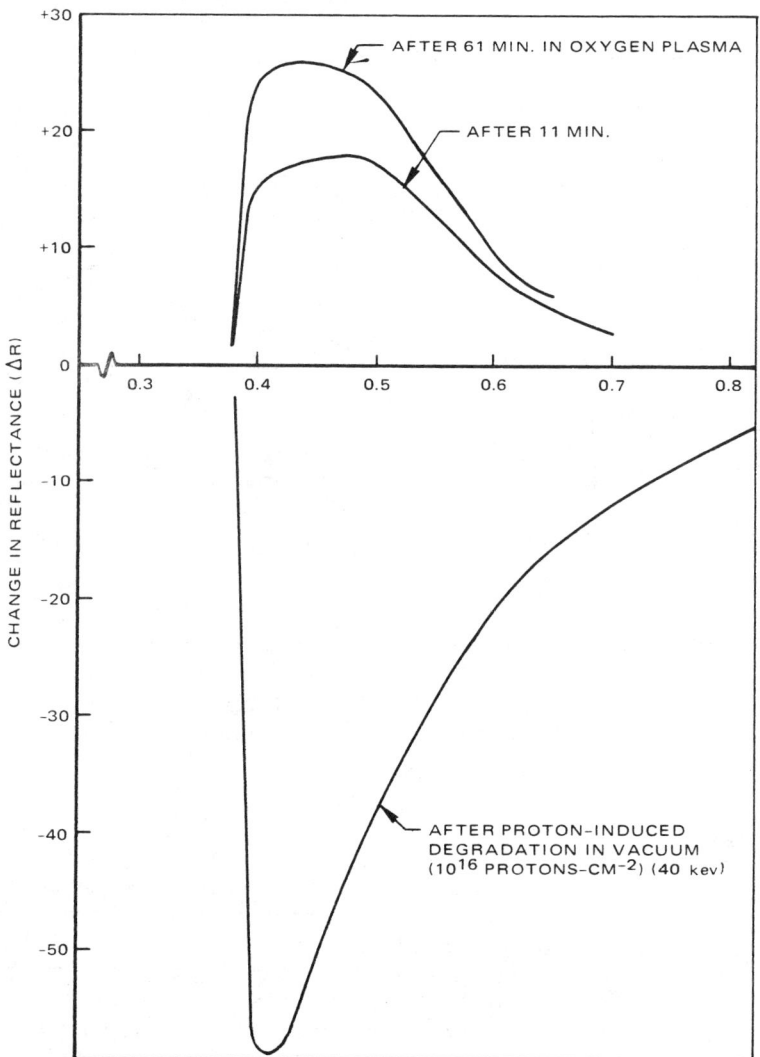

Figure 9  Effect of oxygen plasma on proton-degraded treated ZnO/methyl silicone coating (Goddard SFC type R).

increase in the infrared region where air exposure normally bleaches radiation damage. Up to the present time, however, it has not been possible to obtain meaningful data in the infrared wavelength region because all specimens have been exposed to air between irradiation and plasma treatment.

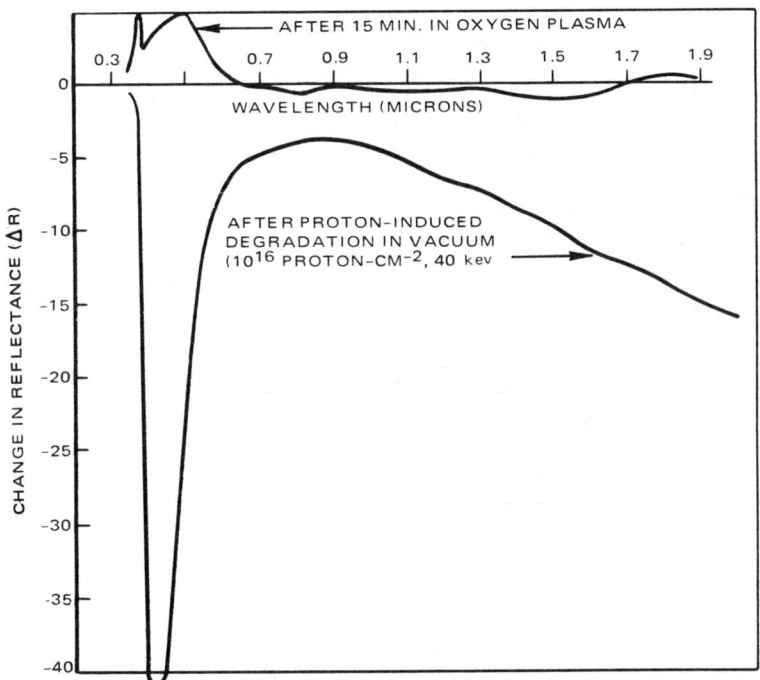

Figure 10 Effect of oxygen plasma on proton-degraded ZnO/potassium silicate (Z-93).

The fact that the short-wavelength absorption band is partially or completely eliminated by oxygen plasma treatment suggests that the band is originally formed as a result of oxygen depletion during irradiation in vacuum. Thus, the experimental results of this work lend credence to the degradation mechanisms proposed by Blakemore[9,10] and Collins and Thomas[11] which are summarized as follows for ZnO irradiated with ultraviolet. A photon with an energy greater than about 3.2 ev (3850 Å), the fundamental band gap of ZnO, creates a photo-hole and a photo-electron in a microcrystal of ZnO. The photo-holes then diffuse to the crystal surface, where they become trapped and combine with $O^{-2}$ lattice ions and adsorbed oxygen atoms. Photo-hole lifetimes of $10^{-9}$ sec and a hole diffusion constant of 1.6 $cm^2$/sec insure that a considerable fraction of the holes will reach the microcrystal surface. The diffusion length resulting from these constants is on the order of $0.4\mu$, larger than the usual microcrystal diameter (0.1 to $0.3\mu$). The resulting oxygen atoms then combine to form oxygen molecules that are thermally desorbed into high vacuum or are desorbed directly

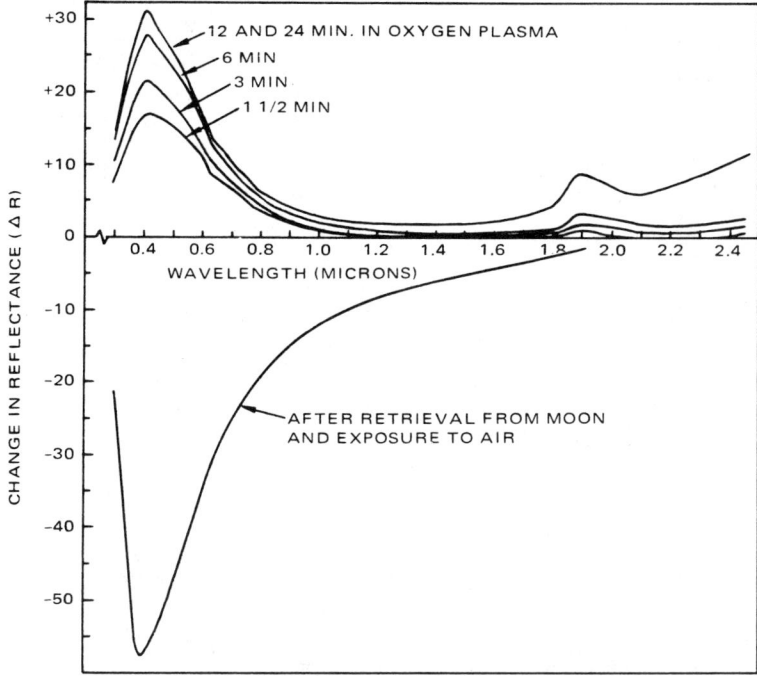

Figure 11  Effect of lunar exposure and subsequent oxygen plasma treatment on surveyor plasmo-clay coating.

as atomic oxygen. If the desorbed oxygen atoms are from lattice oxygen ions, an excess of zinc atoms occurs near the surface. The diffusivity of interstitial zinc is sufficiently high at room temperature so that this excess zinc migrates inward. Migration of the surface zinc inward removes the surface barrier of positively charged $Zn^{+2}$ ions which would eventually limit the diffusion of holes to the surface. Color centers are then formed by trapping of electrons at either the oxygen ion vacancies or interstitial zinc ions. These color centers absorb light in characteristic wavelength regions, depending on the energy of the trapped electron.

Although all of the details of the uv degradation model have not yet been verified, experimental evidence indicates that photo-desorption of oxygen is a critical part of the reaction. Numerous experiments have shown that the optical degradation produced by the preceding mechanisms can be substantially eliminated by exposure to air after irradiation. This effect has

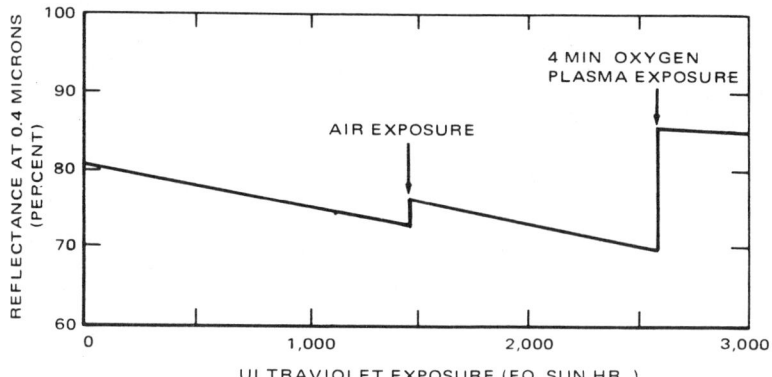

Figure 12  Effect of oxygen plasma on reflectance of ultraviolet-irradiated zinc oxide/potassium silicate at 0.4μ wavelength.

been attributed to readsorption of oxygen, although recent results by Gilligan and Zerlaut[12] have shown that partial elimination of damage occurs when irradiated specimens are exposed to high-purity $N_2$ and $CO_2$. Fogdall et al.[7] have shown that partial recovery of reflectance can occur even in high vacuum ($\sim 10^{-8}$t). It is generally observed in ZnO/methyl silicone coatings that rapid and nearly complete recovery occurs in the near-infrared wavelength region, and slower recovery occurs in the visible region.

Based on the preceding theoretical model and typical experimental data, it is logical to assume that treatment of similar radiation-degraded coatings with reactive oxygen will result in a substantial increase in reflectance. Collins and Thomas[11] have pointed out that the rate-determining step in the oxygen readsorption process may be the breaking up of oxygen molecules during adsorption. If this is true, exposure of the surface to atomic oxygen or energetic oxygen ions, rather than molecules, could greatly enhance the efficiency of the recovery or restoration process. The result would be a much more rapid recovery of reflectance in a plasma vs an air environment.

Generally, re-evacuation after "air-expsoure" recovery induces only a slight degradation with time after re-evacuation. This effect may be attributed to weakly bound oxygen thermally desorbing into high vacuum. It would not be anticipated from theory that tightly bound oxygen (such as lattice atoms) would be desorbed by mere evacuation. Thus, any recovery experienced during oxygen plasma treatment in high vacuum should be semipermanent (ignoring effects of subsequent irradiation). This indeed was the case, as indicated by the data in Fig. 12 for a $ZnO/K_2SiO_3$ specimen irradiated with ultraviolet.

The possibility exists that specimen heating is a contributing factor to restoration experienced during plasma treatment. Experiments to separate plasma-gas effects from temperature effects have not yet been conducted; however, specimen temperature during treatment has been measured. The temperature of a ZnO/methyl silicone specimen as a function of time during oxygen plasma exposure is given in Table 2.

Table 2  Temperature of ZnO/methyl silicone during plasma exposure

| Exposure time, min | Temperature, °F |
|---|---|
| 0 | 75 |
| 1 | 87 |
| 2 | 102 |
| 4 | 125 |
| 8 | 142 |
| 16 | 155 |

It is apparent from the data that an equilibrium temperature in the order of 160°F would be reached for exposure times longer than 16 min. It should also be noted that, for times less than a few minutes (where substantial recovery occurred on some specimens), temperatures were less than 100°F. Thus, oxygen-plasma effects rather than temperature effects probably induced the restoration phenomena presented.

## Conclusions

The following conclusions can be made from the foregoing experiments:

1) The reflectance of most typical radiation-degraded white thermal coatings can be substantially increased by exposure to a low-temperature oxygen plasma.  2) The fact that the short-wavelength absorption band ($\sim 0.3$ to $0.7\mu$) can be partially or completely eliminated by this technique suggests that is was formed during irradiation by depletion of oxygen from the pigment crystal lattice.  3) Results indicate that treatment of coatings with an oxygen plasma may be a useful technique for prolonging coating lifetime in space.

## References

[1] Gillette, R. B. and Kenyon, B. A., "Proton-Induced Contaminant Film Effects on Ultraviolet Reflecting Mirrors," Journal of Optical Society of America, Vol. 10, No. 3, March 1971, pp. 545-551.

[2] Gillette, R. B., Hollahan, J. R., and Carlson, G. L., "Restoration of Optical Properties of Surfaces by Radiofrequency - Excited Oxygen," Journal of Vacuum Science and Technology, Vol. 7, No. 5, Sept./Oct. 1970, pp. 534-537.

[3] Hollahan, J. R., "Analytical Applications of Electrodelessly Discharged Gases," and "Research with Electrodelessly Discharged Gases," Journal of Chemical Education, Vol. 43, Nos. 5 and 6, May and June 1966, pp. A497.

[4] Bell, A. T., "The Physical Characteristics of Electric Discharges," The Application of Plasma to Chemical Processing, edited by Baddour and Timmins, M.I.T. Press, Cambridge, Mass., 1967, pp. 1-12.

[5] Mearns, A. M. and Morris, A. J., "Use of the Nitrogen Dioxide Titration Technique for Oxygen Atom Determination at Pressures Above 2 Torr," Journal of Physical Chemistry, Vol. 74, No. 22, 1970, pp. 3999-4001.

[6] Brown, R. R., Fogdall, L. B. and Cannaday, S. S., "Electron-Ultraviolet Radiation Effects on Thermal Control Coatings," AIAA Progress in Astronautics and Aeronautics: Thermal Design of Spacecraft and Entry Bodies, edited by J. T. Bevans, Vol. 21, Academic Press, New York, 1969, pp. 697-724.

[7] Fogdall, L. B., Cannaday, S. S. and Brown, R. R., "Electron Energy Dependence for In-Vacuum Degradation and Recovery in Thermal Control Surfaces," AIAA Progress in Astronautics and Aeronautics: Applications to Thermal Design of Spacecraft, edited by J. T. Bevans, Vol. 23, Academic Press, New York, 1970, pp. 219-248.

[8] "Test and Evaluation of the Surveyor III Television Camera Returned from the Moon by Apollo 12," Vols. I, II, and III, Dec. 1970. Hughes Aircraft Co.

[9] Blakemore, J. S., "Solar-Radiation Induced Damage to Optical Properties of ZnO-Type Pigments," NAS 8-11266, Sept. 1965, Lockheed Palo Alto Research Laboratories, Technical Summary Report for Period June 1964 to June 1965.

[10]Blakemore, J. S., "A Model for Extraterrestrial Solar Degradation of Zinc Oxide," IEE Transactions on Aerospace and Electronic Systems, May 1966.

[11]Collins, R. J. and Thomas, D. G., "Photoconduction and Surface Effects with Zinc Oxide Crystals," Physics Review, Vol. 112, Oct. 1958, pp. 388-395.

[12]Gilligan, J. E. and Zerlaut, G. A., "A Study of In-Situ Degradation of Thermal Control Surfaces," IITRI-U6061-17, Sept. 1969, IIT Research Institute.

# I Surface Radiation Properties
## I.3 Space Flight Effects

# REPORT ON THE FLIGHT PERFORMANCE OF THE Z-93 WHITE PAINT USED IN THE SERT II THERMAL CONTROL SYSTEM

N. John Stevens[*] and George R. Smolak[*]

NASA Lewis Research Center, Cleveland, Ohio

## Abstract

The change in absorptance of the 1.8 $m^2$ of Z-93 white paint applied as the primary component of a passive thermal control system is evaluated by varying the paint absorptance values used in a calibrated computer model of the satellite until the resulting temperatures match the flight data. This absorptance change is determined for a 6200 hr, constant sunlight period in a 1000 km orbit and found to have increased by 0.04 as expected. It has been concluded that inadvertent contamination has caused higher degradation of one area. This area of higher degradation has not caused thermal problems. Hence, the white paint system is functioning satisfactorily.

## Introduction

The second Space Electric Rocket Test satellite (or SERT II) is a long duration, space environment test vehicle for a 1-kw, mercury bombardment, ion thruster. In its orbiting configuration the SERT II satellite consists of the Agena D booster, a spacecraft support unit (SSU), and a spacecraft section which houses the two ion thrusters and associated experiments (Fig. 1). After reaching a stable orbit, the Agena was shut down and has remained dormant. Power for the satellite is supplied by two solar array wings attached to the Agena.

The satellite was placed in a 1000-km (540-naut-mile), circular, constant sunlight polar orbit on February 3, 1970. The orientation is with the ion thrusters pointing towards the

---

Presented as Paper 71-455 at the AIAA 6th Thermophysics Conference, Tullahoma, Tenn., April 26-28, 1971.
[*]Aerospace Engineer, Spacecraft Technology Division.

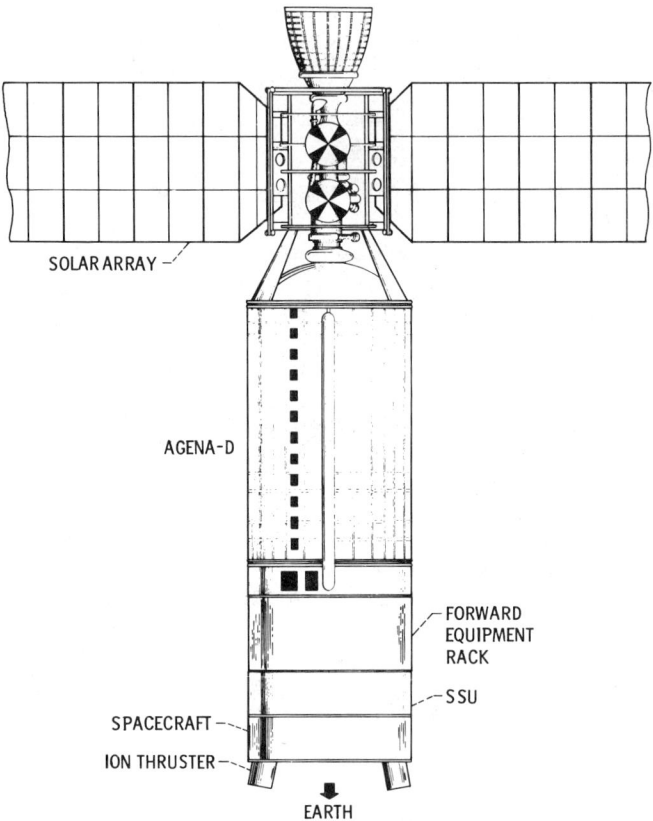

Fig. 1 SERT II satellite.

Earth and the solar cells facing the sun. The satellite is gravity gradient stabilized and does not spin. The oblateness of the Earth has caused the orbit plane to precess so that the satellite remained in constant sunlight.

A passive thermal control system is being used on this satellite.[1] This system consists of Z-93 white paint, aluminum surfaces and black paint. The Z-93 white paint is a zinc oxide-potassium silicate mixture developed at the Illinois Institute of Technology Research Institute.[2] Due to its short shelf life, the paint was formulated at Lewis Research Center especially for the SERT II project. The Z-93 white paint was chosen as the solar reflector coating because of its reported stability in sunlight in near Earth orbits[3-6] where particle induced damage would be minimized.

The purpose of this report is to present the flight performance evaluation of the Z-93 white paint. This evaluation will

## FLIGHT PERFORMANCE OF Z-93 PAINT

cover the period from launch until the premature shutdown of the second ion thruster on October 17, 1970, a total time of 6200 hr in a constant sunlight, space environment. This is believed to be the first such evaluation of Z-93 white paint applied as part of a passive thermal control system over this extended period of time in space.

The evaluation of the paint performance in this mission is complicated by the fact that the coated panels are integral parts of a structure and are not thermally isolated. Changes in temperature, therefore, cannot automatically be correlated to changes in optical properties of the coating. The analytical thermal model, using the design optical properties of the coatings, was subjected to the space environmental heating factors and internal heat dissipation at a given time in the mission and these analytical results were compared to the flight data. Differences between the flight data and the analytical model results, then, were proportional to changes in the solar absorptance of the coating.

Since the analytical thermal model is such a necessary tool in this evaluation, it will be discussed in the next section along with the other components of the SERT II thermal system. Then, the flight data will be discussed and the resulting solar absorptance curve for the 6200 hr of flight will be presented.

### Discussion of the Thermal System

#### Analytical Model

The basic tool used in this evaluation of the coating performance is the analytical thermal model of the satellite. The computer program chosen to solve the analytical thermal network was the Chrysler Improved Numerical Differencing Analyzer or CINDA.[7]

The model consists of a 109-node representation of the spacecraft section, an 81-node representation of the SSU and the Lockheed analytical model of the forward equipment rack for a total of 534 nodes. Heat conduction tests on the shims used between the SSU and Agena rack had indicated that the SSU would be essentially thermally isolated from the Agena. Hence, these sections were sufficient to treat the thermal considerations for orbiting conditions; the Agena rack was used to satisfy the radiation boundary conditions for the two sections containing all of the heat dissipating components.

The majority of the nodes are assigned to the structure. Components on each tray are generally lumped into a single node whereas the tray itself is assigned a separate node. The two ion thrusters which are essentially thermally isolated are assigned one node each. A joint thermal conductance value of about 316 w/m$^2$ was experimentally obtained and incorporated into the model for all structural joints. Internal radiation interchange factors were obtained either by hand calculations or by use of CONFAC II.[8]

## Calibration of the Analytical Model

The SERT II analytical model was calibrated against temperature data obtained in a test program. This calibration was accomplished by instrumenting the prototype spacecraft, SSU and simulated Agena forward rack with 144 thermocouples and collecting data for a total of 160 hr under a variety of external heating conditions and internal power levels. As a boundary condition on the computer model calculations the Agena rack temperatures and the heat receiving panels of the spacecraft and SSU were all held at the experimentally determined temperatures. By a laborious iterative process involving changes in solid conductors, radiation interchange and heat storage capacities the agreement between the computer model results and the experimentally obtained temperatures was made excellent for both transient and steady state results. The details of the test procedure can be found in Ref. 9. The results for the areas of interest in this report for 100 hr of the test program are shown in Figs. 2a to c. As can be seen from the curves the agreement is within 4° C with the predictions generally colder than the thermistor readings. Hence, the analytical thermal model was calibrated for internal heat exchange and radiation losses to the environment provided the heat receiving panel temperatures are known.

## Thermal Control System

The analytical thermal model was used to determine the thermal control pattern that would satisfy the component temperature constraints for the mission. This pattern is shown in Fig. 3. The Z-93 white paint is used on the six sunside panels; an area of about 1.8 m$^2$. The main sun axis panels of both the spacecraft and SSU (bay 8) are completely covered with Z-93 paint while the other four panels are coated with a pattern of three-quarters white paint and one-quarter polished aluminum. The optical properties of the Z-93 paint used in designing the thermal pattern were: solar absorptance of 0.16±0.03; and emittance of 0.90±0.05.

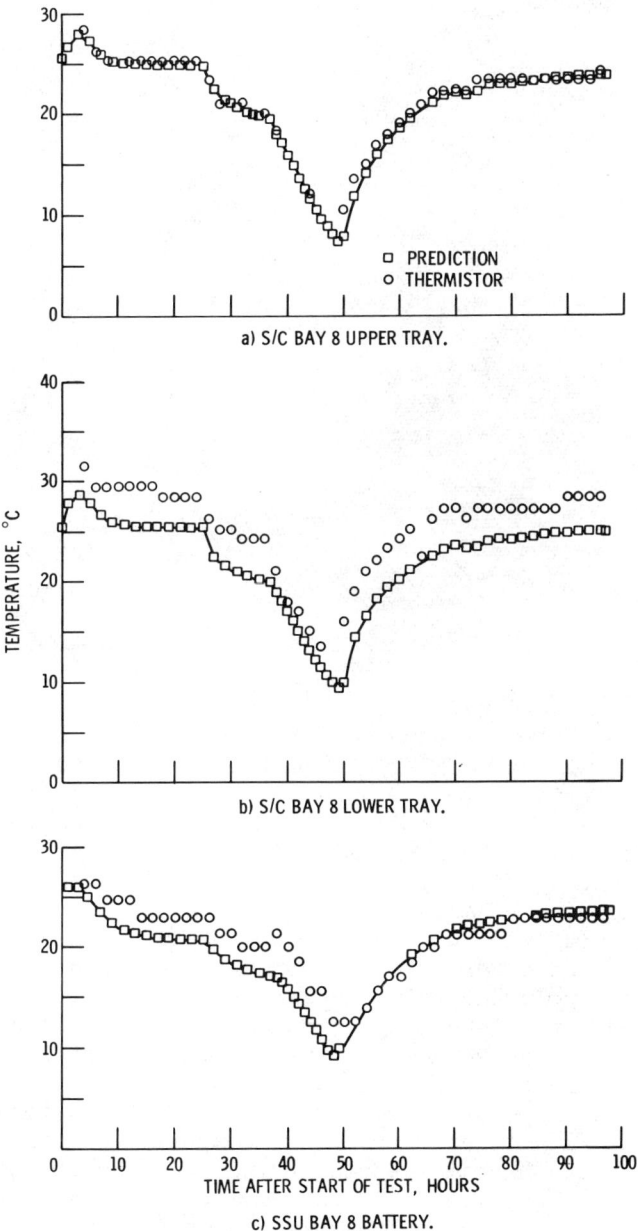

Fig. 2  Comparison of experimental and analytical model temperatures-calibration test.

Fig. 3 Side view of SERT II thermal control pattern. Pattern symmetric.

It was felt that the Z-93 paint disadvantages of easy contamination and difficulty of application could be overcome. This confidence was based on the knowledge that the panels to be painted were removable ones that could be replaced if contaminated. Also an experimental program was conducted to learn how to apply the paint.[10] The paint was kept from being contaminated until launch by maintaining two complete sets of panels; one set which underwent all of the flight acceptance testing with the structure and a flight set which were attached to the satellite just prior to shroud installation. The paint on the flight set of panels was applied within a week of the test set by the same people, using the same procedures as the test set. The flight panels were subjected only to a thermal-vacuum test in an ion-pumped facility and then packaged together for shipment to the launch site. Paint evaluation samples that were painted at the same time as the flight panels were monitored for optical properties at various stages of the flight panel test, packaging and final installation.

FLIGHT PERFORMANCE OF Z-93 PAINT

There were no changes detected in the optical properties of these paint samples.

## Discussion of Flight Results

### Heat Dissipation

The direct solar flux is the most important of the external heating fluxes. The angular variation of the orbit plane with respect to the Earth-Sun line was known for the mission life. This angular dependence coupled with the seasonal variation of the solar flux was sufficient to compute the incident solar flux perpendicular to the orbit plane. Average values for Earth thermal radiation and albedo were obtained from the literature.[11]

The SERT II orbit is fixed with respect to the Earth at a 9° inclination. The satellite is always oriented so that the ion thrusters face Earth and the solar array points toward the Sun. With the satellite in this configuration, there are two oscillations of the incident solar energy over each orbit (Fig. 4). There is a longitudinal oscillation in the solar

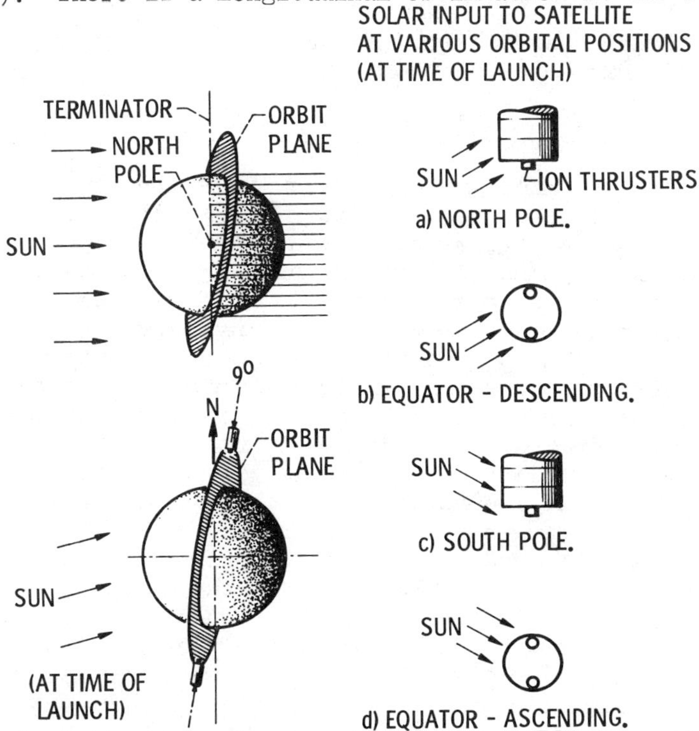

Fig. 4 Variation in solar heating due to orbital position of satellite.

energy incident to the Earth facing surface of the satellite over the polar regions (Figs. 4a and 4c). With the Sun in the southern hemisphere (as it was at the time of launch), the Earth facing surface is sunlit over the North Pole and shaded over the South Pole. The situation is reversed when the Earth moves so that the Sun is in the northern hemisphere. In addition there is a lateral oscillation of incident sunlight on the sides of the satellite over the equator (Figs. 4b and 4d). This oscillation of incident flux has caused the flight temperatures of the four polished metal Z-93 painted panels to vary over a wide range complicating the analysis of this temperature data. The two, fully coated panels were centered on the main axis of the satellite and remained in the orbit plane. Hence, neither oscillation affected the solar flux incident to these two panels and, therefore, did not cause orbital temperature fluctuations in the flight data. The data from these two panels forms the basis of this analysis. The flight temperatures of the other four panels are used to support the conclusions derived from the analysis of the two main panels. For the analysis the oscillations in the incident solar energy were averaged over the orbit and applied to the appropriate panels as heat inputs.

The internal heat dissipation was about 300W for nearly all of the 6200 hr period considered by this report.

## Comparison of Flight and Analytical Results

The analytical model was used to compute the expected steady-state temperatures for conditions of internal heat dissipation and environmental heating corresponding to 80 hr after launch. Nominal design optical properties were used. These first computations resulted in predicted temperatures that were about $15°$ C below the measured flight temperatures. Further corrections to the environmental heat input were required. The solar heating terms were expanded to account for such items as: interior heating through the holes in the ion-thruster side of the satellite during the times when this surface was in sunlight, heating due to the numerous screwheads, and heating due to uncoated metal in the gaps between the panels and between the spacecraft and SSU. The value used for the solar absorptance of the Z-93 paint was increased to 0.17 which was within the tolerance of the design optical properties. After these factors were added the agreement was improved, but the SSU section was still too cold. An additional heat input to the SSU main panel was required in order to obtain reasonable agreement between the flight data and the analytical model results. After considering the possibilities, it was concluded that the solar absorptance of the

SSU main panel must be increased above the value used for the spacecraft main panel paint. The agreement then became within expectations. This implied that the SSU panel paint has become damaged in some manner. This point will be discussed more fully in the next section.

This evaluation of the Z-93 paint is based on the comparison of flight temperatures and computer model steady-state results at eight points over the mission using the environmental heating and internal heat dissipation values corresponding to the point under consideration. All solar heating term corrections found to be necessary to obtain good agreement in the 80 hr case were modified to account for solar incidence angle and included in the computation. The paint emittance was assumed to be constant.

Parametric computer runs at each of the eight points were made using a range of values of paint adsorptance to determine how the absorptance changed. Agreement of the panel and the internal temperatures was required to increase the confidence in the results. Agreement of the panel temperatures alone was not sufficient to insure the validity of the resulting change in absorptance.

The results of the comparison between the flight temperatures and the analytical model results for the bay 8 panels are shown in Figs. 5a and 5b. The analytical temperatures used in these figures are the best fit values to the flight data and, hence, determine the solar absorptance behavior of the Z-93 paint from the 80 hr point used to calibrate the environmental heating of the satellite. The solid line represents a smooth curve between the eight points used in the analysis. With this excellent panel temperature agreement one would expect that the internal temperature trends would be similar to those obtained in the calibration tests. The comparison of these internal temperatures for the flight conditions is shown in Figs. 6a to 6c. Again, the analytical model curve is drawn through the computed points. The agreement for these cases is also very good. The tendency for the battery temperature to be low in the analytical model results was noted in the calibration tests when the battery was not being charged or discharged. The fit in temperatures, as presented here, is felt to be the best possible within the accuracy of the model and the flight data.

Resultant Z-93 Paint Absorptance

The resultant change in the solar absorptance for the Z-93 paint used on the SERT II satellite is shown in Fig. 7. The

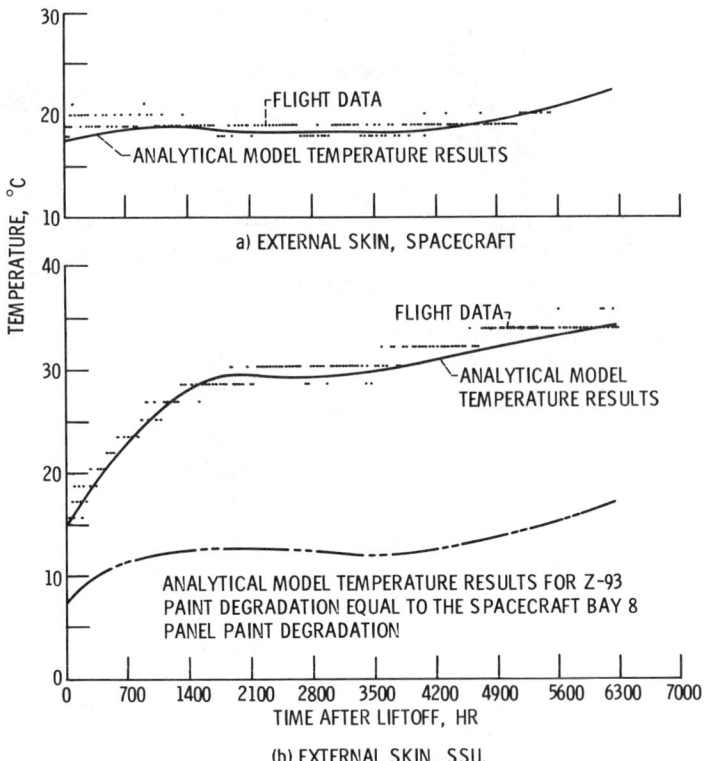

Fig. 5 Comparison of flight temperatures with analytical model results - Bay 8 panel temperatures.

paint on the spacecraft bay 8 has exhibited a slow, continual increase in solar absorptance over the initial value of 0.17. The nominal change for the 6200 hr period is 0.04. The design expectation for absorptance change was a 0.04 increase in 2000 hr based on the work reported in Refs. 12 and 13. The results reported in Ref. 14 are also shown on Fig. 7. The agreement with these literature values is reasonable.

The SSU bay 8 panel Z-93 paint absorptance change however, has shown the characteristics of the Z-93 paint samples flown on ATS I[15] and Mariner IV.[16] This SSU panel paint absorptance curve was generated by assuming that the absorptance of both the spacecraft and SSU panels was initially at the same value. This latter assumption is believed to be logical since both panels were painted within 5 minutes of each other on the same day, by the same man, with the same batch of paint, and cured, tested and packaged together. After the assembly of the flight panels to the structure, both the spacecraft and SSU

# FLIGHT PERFORMANCE OF Z-93 PAINT

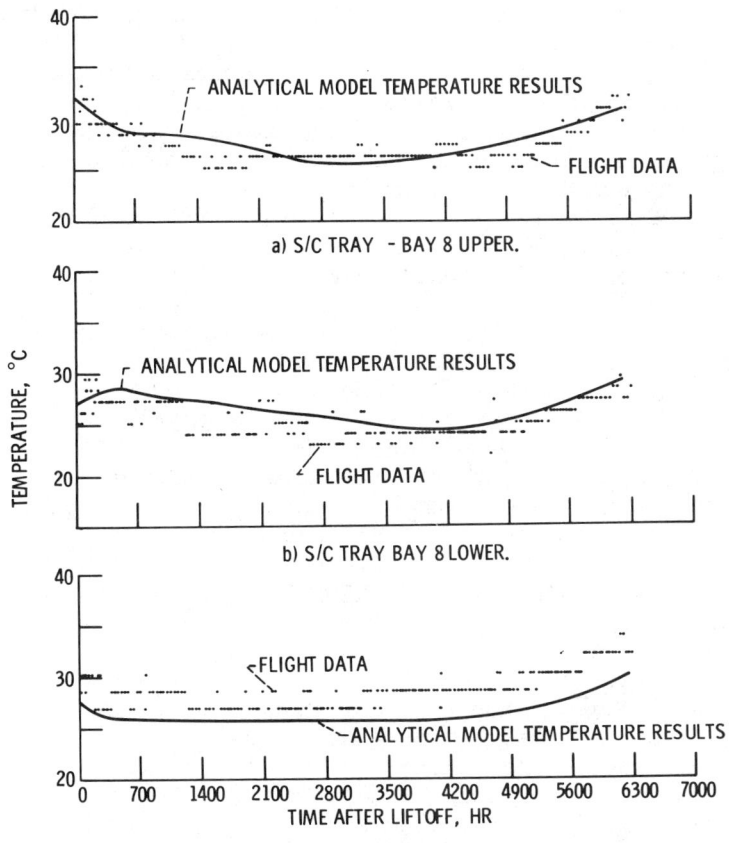

Fig. 6  Comparison of flight temperatures with analytical model results - Bay 8 internal temperatures.

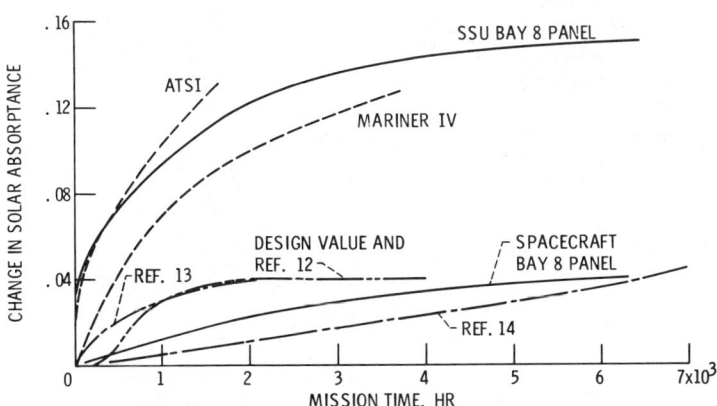

Fig. 7  Change in solar absorptance of SERT II Z-93 paint.

were wrapped in paper to protect the surfaces. The next day the paper was removed and the shroud put in place. With the shroud installed the environment until launch should have been the same for all panels. Paint samples were left in the gantry from the time the panels were removed from their containers until the shroud was installed. The optical properties of these samples were measured and found to be same as the values obtained when the panels were first painted.

The search for possible explanations for the degradation of the Z-93 paint on this SSU bay 8 panel started within a week after launch. Space particle induced damage was considered and rejected since both panels should have experienced the same environment. Degradation due to a charged particle flux from the ion thruster was ruled out since the degradation started before the thrusters were turned on. Cracking and peeling of the Z-93 paint was considered. However, previous experience has indicated that, if the Z-93 paint were going to peel off of the substrate, then it would happen shortly after application. In addition it would be very fortuitous for the paint to crack off in such a fashion that would give the degradation characteristics noted in flight.

After reviewing these and other mechanisms to account for the SSU panel flight temperatures, it was concluded that the paint on this panel had become contaminated in spite of the precautions taken. The source of the contamination is unknown. However, a possible explanation of how this one panel might have been contaminated does exist. The shroud used was a clamshell variety that mounted on a ring immediately below the SSU. The shroud split line was in the center of the bay 8 panels. The clearance between the shroud and the SSU panel was about 2 to 3 cm. The spacecraft bay 8 panel was set back about 1 cm from the SSU so that, in installing the shroud, care had to be taken not to physically damage the paint on the SSU panel. In addition this SSU panel was painted down to the shroud mounting ring. A final connection had to be made at this split line, in front of the white paint, after the shroud was in place. A dry lubricant was also used on the mounting ring to insure that the shroud would fall away. Special precuations were taken to prevent damage to the paint from these causes. However, it is conceivable that contaminations could still have been introduced and localized in the SSU bay 8 panel paint when the shroud was installed.

The absorptance of the Z-93 paint used on the other four panels of the SERT II satellite can only be inferred from the flight panel temperatures because of the orbital fluctuation in solar heating mentioned previously. Based on a comparison of the flight temperatures with the computer model results it is believed that the solar absorptance of this Z-93 paint is following the trends of the paint on the spacecraft bay 8 panel. The data for the SSU bay 1 and 7 panels is shown in Figs. 8a and 8b. Similar data was obtained for the spacecraft bay 1 and 7 panels. The solid line represents the computed temperatures when the Z-93 paint degradation follows the spacecraft bay 8 absorptance. The dashed line represents the analytical results for degradation on these panels corresponding to that of the SSU bay 8 paint. At the minimum solar incidence angle the lateral oscillation of solar energy is minimized reducing the range of temperature variations in

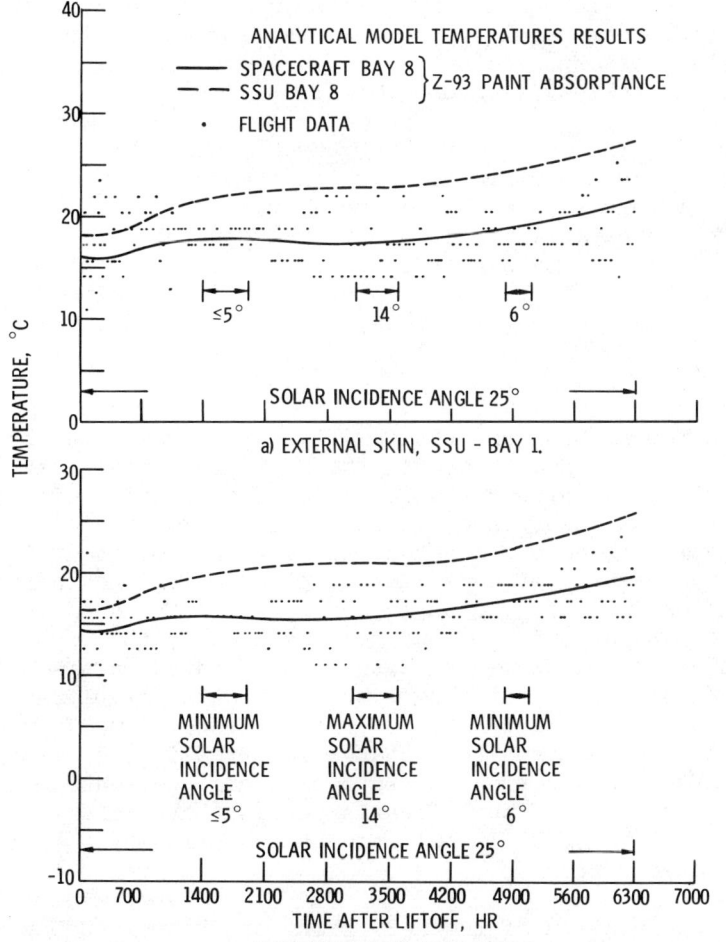

Fig. 8 Comparison of flight temperatures with analytical model results.

these panels. Under this condition the analytical model results should be in good agreement with the flight data. When the angle is larger, the analytical model results should tend to be centered in the fluctuations of the flight data. As shown in Fig. 8, these conditions are met only if the Z-93 paint degradation for these four panels follows the spacecraft bay 8 panel absorptance curve. Hence, it is believed that only one of the six panels have been contaminated. The Z-93 paint on all other panels appears to be following the trends exhibited in laboratory tests for ultraviolet exposure.

In conducting this analysis it was found that a 0.01 change in the solar absorptance value would cause about $1.5°$ C change in the two main panel temperatures and about $2°$ C change in the other four panels. The uncertainity associated with the resultant absorptance curve (Fig. 7) due to telemetry and model uncertainty is about ±0.02.

The 0.15 increase in the SSU bay 8 panel Z-93 paint has caused a panel temperature rise of about $20°$ C. The resulting internal temperature change, however, was only $5°$ C which is within the capability of the thermal control system. Therefore, even with this unexpected degradation of the paint on the SSU bay 8 panel, the Z-93 paint is functioning properly as the solar reflector coating for the SERT II satellite.

## Conclusions

This report has presented a flight performance evaluation of the Z-93 white paint used in the SERT II passive thermal control system. This evaluation is a determination of the change in the solar absorptance of paint for a 6200 hr period. The evaluation is accomplished by a comparison of the flight data with the results of a parametric study using the thermal analytical model. The solar absorptance change of the main spacecraft panel Z-93 paint has degraded by the expected amount but at a slower rate than anticipated. This data correlated well with existing laboratory data for ultraviolet degradation of the paint. Of the increase of 0.04 in 6000 hr of exposure, half of this change occurred in the first 1500 hr. After this period the rate of change decreased. The Z-93 paint on the four off-axis panels (bay 1 and 7 of the spacecraft and SSU) is believed to be following a similar degradation curve.

The degradation of the Z-93 paint on the main SSU panel, however, is far more rapid than that of the spacecraft paint. This increase is about 0.15 in 6000 hr. The rate of change, however, has been continually decreasing for the last 4000 hr of this period. It is believed that the degradation of the Z-93 paint on this panel is due to inadvertent contamination.

The SSU Z-93 paint degradation has caused a panel temperature rise of 20° C. However, the internal bulkhead temperature has increased by only 5° C and there is sufficient margin in the design to tolerate this rise. Hence, the Z-93 paint is functioning satisfactorily in keeping all component temperatures within their respective limits.

## References

[1] Stevens, N. J. and Smolak, G. R , "Design of the Passive Thermal Control System of the SERT II Satellite," *Proceedings of the Symposium on Thermodynamics and Thermophysics of Space Flight*, Western Periodicals, North Hollywood, 1970, pp. 101-120.

[2] Zerlaut, G. A., Harada, Y., and Tompkins, E. H., "Ultraviolet Irradiation of White Spacecraft Coatings in Vacuum," *Symposium on Thermal Radiation of Solids*, NASA SP-55, 1965, pp. 391-420.

[3] Streed, E. R. and Arvesen, J. C., "A Review of the Status of Spacecraft Thermal Control Materials," *Science of Advanced Materials and Process Engineering Series*, Vol. II, Western Periodicals, North Hollywood, 1967, pp. 181-192.

[4] Neel, C. B., "Role of Flight Experiments in the Study of Thermal-Control Coatings for Spacecraft," *AIAA Progress in Astronautics and Aeronautics: Thermophysics of Spacecraft and Planetary Bodies*, Vol. 20, edited by G. B. Heller, Academic Press, New York, 1967, pp. 411-438.

[5] Pearson, B., Jr., "Preliminary Results from the Ames Emissivity Experiment on OSO-II," *AIAA Progress in Astronautics and Aeronautics: Thermophysics of Spacecraft and Planetary Bodies*, Vol. 18, edited by G. B. Heller, Academic Press, New York, 1966, pp. 459-472.

[6] Millard, J. P., "Results of the Thermal Control Coating Experiment on OSO-III," AIAA Paper 68-794, New York, 1968.

[7] Gaski, J. D. and Lewis, D. R., "IBM-7094-II-DCS Computer Program C09945, Chrysler Improved Numerical Differencing Analyzer," April 30, 1966, TN-AP-66-15, Chrysler Corp., New Orleans, La.

[8]Toups, K. A., "A General Computer Program for the Determination of Radiant-Interchange Configuration and Form Factors: CONFAC II," SID-65-1043-2, Oct. 1965, North American Aviation Corp., Downey, Calif.

[9]Smolak, G. R. and Stevens, N. J, "Validation of the SERT II Thermal Analytical Techniques by Thermal Vacuum Testing of the Prototype Satellite." NASA TN D-6421, 1971.

[10]Stevens, N. J., "Application of SERT II Thermal Control Coatings," TM X-2155, Jan. 1971, NASA.

[11]Goetzel, C. G., Rittenhouse, J. B., and Singletary, J. B., eds., Space Materials Handbook, Addison-Wesley, Reading, Pa., 1965.

[12]Cunnington, G. R., Grammer, J. R., and Smith, F. J., "Emissivity Coatings for Low-Temperature Space Radiators," CR-1420, 1969, NASA.

[13]Streed, E. R., "The Influence of Temperature on the Stability of Low Solar Absorptance and Thermal Coatings," Proceedings of Conference on Spacecraft Coatings Development, TM X-56167, 1964, NASA, Washington, D.C.

[14]Zerlaut, G. A., Carroll, W. F., and Gates, D. W., "Spacecraft Temperature-Control Coatings--Selection, Utilization and Problems Related to the Space Environment," SpaceCraft Systems, Vol. 1, edited by Michal Lunc, Gordon and Breach, New York, 1966, pp. 259-313.

[15]Reichard, P. J. and Tiolo, J. J., "Preliminary ATS Thermal Coatings Experiment Flight Data," Proceedings of the Joint Air Force - NASA Thermal Control Working Group, AFML-TR-68-198, AD-841387, Aug. 1968, Air Force Systems Command, Wright-Patterson AFB, Ohio.

[16]Lewis, D. W. and Thostesen, T. O., "Mariner-Mars Absorptance Experiment," AIAA Progress in Astronautics and Aeronautics - Thermophysics and Temperature Control of Spacecraft and Entry Vehicles, Vol. 18, edited by G. B. Heller, Academic Press, New York, 1966, pp. 441-457.

# RADIATION DEGRADATION ANALYSIS OF SURVEYOR III MATERIAL

D.L. Anderson,* B.E. Cunningham,[†] and R.G. Dahms*

NASA Ames Research Center, Moffett Field, Calif.

## Abstract

Results of studies to determine the effects of the lunar environment on the surfaces of selected Surveyor III materials are presented. Two types of surfaces were examined — namely, the thermal-control paint on the television camera and a polished aluminum tube. Spectral reflectance, scanning electron microscopy, and x-ray probe measurements were used in these studies; it was determined that the solar adsorptance ($\alpha_S$) of the white paint degraded from a preflight value of 0.20 to postflight values of from 0.38 to 0.74 and that the $\alpha_S$ of the aluminum changed from 0.15 to values ranging from about 0.26 to 0.75. For the painted surfaces, an analytical model was used to separate the effects of solar radiation and lunar dust. The analysis showed that this dust was the primary cause of reflectance degradation.

## Introduction

During the Apollo 12 mission in November 1969, the astronauts brought back to Earth several pieces of hardware from the Surveyor III spacecraft. The recovery of these parts has provided the scientific community with a unique opportunity to study the effects of the lunar environment on several types of engineering materials. As part of the over-all NASA scientific investigation, a study was conducted at the Ames Research Center to determine the effect of this environment on some of the painted and unpainted exterior surfaces. In this paper are presented results of this study which pertain to the degradation of these surfaces and, hence, to changes in their thermal-control characteristics. These results, combined with analysis, form the basis for a better understanding of the degradation of thermal-control paints and the significance of the lunar micrometeoroid and secondary particle bombardment on thermal-control surfaces.

---

Presented as Paper 71-478 at the AIAA 6th Thermophysics Conference, Tullahoma, Tenn., April 26-28, 1971.

*Research Scientist, Materials Research Branch.
[†]Assistant Chief, Materials Research Branch.

The recovery of Surveyor III was accomplished with few deviations from the original flight plan;[1] however, it will be shown later that in some cases it was necessary to account for the recovery process in the results presented in this paper. During the second extravehicular activity period the astronauts cut the entire television camera and a 19.7-cm length of the Radar Altimeter and Doppler Velocity Sensor (RADVS) support tube from the spacecraft and placed them in separate pockets of their backpacks. During the return flight the parts were subjected to the high "g" loads associated with re-entry and the jolt at splashdown in the Pacific Ocean. In addition, it is believed that the TV camera became dislodged from its stowed position at splashdown, causing two dents in the primary mirror hood.[2] Early in the quarantine period at the NASA Lunar Receiving Laboratory (LRL) in Houston, the parts were taken out of the backpack, photographed, and sealed in polyethylene bags until their release from quarantine on January 7, 1970. The parts subsequently were transferred from the LRL for examination by various investigators. A more complete description of these parts and their handling prior to the investigation reported in this paper can be found in the Hughes Aircraft Company report on Surveyor III materials.[3] This investigation was conducted between July and December 1970. As for the parts examined in this investigation, no specific effort had been made after January 7, 1970, to maintain a vacuum or provide light protection for either the TV camera or the aluminum tube. Thus, the parts have been exposed to a variety of laboratory environmental conditions since their removal from the Surveyor III spacecraft and, therefore, it was expected that the analysis would have to take into consideration such exposure.

## Examination and Analysis

Two types of surfaces were examined in this study: the white thermal-control paint on parts of the television camera (the elevation-drive housing — a small 5.1×7.6×1.3-cm box — and the lower shroud) and the unpainted surface of the RADVS support tube (the two 2.5-cm-long sections "B" and "E" of the 1.3-cm-diameter aluminum tube). Examination of these surfaces was carried out with three techniques: 1) spectral reflectance measurements; 2) energy dispersive x-ray probe analysis; and 3) optical and scanning electron microscopy (SEM). The spectral reflectance of each part was measured in an integrating-sphere reflectometer with the sample located in the center of the sphere whenever possible. Spectral reflectance measurements were made at several locations on each part with special emphasis being given to the cleanest and dirtiest, or most contaminated, areas. The x-ray probe, an accessory to the SEM, was used to obtain the elemental composition of a surface. The microscopy techniques were used to examine surface features at magnifications up to 1,700X and 30,000X, respectively. The normal practice of vapor depositing a gold film over an insulating-paint surface for SEM was not permitted on the Surveyor III parts due to constraints imposed by subsequent

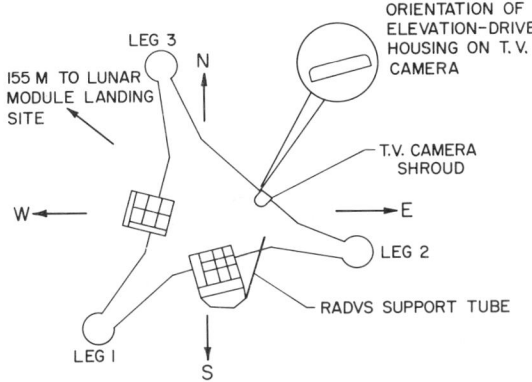

Fig. 1   Schematic diagram of orientation of Surveyor III spacecraft on lunar surface.

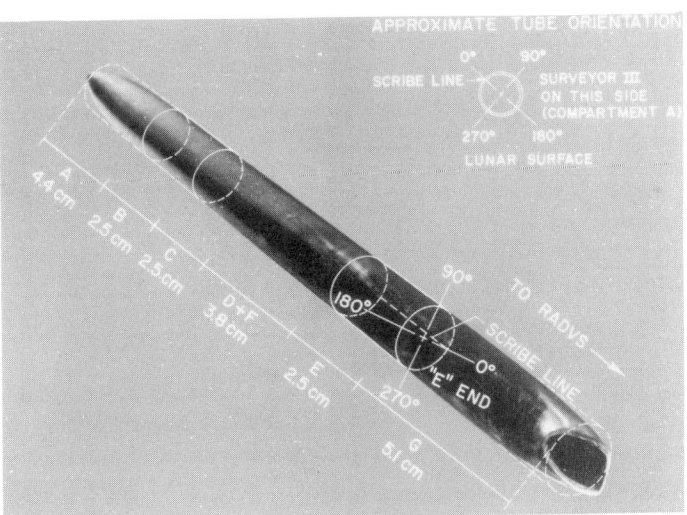

Fig. 2   Photograph of section of RADVS unpainted aluminum support tube as cut from Surveyor III spacecraft.

experiments to be conducted by other investigators. Therefore, some difficulty with charge buildup was encountered which limited the useful magnification for examination of the thermal-control paint to 10,000×.

The locations of the retrieved parts on the Surveyor III spacecraft and their orientation with respect to the lunar surface are shown in Figs. 1, 2, and 3. A schematic diagram (plan view) of the spacecraft at its landing site in the Ocean of Storms[3] and the relative location of the Apollo 12 lunar module (LM) is shown in Fig. 1. The retrieved parts are identified on this figure. The locations of sections "B" and "E" on the RADVS support tube and the

208     D.L. ANDERSON, B.E. CUNNINGHAM, AND R.G. DAHMS

Fig. 3  Photograph of Surveyor III television camera.

television camera parts are shown in Figs. 2 and 3, respectively. The presentation of the results of the examination of these parts is divided into two categories: 1) a discussion of the optical properties of both painted and unpainted surfaces, and 2) a discussion of surface features.

Optical Properties of Painted Surfaces

The major portion of the television camera (Fig. 3) was coated with an inorganic thermal-control paint consisting of a calcined china clay pigment, primarily an aluminum silicate, in a potassium silicate binder. This paint initially had a very white appearance. When the recovered camera was viewed by other investigators[2] at the Lunar Receiving Laboratory in Houston, it had an extensive brown appearance. In order to determine the cause of this change in color, a representative part of the camera, the elevation-drive housing (Fig. 3), was used for measurements of spectral reflectance and surface x-ray analysis. This housing was uniquely suited for these tests due to the orientation of the different sides of the housing relative to the landing site of the LM (Fig. 1), the lunar surface, the overhead path of the Sun, and the other parts of the Surveyor spacecraft. Figure 4 shows this housing after it was sectional for study. The letters indicate measurement locations used in this report.

The distribution of spectral reflectance of the elevation-drive housing was compared with similar measurements on standards prepared by the Surveyor III manufacturer and retained for control purposes. Although the original reflectance of each side of this housing was approximately the same as the

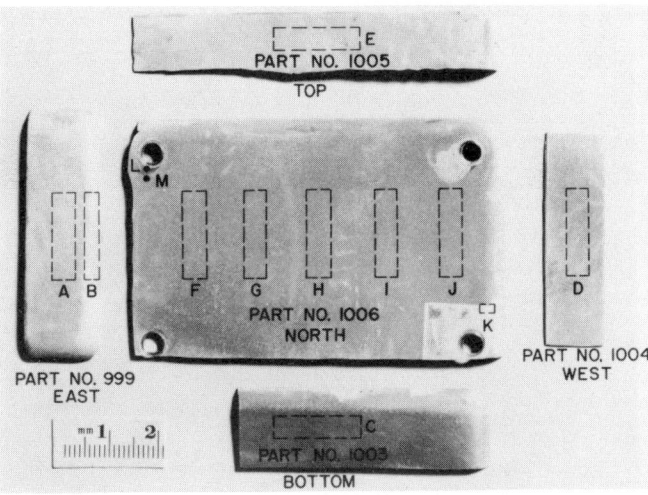

Fig. 4 Photograph of elevation-drive housing showing measurement locations.

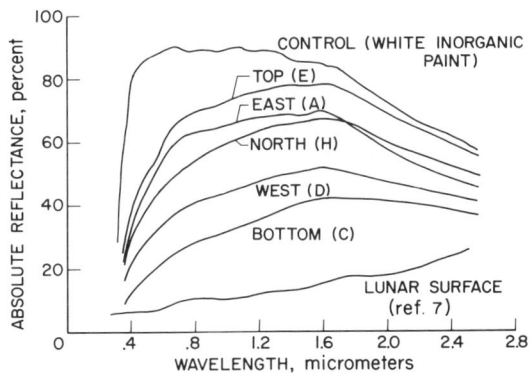

Fig. 5 Spectral distribution of reflectance for the five faces of the elevation-drive housing.

unexposed "control," none of the sides retained their original reflectance or degraded the same amount. Figure 5 shows the spectral reflectance of the five different faces of the housing. These measurements were made on July 29,

Table 1 Reflectance readings

| Part | Side | Measurement location[a] | $\alpha_S$ July 29 1970 | $\alpha_S$ Nov. 24 1970 | $\Delta\alpha_S$ |
|---|---|---|---|---|---|
| 999 | East | A | 0.44 | 0.34 | 0.10 |
|  |  | B | 0.38 | — | — |
| 1003 | Bottom | C | 0.74 | 0.64 | 0.10 |
| 1004 | West | D | 0.63 | 0.54 | 0.09 |
| 1005 | Top | E | 0.38 | 0.38 | 0.00 |
| 1006 | North | F | 0.57 | — |  |
|  |  | G | 0.50 | — |  |
|  |  | H | 0.50 | — |  |
|  |  | I | 0.50 | — |  |
|  |  | J | 0.50 | — |  |
|  |  | K | 0.25 | — |  |

[a]See Fig. 4 for locations.

1970. The significance of this date will be shown later. The values of solar absorptance ($\alpha_S$) corresponding to these reflectance data are shown in Table 1.

In addition to separating the effects of the lunar environment, it became apparent in this study that certain effects of the terrestrial laboratory environment on the data presented in this paper must also be acknowledged. It has been shown[2] that the reflectance of the lunar dust on the camera surfaces increased or "bleached" when exposed to normal laboratory lighting conditions. In general, the exposure of camera parts to light was neither controlled nor recorded. The elevation-drive housing was exposed to various unknown lighting conditions before delivery to Ames so postflight reflectance values shown in Fig. 5 are therefore not the same as would have been recorded immediately upon recovery. To evaluate the effect of this bleaching, a second set of reflectance readings was made four months later, in November 1970. These data also are shown in Table 1. Although the rate of change seems to be relatively slow, the dates of the measurements presented in this report should be considered when comparing results presented in this report with any other investigations.

The effect of angle of surface illumination on reflectance was also investigated. Figure 6 shows these data at a wavelength of 1.34 $\mu$m. The reflectance values are all relatively constant with a slight increase at the high angles of illumination for the more contaminated sides. This increase is not surprising since investigations by others[4] have shown that the reflectance of lunar fines shows this same increase in reflectance at grazing angles of incidence. (The north side, part 1006, could not be measured because of its large size.) The effect of illumination angle was also measured at other

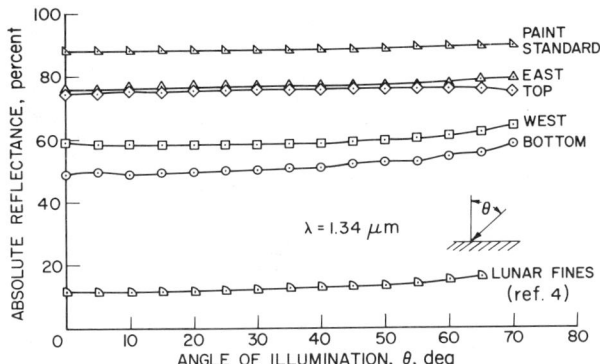

Fig. 6 Effect of angle of surface illumination on reflectance for several faces of the elevation-drive housing.

wavelengths (0.70 and 1.01 μm) with similar results. Since it has been shown by others[5] that these surfaces were bombarded to various degrees by a highly directional spray of lunar dust from the LM, the effect of the rotational orientation of the incident monochromatic light beam was investigated. The measurements were made on part 1004 (west), the side most likely to show the effects of this dust. It was determined that the maximum difference in reflectance values was approximately 3%, a value only slightly outside the normal 2% accuracy of the instrument. The effect of rotational orientation of illumination is, therefore, very small.

Degradation Analysis

Three factors are primarily responsible for the differences between the reflectance values of the respective sides of the elevation-drive housing shown in Fig. 5: 1) solar radiation degradation; 2) surface contamination by lunar dust; and 3) mechanical abrasion of paint during recovery operations. Other environmental factors (electrons, protons, vacuum, and thermal cycling) were considered by other investigators[2] and found to have little, if any, effect on the spectral reflectance of the television camera thermal-control paint. Laboratory tests by other investigators[6] have shown that the reflectance of this inorganic paint could be degraded in the 0.25 to 1.5 μm wavelength range by exposure to uv radiation. The extent of this reflectance degradation is a function of the total sun exposure, therefore, the different faces of the elevation-drive housing should exhibit different reflectances due to radiation damage only. In addition to radiation damage, a coating of lunar dust further modified the reflectance of the paint. For comparison, the spectral reflectance of lunar dust from Ref. 7 is shown. This nongrey reflectance makes the influence of a dust layer wavelength-dependent, with the greatest influence occurring at the short wavelengths.

In order to separate the effects of these two environmental parameters, attempts were made to account for the influence of lunar dust and then to compare the results of this analysis with laboratory data on uv radiation damage. An analysis similar to the one used by the Hughes Co.[2] and Blair et al.[8] was applied to the reflectance data shown in Fig. 5. This analysis assumes that the dust layer will interact with the measurement of reflectance by the following manner: 1) partial absorption of incident energy; 2) partial absorption of reflected energy; and 3) forward and backscatter of incident energy. An expression was derived which considered the monochromatic energy balance of the above factors. The equation below shows the monochromatic energy balance of the above factors.

$$I_m = I_o \rho_d K_1 A_d + I_o \rho_p [(1 - K_2 A_d)(1 - K_3 A_d)]$$

o.

$$\rho_m = \rho_d K_1 A_d + \rho_p (1 - K_2 A_d)(1 - K_3 A_d)$$

where $I_m$ = measured radiant flux (reflected from surface), $I_o$ = radiant flux incident on surface, $\rho_m$ = reflectance = $I_m/I_o$, $\rho_d$ = reflectance of lunar dust, $\rho_p$ = reflectance of paint surface, $A_d$ = fraction of surface area covered by dust, $K_1$, $K_2$, $K_3$ = constants associated with transmission, forward scattering, and geometry of the dust for the incident and leaving beam.

Figure 7 shows the results of this analysis compared with laboratory data[6] on uv damage. The shadowed area shows the range of the uv radiation damage only, without any dust effects, for all the sides. It is significant to note that the reflectance curves for all the sides lie between the laboratory data for 550 and 2800 Equivalent Sun Hours (ESH). Exposure for the housing as

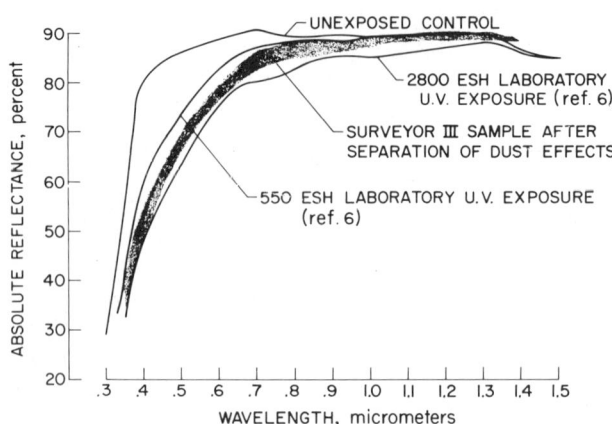

Fig. 7 Results of analysis to separate the solar radiation degradation and the surface contamination by lunar dust on the elevation-drive housing.

calculated from data published by Nickle[9] ranged from 40 ESH on the north side to 3000 ESH for the top. Considering the assumptions that were made in the degradation model for the removal of dust effects, the accuracy of the lunar solar exposure calculations and the limitations of the laboratory data, this degree of agreement between laboratory and flight data was considered to be very good. Also, it should be noted here that other investigators[2] have shown that the abnormal landing maneuvers of the Surveyor spacecraft stirred up enough lunar dust to contaminate some external surfaces. An initial coating of dust at that time would tend to modify the contribution of uv damage. However, the assessment of the relative effects of the lunar dust and radiation damage by this technique seems to be quite reasonable. The contribution of lunar dust was found by this analysis to be by far the major factor in the degradation of reflectance of the television camera thermal-control coating.

The x-ray probe analysis was used to determine the elemental composition of a specimen surface. From this analysis the relative amount of lunar material retained on each of the surfaces was determined. An unexposed paint standard was used to obtain a reference spectrum for the painted surfaces. As expected, aluminum, silicon, and potassium peaks predominate this spectrum. On the elevation-drive housing, additional amounts of calcium, titanium, and iron were evident. This is consistent with the composition of the dust layer found on all parts studied. Analysis of lunar soil by others[10] shows that after silicon and oxygen, iron, at about 14%, is the most abundant element in the lunar samples. Calcium, a major element in pyroxene and plagioclase, represents about 8% of the lunar material. The amount of these two elements was therefore used as an indication of the amount of lunar fines retained on these parts. Table 2 presents these data. Note that the silicon and potassium remained relatively constant on all sides whereas the calcium and iron varied considerably. Although the amount of dust based on calcium does not agree exactly with the iron analysis, the amount of dust can be inferred from these data. The analysis indicates that the dirtiest side was the bottom. This is not a surprising conclusion based on visual observations (Fig. 4), the spectral reflectance data (Fig. 5), and the orientation directly facing the lunar soil. It was estimated that the actual amount of dust on the bottom was

Table 2. X-ray analysis of elevation-drive housing

| Part | Side | Relative amount of element | | | |
| --- | --- | --- | --- | --- | --- |
| | | Si | K | Ca | Fe |
| 1003 | Bottom | 6.00 | 1.40 | 0.97 | 1.02 |
| 1004 | West | 5.90 | 1.50 | 0.30 | 0.25 |
| 1006 | North | 6.10 | 1.44 | 0.44 | 0.30 |
| 1006 | North (Location L) | 6.30 | 1.60 | 0.74 | 0.57 |

$7.4 \times 10^{-5}$ g/cm². An analysis was also made at position L, a location protected from the blast effects of the LM by the head of a mounting screw. The iron and calcium content at this location indicate that the north side was much dirtier prior to the landing of the LM.

Optical Properties of Unpainted Surfaces

The spectral characteristics analysis of the two sections of the unpainted aluminum support tube was compared with a similar analysis of a section of polished aluminum tube made of the same alloy. This tube was polished by the Surveyor manufacturer using the same techniques as were used on the flight hardware. For this "control" specimen, the solar absorptance ($\alpha_S$) was approximately 0.15. The postflight values ranged from an $\alpha_S$ of 0.26 on the "clean" side to 0.75 on the "dirty" side with little variation along the axial length of each tube section. The variation in reflectance around tube section "E" is shown in Fig. 8 for a wavelength of 0.47 μm. It has been judged by the authors and other investigators[2,11] that the portion of the tube with the lowest reflectance (greatest contamination) was oriented toward the lunar surface and slightly toward the spacecraft descent engine number 3. This tends to indicate that this dust was deposited on the tube during the landing maneuver of the Surveyor spacecraft. The reflectance was again measured after the surface was replicated. As indicated in Fig. 8, removal of loose material by replication techniques increased the reflectance on all sides of the tube. Examination of the tube by SEM showed that the aluminum surface under the dust had been eroded. The reflectance after dust removal indicates the extent of this erosion. Figure 9 shows the total spectral distribution of reflectance around the tube. The contamination on the dirty side appears to be primarily of lunar origin or possibly from descent engine exhaust deposits. This contamination is not easily removed, however, since some traces of it remain even after repeated attempts by other investigators to remove it with ultrasonic

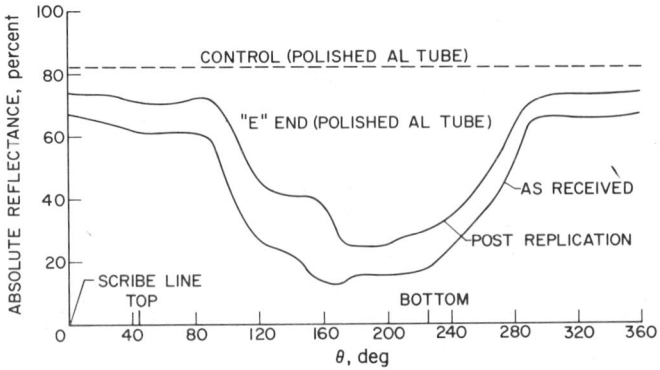

Fig. 8 Spectral reflectance of the unpainted aluminum RADVS support tube, section "E."

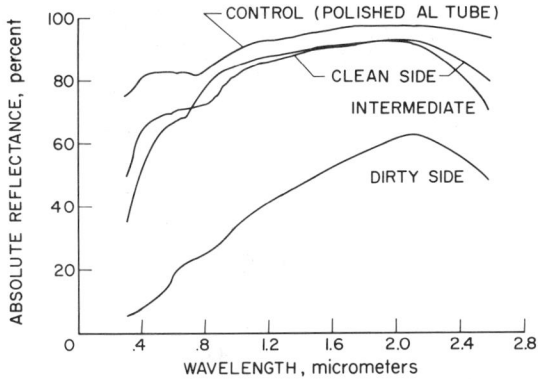

Fig. 9 Spectral distribution of reflectance on various portions of unpainted aluminum RADVS support tube.

cleaning[12] or with replication processes for transmission microscopy experiments.[13]

Surface Features

The features of both painted and unpainted surfaces were examined by optical microscopy and by SEM for impacts, erosion, or other evidence of lunar environmental effects. The painted surfaces of the television camera were examined and compared with the previously mentioned paint standard supplied by the Surveyor III manufacturer. SEM micrographs of these paints are shown in Fig. 10. Note that there are several cracks and small holes in the paint from the lower shroud (Fig. 10a). Similar crack patterns, commonly referred to as "mud cracking," and scattered small holes were also found in the paint standard (Fig. 10b) and also small holes in the elevation-drive housing (Fig. 10c). Although the cracks on the shroud are larger and more pronounced, the existence of similar cracks and small holes on the unexposed standard indicates the possibility that these features existed before flight and are therefore not necessarily a result of exposure to the lunar environment. Other smaller holes were examined at higher magnifications and compared with laboratory-produced hypervelocity impacts in the paint standard. Not one was positively identified as being formed by a micrometeoroid. Examination by optical microscopy showed many holes with very clean white walls — as would be expected if they were formed just before recovery with no time for additional contamination or degradation. It is believed that these holes were made by the previously mentioned shower of lunar dust stirred up by the landing of the LM. Even though the optical properties of the area in and around these craters appeared to differ from the properties shown in Fig. 5, the number of craters and the area of each (Fig. 10) was judged to be too small to significantly affect the total reflection of the spacecraft components.[14]

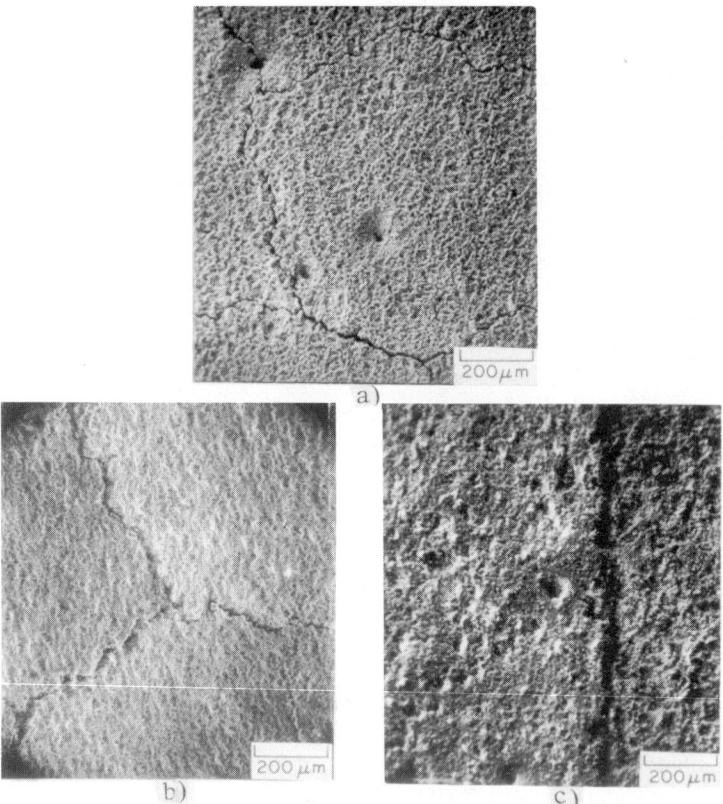

Fig. 10 Scanning electron micrographs of thermal-control paints from Surveyor III television camera [a) and c)] and of a laboratory standard thermal-control paint b).

The surface features of the two portions of the RADVS support tube, sections "B" and "E," were examined. A total of about 200 mm² of the surface section "B" was examined by SEM after removal of contaminants by prior investigators; typical micrographs of the "clean" and "dirty" sides of this section are shown in Fig. 11. The small holes shown in these micrographs are representative of the numerous holes found over the entire surface of the tube by both optical and scanning electron microscopy. Similar holes, although fewer in number, were found on the surface of a "control" specimen – a section of aluminum tube, supplied by the Surveyor III manufacturer, of the same alloy and size as the RADVS support tube. It was therefore concluded that such holes may in some way be characteristic of the tube manufacturing or flight preparation processes.

The question of whether or not the holes are hypervelocity impact sites from micrometeoroid bombardment was also considered by comparing SEM

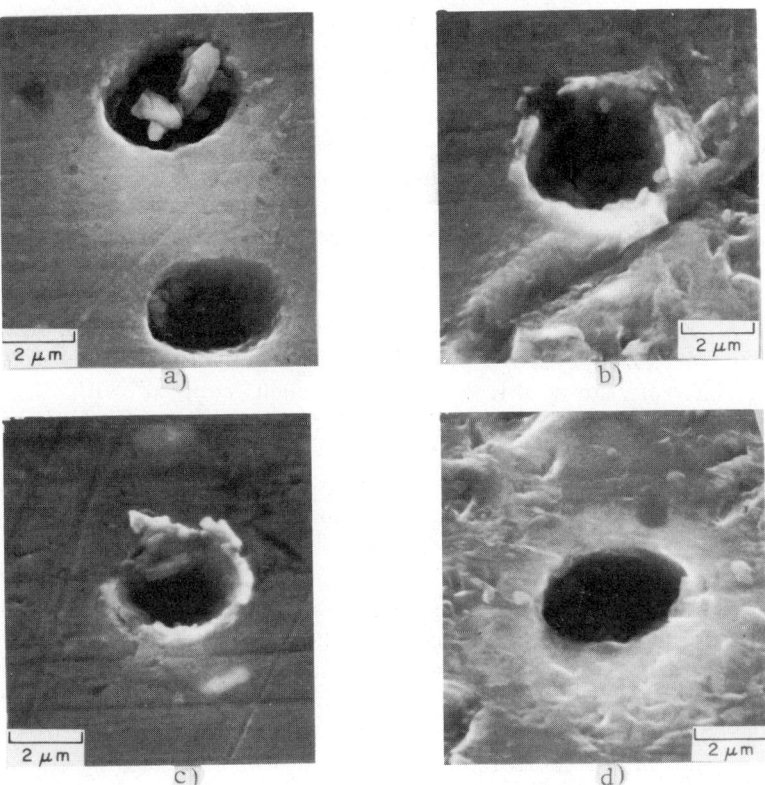

Fig. 11 Scanning electron micrographs of support tube section "B."
a) "Unrimmed" holes on clean side, b) "Semi-rimmed" hole on clean side, c) "Rimmed" hole on clean side, and d) "Unrimmed" hole on dirty side.

micrographs of section "B" with micrographs of hypervelocity impacts produced in laboratory experiments. It was concluded that none of the holes examined by SEM could be definitely characterized as having been caused by hypervelocity impact of primary particles. Based on this conclusion, the known lunar exposure time, and the area examined, an upper limit to the lunar surface meteoroid environment was calculated. For the $10^{-12}$ g regime (the mass applicable to this experiment) the rate of impact was less than $6.2 \times 10^{-5}$ particle/m$^2$ sec. A more thorough discussion of this subject is given in a recent paper by the authors.[14] It is significant to note here that for the type of surfaces studied, this rate of particle bombardment is several orders of magnitude below that which would be detrimental to the surface optical properties of thermal-control coatings similar to that used on the Surveyor III spacecraft.

## Conclusions

The examination of painted and unpainted surfaces of parts retrieved from the Surveyor III spacecraft has provided useful information regarding the effects of the lunar environment on thermal-control surfaces. Furthermore, the opportunity to examine flight hardware and engineering materials after a lengthy exposure to the lunar environment has provided verification of laboratory experiments performed during the design of the Surveyor III spacecraft. With regard to the laboratory techniques used in this investigation, it was found that optical microscopy provided only limited definition of the effects of the lunar environment on the surfaces examined. On the other hand, scanning electron microscopy provided excellent quantitative identification of surface characteristics; this instrument, when coupled with the energy dispersive x-ray probe, enabled the investigators to separate the effects of uv radiation from the effects of contamination by lunar dust. Spectral reflectance measurements provided excellent quantitative data on the degradation of total spectral reflectance of both painted and unpainted surfaces.

It was found that none of the surfaces examined retained their initial optical properties. All the surfaces were found to be coated to some degree with lunar dust such that, in general, those surfaces which were exposed directly to the lunar surface had the greatest amount of dust. It was found that the spectral reflectance of both polished aluminum surfaces and thermal-control paints were affected by the 942-day exposure to the lunar environment. In the case of the polished surfaces, the most significant effect was erosion primarily of lunar origin or possibly from dust stirred up by the Surveyor III descent engine. The postflight values ranged from an $\alpha_S$ of 0.26 on the "clean" side to 0.75 on the "dirty" side with little variation along the axial length of each tube section. The paints, however, were significantly damaged by solar radiation and surface contamination with a resulting change, for a surface facing outer space, in solar absorptance from the preflight value of 0.20 to a postflight value of 0.38. For a surface facing directly toward the lunar surface, the postflight solar absorptance was 0.74. An analytical model was used to separate the effects of solar radiation and lunar dust. This analysis showed that 1) dust contamination was the primary cause of reflectance degradation and 2) laboratory and flight data for uv damage are in reasonable agreement.

The features of both painted and unpainted surfaces were examined for impacts, erosion, or other evidence of lunar environmental effects. It was concluded that no sites were found which could definitely be characterized as being micrometeoroid impact craters. The results of the study, based on a somewhat limited area of examination, indicate an upper-limit value for the micrometeoroid flux on the lunar surface for 1-$\mu$m-diam particles (mass of about $10^{-12}$ g) of less than about $6 \times 10^{-5}$ particle/m$^2$ sec. This rate of particle bombardment is far below that which would have been detrimental to the optical properties of the Surveyor III spacecraft surfaces within the time period of their exposure on the lunar surface.

References

[1] Anon., "Apollo 12 Mission Report," MSC 01855, March 1970, NASA.

[2] Anon., "Test and Evaluation of the Surveyor III Television Camera Returned from the Moon by Apollo 12," Rept. SSD 00545, Vols. I and II, Dec. 1970, Hughes Aircraft Co., Culver City, Calif.

[3] Anon., "Surveyor III Mission Report: Part I. Mission Description and Performance," Rept. 32-1177, Sept. 1967, Jet Propulsion Lab., Pasadena, Calif.

[4] Birkebak, R. C., Cremers, C. J., and Dawson, J. P., "Directional Reflectance of Lunar Fines as a Function of Bulk Density," *Proceedings of the Second Lunar Science Conference,* Vol. 3, 1971, MIT Press, Cambridge, Mass.

[5] Jaffe, L. D., "Blowing of Lunar Soil by Apollo 12: Surveyor III Evidence," *Proceedings of the Second Lunar Science Conference,* Vol. 3, 1971, MIT Press, Cambridge, Mass.

[6] Zerlaut, G. A. and Gilligan, J. E., "Study of In-situ Degradation of Thermal Control Surfaces," IITRI-U6061, Mar. 1969, IIT Research Inst., Chicago, Ill.

[7] Nash, D. B. and Conel, J. E., "Luminescence and Reflectance of Apollo 12 Samples," *Proceedings of the Second Lunar Science Conference,* Vol. 3, 1971, MIT Press, Cambridge, Mass.

[8] Blair, P. M., Jr., Carroll, W. F., Jacobs, S., and Leger, L. J., "Study of Thermal Control Surfaces Returned from Surveyor III," (published elsewhere in this volume).

[9] Nickle, N.L., "Surveyor III Material Analysis Program," *Proceedings of the Second Lunar Science Conference,* Vol. 3, 1971, MIT Press, Cambridge, Mass.

[10] LSPET (Lunar Sample Preliminary Examination Team), "Preliminary Examination of the Lunar Samples from Apollo 12," *Science,* Vol. 167, No. 3923, March 1970, pp. 1325-1339.

[11] Carroll, W. F., Blair, P. M., Jacobs, S., and Leger, L., "Discoloration and Lunar Dust Contamination of Surveyor III Surfaces," *Proceedings of the Second Lunar Science Conference,* Vol. 3, 1971, MIT Press, Cambridge, Mass.

[12] Bühler, F., Eberhardt, P., Geiss, J., and Schwarzmuller, J., "Trapped Solar Wind Helium and Neon in Surveyor III Material," *Proceedings of the Second Lunar Science Conference,* Vol. 3, 1971. MIT Press, Cambridge, Mass.

[13] Buvinger, E. A., "Replication Electron Microscopy on Surveyor III Unpainted Aluminum Tubing," *Proceedings of the Second Lunar Science Conference,* Vol. 3, 1971, MIT Press, Cambridge, Mass.

[14] Anderson, D. L., Cunningham, B. E., Dahms, R. G., and Morgan, R. G., "X-ray Probe, SEM, and Optical Property Analysis of the Surface Features of Surveyor III Materials," *Proceedings of the Second Lunar Science Conference,* Vol. 3, 1971, MIT Press, Cambridge, Mass.

STUDY OF THERMAL CONTROL SURFACES
RETURNED FROM SURVEYOR III

Paul M. Blair Jr.*

Hughes Aircraft Company, Culver City, Calif.

W. F. Carroll[†]

Jet Propulsion Laboratory, Pasadena, Calif.

and

S. Jacobs[‡] and L. J. Leger[‡]

NASA Manned Spacecraft Center, Houston, Texas

Abstract

Surveyor III parts were returned to earth by Apollo 12 astronauts after 2 1/2 years on the lunar surface. Changes found in the thermal optical properties of the surface finishes of these parts are attributed to expected solar radiation damage (ultraviolet and solar wind protons) and to greater than expected lunar dust deposition. The dust was

---

Presented as Paper 71-479 at the AIAA 6th Thermophysics Conference, Tullahoma, Tenn., April 26-28, 1971. This paper presents results of one phase of research carried out at Hughes Aircraft Company, Culver City, Calif. under Contracts JPL 952792 and NAS 9-10492; at the Jet Propulsion Laboratory, California Institute of Technology, under Contract NAS7-100; and at NASA Manned Spacecraft Center, Houston, Texas, sponsored by the National Aeronautics and Space Administration. Numerous individuals performed tasks supporting the studies described in this paper. The authors acknowledge the efforts of those people at Hughes, Jet Propulsion Laboratory, NASA Manned Spacecraft Center, and NASA Headquarters who made this program possible. The work of E. E. Luedke and W. D. Miller at TRW Systems in performing some of the optical measurements is also acknowledged.
*Section Head.
†Member of Technical Staff.
‡Aerospace Technologists.

generated by the Surveyor landing and the Lunar Module landing. Surveyor surfaces were significantly altered by the Lunar Module dust even though the Lunar Module landed more than 500 ft from the Surveyor. Condensed outgassed materials do not appear to be significant.

## Introduction

The prediction of the effects of the space environment on materials is a problem faced in all space programs. The problem of predicting the effect of radiation interaction (in vacuum) with materials is complex and is difficult to solve. Experimentally obtained data are limited because of inadequate simulation techniques. The best engineering information available today on the effects of the space environment on thermal control surfaces is based on flight experiments. However, such data are limited to the accuracy of temperature measurements and may be significantly biased by unknown events that occurred prior to or during the mission. Ideally, placing specimens in space and measuring in place or returning them to earth for measurement would produce valid information; the latter only if the specimens could be returned in a controlled environment. Except for short duration missions this has not been possible to date.

An opportunity to obtain valuable scientific and engineering data presented itself during the Apollo 12 mission when on Nov. 19, 1969 a precision landing was made near Surveyor III, a spacecraft that had soft landed on the moon 2 1/2 yr earlier. Selected parts of the Surveyor III spacecraft were retrieved and returned to earth for study. The data obtained from these parts are expected to be useful for all future missions although the parts could not be returned under controlled conditions.

An extensive evaluation of the returned parts was made in an attempt to determine the effect of the 2 1/2-yr lunar stay.[1] This paper will review the results obtained to date on the thermal control finishes.

## Surveyor III Parts Return

Surveyor III, typical of most spacecraft, was exposed to a wide range of prelaunch and flight environments. A knowledge of these environments is important to the correct interpretation of the data collected on the returned parts. The environments to which Surveyor III was exposed are shown in Table 1. The exposure has been divided into six periods, and the critical environments of each period have been identified.

Table 1  Environments of Surveyor III critical
to thermal control surfaces

| Period of Exposure | Environment |
|---|---|
| Prelaunch | Manufacturing<br>Testing (including thermal vacuum)<br>Thermal surface rework |
| Transit to moon (66 hr) | Vacuum<br>Solar electromagnetic radiation<br>Particulate radiation<br>Materials outgassing<br>Vernier engine exhaust |
| Landing | Retro rocket exhaust<br>Vernier engine exhaust<br>Lunar dust |
| Post landing (2 1/2 yr) | Meteoroids<br>Dust from meteoroids<br>Solar electromagnetic radiation<br>Solar wind protons and electrons<br>Extreme temperature variations<br>Vacuum<br>Materials outgassing<br>Dust from LM landing |
| Recovery and return | Dust from astronauts approach<br>Handling<br>Packaging<br>Air exposure<br>Light exposure<br>Earth landing impact |
| Laboratory study | Handling<br>Light exposure |

Each of these environments possibly could have altered the surfaces of Surveyor III and therefore influenced the reflectance measurements made on thermal control surfaces of the returned parts.

The parts of Surveyor III returned to earth by the Apollo 12 crew were the television camera, portions of the cable attached to the camera, the soil mechanics/surface sampler (SM/SS) scoop, a section of a polished aluminum support tube and a section of a white painted support tube. Table 2 describes the thermal control surfaces used on the returned

Table 2  Description of thermal control surfaces on the returned Surveyor III parts

| Thermal control surface | Typical initial thermal optical properties | | Description |
|---|---|---|---|
| | Solar absorptance $a_s$ | Total normal emittance (80°F) $\epsilon_N$ | |
| Anodized aluminum | | | 6061-0 Al alloy anodized per MIL-A-8625, Type II |
| Clear | 0.20-0.30 | 0.90 | Class 1, clear |
| Dyed black | 0.70 | 0.90 | Class 2, dyed black |
| Polished aluminum | 0.19 | 0.04 | Aluminum alloy, 6061, 2024 |
| Fiberglass | not measured (white) | not measured | MIL-Y-1140, Form 1, Class C ECG 150-4/3 yarn 0.024-in. diam. |
| SiO over aluminized first surface mirror | not measured | not measured | Beryllium mirror blank, electroless nickel coated and polished, vapor deposited Al, vapor deposited SiO. |
| Teflon FEP insulation | not measured | not measured | Clear Teflon FEP wire insulation, 0.009 in. thick. |
| White inorganic paint | 0.17 | 0.92 | Calcined china clay in a water solution of potassium silicate. Clay is predominantly aluminum silicate. |
| White organic paint | 0.22 | 0.91 | Titanium dioxide in an acrylic binder, 3M White Velvet 202-A10. |
| Black organic paint | 0.95 | 0.92 | Carbon black in silicone alkyd binder, 3M Black Velvet 101-C10 |
| Blue inorganic paint | 0.31 | 0.92 | China clay/potassium silicate paint with a small percentage of a blue ceramic frit added for color. |
| Aluminized Teflon FEP, second surface mirror | 0.15 | 0.68 | Teflon FEP second surface mirror, Teflon 2 mils thick, Type A (untreated) |

Surveyor parts. Typical preflight thermal optical properties are given for each surface where known. The preflight properties represent average values for freshly processed surfaces and do not reflect any change introduced into the surfaces during testing or rework prior to launch.

The camera, cable, scoop, and unpainted tube were returned in the astronaut's backpack with no attempt made to protect them from Earth's environment. When these parts reached the Lunar Receiving Laboratory at the Manned Spacecraft Center, they were removed from the backpack, photographed, and then heat sealed in polyethylene bags and stored during the quarantine period. The surfaces were exposed to light, oxygen, and handling.

A different method of return was used for a small section of cable and the white painted support tube from the TV camera. The cable section was to be used in the search for microorganisms which might have survived the 2 1/2 yr in the lunar environment. It was necessary to protect these parts from terrestrial biological contamination during their return. A small canister, vacuum and light tight, the Sample Environmental Sealed Container (SESC), was used to return the cable section since it was designed to provide the necessary protection. The section of the white painted tube also was returned in the SESC since it was desired to return at least one sample of white paint in a condition close to that in which it had existed on the lunar surface.

The cable section and the section of white painted tube were removed from the spacecraft and placed in the SESC. The container was closed on the lunar surface, returned to Earth, and stored for several months. After this storage time and prior to its opening, the SESC was checked for leaks using $SF_6$ as a tracer gas. The test results indicated an apparent gross leak in the vacuum seal. It is assumed that the leak was present upon return to Earth, and thus the parts were exposed to air (and oxygen) during the several months prior to opening. However, the parts were not exposed to light, and therefore this method of return eliminated possible light bleaching of radiation damage of the optical surfaces.

## Visual Examination of Surveyor Parts

Upon termination of the mission quarantine (6 weeks after return from the lunar surface), the Surveyor III parts, except for those in the SESC, were visually examined at the Lunar Receiving Laboratory (LRL).

During the recovery of the parts from Surveyor III, the Apollo 12 astronauts had observed that the preflight color of the spacecraft paint, white, had changed to light tan. This change in color was anticipated due to radiation damage in the white coating. However, when viewed at the LRL the color changes were found to be much greater than expected with painted surfaces of the television camera varying from a yellowish brown to a dark gray.

In addition to the over-all discoloration, sharply defined regions of much darker discoloration, or "shadow patterns," were observed. In all cases these sharply defined darker regions were found on the side of the camera that faced northwest, toward the LM landing site. Each shadow pattern was found to be associated with a protruding or raised surface located on the camera and near the dark region. The sharply defined discoloration patterns on the camera surface were not caused by solar exposure.

In studying the discoloration patterns it was possible to assume a unique viewing position such that all dark regions disappeared from view behind the protruding surface associated with that region; a wire, a raised cover, a support strut, etc. This viewing position was found to be in the direction of the LM landing site. Therefore, it is concluded that the landing of the LM stirred a significant amount of lunar dust which impinged on the Surveyor III surfaces, except where protected by the protruding surface. The impinging dust "sandblasted" the exposed area of the camera, removing darker surface material and forming the well-defined patterns.

The first surface mirror on the camera was found to have a diffuse appearance and a light tan color. This change is attributed to deposited lunar dust, a portion of which was deposited during the Surveyor landing[1]. In regions where the deposit had been wiped off, the aluminum layer overcoated with silicon monoxide appeared bright.

Two cable sections were found still attached to the camera upon its return. One was covered with a fiberglass sleeve and the other cable was overwrapped with second surface aluminized Teflon FEP. The fiberglass sleeve was gray in color (originally white) but its physical properties appeared to be unchanged. The Teflon FEP was yellowish in color where its surface was exposed to solar radiation. In addition there appeared to be a brown deposit on some surfaces of the Teflon that was assumed to be lunar dust. A Nylon cord used to secure the Teflon wrap was severely discolored where it was directly exposed to solar radiation.

The camera was coated with inorganic white paint (Table 2). Physically, the paint had survived the lunar environment except for some cracking, but no loss of adhesion or cohesion was noted. Inorganic paint also was used on the SM/SS scoop over aluminum and fiberglass surfaces. The paint over the aluminum did not exhibit any cracking when returned to Earth. The paint over the fiberglass was severely cracked; however, except for a few small chips at the corners, the paint was adhering to the fiberglass when returned.

The SM/SS scoop was found to be heavily contaminated with lunar dust. The majority of this contamination apparently occurred during return of the scoop. Several grams of lunar soil that had remained in the scoop fell out through openings in the sides. This loose soil was redistributed over the surface of the SM/SS scoop both in the backpack and later in the polyethylene bag. This happenstance reduced the engineering interest in making optical measurements on the surfaces of the SM/SS scoop since this induced contamination was not a direct lunar effect. Measurements made on the SM/SS scoop by Dr. Ronald F. Scott, reported at the Lunar Science Conference[2] in Houston, gave an excellent indication of the affinity of the dust for the various surfaces.

In general, the blue inorganic painted surfaces of the SM/SS scoop that were directly exposed to solar radiation appear to have faded. No change in color of the black organic paint was noted; however, lunar dust may have masked radiation-caused optical changes. Anodized aluminum surfaces, described in Table 2, did not appear to have changed color as a result of lunar exposure. No optical measurements were made on these surfaces because of size and configuration. Transparent Teflon FEP wire insulation had discolored and appeared yellowish after an estimated 3000 hr of solar radiation.

The polished aluminum support tube was found to be severely discolored on one side, with the degree of discoloration varying from heavy at one end to light at the opposite end. The discoloration, a dark tan, is due to a coating of material, which was later determined (by microprobe analysis) to be lunar dust. The coating was very adherent and was not easily rubbed off or redistributed. The other side of the polished aluminum tube was shiny and appeared to be free of any lunar effects. Reflectance measurements made on this tube indicated that the shiny side also had a light layer of lunar dust.

All other polished aluminum surfaces were found to be hazy because of light coatings of lunar dust. In regions where the dust had been removed as a result of handling, the aluminum remained shiny.

## Experimental Results and Discussion

The discoloration observed on the various Surveyor III thermal control surfaces is due to a combination of several environmental factors. Solar radiation damage (ultraviolet and low-energy protons) and lunar dust are primarily responsible, but their relative contributions vary. The origin of the dust as related to both the Surveyor landing and the LM landing was described in detail at the Apollo 12 Lunar Conference.[3] The experimental results presented in this paper relate to reflectance measurements made on the returned parts and describe a technique used to separate the optical effects of the dust and the radiation damage in the white inorganic paint.

### White Inorganic Paint

The reflectance spectrum of the white inorganic paint (0.3 µm to 2.6 µm) was measured at several positions on the TV camera two months after sample release from quarantine. As a result of these and later measurements, it was concluded that the optical effects of the dust could be separated from the measured reflectance spectrum. The reflectance measurements made two months after release were done with an integrating sphere with the sample (TV camera) located at the wall of the sphere. All other measurements were made by cutting 1 cm by 2 cm samples from the camera and inserting these into a Gier-Dunkle integrating sphere. A control surface was measured using both techniques so that any errors associated with the sample at the wall method could be eliminated.

Samples of the returned white inorganic paint were found to have decreased in reflectance at wavelengths greater than 1.0 µm as well as at shorter wavelengths. Laboratory tests were conducted and show that neither ultraviolet radiation nor low-energy protons (1 kev) cause optical damage in this paint at these longer wavelengths. Thus, the observed reduction in the reflectance of the Surveyor surfaces is attributed to the presence of lunar dust. The dust effect can be separated from the measured Surveyor III spectra with fair accuracy by comparing the reflectance spectrum, at 1.5 µm, of lunar dust on white paint to the returned Surveyor samples. An expression developed for this purpose and shown by the Eq. (1) recognizes that the dust acts as a filter, and light incident on the paint surface must pass through the filter two times to be reflected.

$$\rho_{m_\lambda} = R_\lambda (1 - F_\lambda)^2 \tag{1}$$

where $\rho_{m_\lambda}$ = measured reflectance at wavelength $\lambda$; $R_\lambda$ = reflectance of the paint surface due to radiation changes only ($R_\lambda = \rho_{o_\lambda}$ if no radiation damage has occurred where $\rho_{o_\lambda}$ is the initial reflectance); and $F_\lambda$ = factor relating area of dust coverage, reflectance of dust, and scattering of dust at wavelength $\lambda$. The effect of applying Eq. (1) is shown in Fig. 1. A reflectance measurement was made on one part of the camera known to have seen significant solar radiation, estimated to be about 2000 equivalent solar hours[4] (Fig. 2, position 2). This measurement is shown as curve B of Fig. 1. Curve A in Fig. 1 is a typical preflight reflectance spectrum of the inorganic white paint. Curve C in Fig. 1 shows the reflectance of the painted surface with the effect of the lunar dust removed. This was done, using Eq. (1), to calculate $R_\lambda$ after first evaluating $F_\lambda$ for lunar dust only on white paint. The

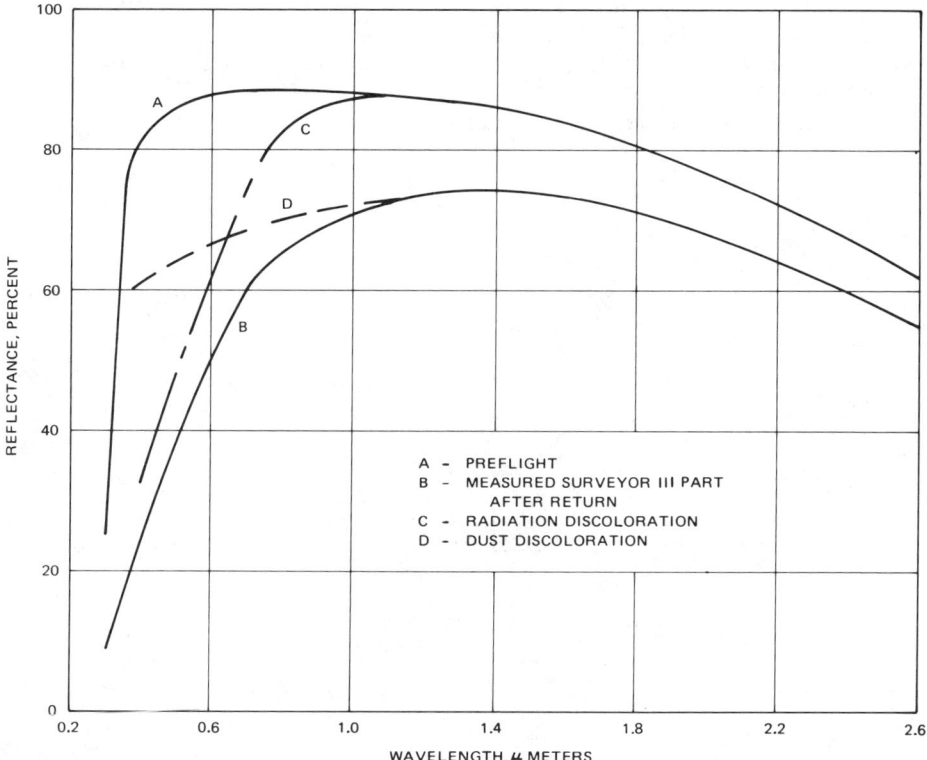

Fig. 1 Separation of spectral reflectance effects due to dust and radiation.

resulting data (Curve C) are characteristic of the radiation damage observed in laboratory tests of the inorganic white paint both in curve shape and magnitude[5]. Curve D in Fig. 1 is a dust only curve with the effect of the radiation damage removed from the measured reflectance spectrum.

This type of analysis was applied to other surface measurements and, as expected, the radiation damage patterns were detected. That is, the portions of the camera facing lunar east, which received the most solar illumination, show the major radiation induced discoloration; the surfaces facing west, which were to a significant degree shielded from the sun by various portions of the spacecraft structure, showed much less radiation damage. Similar results were found by Anderson[6] and co-workers on Surveyor III parts.

As mentioned earlier, the spectral reflectance at several different positions on the camera was measured about two months after release from quarantine. Fig. 2 shows the position at which these measurements were made, and Table 3 lists the calculated solar absorptance. The absolute values of the foregoing results have to be qualified, since the radiation discoloration of the inorganic white paint was bleached by exposure of the camera to light. Oxygen bleaching

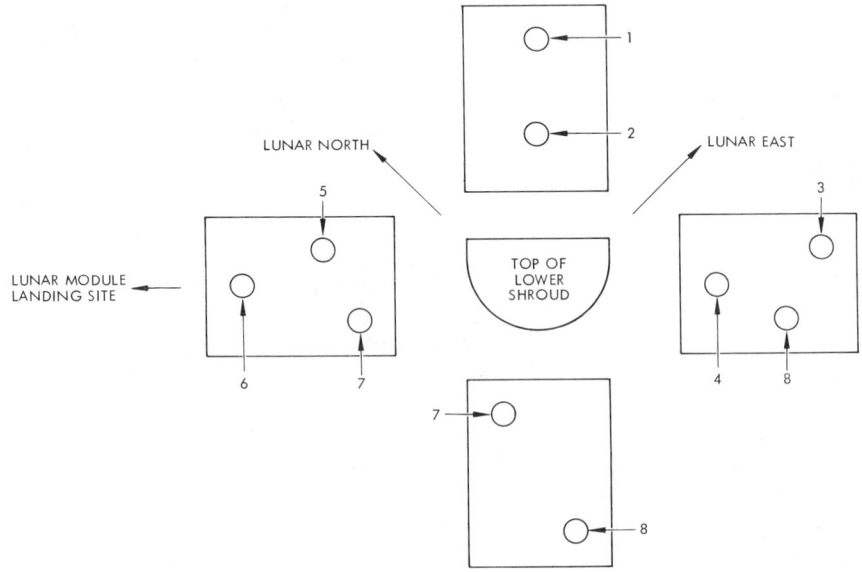

Fig. 2 Reflectance measurement positions and lunar orientation of the TV camera lower shroud.

Table 3 Solar absorptance of the inorganic white paint measured at several locations on the TV camera (Fig. 2)

| Measurement position lower shroud, TV camera | Position description | Solar absorptance $a_s$ |
|---|---|---|
| 1 | Covered by close fitting bracket while on lunar surface | 0.14 |
| 2 | Surface faced northeast[a] | 0.46 |
| 3 | Partially protected by Teflon wrapped cable, faced southeast away from LM landing site | 0.23 |
| 4 | Same as 3 except no protection | 0.32 |
| 5 | Faced northwest, toward LM landing site | 0.51 |
| 6 | Same as 5 except in partially cleaned area | 0.36 |
| 7 | Faced west, partially viewed LM landing site | 0.61 |
| 8 | Surface faced south, away from LM landing site | 0.51 |

[a]All directions are at Surveyor III location on moon.

of radiation damage does not occur in this inorganic white paint[7]. Several months after sample release from quarantine, this effect of light bleaching was measured, using the camera cable bracket. This bracket, which was discolored identical to the camera, had been removed at the time of sample release and stored in the dark. Five months later this bracket was removed from dark storage. A comparison between the bracket and the camera revealed that the camera had changed color to a dirty white while the bracket maintained its brownish cast. Reflectance measurements established that, at 1.5 $\mu$m, no reflectance change had occurred in the camera coating, indicating that very little lunar dust had been lost from the camera surface. The change in color of the camera was due to exposure of the radiation damaged surface to fluorescent lights. These reflectance measurements, summarized in Table 4, show that continued storage of the bracket in the dark has no effect on its reflectance. However, 17 days of exposure of a sample from this bracket to fluorescent lights caused a significant increase in the visible reflectance spectrum. These results are not surprising since this mechanism of bleaching of induced optical damage is well known[8].

The authors would like to emphasize that the observed effect was real and must be considered when planning the return to

Table 4  Paint reflectance changes after return to earth

Lower shroud exposed continuously to flourescent lights (% reflectance)

| Wavelength, $\mu$m | April 1970 | July 1970 | Oct 1970 |
|---|---|---|---|
| 0.4 | 27 | 36 | 37 |
| 0.6 | 46 | 48 | 51 |
| 1.0 | 60 | 63 | 64 |
| 1.5 | 68 | 68 | 70 |

Cable bracket (% reflectance)

| Wavelength, $\mu$m | May 1970 (dark storage) | Aug. 1970 (dark storage) | Nov. 1970 (dark storage) | Nov. 1970 17-day fluorescent light exposure |
|---|---|---|---|---|
| 0.4 | 25 | 24 | 25 | 39 |
| 0.6 | 50 | 49 | 49 | 62 |
| 1.0 | 70 | 70 | 70 | 74 |
| 1.5 | 74 | 74 | 74 | 75 |

Earth of other hardware that has been exposed to space.
Shielding from light and heat (thermal exposure to 400°F produced the same phenomena as the light bleaching), as well as air, is required to assure that no optical changes occur.

Additional data were obtained on bleaching effects by making reflectance measurements on the white painted support tube which had been returned in the SESC and protected from light but not oxygen. The SESC was opened within a sterile glove box several months after sample release from quarantine. The glove box was purged and filled with argon prior to opening. The opening and transfer of the tube was accomplished under low-level, red-light illumination. No other light was incident on the tube until a controlled light exposure test was conducted (except from the reflectance spectrophotometer). The argon atmosphere transfer was planned prior to parts return so as not to expose the white painted tube to oxygen until a later controlled test could be conducted. Although a leak was found in the SESC, as mentioned earlier, it was decided to continue with the tests as if no leak occurred.

The white painted support tube was placed in a quartz tube attached to a vacuum valve. The quartz tube was covered with a light tight shield, and the entire test chamber was removed from the glove box. The chamber was then attached to an ion pump and, using sorption rough pumping, the argon pressure was reduced to $10^{-4}$ torr. During this operation the quartz chamber was maintained in a light tight condition. The optical reflectance of the white painted tube was measured after darkening the room, removing the light tight cover from the quartz vacuum tube, and inserting into a Gier-Dunkle integrating sphere[9]. The reflectance was measured over the spectral range of 0.3 $\mu$m to 2.6 $\mu$m. Table 5 describes the complete sequence of tests.

The spectral reflectance of the inorganic white painted tube remained unchanged through the first four measurements. Following the exposure to light, the reflectance increased at wavelengths shorter than 1 $\mu$m. This again demonstrates the bleaching of radiation damage by light. The reflectance of the white painted tube before and after these tests is shown in Fig. 3.

Black Paint
---

The organic black painted surface chosen for measurement was from inside the mirror hood. The black painted surface on the SM/SS scoop were not measured because of limited area

Table 5  Measurement conditions for the white painted tube returned in the SESC

| Measurement sequence | Chamber pressure | Sample light exposure |
|---|---|---|
| 1 | $4 \times 10^{-4}$ torr | Dark |
| 2 | $3 \times 10^{-1}$ torr | Dark |
| 3 | 1 atmosphere (air) | Dark |
| 4 | 3 days in air | Dark |
| 5 | Air | 48-hr exposure to 214 mw/cm$^2$ of white light (Xenon arc with UV wavelengths removed) |

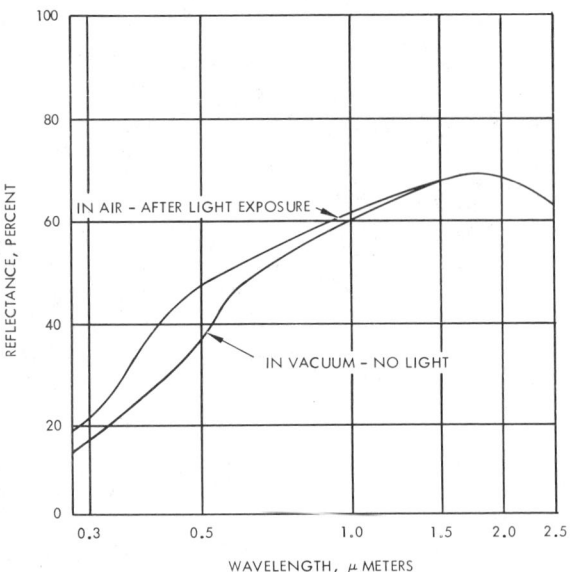

Fig. 3  Reflectance of white painted tube returned in SESC.

available and because of the heavy lunar dust comtamination, previously described.

The painted area inside the hood saw direct solar radiation only during sunrise and early morning. The total irradiance is estimated to have been less than 1000 hr. The painted surface had a brownish cast which indicated the presence of lunar dust. Samples (1 cm x 2 cm) were cut from the hood, and the reflectance was measured using a Gier-Dunkle integrating sphere. The reflectance of the samples was found to average about 4-5% compared to a preflight value of about 3% in the spectral region of 0.3 $\mu$m to 2.6 $\mu$m. The slight increase in reflectance is attributed to the layer of lunar dust and solar bleaching.

Polished Aluminum

The reflectance of polished aluminum was measured on samples taken from two different locations, the bottom of the TV camera and the polished aluminum support strut. The measurement made on the bottom of the camera is not as significant because of the handling the camera received. This surface was coated with gray dust when initially viewed at the LRL. The amount of adhering dust was reduced by handling when the lower shroud was removed for microbiological sampling. The spectral reflectance of the lower shroud surface was not measured until two months after release of parts from the LRL, and several months later the reflectance was again measured. The result of this second measurement indicated some additional loss of lunar material. A comparison of these two measurements is shown in Fig. 4.

The polished aluminum support strut provides a surface of greater interest. As described earlier, one side of this

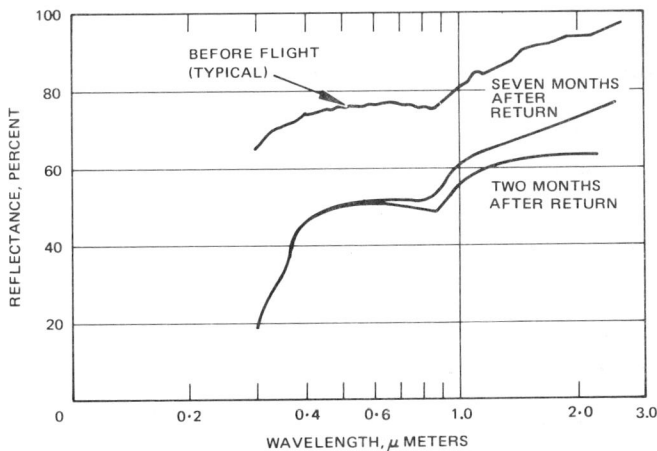

Fig. 4 Reflectance of polished aluminum surface from the TV camera of Surveyor III.

aluminum tube was coated with a brownish deposit identified by microprobe and IR measurements as lunar material. A significant difference was noted between the deposit on the polished tube and that on the aluminum surface of the camera; the deposit on the tube was more strongly adherent than that on the camera. Because of the location of the deposit on the tube, and since it had been identified as lunar soil, it was concluded that the contamination was deposited during the Surveyor III landing. The lunar material was disturbed by the exhaust from one of the vernier engines of Surveyor.

The reflectance of the opposite sides of the polished aluminum support tube (areas of heaviest and lightest contamination) was measured over the spectral region between 0.3 $\mu$m and 2.6 $\mu$m using a Gier-Dunkle integrating sphere. The reflectance data are shown in Fig. 5, and calculated solar absorptance values are given in Table 6. The absolute reflectance of the tube was not measured at wavelengths greater than 2.6 $\mu$m. However, a relative reflectance was determined for the contaminated surface of the tube compared to a clean polished tube from 3 $\mu$m to 14 $\mu$m. This measurement revealed the presence of a strong absorption band in the contaminant on the Surveyor III tube between 8 and 14 $\mu$m, centered at about 10.5 $\mu$m. No other thermally significant absorption bands were noted. The emittance of the tube could not be calculated using this relative reflectance data. However, the absorptance in the infrared region indicates a greater than normal emittance for this tube.

Aluminized Teflon

Reflectance values were measured on samples of aluminized Teflon FEP taken from the outer thermal control wrap of the cables. One set of measurements was made on samples removed

Table 6. Solar absorptance of the Surveyor III polished aluminum tube

|  | Solar absorptance, $a_s$ |
|---|---|
| Polished aluminum (typical preflight) | 0.18 |
| Polished tube from Surveyor, heaviest contamination | 0.76 |
| Polished tube from Surveyor, side opposite from heavy contamination | 0.34 |

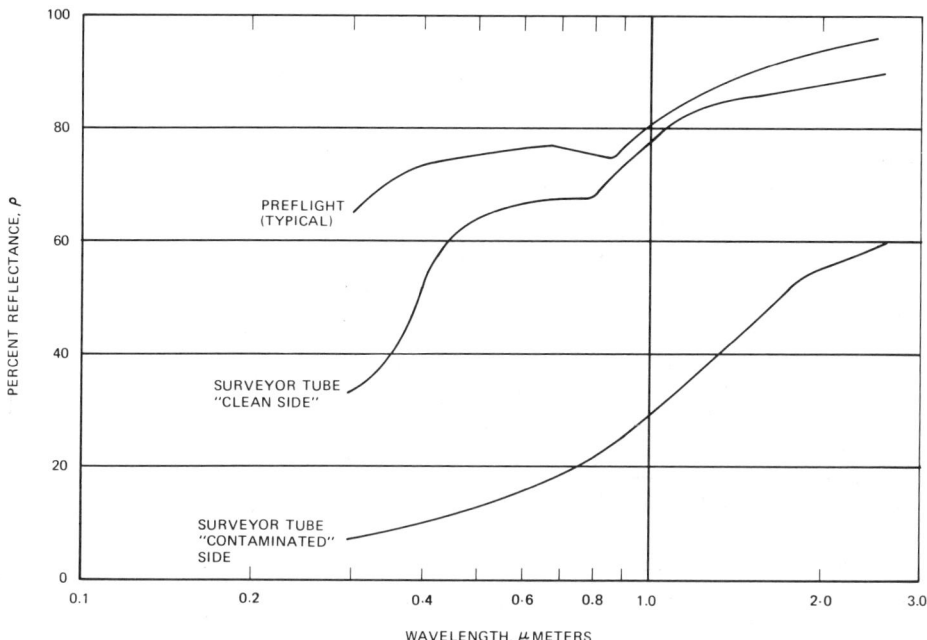

Fig. 5   Reflectance of polished aluminum support tube returned from Surveyor III.

from the cable attached to the TV camera. This material was exposed to the general laboratory environment prior to these measurements and to handling. The second set of samples was taken from the cable section returned in the SESC. As described earlier, these surfaces were protected from light exposure and were not handled directly until several months after sample return. The samples in the SESC were exposed to air, however, because of a leak in the vacuum seal.

The aluminized Teflon FEP taken from the cable returned with the camera was found to have a nonuniform coating of aluminum which was quite thin in some areas. There is the possibility that this was caused by the lunar environment exposure. However, it is believed that a thin layer of aluminum probably existed prior to flight. The Teflon FEP in question is produced in rolls, several hundred feet long, and it is not unusual to find sections within these rolls that have a thinner than normal coating of aluminum.

The second set of reflectance measurements was made on Teflon FEP cable wrap removed from the cable contained in the SESC. A test specimen was cut and mounted with the Teflon side out. The cable wrap was 1 in. wide and was overlapped about 1/2 in. on each turn. The Teflon FEP sample was measured in a region directly exposed to the lunar environment and where protected from the environment by an overlay of Teflon. Reflectance measurements were made as reported in the painted tube section. Fig. 6 is a plot of the spectral reflectance of the Teflon FEP from the SESC measured before any light or air exposure. Fig. 6 also shows a typical preflight reflectance curve for this material. The reduction in reflectance observed in the protected sample as compared to the preflight sample is attributed either to lunar material deposits or to a nonuniform aluminum coating. This sample is still contained in the test chamber, and a close visual study of its surface has not been made. The measured decrease in reflectance of the exposed portion of the Teflon is attributed to two causes, heavy lunar material deposition and radiation damage. The change in the slope of the reflectance curve between 0.3 $\mu$m and 0.7 $\mu$m suggests that discoloration due to radiation damage is significant.

Following the initial vacuum measurements, the Teflon FEP was tested as described in Table 5. After each new condition, a reflectance measurement was made with the sample still in the quartz tube. No significant changes in the reflectance spectrum were noted.

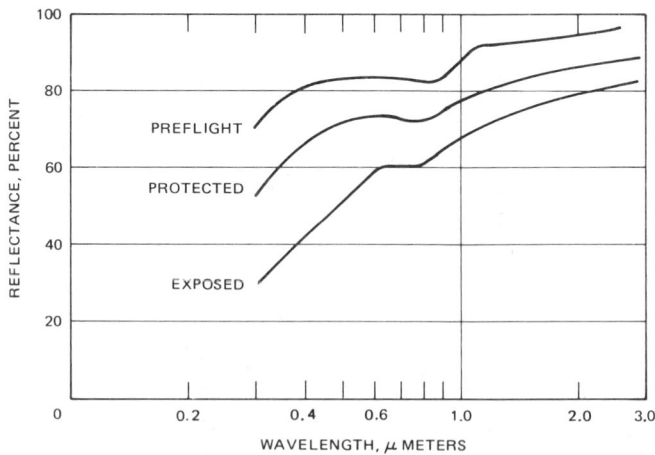

Fig. 6  Reflectance of aluminized teflon FEP film (second surface, 2 mils thick) from Surveyor III returned in SESC.

## Conclusions

This paper has discussed the visually observed effects of the lunar environment on typical thermal control surfaces used on the Surveyor III spacecraft and results of the measurement of the spectral reflectance of several of these surfaces. These results must be carefully reviewed in light of the complex environment experience by the Surveyor III parts.

Not described in this paper was a series of experiments on the analysis of surface contaminants on samples from several locations on the returned hardware. The results of these tests[1] indicate a significant factor from the thermal control aspect: organic material outgassed from the Surveyor III during its mission, and organic material deposited prior to flight (from vacuum pumps and airborne sources), did not significantly effect the radiation stability of the thermal control surfaces. It is emphasized that the foregoing statement applies only to the engineering property of solar absorptance and not to critical optical surfaces such as mirrors or lenses. Traces of organic material were found in these tests, and additional studies by others are continuing to define the effect on optical surfaces.

An over-all assessment of the experimental results of the surface discoloration studies has led to a proposed damage model. The proposed model states that the discoloration of the surfaces of the returned Surveyor III hardware is attributable to a combination of two dominant effects: radiation damage and lunar dust, the latter from both the original Surveyor landing and the landing of the Apollo lunar module. There is some still inconclusive evidence of organic contamination, but its contribution to the total discoloration is believed to be minor. The model has yet to be fully tested for all of the surfaces to synthesize the total and complex discoloration pattern.

The significance of the lunar dust found on the various parts is of particular importance. Some areas of the TV camera were heavily coated with dust during the Surveyor III landing. The second source of dust, the LM landing, was also of major importance even though the LM landed more than 500 ft from the Surveyor. These two conclusions must be considered when planning future soft landings on nonterrestrial surfaces.

The return of the Surveyor hardware has demonstrated quite clearly the need to provide protection for parts returned from space if future studies of optical and thermal control

This conclusion dictates that all surfaces must be returned in vacuum tight, light tight containers. The containers must be opened, the parts handled, and the experiments conducted under carefully controlled conditions.

## References

[1] NASA Special Publication, Surveyor Parts Retrieval, In Publication.

[2] Scott, R.F., "Surface Examination of Returned Surveyor III Scoop; Some Preliminary Results," Proceedings of Second Lunar Science Conference, Vol 3, MIT Press, 1971.

[3] Carroll, W. F. and Blair, P.M., "Discoloration and Lunar Dust Contamination of Surveyor III Surfaces," Proceedings of Second Lunar Science Conference, Vol 3, MIT Press, 1971.

[4] Nickle, N. L., "Surveyor III Material Analysis Program," Proceedings of Second Lunar Science Conference, Vol 3, MIT Press, 1971.

[5] Gilligan, J. E. and Zerlaut, G. A., "Study of In-situ Degradation of Thermal Control Surfaces," NAS 8-21074, March 17, 1969.

[6] Anderson, D. L., Cunningham, B. E., Dahms, R.G., and Morgan, R. G., "Thermal Radiation Degradation Analysis of Surveyor III Material" Progress in Aeronautics and Astronautics, Vol 29, Academic Press, 1971.

[7] Blair, P. M., Pezdirtz, G. F., and Jewell, R. A., "Ultraviolet Stability of Some White Thermal Control Coatings Characterized in Vacuum," AIAA Paper 67-345, New Orleans, 1967.

[8] Schulman, James E. and Compton, W. D., Color Centers in Solids, The Macmillan Company, New York, 1962.

[9] Miller, W. D. and Leudke, E. E., "In Situ Solar Absorptance Measurement, an Absolute Method," SAMPE Proceedings, Vol 11, April 20, 1967, pp. 75-84.

# II Thermal Analysis

DIRECTIONAL PROPERTY EFFECTS ON RADIANT
HEAT TRANSFER AND EQUILIBRIUM TEMPERATURE

A. F. Houchens[*]
Oakland University, Rochester, Mich.

and

R. G. Hering[/]
University of Illinois at Urbana-Champaign, Urbana, Ill.

Abstract

An analysis is presented for radiative transfer between adjoint planes uniformly irradiated by a collimated radiative flux. The analysis employs a semigrey spectral model with a diffuse plus specular reflection model and direction dependent emittance and absorptance. Numerical results are presented for local and over-all heat transfer in the absence of an external source and for local equilibrium temperature of radiatively adiabatic surfaces in the presence of a solar flux. Directional effects on local heat transfer are nearly independent of whether the material is specularly or diffusely reflecting and are significant for low and high emittance materials at locations of high radiant interaction. Directional property effects on over-all heat transfer are small. Temperature errors which arise when directional property effects for surface emitted radiation are ignored are small. Directional solar property effects on local temperature are most important for low solar absorptance materials which are specularly reflecting to solar radiation.

Introduction

Radiation properties of engineering materials often deviate substantially from simple property models commonly employed in

---

Presented as Paper 71-76 at the 9th Aerospace Sciences Meeting, New York, N.Y., January 25-27, 1971. This paper presents the results of research supported in part by the Jet Propulsion Laboratory, California Institute of Technology, Contract 951661.
[*]Assistant Professor of Engineering.
[/]Professor, Dept. of Mechanical Engineering; presently Professor and Chairman, Dept. of Mechanical Engineering, The University of Iowa, Iowa City, Iowa.

analysis of radiative transfer among interacting opaque surfaces. The directional dependence of radiation properties is one important characteristic which is commonly ignored and may lead to significant discrepancy between predicted and observed radiative energy exchange rates. Little is known about directional property effects on radiative heat transfer and the accuracy attainable with analysis using simple property models.

Effects of directional dependent emittance and absorptance on local and over-all radiant heat transfer for simple systems of interacting specularly reflecting surfaces have been reported.[1-3] Generally, directional property effects on over-all heat transfer were small. Hering[1] reported discrepancies greater than a factor of two in calculated local heat transfer when directional property effects were ignored. Although engineering materials generally exhibit angle dependent emittance and absorptance, they do not generally reflect radiation specularly. The objective of this study is to obtain insight into the importance of directional property effects for surfaces which are not specularly reflecting.

Directional property effects on radiative transfer are evaluated for a representative system of simple geometry. The system of interacting surfaces selected for study is shown in Fig. 1 and consists of identical semi-infinite plates sharing a common edge. The surfaces are uniformly irradiated with a collimated solar flux directed along the bisector of the included angle $\gamma$. The surfaces are opaque and have uniform, temperature independent surface properties. Heat exchange occurs only by radiant transport through the transparent intervening media of unit refractive index, and polarization effects are ignored. This system was selected because it includes phenomena common in temperature control components such as inter-reflections and an external radiation field. Furthermore, extensive heat transfer and temperature results are available for certain simple surface property models[4] as well as more detailed ones.[1] Local and over-all heat transfer rates are determined for surfaces of specified uniform temperature in the absence of an external source of radiant energy, and equilibrium surface temperature distribution is evaluated for radiatively adiabatic surfaces illuminated by a uniform collimated solar flux.

A semigrey spectral model[5] is used to approximate the differences in surface property values for solar and surface emitted radiation. The directional dependence of emittance and absorptance is described by the relations of electromagnetic theory.[6] Although these relations strictly apply only to specularly reflecting materials, it appears[7] they may be representative of those for slightly rough surfaces which

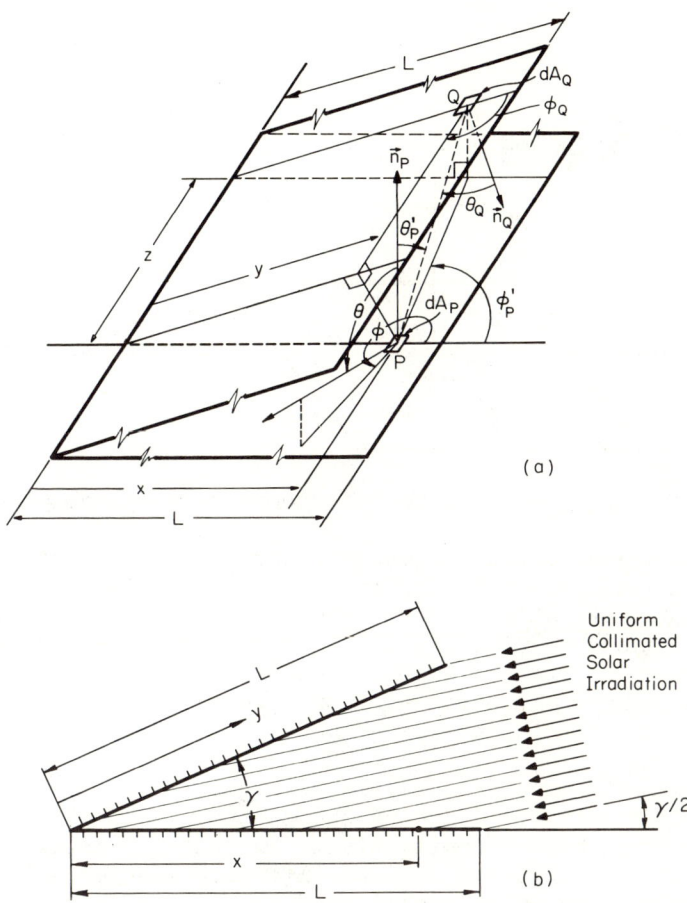

Fig. 1  Adjoint plate system.

strongly alter the spatial distribution of reflected radiation. A reflection model with a directional independent apportionment of reflected radiation to specular and diffuse components is employed for both solar and surface emitted radiation.

## Analysis

### Equations of Radiative Transfer

In this section the fundamental principles of radiative transfer are applied to the selected system, and the equations governing heat flux and temperature are developed incorporating sufficient detail to include directional dependent properties. In the analysis symbols x and y denote distances measured normal to the common edge of the plates along the

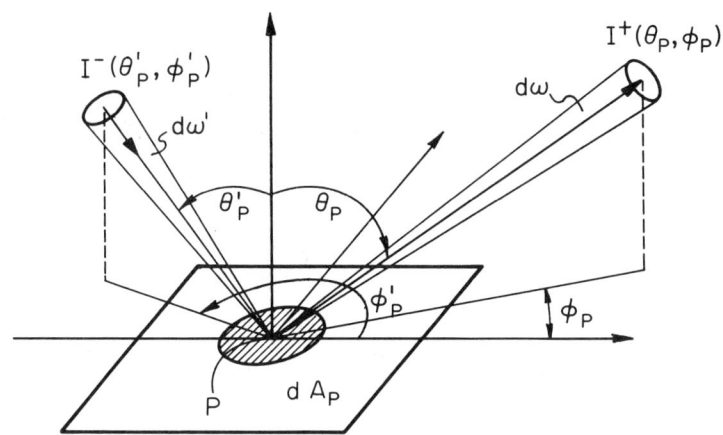

Fig. 2  Local spherical coordinate system.

lower and upper surfaces, respectively, while z denotes distance measured perpendicular to an arbitrary plane normal to the common plate axis. Local spherical coordinates ($\theta'$, $\phi'$) and ($\theta,\phi$) shown in Fig. 2 define the direction of incident (prime superscript) and emergent (no superscript) energy. Later the spherical coordinates are expressed as functions of the position coordinates x, y, and z.

The complexity of the analysis is significantly reduced because of the infinite width of the plates, the geometrical symmetry, and the uniformity of the surface properties and external radiation field. Radiant intensity leaving an element of either surface in a prescribed direction defined by polar and azimuthal angles ($\theta,\phi$) depends only on the position of the element measured normal to the common axis of the plates. For specified polar angle ($\theta$) intensity is azimuthally symmetric about a plane normal to the common plate edge. Furthermore, radiant intensity distributions are identical on the two surfaces at equal distances (x=y) from the apex.

Net heat flux q(x) at a typical element of the lower surface located at point P(x,z=o) is evaluated as the difference between the rate of emission and absorption of radiant energy.

$$q(x) = \varepsilon_H \sigma T^4(x) - \int_{y=o}^{L} \int_{z=-\infty}^{\infty} \alpha_d(\theta_P') I_t^-(x,z=o,\theta_P',\phi_P')$$

$$K(x,y,z)dzdy - q_{a,so}(x) \qquad (1)$$

The geometry kernel $K(x,y,z)$ is given by the relation

$$K(x,y,z) = xy \sin^2(\gamma)/D^4 \qquad (2)$$

where

$$D = (x^2 + y^2 - 2xy \cos(\gamma) + z^2)^{1/2} \qquad (3)$$

and $\theta_P'$ and $\phi_P'$ are expressed in terms of the position coordinates as

$$\cos(\theta_P') = y \sin(\gamma)/D \qquad (4)$$

$$\tan(\phi_P') = z/[y \cos(\gamma) - x] \qquad (5)$$

Local absorbed solar radiation is denoted $q_{a,so}(x)$. Directional absorptance for energy incident at polar angle $\theta_P'$ and hemispherical emittance are denoted by the symbols $\alpha_d(\theta_P')$ and $\varepsilon_H$, respectively. Directional absorptance is taken azimuthally independent. The symbol $I_t^-(x,z=0, \theta_P', \phi_P')$ denotes the intensity function for surface (thermal) radiation incident at point $P(x,z=0)$ from the $(\theta_P',\phi_P')$ direction, while later a superscript plus (+) will signify thermal radiation intensity leaving a surface. Local absolute surface temperature is denoted $T(x)$ and $\sigma$ is the Stefan-Boltzmann constant. The terms on the right of Eq. (1) represent in order of appearance: emissive power, surface radiation incident from the adjacent plate which is absorbed, and absorbed solar irradiation. Since the intensity at P from the $(\theta_P',\phi_P')$ direction $I_t^-(x,z=0,\theta_P',\phi_P')$ equals the intensity emergent from a point $Q(y,z)$ on the adjacent plate in the $(\theta_Q,\phi_Q)$ direction $I_t^+(y,z,\theta_Q,\phi_Q)$, Eq. (1) can be written

$$q(x) = \varepsilon_H \sigma T^4(x) - \int_{y=0}^{L} \int_{z=-\infty}^{\infty} \alpha_d(\theta_P') I_t^+(y,z,\theta_Q,\phi_Q)$$

$$K(x,y,z) \, dz \, dy - q_{a,so}(x) \qquad (6)$$

where $\theta_Q$ and $\phi_Q$ are expressed in terms of the position coordinates as

$$\cos(\theta_Q) = x \sin(\gamma)/D \qquad (7)$$

$$\tan(\phi_Q) = z/[x \cos(\gamma) - y] \qquad (8)$$

To evaluate local heat flux (temperature) when local temperature (flux) and the radiative properties for surface radiation are specified it is first necessary to determine the local thermal intensity function and local absorbed solar radiation.

The intensity function for surface radiation leaving P in the arbitrary $(\theta_P, \phi_P)$ direction consists of emitted radiation and radiation incident from the adjacent surface which is reflected into the $(\theta_P, \phi_P)$ direction. It satisfies the following equation:

$$I_t^+(x,\theta_P,\phi_P) = \varepsilon_d(\theta_P) \frac{\sigma T^4(x)}{\pi} + \int_{y=0}^{L} \int_{z=-\infty}^{\infty} \rho_{bd}(\theta_P', \phi_P'; \theta_P, \phi_P)$$
$$I_t^+(y,z,\theta_Q,\phi_Q) \, K(x,y,z) \, dz \, dy \qquad (9)$$

The symbol $\varepsilon_d(\theta_P)$ denotes azimuthally independent directional emittance for polar emission angle $\theta_P$. The remaining surface property in Eq. (9) is bidirectional reflectance $\rho_{bd}(\theta_P', \phi_P'; \theta_P, \phi_P)$ defined as the ratio of the intensity of reflected energy $dI_{t,r}^+(x; \theta_P', \phi_P'; \theta_P, \phi_P)$ in the $(\theta_P, \phi_P)$ direction due to energy incident from the $(\theta_P', \phi_P')$ direction within elemental solid angle $d\omega'$ to the radiant power per unit surface area incident from the $(\theta_P', \phi_P')$ direction.

$$\rho_{bd}(\theta_P', \phi_P'; \theta_P, \phi_P) = \frac{dI_{t,r}^+(x; \theta_P', \phi_P'; \theta_P, \phi_P)}{I_t^-(x; \theta_P', \phi_P') \cos(\theta_P') \, d\omega'} \qquad (10)$$

Because of the system symmetry the intensity function on the left and that within the integral operator of Eq. (9) denote physically identical quantities. Thus, Eq. (9) is a linear integral equation for the local intensity function for surface radiation.

Local absorbed solar radiation at P consists of the sum of directly incident solar energy which is absorbed and inter-reflected solar radiation emergent from the adjacent surface which is absorbed.

$$q_{a,so}(x) = \alpha_d^*(\theta_{so}') \, S \, \sin(\gamma/2) + \int_{y=0}^{L} \int_{z=-\infty}^{\infty} \alpha_d^*(\theta_P')$$
$$I_{so}^+(y,z,\theta_Q,\phi_Q) K(x,y,z) dz dy \qquad (11)$$

In Eq. (11) the polar angle of incidence for directly incident solar radiation is

$$\theta_{so}' = (\pi-\gamma)/2 \qquad (12)$$

Subscript (so) denotes the radiant intensity function and geometrical factors associated with solar radiation, while the asterisk superscript distinguishes surface properties for short wavelength solar radiation. The external solar energy flux value is denoted S. Evaluation of absorbed solar radiation from Eq. (11) requires that the local intensity function for solar radiation be available.

Within the framework of the semigrey approximation, solar intensity leaving P in the direction $(\theta_P, \phi_P)$ consists entirely of reflected radiation.

$$I_{so}^+(x,\theta_P,\phi_P) = \rho_{bd}^*(\theta_{so}', 0; \theta_P, \phi_P) \, S \sin \gamma/2 + \int_{y=0}^{L} \int_{z=-\infty}^{\infty} \rho_{bd}^* (\theta_P', \phi_P'; \theta_P, \phi_P) \, I_{so}^+(y,z,\theta_Q,\phi_Q) K(x,y,z) \, dz \, dy \quad (13)$$

The first term on the right of Eq. (13) represents reflected intensity in the $(\theta_P, \phi_P)$ direction due to directly incident solar radiation, and the integral term is the reflected intensity due to solar energy emerging from the adjacent surface. Equation (13) is a linear integral equation for the local intensity function for solar radiation.

Consider a system of surfaces with known radiation properties for the solar spectral interval irradiated by a specified collimated source. The local solar intensity function can be evaluated from Eq. (13) and, in turn, local absorbed solar radiation follows from Eq. (11). For specified surface properties for the thermal spectral interval and surfaces of prescribed temperature, the local thermal intensity function may be calculated from Eq. (9) and local heat transfer follows from Eq. (6). When radiative flux is prescribed it is convenient to eliminate the unknown temperature between Eqs. (6) and (9) and to evaluate the local thermal intensity function from the resulting integral equation. For radiatively adiabatic surfaces [$q(x) \equiv 0$] this procedure gives

$$I_t^+(x,\theta_P,\phi_P) = \frac{\varepsilon_d(\theta_P)}{\pi \varepsilon_H} q_{a,so}(x) + \int_{y=0}^{L} \int_{z=-\infty}^{\infty} \left[ \rho_{bd}(\theta_P', \phi_P'; \theta_P, \phi_P) + \frac{\varepsilon_d(\theta_P) \alpha_d(\theta_P')}{\pi \varepsilon_H} \right] I_t^+(y,z,\theta_Q,\phi_Q) K(x,y,z) \, dz \, dy \quad (14)$$

Once the thermal intensity function is available the temperature distribution follows from Eq. (6) with radiative flux equal to zero.

$$T^4(x) = \frac{q_{a,so}(x)}{\varepsilon_H \sigma} + \frac{1}{\varepsilon_H \sigma} \int_{y=0}^{L} \int_{z=-\infty}^{\infty} \alpha_d(\theta_P') I_t^+(y,z,\theta_Q,\phi_Q) \, K(x,y,z) \, dz \, dy \quad (15)$$

Accounting for symmetry and introducing the variables

$$H_{so} = \pi I_{so}^+/S, \quad \Theta^4 = \sigma T^4/S, \quad H_t = \pi I_t^+/S, \quad \Psi = q_{a,so}/S$$

$$\eta = x/L, \quad \xi = y/L \quad \text{and} \quad \zeta = z/L$$

the governing equations may be expressed in dimensionless form. For radiatively adiabatic surfaces the dimensionless temperature distribution is governed by

$$\Theta^4(\eta) = \frac{\Psi(\eta)}{\varepsilon_H} + \frac{2}{\pi \varepsilon_H} \int_{\xi=0}^{1} \int_{\zeta=0}^{\infty} \alpha_d(\theta_P') \, H_t(\xi,\zeta,\theta_Q,\phi_Q) \, K(\eta,\xi,\zeta) \, d\zeta \, d\xi \quad (16)$$

with the dimensionless thermal intensity function evaluated from the integral equation

$$H_t(\eta,\theta_P,\phi_P) = \frac{\varepsilon_d(\theta_P) \Psi(\eta)}{\varepsilon_H} + \int_{\xi=0}^{1} \int_{\zeta=0}^{\infty} \left[ \rho_{bd}(\theta_P',\phi_P';\theta_P,\phi_P) + \right.$$

$$\left. \rho_{bd}(\theta_P', 2\pi-\phi_P'; \theta_P,\phi_P) + \frac{2\varepsilon_d(\theta_P) \alpha_d(\theta_P')}{\pi \varepsilon_H} \right]$$

$$\cdot H_t(\xi,\zeta,\theta_Q,\phi_Q) \, K(\eta,\xi,\zeta) \, d\zeta \, d\xi \quad (17)$$

Dimensionless absorbed solar radiation is given by

$$\Psi(\eta) = \alpha_d^*(\theta_{so}') \sin(\gamma/2) + 2/\pi \int_{\xi=0}^{1} \int_{\zeta=0}^{\infty} \alpha_d^*(\theta_P') \, H_{so}(\xi,\zeta,\theta_Q,\phi_Q)$$

$$K(\eta,\xi,\zeta) \, d\zeta \, d\xi \quad (18)$$

with the dimensionless solar intensity function determined from the following integral equation

$$H_{so}(\eta,\theta_P,\phi_P) = \rho_{bd}^*(\theta_{so}',0;\theta_P,\phi_P)\sin(\gamma/2) + \int_{\xi=0}^{1} \int_{\zeta=0}^{\infty} \left[ \rho_{bd}^*(\theta_P', \right.$$

$$\left. \phi_P';\theta_P,\phi_P) + \rho_{bd}^*(\theta_P',2\pi-\phi_P';\theta_P,\phi_P) \right] H_{so}(\xi,\zeta,\theta_Q,\phi_Q)$$

$$K(\eta,\xi,\zeta) \, d\zeta \, d\xi \quad (19)$$

Note that the integration limits on $\zeta$ are now 0 to $\infty$, and an additional term has been introduced in the integrands of the integral Eqs.(17) and (19) to account for the symmetry of the radiant intensity functions about $\zeta = 0$ for specified values of $\xi$, $\theta_Q$ and $\phi_Q$.

For specified temperature surfaces in the absence of an external source introduce

$$H = I_t^+ / (\sigma T^4 / \pi) \text{ and } T = q / \varepsilon_H \sigma T^4$$

Since local absorbed solar radiation and local solar intensity are identically zero, dimensionless radiative flux distribution is evaluated from

$$T(\eta) = 1 - \frac{2}{\pi \varepsilon_H} \int_{\xi=0}^{1} \int_{\zeta=0}^{\infty} \alpha_d(\theta_P') H(\xi,\zeta,\theta_Q,\phi_Q) K(\eta,\xi,\zeta) \, d\zeta d\xi \quad (20)$$

The governing integral equation for the local thermal intensity function is

$$H(\eta,\theta_P,\phi_P) = \varepsilon_d(\theta_P) + \int_{\xi=0}^{1} \int_{\zeta=0}^{\infty} \left[ \rho_{bd}(\theta_P',\phi_P';\theta_P,\phi_P) + \rho_{bd}(\theta_P',2\pi-\phi_P';\theta_P,\phi_P) \right] H(\xi,\zeta,\theta_Q,\phi_Q) K(\eta,\xi,\zeta) \, d\zeta d\xi \quad (21)$$

Dimensionless over-all heat transfer per unit length from each plate is obtained by multiplying Eq. (20) by $\varepsilon_H$ and integrating over the plate length

$$\frac{Q/L}{\sigma T^4} = \varepsilon_H \int_{\eta=0}^{1} T(\eta) \, d\eta \quad (22)$$

The geometrical relations given by Eqs. (2-5) as well as (7) and (8) continue to apply if x, y, and z are replaced by dimensionless variables $\eta, \xi$, and $\zeta$, respectively.

Radiation Property Models

The radiation property model selected for both surface emitted and solar radiation incorporates directional dependent emittance and absorptance with a specular plus diffuse reflection model. Bidirectional reflectance is expressed as the sum of a specular and a diffuse component.

$$\rho_{bd}(\theta',\phi';\theta,\phi) = 2\rho_{sp}(\theta')\delta[\sin^2(\theta') - \sin^2(\theta)]\delta[\phi' - (\phi \pm \pi)] + [\rho_D(\theta')/\pi] \quad (23)$$

In Eq. (23) $\delta(\eta)$ is the Dirac delta function with argument $\eta$. Specular reflectance $\rho_{sp}(\theta')$ is defined as the ratio of reflected radiant intensity in a specular reflection to the intensity of radiation incident from polar angle $\theta'$. Diffuse reflectance $\rho_D(\theta')$ is the fraction of radiant energy incident within solid angle $d\omega'$ around the $(\theta',\phi')$ direction which is diffusely reflected throughout hemispherical space.

Directional-hemispherical reflectance $\rho_d(\theta')$ is obtained by multiplying bidirectional reflectance with $[\cos(\theta)\sin(\theta)d\theta\,d\phi]$ and integrating over all directions for reflected energy $(0 \leq \theta \leq \pi/2;\ 0 \leq \phi \leq 2\pi)$. The result is

$$\rho_d(\theta') = \rho_{sp}(\theta') + \rho_D(\theta') \tag{24}$$

The apportionment of reflected radiation between specular and diffuse components is taken independent of direction. Hence, the specularity ratio defined as the ratio of specular to directional hemispherical reflectance $\rho_{sp}(\theta')/\rho_d(\theta')$ is direction independent and will hereafter be written without the argument $(\theta')$. Substituting Eq. (24) into Eq. (23) gives

$$\rho_{bd}(\theta',\phi';\theta,\phi) = \rho_d(\theta') \left\{ 2(\rho_{sp}/\rho_d)\delta[\sin^2(\theta')-\sin^2(\theta)] \right.$$
$$\left. \delta[\phi'-(\phi\pm\pi)] + [1-(\rho_{sp}/\rho_d)]/\pi \right\} \tag{25}$$

Bidirectional reflectance is completely determined by specifying directional-hemispherical reflectance and the specularity ratio. For opaque materials directional-hemispherical reflectance and directional absorptance are related by the equation

$$\rho_d(\theta') = 1 - \alpha_d(\theta') \tag{26}$$

and Kirchhoff's law equates directional emittance and absorptance giving

$$\varepsilon_d(\theta=\theta') = \alpha_d(\theta') = 1 - \rho_d(\theta') \tag{27}$$

Directional emittance, directional absorptance, and directional-hemispherical reflectance are evaluated from the Fresnel relations[6] of electromagnetic theory. To facilitate direct comparison of heat transfer results with reported results the directional distributions selected by Hering[1] are used. Directional emittance distributions for the selected optical indices are shown in Fig. 3. Values of optical indices were selected to give hemispherical emittance values and directional emittance distributions characteristic of a metal ($\varepsilon_H$ = 0.1; n = 23.452, k = 1.0), a dielectric ($\varepsilon_H$ = 0.9; n = 1.5565, k = 0.0), and a material with characteristics intermediate to a metal and a dielectric ($\varepsilon_H$ = 0.5; n = 6.1038, k = 0.0).

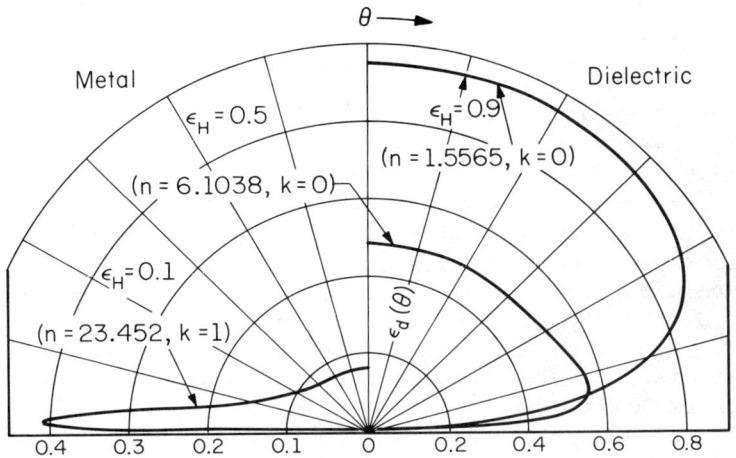

Fig. 3  Directional emittance distribution.

This model and results derived from its use are denoted DSD (Direction dependent Specular plus Diffuse property model.) When directional reflectance is taken independent of $\theta'$, Eqs. (23-27) reduce to well-known expressions for the grey property, diffuse emission, direction independent specular plus diffuse reflection model. Results derived using this constant property model are designated CSD (Constant Specular plus Diffuse property model.) The CSD model includes as limiting cases direction independent specular and diffuse reflection.

## Method of Solution

The complexity of the governing equations appears to preclude analytical solution methods, and numerical techniques were employed. The numerical methods used to solve the integral equations governing the dimensionless solar and surface radiation intensity functions as well as to evaluate local absorbed solar radiation, equilibrium temperature distribution, and heat flux distribution are described in detail elsewhere[9].

The numerical procedure was verified by calculating heat flux and equilibrium temperature distribution with selected property models and comparing the results with those available in the literature. For both values of included angle considered in this study, heat flux was calculated for materials with properties represented by the CSD model for selected values of surface specularity. Only for a low emittance, diffusely reflecting material and 45° included angle at points near the apex were the results distinguishable from published results. The maximum error then was no more than 2%. A more critical test of the numerical method was provided by comparing heat flux results obtained using the DSD model for specularly reflecting surfaces with those reported by Hering[1]. Generally, the results were in excellent agreement. The largest differences occurred for the smaller included angle and high emittance material at surface elements near the apex. The discrepancy however, was less than 3%. Equilibrium temperature distributions for properties represented by the semigrey CSD property model were compared to those reported in Ref. 4. Again, excellent agreement was obtained with the largest discrepancies occurring near the apex. These differences were limited in magnitude to about 1%. No temperature results for direction dependent properties were available for comparison.

## Results and Discussion

### Radiant Heat Transfer

Dimensionless local and over-all heat transfer rates are presented for uniform temperature adjoint plates with direction dependent surface properties represented by the DSD model in the absence of an external radiation source. The influence of angular property dependence on heat transfer for markedly different spatial distributions of reflected energy is assessed.

The general dependence of heat flux on position, included angle, hemispherical emittance, and reflection model for adjoint plates with direction independent properties has been fully discussed elsewhere[4], and therefore is only briefly summarized here. For prescribed uniform temperature plates, radiant heat flux is minimum at the apex and continuously increases to the outer edge. The level of the heat transfer increases as either hemispherical emittance or included angle is increased. For all other conditions fixed, local heat transfer is greater for specularly reflecting surfaces than for diffusely reflecting surfaces. Discussion of these general trends is not repeated in this paper. Instead, attention is directed to evaluation of variable property effects.

## Local Heat Transfer

Dimensionless heat flux distributions are shown in Figs. 4-7 for surfaces with properties represented by the DSD and CSD models. Results are shown for specularity ratio values of 0.0, 0.5, and 1.0. Figs. 4-6 are for included angle $\gamma = 45°$ and hemispherical emittance values equal to 0.1, 0.5, and 0.9, respectively. Fig. 7 illustrates results for $\gamma = 90°$ for each of the three emittance values but only for the limiting values for specularity parameter. Following Hering[1], results obtained using directional property and constant property models are denoted by the acronym DP and CP, respectively. Hering[1] compared results from DP and CP analysis for specularly reflecting surfaces. The objective here is to investigate directional property effects when the participating surfaces are not specular reflectors.

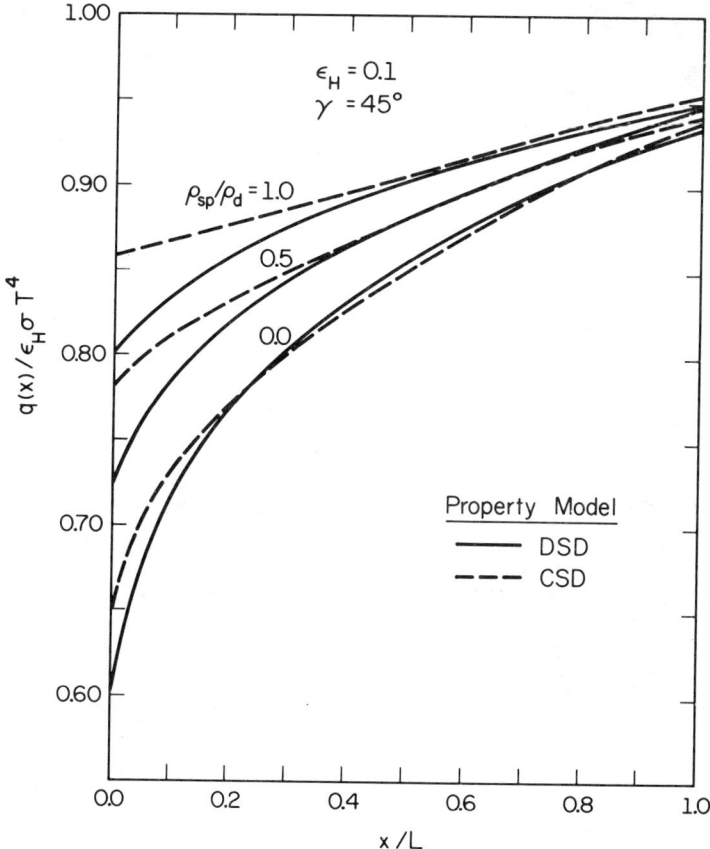

Fig. 4  Angular dependent property effects on radiant heat flux ($\varepsilon_H = 0.1$, $\gamma = 45°$).

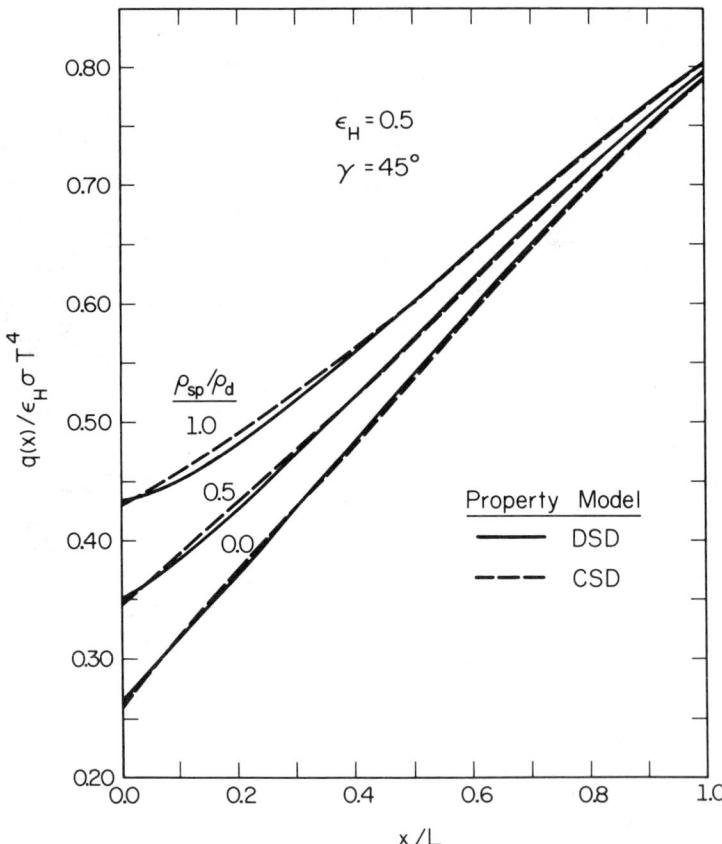

Fig. 5 Angular dependent property effects on radiant heat flux ($\epsilon_H$ = 0.5, $\gamma$ = 45°).

It is important to note the magnitude of the differences between results obtained with DP specular and DP diffuse reflection analysis. The largest differences occur at the apex. For $\gamma$ = 45° and hemispherical emittance values of 0.1, 0.5, and 0.9, DP specular reflection results are higher than those obtained from DP diffuse reflection analysis by approximately 32, 67, and 14%, respectively. For $\gamma$ = 90° the respective differences are reduced by a factor of five. These differences are nearly identical to those previously reported[10] for CP analysis. It is clear that for systems in which multiple reflections are important, knowledge of the spatial distribution of reflected energy is essential for applications requiring accurate heat transfer predictions even when the directional dependence of emittance and absorptance are accounted for.

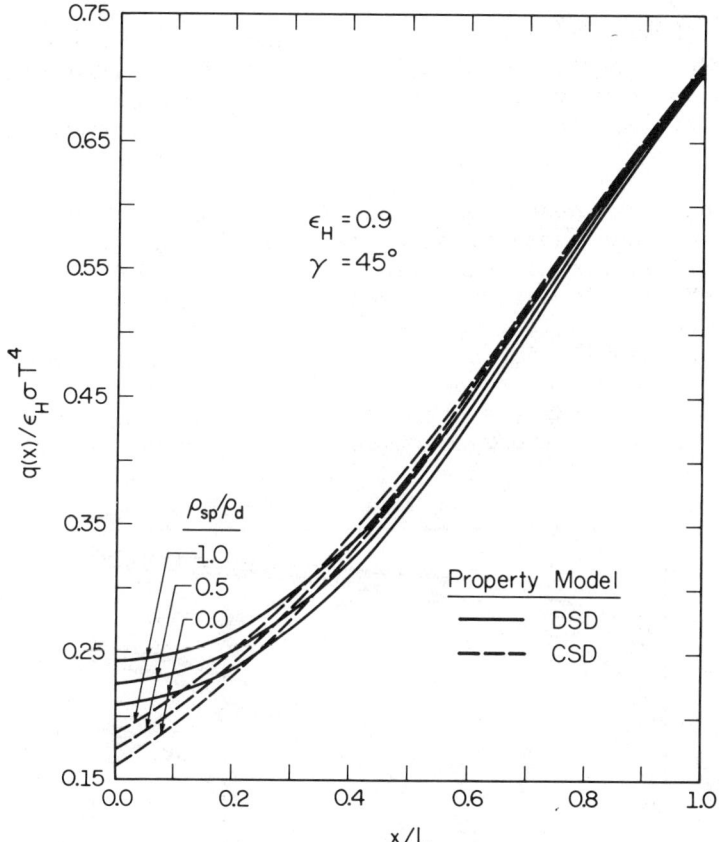

Fig. 6 Angular dependent property effects on radiant heat flux ($\epsilon_H = 0.9$, $\gamma = 45°$).

Several trends are common to results for the two included angles. For the intermediate emittance value ($\epsilon_H = 0.5$), CP results are in excellent agreement with DP results along the entire plate length for each value of the specularity parameter. The largest difference between DP and CP heat flux values is approximately 2%. Consequently, attention is directed to the results for low ($\epsilon_H = 0.1$) and high ($\epsilon_H = 0.9$) emittance surfaces. For each value of the specularity parameter, the largest difference between results obtained using DP and CP analysis occurs for surface elements near the apex. This is attributed to the fact that these elements receive major contributions to their irradiation from energy incident at large angles of illumination. Furthermore, a significant portion of the irradiation of these elements is due to energy which is emitted by the adjacent surface at large angles of emission. These factors are important because the largest differences between direction dependent and direction independent properties occur for polar angles greater than about 60° (Fig. 3).

Near the apex, heat flux values obtained from CP analysis are greater than those from DP analysis for low-emittance surfaces and less for high-emittance surfaces. For $\gamma$ = 90° these relationships exist along the entire plate length. The discrepancy between DP and CP results is nearly independent of surface specularity for both low and high emittance surfaces. For low-emittance surfaces the maximum differences are approximately 8 and 5% for $\gamma$ = 45° and 90°, respectively. For high emittance surfaces the respective maximum discrepancies are 25 and 12%.

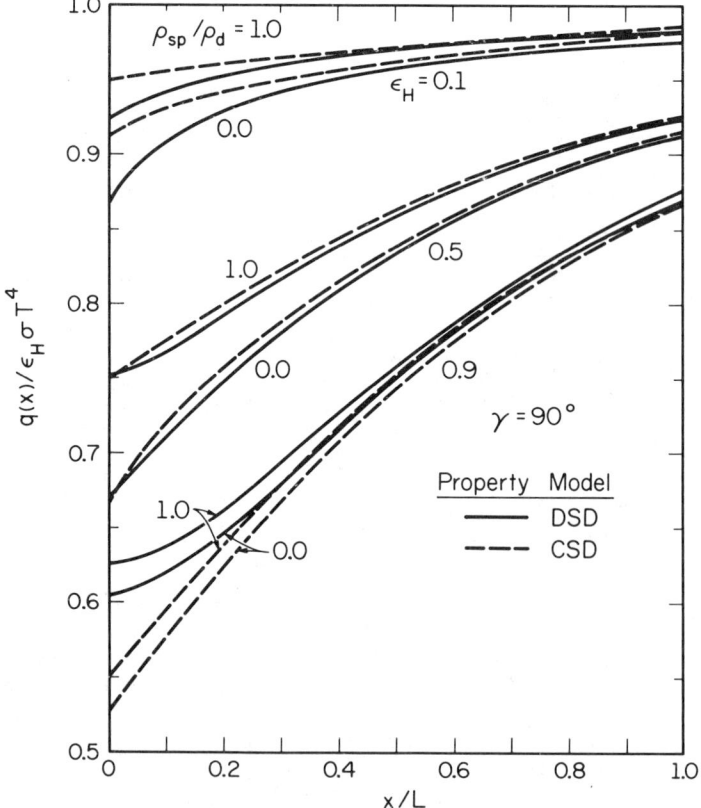

Fig. 7   Angular dependent property effects on radiant heat flux ($\gamma$ = 90°).

The major difference between results for $\gamma = 45°$ and $\gamma = 90°$ is that the discrepancy between DP and CP results is generally larger for the former. This is evidence of the greater importance of interreflection phenomena and of irradiation at large angles of illumination for $\gamma = 45°$.

An interesting comparison may be made between present results and those reported by Schoenhorst and Viskanta[2] for heat flux on identical, equal temperature, semi-infinite, parallel plates. For emittance values less than 0.34 and a plate spacing to-width ratio value of 0.5, these investigators reported discrepancies between DP and CP heat flux results which were much larger for specular than for diffuse reflection (in some cases more than ten times larger). For the adjoint plate system directional property effects on local heat transfer are less sensitive to the specularity of the surfaces.

Over-all Heat Transfer

Dimensionless over-all heat transfer per unit width is presented in Table 1.

Table 1  Over-all heat transfer $(Q/L)/\sigma T^4$

| | | Specularity Parameter $\rho_{sp}/\rho_d$ | | | | | |
|---|---|---|---|---|---|---|---|
| | | 1.0 | | 0.5 | | 0.0 | |
| $\gamma$ | $\varepsilon_H$ | DSD | CSD | DSD | CSD | DSD | CSD |
| 45° | 0.1 | 0.0895 | 0.0906 | 0.0869 | 0.0877 | 0.0835 | 0.0838 |
| | 0.5 | 0.304 | 0.304 | 0.285 | 0.286 | 0.268 | 0.267 |
| | 0.9 | 0.378 | 0.376 | 0.369 | 0.370 | 0.359 | 0.364 |
| 90° | 0.1 | 0.0965 | 0.0971 | --- | --- | 0.0947 | 0.0958 |
| | 0.5 | 0.425 | 0.427 | --- | --- | 0.410 | 0.412 |
| | 0.9 | 0.677 | 0.663 | --- | --- | 0.668 | 0.655 |

The influence of the reflectance model on over-all heat transfer in the adjoint plate system is not very great. This is attributed to the dominance of emission over absorbed incident radiation for a major portion of the plate surface. For the smaller included angle, over-all heat transfer evaluated using DP analysis and a specular reflection model is greater than that calculated with a diffuse reflection model by 13% for the intermediate emittance surface. The same comparison for low and high emittance surfaces reveals that the over-all heat transfer differences are reduced to 7 and 5%, respectively. For the larger included angle the aforementioned differences are reduced by approximately a factor of three. The influence of the reflection model on over-all heat transfer is not significantly different from that reported[10] for constant property analysis.

Inspection of the over-all heat transfer results presented in Table 1 shows that CP analysis yields results which are in excellent agreement with those obtained using DP analysis. The largest difference is less than 3% and is observed for specularly reflecting surfaces of high emittance with the larger value for included angle. Thus, CP analysis accurately predicts over-all heat transfer in the adjoint plate system.

Equilibrium Temperature

Equilibrium temperature distributions are presented for perpendicular adjoint plates illuminated by a uniform collimated solar field. The surfaces are taken specularly reflecting to surface emitted thermal radiation. Directional dependent semigrey surface properties for solar and surface radiation

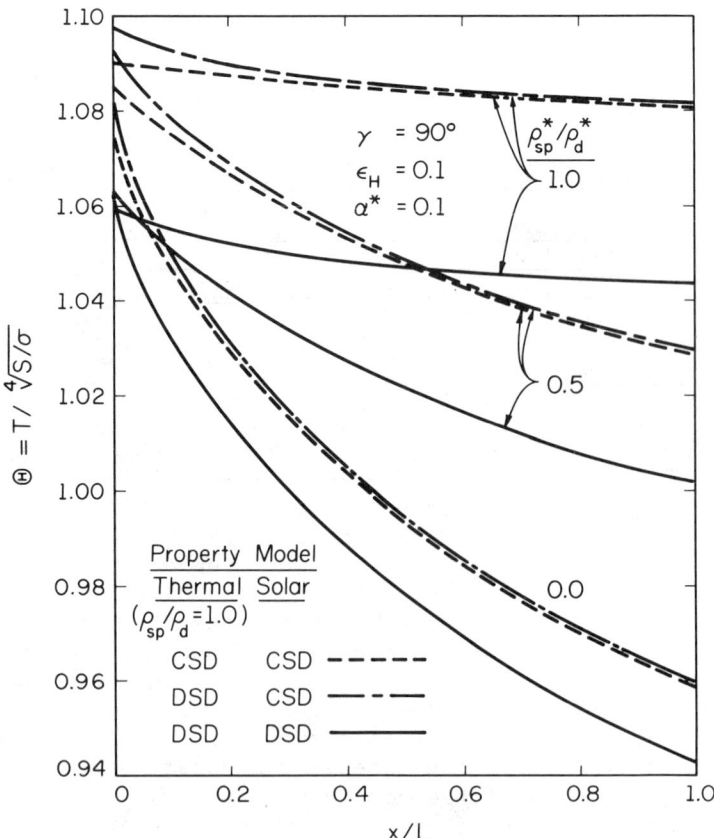

Fig. 8  Angular dependent property effects on local equilibrium temperature ($\varepsilon_H = 0.1$, $\alpha^* = 0.1$).

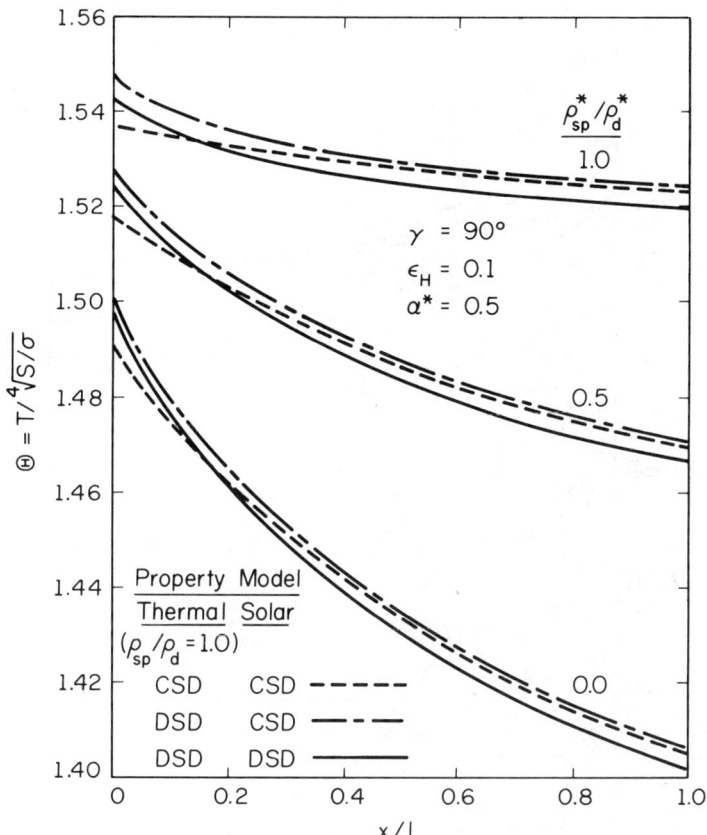

Fig. 9 Angular dependent property effects on local equilibrium temperature ($\varepsilon_H = 0.1$, $\alpha^* = 0.5$).

are represented by the DSD property model. The influence of angular property dependence on equilibrium temperature is assessed for markedly different spatial distributions of reflected solar radiation.

The general dependence of local equilibrium temperature in the adjoint plate system on position, included angle, emittance, solar absorptance, and surface specularity for surface and solar radiation was thoroughly investigated by Hering[4] for direction independent surface property models. These general trends are only briefly summarized here. For uniformly irradiated perpendicular plates with semigrey, direction independent surface properties local equilibrium temperature is maximum at the common edge and decreases continuously to the outer edge. With all other parameters fixed, equilibrium temperature in-

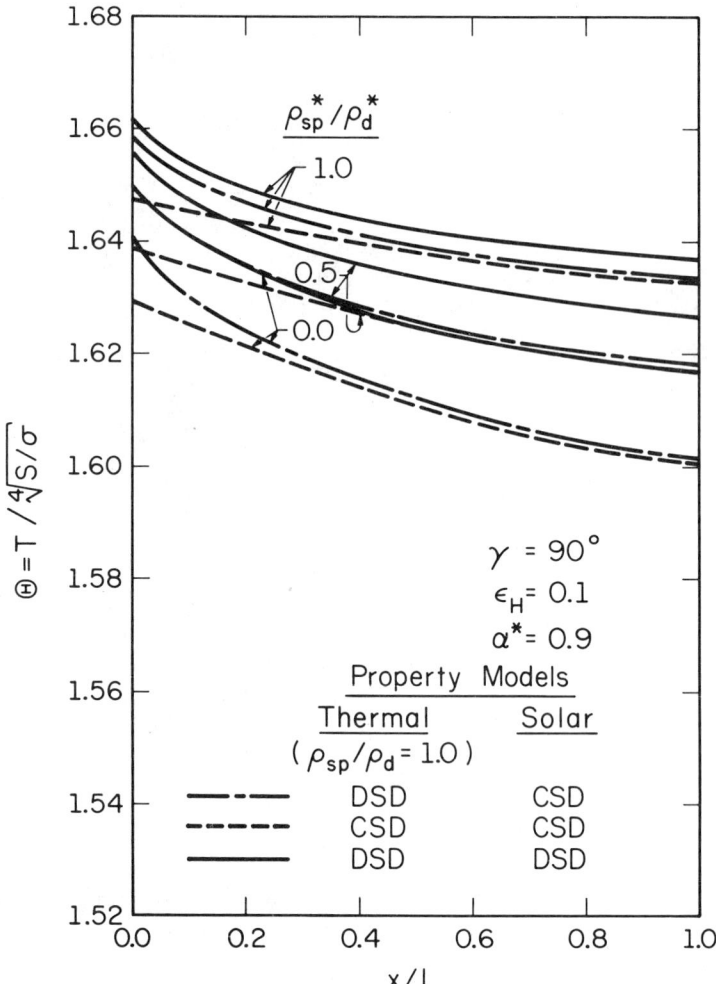

Fig. 10 Angular dependent property effects on local equilibrium temperature ($\epsilon_H = 0.1$, $\alpha^* = 0.9$).

creases as solar absorptance becomes larger or emittance takes on smaller values. Surface specularity for solar radiation has a major effect on equilibrium temperature causing it to decrease with decreasing specularity. Discussion of these trends is not repeated in this paper. Instead, attention is directed to the influence of directional dependent properties on equilibrium temperature distribution.

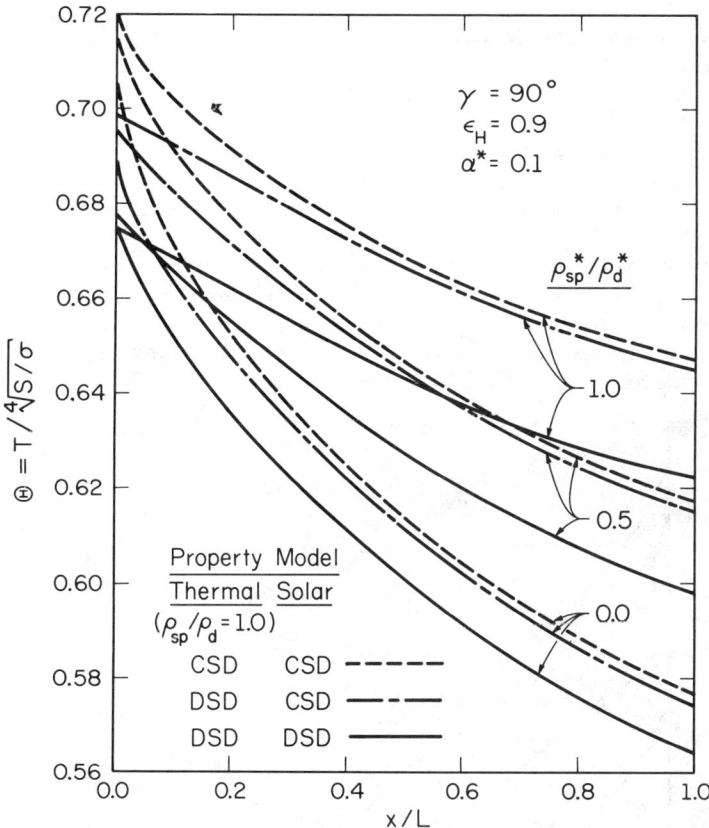

Fig. 11 Angular dependent property effects on local equilibrium temperature ($\varepsilon_H = 0.9$, $\alpha^* = 0.1$).

## Directional Dependent Thermal Properties

Dimensionless equilibrium temperature distributions are presented in Figs. 8-10 for surfaces with hemispherical emittance ($\varepsilon_H$) equal to 0.1 and solar absorptance values ($\alpha^*$) of 0.1, 0.5, and 0.9, respectively. Results for a hemispherical emittance value of 0.9 and solar absorptance values of 0.1 and 0.5 are shown in Figs. 11 and 12, respectively. Surface properties for long wavelength thermal radiation are represented by the DSD model for the limiting case of specular reflection. Surface reflection for solar radiation is represented by the direction independent CSD model (designated by broken lines) and the direction dependent DSD model (designated by solid lines) for specularity ratio values 0.0, 0.5, and 1.0. For purposes of comparison, results of analysis using direction independent thermal and solar properties represented by the CSD model are included and shown as dashed lines.

The effect of direction dependent thermal properties is observed by comparing temperature distributions obtained with DSD (broken lines) and CSD (dashed lines) thermal property models when solar properties are represented by the CSD model. For reasons which were discussed in a previous section of this paper, the influence of directional dependent thermal properties is most pronounced at critical surface elements near the common edge, and continuously decreases with distance from the apex. For a low emittance material ($\varepsilon_H$ = 0.1), CP model

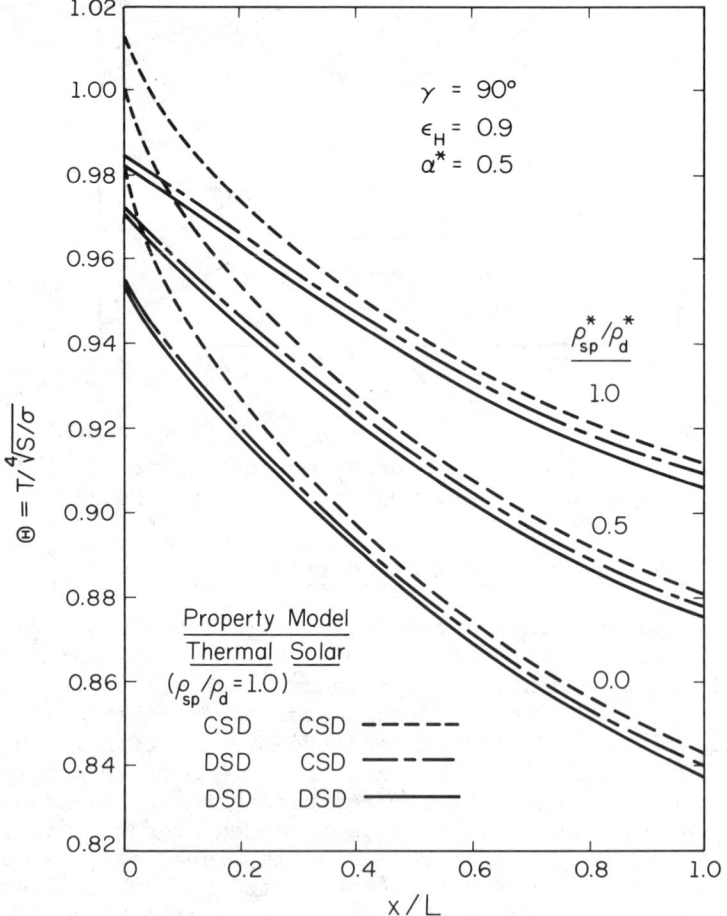

Fig. 12 Angular dependent property effects on local equilibrium temperature ($\varepsilon_H$ = 0.9, $\alpha^*$ = 0.5).

results yield lower temperatures than those evaluated with analysis which includes angular dependent thermal properties. However, the discrepancy is less than 1%, and it is insensitive to the value for solar absorptance and solar specularity. For the high emittance surface ($\varepsilon_H = 0.9$), local temperature is higher from the CP model than that evaluated with direction dependent thermal properties, but the difference is less than 3%. Again, the error is insensitive to the value of solar absorptance and solar specularity.

It is interesting to note that the percentage error in local dimensionless equilibrium temperature associated with ignoring directional dependence of thermal properties is less than that for local radiant heat transfer discussed earlier for the same emittance value by a factor of 4 to 5.

## Directional Dependent Solar Properties

The effect of angle dependent solar properties may be observed by comparing temperature distributions in Figs. 8-12 for solar properties represented by the direction dependent DSD model (solid lines) and the direction independent CSD model (broken lines). Thermal properties are represented by the DSD model for specular reflection.

The difference between results for direction dependent and independent solar properties is nearly uniform over the entire plate length. For both low ($\varepsilon_H = 0.1$) and the high ($\varepsilon_H = 0.9$) emittance materials, local temperature is higher from analysis with direction independent solar properties for low ($\alpha^* = 0.1$) and intermediate ($\alpha^* = 0.5$) solar absorptance values. However, for surfaces of high solar absorptance ($\alpha^* = 0.9$), constant solar property analysis yields low temperatures. For intermediate and high solar absorptance values, the difference between local temperature values for direction dependent and independent solar property models is less than 1%. For low solar absorptance the error is less than 4% and decreases as the surface changes from specularly reflecting to diffusely reflecting to solar energy.

For a surface of low emittance ($\varepsilon_H = 0.1$) and low to intermediate solar absorptance, the effects of direction dependent thermal and solar properties are partially compensating. Therefore, results obtained with analysis for direction independent thermal and solar properties show slightly better agreement with results for direction dependent thermal and solar properties than do those which account only for angular variation of thermal properties. However, for the high emittance surface, results from analysis employing constant thermal and solar properties are too high by 7 and 3% for $\alpha^* = 0.1$ and 0.5, respectively.

It is important to note the uncertainty in local equilibrium temperature which results from inadequate knowledge of the spatial distribution of reflected solar radiation. Consider the results for direction dependent thermal and solar properties. The difference between results for specular ($\rho^*_{sp}/\rho^*_d = 1.0$) and diffuse ($\rho^*_{sp}/\rho^*_d = 0.0$) reflection of solar radiation is largest for a surface of low solar absorptance ($\alpha^* = 0.1$) and decreases continuously with increasing values for $\alpha^*$. The uncertainty is 10, 8, and 1% for $\alpha^* = 0.1, 0.5$, and 0.9, respectively. These values are nearly the same as those reported earlier[4] for direction independent surface properties.

## Conclusion

Basic principles of radiative transfer were employed to formulate the equations governing radiant heat transfer between identical, uniform property adjoint plates uniformly irradiated by a collimated radiative flux. The formulation accounted for direction dependent semigrey surface properties. Numerical results for local and over-all heat transfer were obtained in the absence of external radiative flux, and equilibrium temperature distributions were evaluated for radiatively adiabatic surfaces in the presence of a solar flux. The influence on heat transfer and temperature of direction dependent emittance, absorptance, and reflectance was evaluated for surfaces with directional characteristics represented by the Fresnel relations and markedly different spatial distributions of reflected radiation.

Radiant heat flux distributions evaluated for plates of specified uniform temperature demonstrate that the spatial distribution of reflected energy can significantly influence heat flux values. The use of a representative reflectance model for a given surface is particularly important for surface elements of intermediate and low emittance materials which experience large radiant interaction with other elements and receive major contributions to their irradiation from energy incident at large angles of incidence. For the system studied, the directional dependence of surface properties was unimportant for surfaces of intermediate emittance and analysis which completely ignores directional property dependence is exceptionally accurate. For the critical surface elements mentioned above, disregard of directional property dependence yields high flux values for low emittance surfaces and low flux values for high emittance surfaces. Directional property effects were most pronounced for the high emittance surface where the lack of accounting for surface property detail gave a 25% error in heat flux. The error incurred by analysis which ignores directional property dependence was nearly independent of the specularity of the surface, that is, whether the surface was specularly or diffusely reflecting. Directional dependence of surface properties is not important for evaluating over-all heat transfer and constant property analysis gave accurate results.

The equilibrium temperature results presented suggest several conclusions concerning the importance of directional dependent thermal and solar radiation properties. Directional thermal property effects are most important for the critical surface elements near the common edge, and are insensitive to the value of solar absorptance and whether the surfaces are specularly or diffusely reflecting to solar energy. Analysis which ignores directional dependence of thermal properties yields equilibrium temperature which is high for high surface emittance and low for low surface emittance. However, the resulting temperature errors are small being four to five times less than the corresponding errors in heat flux for uniform temperature plates. Directional dependent solar property effects were found to be nearly uniform over the extent of the surface. For both high and low emittance materials, analysis which ignores directional dependence of solar properties yields equilibrium temperature which is high for low and intermediate values of solar absorptance and temperature which is low for high solar absorptance. Directional property effects are most important for low solar absorptance values, and with all other parameters fixed they decrease in importance as the surface changes from specularly to diffusely reflecting for solar radiation. Directional independent thermal and solar property analysis was found to yield largest temperature errors for the high emittance surface. However, the error was limited to less than 7%. Finally, it was noted that the spatial distribution of reflected solar radiation can have an important effect on equilibrium temperature particularly for surfaces with low and intermediate values of solar absorptance.

## References

[1] Hering, R. G., "Radiative Heat Exchange Between Specularly Reflecting Surfaces with Directional Dependent Properties," Proceedings of the Third International Heat Transfer Conference, American Institute of Chemical Engineers, New York, Vol. V, 1966, pp. 200-206.

[2]Schornhorst, R. J. and Viskanta, R., "Effect of Direction and Wavelength Dependent Surface Properties on Radiant Heat Transfer," AIAA Journal, Vol. 6, No. 8, Aug. 1968, pp. 1450-1455.

[3]Toor, J. S. and Viskanta, R., "A Numerical Experiment on Radiant Heat Interchange by the Monte Carlo Method," International Journal of Heat Mass Transfer, Vol. 11, No. 5, May 1968, pp. 883-897.

[4]Hering, R. G., "Radiative Heat Exchange and Equilibrium Surface Temperature in a Space Environment," Journal of Spacecraft and Rockets, Vol. 5, No. 1, Jan. 1968, pp. 47-54.

[5]Bobco, R. P., "Radiation Heat Transfer in Semigrey Enclosures with Specularly and Diffusely Reflecting Surfaces," Journal Heat Transfer, Transactions ASME, Ser. C, Vol. 86, No. 1, Feb. 1964, pp. 123-130.

[6]Hering, R. G. and Smith, T. F., "Surface Radiation Properties from Electromagnetic Theory," International Journal Heat Mass Transfer, Vol. 11, 1968, pp. 1567-1571.

[7]Edwards, D. K., "Radiative Transfer Characteristics of Materials," Journal Heat Transfer, Transactions ASME, Ser. C, Vol. 91, No. 1, Feb. 1969, pp. 1-15.

[8]Hering, R. G., "Selected Topics on Radiative Heat Transfer," Advanced Heat Transfer, edited by B. T., Chao, University of Illinois Press, Urbana, Ill., 1969, pp. 75-154.

[9]Houchens, A. F., "Real Surface Effects on Radiative Transfer," Ph.D. dissertation, 1970, Dept. of Mechanical and Industrial Engineering, University of Illinois at Urbana-Champaign, Urbana, Ill.

[10]Eckert, E. R. G. and Sparrow, E. M., "Radiative Heat Exchange Between Surfaces with Specular Reflection," International Journal Heat Mass Transfer, Vol. 3, Aug. 1961, pp. 42-54.

# RADIANT HEAT TRANSFER BETWEEN NONGRAY SURFACES

R. G. Hering[*] and W. D. Fischer[†]

University of Illinois at Urbana-Champaign, Urbana, Ill.

## Abstract

Nongray surface property effects on local and over-all radiant heat transfer are studied for a simple system of interacting opaque surfaces. Detailed wavelength and temperature dependence of spectral emittance is included in the analysis for equal and unequal temperature surfaces which vary from diffusely to specularly reflecting. Tungsten is employed as a representative metal and Robert's model is used to describe the wavelength and temperature dependence of its optical parameters. Numerical results establish that gray analysis adequately predicts general trends of nongray analysis and generally yields acceptable local and over-all heat transfer for small differences in plate temperature. Gray theory results for local and over-all heat transfer of the low-temperature surface were often grossly in error when large temperature differences were present. A gray model motivated by physical considerations gave results in exceptional agreement with those determined from nongray analysis.

## Nomenclature

| | |
|---|---|
| B | = radiosity |
| c | = speed of light in vacuum |
| $e_b$ | = emissive power of a black body |
| $f, f^o$ | = functions defined in Eqs. (5) and (11) |
| H | = irradiation |

---

Presented as Paper 71-464 at the AIAA 6th Thermophysics Conference, Tullahoma, Tenn., April 26-28, 1971. This paper presents results of research supported in part by Jet Propulsion Laboratory, California Institute of Technology, Contract 951661.

[*]Professor, Department of Mechanical Engineering; presently Professor and Chairman, Department of Mechanical Engineering, The University of Iowa, Iowa City, Ia.

[†]Research Assistant, Department of Mechanical Engineering.

$K, K^o$ = geometry kernel, geometry kernel at the corner
$k$ = absorption index
$n$ = refractive index
$p$ = integer
$N, P$ = integers defined in Eq. (6)
$q$ = local heat transfer rate per unit area
$Q$ = over-all heat transfer rate per unit width
$T$ = temperature
$x, x'$ = coordinates

$\left.\begin{array}{l} \beta_{om} \\ \delta_m \\ \lambda_{sm} \\ \lambda_{\tau n} \\ \sigma_n \end{array}\right\}$ = constants used in Eq. (15)

$\Delta$ = function defined in Eq. (10)
$\varepsilon$ = hemispherical emittance
$\varepsilon_o$ = permittivity of free space
$\gamma$ = included angle
$\lambda$ = wavelength
$\rho, \rho^D, \rho^S$ = reflectance, diffuse reflectance, specular reflectance

Subscripts

$i, j$ = surface

## Introduction

Engineering materials exhibit radiation characteristics which often significantly differ from property models employed in analysis. Such real surface characteristics may be broadly classified as wavelength and directional dependence. The influence of directional dependence of surface properties on radiant heat transfer and equilibrium temperature has received considerable attention.[1-5] On the other hand, definitive information concerning the effect of spectral and temperature dependence of surface properties is somewhat limited. It is the purpose of this study to investigate the influence of spectral and temperature dependence of surface properties (nongray property effects) on radiant heat transfer and to evaluate the magnitude of the discrepancy incurred when simpler gray analysis is employed. This study is an initial step in determining the relative importance of spectral and directional property dependencies on radiant heat transfer.

Reported studies which consider the spectral dependence of surface properties are confined to evaluating the influence of nongray effects on over-all radiant heat exchange. Goodman[6] and Branstetter[7,8] studied heat transfer between infinite parallel plates using a diffuse reflection model, while Holt and Grosh[9] employed a specular reflection model and included both directional and wavelength dependence of surface properties to study the same system. Chupp and Viskanta[10] reported on nongray effects for concentric spheres and coaxial cylinders. The above investigators found differences between nongray and gray heat exchange rates as large as a factor of two.

The system of interacting opaque surfaces selected for this study consists of unit length plates sharing a common edge and including angle $\gamma$ (Fig. 1). This system was chosen for a number of reasons. First, extensive results are available for radiant heat transfer and for equilibrium temperature using gray or semigray spectral models for both direction dependent[2] and direction independent[11-13] property models with both specular and diffuse reflection. Second, unlike geometries previously studied, the system can readily accommodate an external radiant flux and, hence, the analysis may be extended to include an essential feature of the space environment; namely, a collimated solar flux. Also, in contrast to earlier studies, nongray effects on local radiant flux, as well as on over-all heat transfer, may be evaluated for grossly different distributions of reflected energy. For the purposes of this study, each surface has a uniform temperature and uniform properties, external sources of radiant energy are absent, and the intervening media is radiatively nonparticipating. Both surfaces

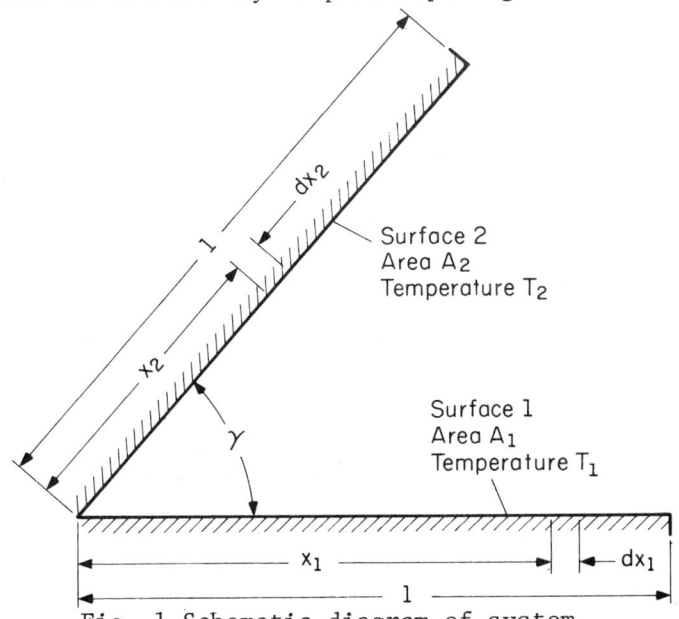

Fig. 1 Schematic diagram of system.

are considered diffusely emitting and a direction independent specular plus diffuse model is employed to describe the spatial distribution of reflected energy. Energy transfer mechanisms other than radiative transport are assumed negligible.

## Analysis

### Nongray Analysis

Local monochromatic heat flux for a typical surface area element $dA_i$ of finite surface $A_i$ ($i = 1,2$), $q_{i,\lambda}(x_i)$, is the difference between emission rate and rate of absorption of incident energy with wavelengths between $\lambda$ and $\lambda + d\lambda$. Thus, on a per unit wavelength interval basis

$$q_{i,\lambda}(x_i) = \varepsilon_{i,\lambda} e_{bi,\lambda} - \varepsilon_{i,\lambda} H_{i,\lambda}(x_i) \quad (i = 1,2) \tag{1}$$

In Eq. (1), monochromatic absorptance of surface $i$ has been taken equal to monochromatic hemispherical emittance at the surface temperature $T_i$; that is, $\varepsilon_{i,\lambda}$. The symbol $e_{bi,\lambda}$ denotes monochromatic emissive power of a black surface at $T_i$ which is available from Planck's law[14] and $H_{i,\lambda}(x)$ represents the monochromatic irradiation function for surface $i$ at position $x_i$. The spectral irradiation functions may be expressed in terms of spectral radiosity functions $B_{i,\lambda}(x_i)$ by the relation[15]

$$H_{i,\lambda}(x_i) = \int_0^1 B_{i,\lambda}(x'_i) K_{ij,\lambda}(x_i, x'_i) dx'_i + \int_0^1 B_{j,\lambda}(x_j) K_\lambda(x_j, x_i) dx_j$$

$$(i = 1,2) \tag{2}$$

where $j = 3 - i$. The spectral radiosity functions satisfy the following pair of simultaneous linear integral equations

$$B_{i,\lambda}(x_i) = \varepsilon_{i,\lambda} e_{bi,\lambda} + \rho^D_{i,\lambda} H_{i,\lambda}(x_i) \quad (i = 1,2) \tag{3}$$

where $\rho^D_{i,\lambda}$ is the diffuse component of spectral reflectance $\rho_{i,\lambda} (= 1 - \varepsilon_{i,\lambda})$. In Eq. (2), the kernel functions are [15]

$$K_{ij,\lambda}(x_i, x'_i) = \sum_{k=1}^{N} (\rho^S_{i,\lambda})^{k-1} (\rho^S_{j,\lambda})^k f_{2k\lambda}(x_i, x'_i)$$

$$\tag{4}$$

$$K_\lambda(x_i, x_j) = \sum_{k=0}^{P} (\rho^S_{i,\lambda} \rho^S_{j,\lambda})^k f_{(2k+1)\gamma}(x_i, x_j)$$

with

$$f_{p\gamma}(x_i,x_j) = \frac{\sin^2 p\gamma}{2} \frac{x_i x_j}{[x_i^2 + x_j^2 - 2x_i x_j \cos p\gamma]^{3/2}} \tag{5}$$

and

$$N = \{\pi/2\gamma\}, \quad P = \{(\pi-\gamma)/2\gamma\}. \tag{6}$$

(The notation $\{X\}$ denotes the operation of taking the integer part of X. If X is precisely an integer, then $\{X\}$ is that integer minus 1.) In Eqs. (4), $\rho_{i,\lambda}^s$ is the specular component of spectral reflectance.

Once the spectral radiosity functions are available, the spectral irradiation functions may be evaluated from Eq. (2) and then monochromatic flux follows from Eq. (1). Local radiant heat flux $q_i(x_i)$ is obtained by integration over wavelength of the spectral flux.

$$q_i(x_i) = \int_0^\infty q_{i,\lambda}(x_i) d\lambda \qquad (i = 1,2) \tag{7}$$

Over-all heat transfer $Q_i$ is determined by integration of local flux over plate length.

$$Q_i = \int_0^1 q_i(x_i) dx_i \qquad (i = 1,2) \tag{8}$$

In order to evaluate local heat flux and over-all heat transfer, it is generally necessary to solve the integral equations governing the radiosity functions at a sufficient number of wavelengths to ensure a reasonably accurate integration over wavelength as indicated in Eq. (7). The necessity for repeatedly solving the integral equations is curcumvented when local radiant flux at the apex ($x_i = 0$, $i = 1,2$) is evaluated. By using physical arguments or techniques such as those employed in Ref. 16, it may be shown that the corner values for the spectral radiosity functions $B_{i,\lambda}(0)$ are

$$B_{i,\lambda}(0) = (1/\Delta_\lambda)[\varepsilon_{i,\lambda}(1-\rho_{j,\lambda}^D K_{ji,\lambda}^o)e_{bi,\lambda} + \varepsilon_{j,\lambda}\rho_{i,\lambda}^D K_\lambda^o e_{bj,\lambda}] \tag{9}$$

where

$$\Delta_\lambda = (1 - \rho_{i,\lambda}^D K_{ij,\lambda}^o)(1 - \rho_{j,\lambda}^D K_{ji,\lambda}^o) - \rho_{i,\lambda}^D \rho_{j,\lambda}^D (K_\lambda^o)^2 \tag{10}$$

The functions $K^o_{ij,\lambda}$ and $K^o_\lambda$ are given by Eq. (4), except for the replacement of $f_{p\gamma}(x_i,x_j)$ with $f^o_{p\gamma}$, where

$$f^o_{p\gamma} = (1+\cos_{p\gamma})/2 \tag{11}$$

The spectral irradiation function values at the apex $H_{i,\lambda}(0)$ follow as

$$H_{i,\lambda}(0) = B_{i,\lambda}(0) K^o_{ij,\lambda} + B_{j,\lambda}(0) K^o_\lambda \qquad (i = 1,2) \tag{12}$$

Corner values for local flux $q_i(0)$ are evaluated by integration of $q_{i,\lambda}(0)$.

The local radiant flux value depends on a number of parameters. These include temperature of both surfaces $(T_1,T_2)$, included angle $(\gamma)$, apportionment of reflected energy between the specular and diffuse components $(\rho^s_{i,\lambda}/\rho_{i,\lambda})$, and spectral and temperature dependence of plate emittances $(\varepsilon_{i,\lambda})$. For the purposes of this study the specularity parameter $\rho^s_{i,\lambda}/\rho_{i,\lambda}$ is taken wavelength independent. The spectral emittance model employed to obtain quantitative results is briefly discussed later.

Gray Analysis

Gray analysis ignores spectral dependence of radiation properties and yields expressions identical to those of Eqs. (2)-(6) except for the replacement of all spectral quantities with corresponding total (integrated over wavelength) values. For the purpose of distinguishing gray results presented later, the expression for total heat flux is written

$$q_i(x_i) = \varepsilon_i e_{bi} - \alpha_i H_i(x_i) \qquad (i = 1,2) \tag{13}$$

where $\varepsilon_i$, $e_{bi}$, and $H_i(x_i)$ are total hemispherical emittance at temperature $T_i$, total emissive power of a black surface at $T_i (=\sigma T_i^4)$, and total irradiation function for surface i, respectively. Symbol $\alpha_i$ denotes total absorptance of surface i. Values employed for $\alpha_i$ and the corresponding total reflectance $\rho_i (=1 - \alpha_i)$ in the equations of gray analysis distinguish the various gray theory results presented later. Surface absorptances were approximated by evaluating the expression

$$\alpha_i = \frac{1}{\tilde{H}} \int_0^\infty \varepsilon_{i,\lambda} \tilde{H}_{i,\lambda} \, d\lambda \tag{14}$$

with $\tilde{H}_{i,\lambda}$ a representative spatially uniform spectral irradiation function and $\tilde{H}_i$ the corresponding wavelength integrated value. It is also convenient at this point to let surface 1 be the higher temperature surface when plate temperatures are unequal.

Results are presented later for four different gray calculations corresponding to different choices for the values of plate absorptances. Gray models A,B, and C employed total emittance($\varepsilon_1$) for total absorptance($\alpha_1$) of the high-temperature plate. This corresponds to the use of the spectral emissive power of a black surface at $T_1$ for $H_{1,\lambda}$ in Eq. (14). Models A,B, and C also employed black body spectral emissive power for $H_{2,\lambda}$ in Eq. (14) but differed in the value for temperature at which the emissive power was introduced. Model A utilized the temperature of the colder surface($T_2$); model B employed the geometric mean of the surface temperatures($\sqrt{T_1 T_2}$); and, model C used the temperature of the hot surface ($T_1$). Model A completely ignores the difference between emittance and absorptance and is valid for spatially uniform irradiation when the monochromatic emittances are wavelength independent. Model B is suggested by the fact that for highly reflecting metal surfaces such as are considered later, the dominant energy interreflecting within the system is that of the high-temperature surface. Furthermore, electromagnetic theory predicts that at sufficiently long wavelengths absorptance of metal surfaces may be approximated by emittance evaluated at the geometric mean of the temperatures of the source and the absorbing surface.[17] Model C is motivated by the observation that for highly reflecting metal surfaces, the spectral variation of irradiation absorbed by the surfaces, if approximated by a black surface, is probably most closely represented by the distribution corresponding to the temperature of the hot plate. Finally, model D is an improvement over model C since the absorptance of both surfaces are based on the spectral emissive power of the higher temperature surface. These absorptances correspond to use of $\tilde{H}_{1,\lambda} = \tilde{H}_{2,\lambda} = \varepsilon_{1,\lambda} e_{b1,\lambda}$ and $\tilde{H}_1 = \tilde{H}_2 = \varepsilon_1 e_{b1}$ in Eq. (14). Generally, for the system considered, gray theory results are expected to approximate nongray values more closely as the gray models progress from model A to model D.

## Spectral Surface Properties

One of the most uncertain aspects of nongray radiative heat-transfer calculations is an accurate representation for radiative properties of engineering materials. Branstetter[8,9] employed experimental data. Although this is, no doubt, the best technique, it is limited to materials and temperatures for which data are available. Property measurements of sufficient scope in wavelength and temperature are not readily available for many materials of interest and it is useful to employ theoretical models which exhibit the general characteristics observed for classes of engineering materials.

Several investigators have used classical mechanics to develop expression for the emittance of metals which include both spectral and temperature dependence. The most widely known result is that from Drude single electron theory.[18] The greatest attribute of the Drude theory results from an engineering standpoint is probably the simplicity of the expression for spectral emittance. The Drude theory result for spectral emittance, however, does not agree well with experimental data for metals for wavelengths less than about 5μ. Although models based on quantum theory have been developed, the complexity of these relationships generally precludes their use in engineering calculations. Roberts[19] extended Drude theory and incorporated some of the phenomena predicted by quantum theory. By fitting data and using some physical constraints, Roberts proposed, for the optical indices $n_\lambda$ and $k_\lambda$ of a metal,

$$n_\lambda^2(1- ik_\lambda) = 1 + \sum_m [(\beta_{om}\lambda^2)/(\lambda^2-\lambda_{sm}^2 + i\delta_m \lambda_{sm}\lambda)] - (\lambda^2/2\pi\varepsilon_o c) \sum_n [\sigma_n/(\lambda_{\tau n} - i\lambda)] \quad (15)$$

where $\varepsilon_o'$ and c are permittivity of free space and speed of light in vacuum, respectively, and $i = \sqrt{-1}$. The symbols $\beta_{om}$, $\lambda_{sm}$, $\delta_m$, $\sigma_n$, and $\lambda_{\tau n}$ denote emperical parameters which are generally temperature dependent and are determined from experimental data for application to a particular metal. Once the optical parameters $n_\lambda$ and $k_\lambda$ are evaluated, spectral hemispherical properties for optically smooth uncontaminated surfaces may be determined by employing the relationships of electromagnetic theory.[20]

Tungsten is a convenient material to employ for investigating nongray property effects on radiative heat transfer because its spectral properties are well documented over a wide range of temperature and wavelength. Furthermore, Roberts[19] has reported values for the empirical parameters of Eq. (15) for tungsten. These values were used in the relations of electromagnitic theory[20] to evaluate spectral hemispherical emittance for tungsten. In Fig. 2, spectral hemispherical emittance calculated using the above procedure is presented with data for tungsten. Although the temperature for the calculated emittance values and the data are not identical, the results evaluated with the model generally exhibit the trends observed in the data.

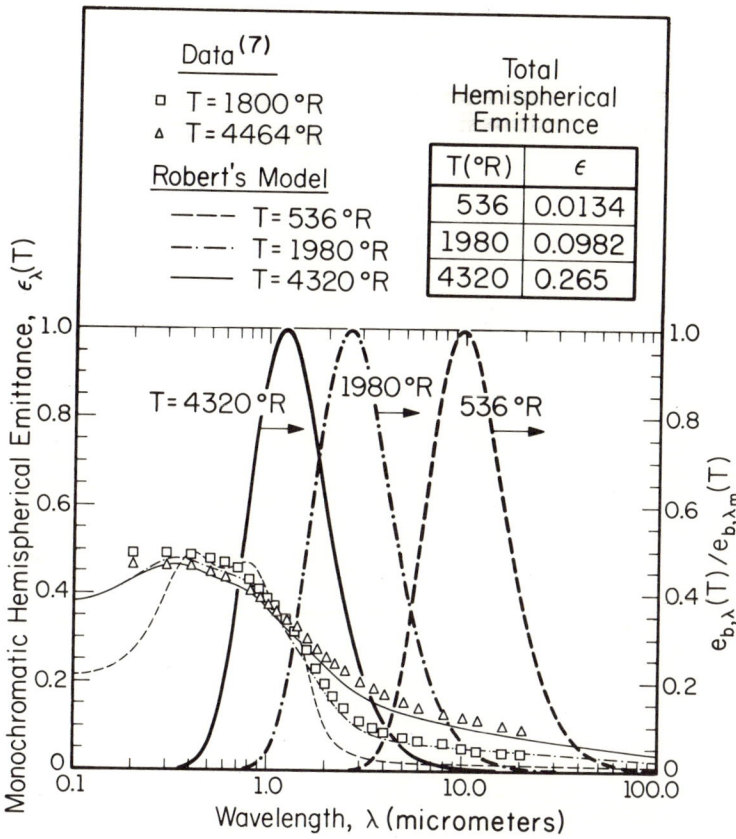

Fig. 2 Comparison of Roberts' property model to data for tungsten.

Results

Nongray property effects on local and total radiant heat transfer were investigated for equal and unequal temperature tungsten plates for an included angle ($\gamma$) of $45°$. All calculations employed identical values for the specularity parameter of both surfaces, that is, $\rho_1^s/\rho_1 = \rho_2^s/\rho_2 = \rho^s/\rho$. The general character of local flux distribution and the over-all influence of emittance value as well as reflectance model on this distribution has been discussed in detail for gray surfaces elsewhere[11] and is not repeated here. Since nongray effects are not expected to significantly alter the general character of trends predicted with gray theory, emphasis is placed on the differences between nongray and gray heat-transfer results and the error incurred in using gray analysis.

## Equal Temperature Surfaces

Dimensionless radiant flux distributions, $q(x)/\epsilon\sigma T^4$, are illustrated in Fig. 3 for equal temperature tungsten plates at temperatures of 536°R, 1980°R, and 4320°R. Over this temperature range emittance increases twenty-fold from a value of 0.013 at 536°R to a value of 0.26 at 4320°R. (See insert on Fig. 2) Results are presented for both specularly reflecting (solid curves), $\rho^s/\rho = 1.0$) and diffusely reflecting (dashed curves, $\rho^s/\rho = 0.0$) surfaces. Distributions based on both nongray and gray theory are shown. Gray theory models A, B, and C degenerate to identical models for equal temperature

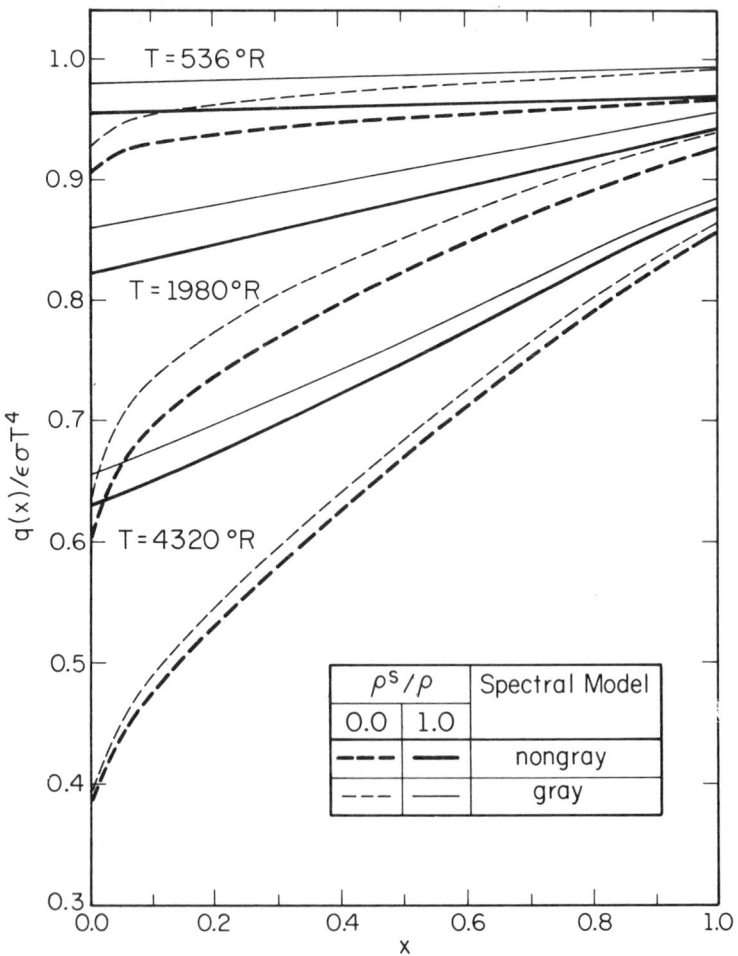

Fig. 3 Gray and nongray dimensionless radiant flux distributions for equal temperature plates ($\gamma = 45°$).

plates and are simply denoted as gray results in Figs. 3 and 4 as well as in Table 1.

As expected, the general character of the flux distributions is not significantly influenced by the spectral and temperature dependence of surface properties. Furthermore, in accordance with the predicitons for gray surfaces,[11] local flux for a nongray surface of specified emittance is greater for a specularly reflecting than for a diffusely reflecting surface and local flux per unit emissive power diminishes with increasing emittance. Thus, gray theory adequately predicts the general character of the real surface results.

Flux distributions computed with gray theory exceed those which account for detailed spectral and temperature dependence of surface properties when all other influencing parameters are identical. This situation may be explained by the following considerations. Local flux is the difference between emission rate and rate of absorption of incident radiation. Since the emission rate calculated using gray and nongray theory is identical, the higher flux value from gray theory must be attributed to a lower value for absorbed irradiation calculated from gray theory than that evaluated considering the detailed property dependence in nongray analysis. Now it may be shown that the absorptance (emittance) used in the gray calculations is less than that evaluated for the real surfaces. However, since the absorbed irradiation per unit emissive power decreases when the emittance is reduced, it follows that the gray theory calculations which employ a low value for absorptance will yield smaller values for absorbed irradiation and, therefore, larger local heat flux per unit emissive power than nongray analysis.

Gray theory local flux results have their largest error at the corner ($x = 0$) where reflection phenomena are most important in this system. Gray theory, however, yields surprisingly accurate values with a maximum flux error of less than 6% for a diffusely reflecting surface of intermediate temperature. Since corner flux values exhibit all trends observed at other locations, a more extensive study was made of corner heat flux values. The results are presented in Fig. 4 over a large temperature range for the reflectance models previously employed as well as for a surface with an equal apportionment between specular and diffuse reflectance components. Generally, the error in gray theory results is a weak function of the reflectance model employed. As should be expected, gray analysis is least accurate at intermediate temperature levels. At low plate temperatures, the spectral emittance in the important long wavelength region is nearly independent of wavelength (Fig. 2) and gray analysis is accurate. At high plate temperatures, emission dominates local radiant flux and hence gray theory errors diminish.

A comparison of dimensionless over-all heat transfer, $Q/\sigma T^4$, based on gray and nongray analysis is presented in Table 1 for equal temperature tungsten plates. In view of the relatively small influences of nongray property behavior on local flux values for equal temperature surfaces, it follows that over-all heat-transfer results from gray and nongray analysis should not be significantly different. The results of Table 1 confirm the accuracy of gray theory with a maximum discrepancy of 3% occurring at the intermediate temperature.

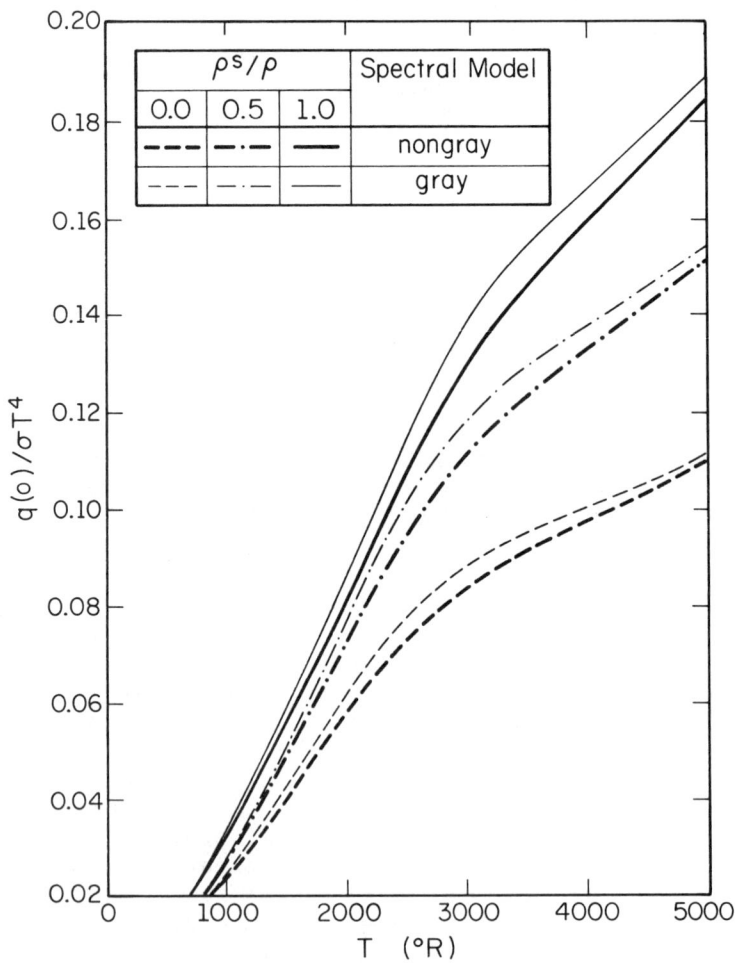

Fig. 4 Gray and nongray dimensionless corner heat flux values for equal temperature tungsten plates ($\gamma = 45°$).

Table 1 Comparison of over-all dimensionless radiant heat transfer for equal temperature tungsten plates ($\gamma=45°$).

| Temperature($°R$) | Reflectance model, $\rho^s/\rho$ | Spectral Model | | % error of gray model |
|---|---|---|---|---|
| | | Nongray | Gray | |
| 536 | 0.0 | 0.0131 | 0.0134 | 2.29 |
| | 1.0 | 0.0133 | 0.0136 | 2.26 |
| 1980 | 0.0 | 0.0797 | 0.0825 | 3.40 |
| | 1.0 | 0.0866 | 0.0891 | 2.80 |
| 4320 | 0.0 | 0.1748 | 0.1785 | 2.11 |
| | 1.0 | 0.1990 | 0.2309 | 2.46 |

Unequal Temperature Surfaces

Representative results for dimensionless radiant flux distributions are presented in Figs. 5 and 6 for unequal temperature tungsten plates using a specular reflectance model ($\rho^s/\rho = 1.0$) with the high temperature surface at $4320°R$. Results of Figs. 5 and 6 pertain to low temperature surfaces of $536°R$ and $1980°R$, respectively. In addition to the nongray radiant heat flux distributions, gray results for the various models previously discussed are presented in each figure.

It is evident from both Figs. 5 and 6 that gray theory continues to yield the trends observed for the nongray results. Gray analysis underestimates local heat flux for the hotter surface and generally overestimates flux for the colder surface. Although all gray models yield acceptable accuracy for the high-temperature surface(maximum error less than 3%), discrepancies between nongray and gray results as large as a factor of two are clearly evident for the low-temperature surface. Of the gray models considered, model D yields exceptionally accurate results which differ almost imperceptibly from those determined using nongray analysis. This should be expected since local heat flux of the low-temperature surface is dominated by the absorption of energy emitted by the high-temperarue surface. Clearly, the absorptance evaluated with the high-temperature surface spectral emissive power corresponds to the physical situation more closely than evaluation of absorptance in any other manner. Even with the somewhat simpler model C, however, gray theory results may be adequate for some applications since the error for the low-temperature surface flux is limited to about 20%. Gray analysis local flux values using model C are within 0.5% of the nongray results for the high-temperature surface.

Figure 7 presents results for radiant heat flux at the corner as a function of the ratio of cold plate temperature to hot plate temperature $(T_2/T_1)$ for a hot surface temperature of 4320°R. Results are shown for nongray analysis as well as for the gray models previously discussed for specularly reflecting $(\rho^s/\rho = 1.0)$, diffusely reflecting $(\rho^s/\rho = 0.0)$, and specularly plus diffusely reflecting $(\rho^s/\rho = 0.5)$ surfaces. The most significant additional insight obtained from the results in Fig. 7 is that the accuracy of gray analysis deteriorates for the high-temperature surface as the surface changes from specularly reflecting to diffusely reflecting. Although gray model A yields accurate heat flux values for the specularly reflecting high-temperature surface over the entire range of

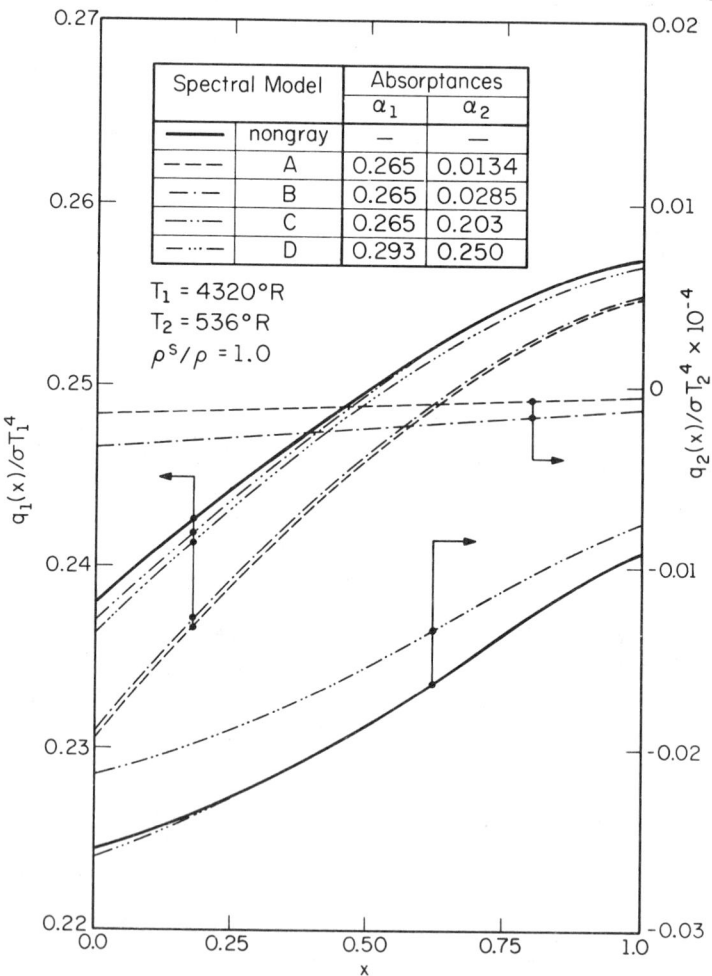

Fig. 5 Gray and nongray dimensionless radiant flux distributions for unequal temperature tungsten plates ($\gamma = 45°$, $\rho^s/\rho = 1.0$, $T_1 = 4320°R$, $T_2 = 536°R$).

temperature ratio, it gives values which are low by 25% at moderate cold plate temperatures for diffusely reflecting surfaces. This phenomenon is attributed to the major role multiply reflected energy attains for diffusely reflecting surfaces. Since the low-temperature surface reflectance at moderate temperatures is very high ($\sim 0.99$ at $T_2 = 536°R$), the total absorbed irradiation is more sensitive to changes in reflectance for diffusely reflecting surfaces which offers the greatest impedance to energy flow out the opening. As observed earlier, model A yields poor results for the low-temperature surface heat flux. Also, although gray model C yields a signifi-

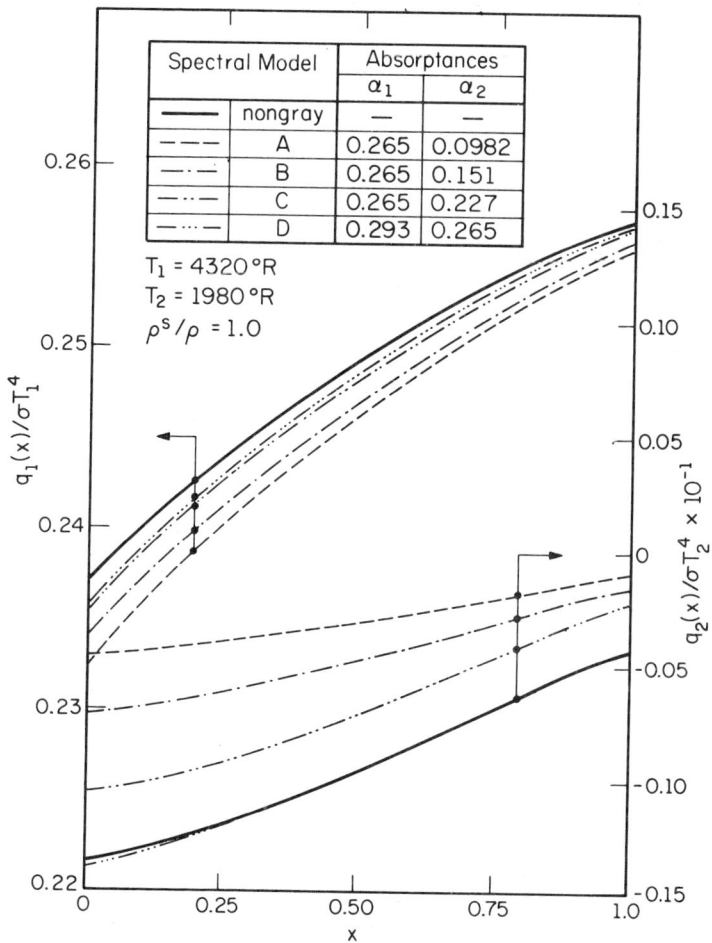

Fig. 6 Gray and nongray dimensionless radiant flux distributions for unequal temperature tungsten plates ($\gamma = 45°$, $\rho^s/\rho = 1.0$, $T_1 = 4320°R$, $T_2 = 1980°R$).

cant improvement in agreement with nongray results, the error is not at an acceptable level for moderate values of plate temperature ratio. Gray model D results are essentially in

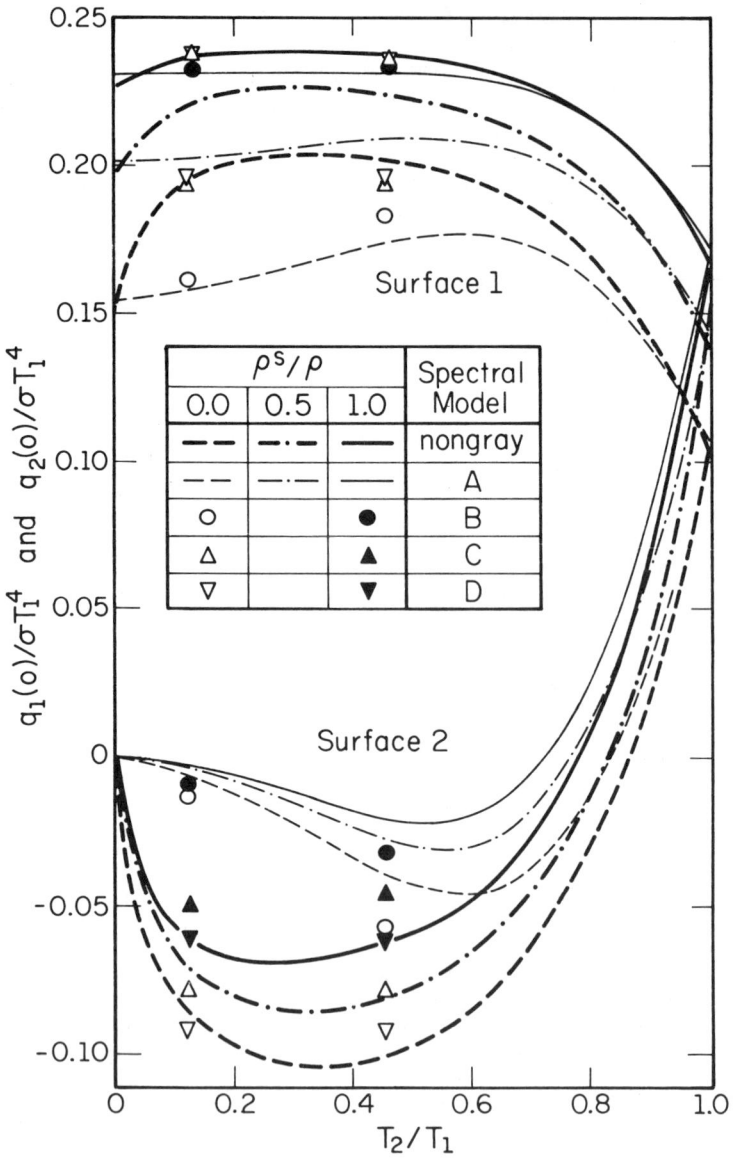

Fig. 7 Gray and nongray dimensionless corner heat flux values for unequal temperature tungsten plates ($\gamma = 45°$).

agreement with nongray results for all values of the influencing parameters.

Table 2 presents dimensionless over-all radiant heat-transfer values for unequal temperature specularly reflecting plates based on nongray and gray analysis for a hot plate temperature of 4320°R. Gray analysis, total heat-transfer values for the high-temperature surface are all within 3% of those determined from nongray analysis. Gray model D yields the best over-all agreement. As expected, the low-temperature surface total heat flux results from gray theory can differ significantly from those derived from nongray analysis. Gray model C predicts total heat transfer for the cold plate with an error of almost 30%. On the other hand, gray model D results differ from the nongray values for the cold surface total heat transfer by less than 1%.

## Conclusions

Analysis and results have been presented for local and over-all radiant heat transfer between interacting opaque tungsten surfaces which include detailed spectral and temperature dependence of radiation surface properties. It has been demon-

Table 2 Comparison of over-all dimensionless radiant heat transfer $Q_1/\sigma T_1^4$ and $Q_2/\sigma T_2^4$ for unequal temperature tungsten plates ($\gamma=45°$, $\rho^s/\rho = 1.0$, $T_1 = 4320°R$)

| $T_2/T_1$ | Nongray | High temperature plate, $Q_1/\sigma T_1^4$ | | | |
|---|---|---|---|---|---|
| | | Gray (percent error) | | | |
| | | A | B | C | D |
| 0.124 | 0.2492 | 0.2450 (1.7) | 0.2454 (1.5) | 0.2490 (0.1) | 0.2483 (0.4) |
| 0.458 | 0.2488 | 0.2460 (1.1) | 0.2467 (0.8) | 0.2477 (0.4) | 0.2478 (0.4) |
| 1.000 | 0.1990 | 0.2038 (2.4) | 0.2038 (2.4) | 0.2038 (2.4) | 0.1984 (0.3) |
| | | Low temperature plate, $Q_2/\sigma T_2^4$ | | | |
| 0.124 | -181.7 | -10.0 (94.5) | -21.3 (88.4) | -150.0 (17.5) | -183.4 (10.9) |
| 0.458 | -0.9438 | -0.2982 (68.5) | -0.4576 (51.6) | -0.6868 (27.2) | -0.9524 (0.9) |
| 1.000 | 0.1990 | 0.2038 (2.4) | 0.2038 (2.4) | 0.2038 (2.4) | 0.1984 (0.3) |

strated that gray theory which ignores such real surface characteristics adequately predicts the general trends observed in nongray heat-transfer results. Furthermore, when the difference in temperature between the surfaces is not large, gray theory results for local and over-all heat transfer are of acceptable engineering accuracy independent of the reflectance model employed. Even for large differences in temperature, gray theory local and over-all heat transfer for the high-temperature surface were sufficiently accurate for engineering purposes when the surfaces were specularly reflecting. Gray theory was generally less accurate when the interacting surfaces were diffusely reflecting. On the other hand, gray theory local and over-all heat transfer for the low-temperature surface were generally grossly in error when the difference in surface temperature was moderate to large. Of the gray models employed, that which evaluated the absorptance of the low-temperature surface with the spectral emissive power at the temperature of the high-temperature surface gave exceptional agreement with the nongray heat-transfer results. Models which were less physically motivated, however, gave local and over-all heat-transfer values for the low-temperature surface which were substantially in error.

## References

[1] Bevans, J. T. and Edwards, D. K., "Radiation Exchange in an Enclosure with Directional Properties", *Transactions of the ASME, Series C, Journal of Heat Transfer*, Vol. 87, No. 3, Aug. 1965, pp. 388-396.

[2] Hering, R. G., "Radiative Heat Exchange between Specularly Reflecting Surfaces with Direction Dependent Properties," *Proceedings of the Third International Heat Transfer Conference*, American Institute of Chemical Engineers, New York, 1966, pp. 200-206

[3] Toor, J. S. and Viskanta, R., "Effect of Direction Dependent Properties on Radiation Interchange," *Journal of Spacecraft and Rockets*, Vol. 5, No. 6, June 1968, pp. 742-743.

[4] Hering, R. G. and Smith, T. F., "Surface Roughness Effects on Radiant Transfer between Surfaces," *International Journal of Heat and Mass Transfer*, Vol. 13, 1970, pp. 725-739.

[5] Hering, R. G. and Smith, T. F., "Surface Roughness Effects on Equilibrium Temperature of Interacting Surfaces," *Journal of Spacecraft and Rockets*, Vol. 8, No. 4, April 1971, pp. 367-373.

[6] Goodman, S., "Radiant-Heat Transfer between Nongray Parallel Plates," Journal of Research of the National Bureau of Standards, Vol. 58, No. 1, Jan. 1957, pp. 37-40.

[7] Branstetter, J. R., "Radiant Heat Transfer between Nongray Parallel Plates of Tungsten, TN D-1088, Aug. 1961, NASA.

[8] Branstetter, J. R., "Formulas for Radiant Heat Transfer between Nongray Parallel Plates of Polished Refractory Metals," TN D-2902, June 1965.

[9] Holt, V., Grosh, R. J., and Geynet, R., "Evaluation of the Net Radiant Heat Transfer between Specularly Reflecting Plates," The Bell System Technical Journal, Vol. 41, Nov. 1962, pp. 1865-1874.

[10] Chupp, R. E. and Viskanta, R., "Radiant Heat Transfer between Concentric Spheres and Coaxial Cylinders," Journal of Heat Transfer, Vol. 88, No. 3, Aug. 1966, pp. 326-327.

[11] Sparrow, E. M., et al., "Analysis, Results, and Interpretation for Radiation between Some Simply Arranged Gray Surfaces," Journal of Heat Transfer, Vol. 83, No. 2, May 1961, pp. 207-214.

[12] Eckert, E.R.G. and Sparrow, E. M., "Radiative Heat Exchange between Surfaces with Specular Reflection," International Journal of Heat and Mass Transfer, Vol. 3, No. 1, Aug. 1961, pp. 193-199.

[13] Hering, R. G., "Radiative Heat Exchange and Equilibrium Surface Temperature in a Space Environment," Journal of Spacecraft and Rockets, Vol. 5, No. 1, Jan. 1968, pp. 47-54.

[14] Sparrow, E. M., and Cess, R., Radiation Heat Transfer, Brooks/Cole Publishing Co., Belmont, Calif., 1966.

[15] Hering, R. G. and Smith, T. F., "Apparent Radiation Properties of a Rough Surface," AIAA Progress in Astronautics and Aeronautics: Thermophysics: Applications to Thermal Design of Spacecraft, edited by J. T. Bevans, Vol. 23, Academic Press, New York, 1970, pp. 337-361.

[16] Heaslet, N. A. and Lomax, H., "Numerical Prediction of Radiative Interchange between Conducting Fins with Mutual Irradiations," TR R-116, 1961, NASA.

[17] Eckert, E.R.G. and Drake, R. M., Heat and Mass Transfer, McGraw-Hill, New York, 1959.

[18] Mott, N. F. and Jones, H., Theory of the Properties of Metals and Alloys, Dover, New York, 1958.

[19] Roberts, S., "Optical Properties of Nickel and Tungsten and Their Interpretation According to Drude's Formula," The Physical Review, Vol. 114, No. 1, April 1959, pp. 104-115.

[20] Hering, R. G. and Smith, T. F., "Surface Radiation Properties from Electromagnetic Theory," International Journal of Heat and Mass Transfer, Vol. 11, Oct. 1968, pp. 1561-1571.

# THERMAL CONTACT CONDUCTANCE OF TURNED SURFACES

M. Michael Yovanovich[*]

University of Waterloo, Ontario, Canada

## Abstract

This paper presents an analytical work performed to determine the thermal resistance to heat transfer at the interface formed by the contact of a hard smooth flat surface with a softer turned surface. The results are valid for surfaces in a vacuum environment when there is negligible radiation heat transfer across the gaps. The thermal analysis was based upon steady heat flow in a two-dimensional heat channel and the contact analysis was based upon plastic deformation of a ridge formed by the turning process. A dimensionless group consisting of contact conductance, harmonic mean thermal conductivity of the contacting surfaces and the distance between adjacent contacting ridges correlates the available data if surface roughness is also taken into account.

## Nomenclature

| | | |
|---|---|---|
| a | = | half-width of contact strips |
| Aa | = | apparent contact area |
| b | = | half-width of heat channel |
| C | = | coefficient in Eq. (35) |
| $d_1$ | = | average diameter of innermost heat channel |
| D | = | diameter of contacting cylinders |
| $h_c$ | = | contact conductance |
| $h_{cr}$ | = | contact conductance due to contact spots |
| $h_{cw}$ | = | contact conductance due to contact strips |
| $J_o$ | = | Bessel function |
| k | = | thermal conductivity |

---

Presented as Paper 71-80 at the 9th Aerospace Sciences Meeting, New York, N.Y., January 25-27, 1971. This work was supported by the National Research Council of Canada. The author also gratefully acknowledges the fact that Professor H. Cordier and his co-workers made available information regarding test procedures and the results of their tests.

[*]Associate Professor of Mechanical Engineering.

$k_m$ = harmonic mean thermal conductivity
$k_m = 2k_1 k_2 / (k_1 + k_2)$
$\ell$ = length of heat channel
$N$ = number of heat channels
$P_a$ = apparent contact pressure
$P_m$ = maximum yield pressure
$P^*$ = dimensionless contact pressure
$Q$ = heat flow rate
$R_c$ = constriction resistance
$T$ = temperature
$T_a$ = average contact temperature, Eq. (13)
$T_c$ = contact strip temperature
$u$ = transformation, Eq. (42)
$x, y$ = Cartesian coordinates
$\alpha$ = contact angle
$\gamma$ = dimensionless area ratio, Eq. (36)
$\delta$ = distance between spirals
$\theta$ = transformation, Eq. (42)
$\sigma$ = root-mean-square of surface roughness
$\eta, \psi$ = elliptic coordinates
$\Psi$ = geometric factor, Eq. (23)

Subscripts

1,2 = solids 1 and 2, respectively
$i$ = ith heat channel

## Introduction

The thermal resistance to steady heat flow across interfaces formed by contacting solids is currently of great interest to aerospace engineers, especially when the interfaces are placed in a vacuum environment. Many investigators have studied (analytically and experimentally) various aspects of this rather complex problem. The complexity is due to the fact that there are essentially two related problems which have to be studied: the thermal and the mechanical. The thermal problem is solved when one is able to predict the interface resistance from a knowledge of certain physical (thermal conductivities of the contacting solids and interstitial substance) and geometric (number, shape, size and placement of the contact spots) characteristics. The mechanical problem is solved when one is capable of predicting the required geometric characteristics from a knowledge of the geometry (surface roughness and waviness) of the contacting surfaces and certain physical characteristics (modulus of elasticity, maximum yield pressure and apparent contact pressure).

Several important problems dealing with nominally flat rough surfaces and smooth wavy surfaces have been extensively studied and the solutions can be found in the open literature.[1-6] Some aspects of these results will be applied in this study which will consider the heat transfer across an interface formed when a hard smooth flat surface contacts a softer turned surface. Certain investigators[3,7,11] have presented theoretical works dealing with an ideal two-dimensional heat channel, but, to-date, no one has examined the complete thermal-mechanical problem. This paper will be limited to the study of thermal contact conductance in a vacuum only, since it serves as a logical starting point for the more difficult analysis required to handle interstitial substances such as fluids.

## Statement of the problem

A solid metallic cylinder whose surface has first been made flat and then turned on a lathe is brought into contact with a second metallic cylinder of different material whose surface is smooth and flat. A contact area (consisting of individual contact spots) ressembling a long spiral of effective width 2a (Fig.1) is formed as a result of the plastic deformation of the ridge formed during the turning process (Fig. 2). It is the softer turned surface which undergoes the plastic deformation. The distance between adjacent ridges (or spirals) $\delta$ will depend upon the turning rate and the rate of tool advance. This dimension will be much larger than the effective width of the spiral which actually consists of differently shaped contact spots. Some of the spots are relatively far apart while others are very close to each other. Those close together will merge to form larger ones as the contact load is increased. In fact at very high loads, the majority of the contact spots will have merged to form a quasi-continuous strip of width 2a.

When the interface described above is placed in a vacuum and steady linear heat flow occurs in both cylinders, a pseudo-tem-

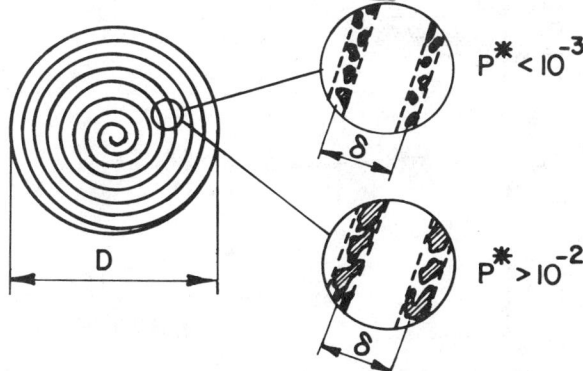

Fig. 1 Contact areas for turned surfaces.

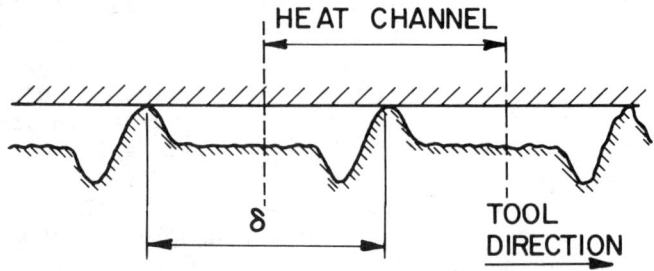

Fig. 2  Profiles of contacting surfaces.

perature drop appears at the interface (Fig. 3). This temperature drop is a direct measure of the thermal resistance occurring at the imperfect interface. The objective, then, is to predict this thermal resistance or its reciprocal, the thermal conductance. For interfaces placed in a vacuum and negligible radiation heat transfer across the gaps, there will be one thermal path available for heat transfer across the interface and that is conduction through the contact spots.

The thermal problem will be modeled as N concentric circular heat channels thermally connected in parallel contacting N other concentric circular heat channels. The first set of heat channels are thermally connected in series with the second set of heat channels. Once the thermal constriction resistance of a typical heat channel has been determined, it is relatively simple to determine the total resistance of the first set of heat channels and finally the total resistance of the two sets together.

Analytical solution

<u>Constriction Resistance of a Single Heat Channel</u>

Fig. 4 shows the model used in this analysis as well as the model used by other investigators.[3,7] Only one half of a typ-

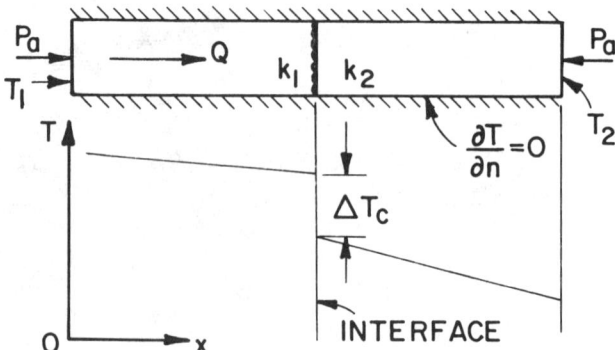

Fig. 3  Temperatures in contacting solids.

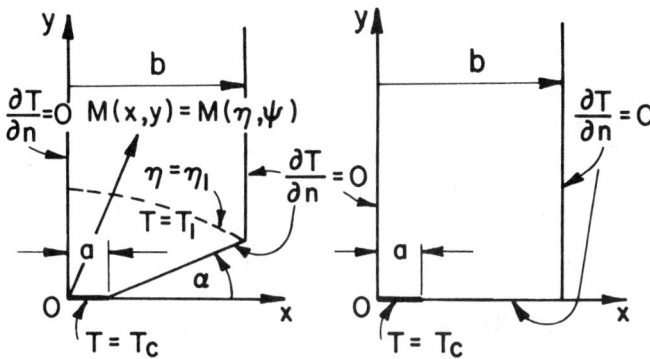

Fig. 4 Heat channel models.

ical heat channel is shown because of temperature symmetry about the oy-axis. The required temperature field must satisfy the following differential equation in Cartesian coordinates as well as the boundary conditions:

$$(\partial^2 T/\partial x^2) + (\partial^2 T/\partial y^2) = 0 \qquad (1)$$

$$T = T_c \quad y = 0 \quad 0 < x < a \qquad (2)$$

$$\partial T/\partial n = \cos\alpha(\partial T/\partial y) - \sin\alpha(\partial T/\partial x) = 0, \quad 0 < y < (b-a)\tan\alpha$$

$$a < x < b \qquad (3)$$

$$\left.\begin{array}{l} \partial T/\partial x = 0 \quad x = 0 \\ \\ \partial T/\partial x = 0 \quad x = b \end{array}\right\} \quad 0 < y < \infty \qquad (4)$$

$$\partial T/\partial y \to -Q/k2b \quad y \to \infty \qquad (5)$$

Equations (2) and (3) are the mixed boundary conditions specified over the contact area and the surface outside the contact area. The uniform temperature is prescribed over the contact, while a zero heat flux in the normal direction is prescribed over the remainder of the apparent contact area. Both Mikic[3] and Veziroglu[7] analyzed the simple case where the contact angle is zero ($\alpha = 0$). The theory presented in this paper can treat the more general case of $\alpha \ne 0$ and yields an expression for the constriction resistance for $\alpha = 0$ which is superior to those expressions developed by Mikic and Veziroglu.

The thermal problem as stated above can be reformulated in terms of elliptic-cylinder coordinates ($\eta,\psi$) if one uses the following transformation equations:

$$x = a \cosh\eta \cos\psi \quad y = a \sinh\eta \sin\psi \tag{6}$$

where a is the half-width of the contact area assumed to be a very long rectangular strip. The parameter $\eta$ determines the elliptic isothermal surfaces in the neighborhood of the contact area, and $\eta = 0$ represents the isothermal contact area while $\eta = \eta_1 > 0$ represents the elliptic isothermal surface located far from the contact area, (Fig. 4). The parameter $\psi$ is an angular measure determined from $\psi = \arctan(y/x)$. Employing the transformations of Eqs. (6), Eq. (1) transforms to

$$d^2T/d\eta^2 = 0 \tag{7}$$

where the temperature field depends only upon one parameter $\eta$. The original differential equation has been transformed into a much simpler equation and its solution is

$$T = C_1\eta + C_2 \tag{8}$$

and the corresponding boundary conditions are

$$T = T_c \quad \eta = 0 \tag{9}$$

$$T = T_1 < T_c \quad \eta = \eta_1 \tag{10}$$

The mixed boundary conditions which are difficult to satisfy when Cartesian coordinates are used, are automatically satisfied when elliptic-cylinder coordinates are employed.

Upon substitution of Eqs. (9) and (10) into Eq. (8) and after evaluating the two constants of integration, one obtains the following expression for the temperature distribution in the region of interest:

$$(T_c - T)/(T_c - T_1) = \eta/\eta_1 \tag{11}$$

This is a simple linear temperature distribution in terms of $\eta$, but a complex two-dimensional field in terms of x and y. Equation (11) can also be written in closed form as

$$\frac{T_c - T}{T_c - T_1} = \frac{\cosh^{-1}(x/a \cos\psi)}{\cosh^{-1}(b/a \cos\psi)} \tag{12}$$

where the first expression of Eq. (6) was used to obtain the relationship between $\eta$ and $\psi$.

Equation (12) will now be used to evaluate the average interface temperature which is defined as

$$T_a = \frac{1}{b}\int_0^a T dx + \frac{1}{b}\int_a^b T dx \qquad (13)$$

After substitution of Eq. (12) in Eq. (13) the average interface temperature is found to be

$$T_a = T_c - \frac{(T_c - T_1)}{b\cosh^{-1}(b/a\cos\psi)}\int_a^b \cosh^{-1}(x/a\cos\psi)dx \qquad (14)$$

or

$$T_a = T_c - \frac{(T_c - T_1)a\cos\psi}{b\cosh^{-1}(b/a\cos\psi)}\int_{1/\cos\psi}^{b/a\cos\psi} \cosh^{-1} u\, du \qquad (15)$$

where $u = x/a\cos\psi$. It should be noted that the average interface temperature depends upon the contact area temperature, the temperature difference between the contact area and an isothermal surface located at the far boundary of the region of interest as well as certain geometric characteristics.

The effective temperature difference which is required to overcome the thermal constriction resistance will be defined as

$$T_c - T_a = \frac{(T_c - T_1)a\cos\psi}{b\cosh^{-1}(b/a\cos\psi)}\int_{1/\cos\psi}^{b/a\cos\psi} \cosh^{-1} u\, du \qquad (16)$$

The steady heat flow rate through the heat channel can be evaluated at the contact area

$$Q = 2\int_0^\ell \int_0^a -k\frac{\partial T}{\partial y}(x,0)\, dx dz \qquad (17)$$

where $\ell$ is the effective length of the heat channel measured into the paper. By means of Eq. (6), Eq. (17) can be transformed into an expression dependent upon $\eta$ and $z$:

$$Q = 2\int_0^\ell \int_\alpha^{\pi/2} -k\frac{\partial T}{\partial \eta}(\eta = 0)\, d\psi dz \qquad (18)$$

where $\alpha$ is the contact angle, Fig. 4. Eq. (18) can be integrated to yield an expression for the total heat flow rate

$$Q = \frac{2k\ell(\pi/2 - \alpha)(T_c - T_1)}{\cosh^{-1}(b/a\cos\alpha)} \qquad (19)$$

The thermal constriction resistance of a heat channel is defined as the effective temperature difference divided by the total heat flow rate. Therefore,

$$R_c = (T_c - T_a)/Q \tag{20}$$

and after substitution of Eqs. (16) and (19) into (20) one obtains

$$R_c = \frac{(a\ \cos\alpha/b)}{k\ell\pi(1 - 2\alpha/\pi)} \int_{1/\cos\alpha}^{b/a\ \cos\alpha} \cosh^{-1} u\ du \tag{21}$$

where k is the thermal conductivity and the geometric characteristics of the interface are $a, b, \alpha$, and $\ell$. Equation (21) can be written as

$$R_c = \Psi/k\ell \tag{22}$$

where the dimensionless geometric factor $\Psi$ defined as

$$\psi = \frac{(a\ \cos\alpha/b)}{\pi(1 - 2\alpha/\pi)} \int_{1/\cos\alpha}^{b/a\ \cos\alpha} \cosh^{-1} u\ du \tag{23}$$

is seen to depend only upon $\alpha$ and $a/b$.

Equation (22) is the total thermal constriction resistance of a typical heat channel. The resistance is a function of the thermal conductivity as well as certain geometric characteristics.

## Thermal Resistance of Multiple Channels in Parallel

The total thermal resistance of the spiral arrangement shown in Fig. 1 will now be modelled as a number of concentric circular heat channels thermally connected in parallel. The constriction resistance of the ith channel is

$$R_{ci} = \Psi_i/k_1 \ell_i \tag{24}$$

where $k_1$ is the thermal conductivity of the solid which has been separated into N heat channels. $\ell_i$ is the effective length of the heat channel and $\Psi_i$ is the corresponding geometric factor. The total resistance for solid 1 is, therefore,

$$1/R_{tc1} = \sum_{i=1}^{N} 1/R_{ci} = k_1 \sum_{i=1}^{N} \ell_i/\Psi_i \tag{25}$$

when N heat channels are connected in parallel. If the heat channels are geometrically similar, the geometric factor $\Psi_i$ will be the same for each heat channel, and Eq. (25) becomes

$$1/R_{tc1} = (k_1/\Psi_1) \sum_{i=1}^{N} \ell_i \qquad (26)$$

where $\Psi_1$ is now the geometric factor corresponding to solid 1 and

$$\sum_{i=1}^{N} \ell_i$$

is the total effective length of the contact.

### Effective Length of Contact

In order to evaluate the total resistance given by Eq. (26), it is necessary to relate the total effective length of the contact to the apparent contact area and the spacing between adjacent spirals. Since the contact has been modeled as concentric circular heat channels with common spacing $\delta$, (Fig. 5) the effective length of each channel, starting from the center of the contact area, is

$$\begin{aligned} \ell_1 &= \pi d_1 \\ \ell_2 &= \pi d_1 + 2\pi\delta \\ &\vdots \\ \ell_i &= \pi d_1 + 2\pi(i-1)\delta \end{aligned} \qquad (27)$$

where $d_1$ is the average diameter of the innermost spiral and it is of the order of $\delta$. The total effective length of the contact is the sum of all the heat channel lengths

$$\sum_{i=1}^{N} \ell_i = \pi\{Nd_1 + 2\delta \sum_{i=1}^{(N-1)} i\} \qquad (28)$$

The second term within the bracket can be summed and it is equal to

$$\delta N(N-1) \qquad (29)$$

The total number of heat channels, to a first approximation, is $D/2\delta$ where $D$ is the diameter of the apparent contact area.

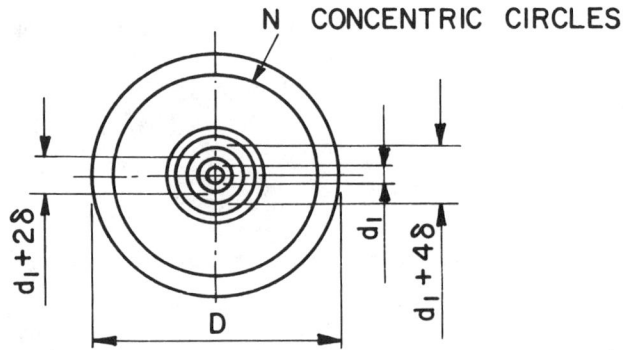

Fig. 5  Contact model for conductance theory.

Upon substitution of Eq. (29) in Eq. (28) we obtain as the total effective length of contact

$$\sum_{i=1}^{N} \ell_i = \pi[Nd_1 - N\delta + \frac{D^2}{4\delta}] = \frac{\pi D^2}{4\delta} \qquad (30)$$

where $\pi D^2/4$ is the apparent contact area.

## Thermal Contact Conductance of the Spiral

The total contact resistance, Eq. (26), can now be written as

$$1/R_{tc1} = (k_1/\Psi_1 \delta)(\pi D^2/4) \qquad (31)$$

The definition of contact resistance, $R_c = \Delta T_c/Q$, and that of contact conductance, $h_c = \Delta T_c/(Q/A_a)$, allows one to equate the resistance and conductance in the following manner:

$$h_c = 1/(R_c A_a) \qquad (32)$$

Thus Eq. (32) with Eq. (31) shows that the thermal contact conductance of N heat channels in solid 1 can be written as

$$h_{c1} = k_1/\Psi_1 \delta \qquad (33)$$

A similar expression can be written for the second set of N heat channels in solid 2.

The over-all contact conductance for two sets of N dissimilar heat channels thermally connected in series is

$$1/h_{cw} = 1/h_{c1} + 1/h_{c2} = \delta[\Psi_1/k_1 + \Psi_2/k_2] \qquad (34)$$

where $\Psi_1$ and $\Psi_2$ are the geometric factors, defined by Eq. (23), corresponding to solid 1 and solid 2, respectively. The additional subscript w will be used to denote thermal conductance due to the spirals under ideal conditions, i.e., when the spirals or strips are continuous (all the contact spots have merged).

## Contact Model

If one is to be able to use Eq. (34) to predict the thermal conductance of an ideal turned surface, it is necessary to have a relationship between the geometric parameter a/b, the apparent contact pressure and the maximum yield pressure of the softer turned surface. If one assumes that during the first compression cycle the contacting ridges undergo plastic deformation, then one can write the following simple expression which is a force balance at the interface:

$$2aLP_m = (\pi D^2/4)P_a \qquad (35)$$

where 2aL is the contact area formed because of plastic deformation of the spirals, $P_m$ is the maximum yield pressure and $P_a$ is the apparent contact pressure. This simple model will be a good approximation of the real situation for the first loading cycle.

Replacing L, the total effective length of the contact area, by Eq. (30), one obtains

$$a/b = P^* \qquad (36)$$

where $P^* = P_a/P_m$. The effective contact width may be smaller than that predicted by Eq. (36), and in order to take this into account, write

$$a/b = P^*/\gamma \qquad (37)$$

where $\gamma \geq 1$ depending upon how closely the contact model agrees with the real situation. Equation (37) can now be substituted into Eq. (23) to obtain a relationship between conductance, applied load and material strength.

## Over-all Conductance Including Surface Roughness

All surfaces possess surface roughness prior to the turning process. This roughness will not change during the turning process or will be altered somewhat (increased). The ridges formed during turning will, therefore, not be perfect and will not form a continuous line during the initial contact. The contact will consist of discrete contact spots formed on a line corresponding to the lay of the ridges when the load is very light. The discrete contact spots will be relatively far apart and, therefore, the total conductance will be influenced by the surface roughness. The presence of the contact spots within the spiral contact and their influence on the over-all conductance can be taken into consideration by superposing[8] upon Eq. (34) the conductance due to surface roughness. For light and moderate loads an expression such as the one developed by Cooper, Mikic and Yovanovich[5] is recommended.

$$h_{cr} = C(k_m/\sigma)(P^*)^{0.985} \qquad (38)$$

In Eq. (38), C is a geometric parameter depending upon the slope of the contacting asperities (surface roughness). Typical values of $C^{12}$ are 0.036, 0.175 and 0.290 for lapped surfaces, average rough and very rough flat surfaces, respectively. The other parameters appearing in Eq. (38) are: $\sigma$, the standard deviation of the profile heights of the asperities; $k_m$, the harmonic mean thermal conductivity; and $P^*$, the dimensionless contact pressure.

As the contact pressure becomes very large, more contact spots appear and the characteristic distance between them decreases becoming zero for many contact spots. At these very high loads most of the contact spots will have merged to form a quasi-continuous spiral. At these high contact pressures Eq. (34) should be adequate, but for the lower pressures the overall conductance will be more accurately predicted by

$$1/h_c = (1/h_{cw}) + (1/h_{cr}) \qquad (39)$$

provided that Eq. (38) can predict the roughness conductance at moderately high pressures.

Comparison of Theory and Some Experimental Data

The theory presented in this paper will be compared with the results of the experimental work preformed at the University of Poitiers, France. Under the direction of Cordier, Roiron,[9] Bardon,[10] and Fouché[11] systematically studied various aspects of heat transfer across the interface formed by the contact of a smooth flat surface and a softer turned surface. In their experimental work conducted under ambient and vacuum conditions, they sought an empirical correlation for the thermal contact conductance. The one correlation by Fouché[11] developed for ambient as well as vacuum conditions failed to predict the thermal conductance with any degree of accuracy. The vacuum test data of Bardon[10] will be compared with the present theory. In his work Bardon placed stainless steel (18-8) into contact with an alloy of magnesium-zirconium (0.7% zirconium). The stainless steel cylinder was prepared in the following manner: A collar fabricated from identical material was placed around the cylinder; then both were made as flat as possible by turning. After turning, the pieces were lapped and finally polished. After these operations the collar was removed and the stainless steel surface was observed to be smooth and optically flat. The softer material was made flat by turning it on a lathe such that the depth of cut was just sufficient to eliminate all the high spots. During this operation the tool advance was 0.02 mm. The surface roughness was measured to be about $12 \times 10^{-6}$ inches (rms). After this initial flattening process the alloy material was again turned on a lathe using a tool steel cutter having an angle of 60°. The cylinders were turned at 250 rpm. The tool advance and depth of cut were: a) 0.5 mm and 0.05 mm; b) 0.25 mm and 0.05 mm; and finally c) 0.125 mm and 0.025 mm, respectively, for the three interfaces studied. The diameter of all these pieces was 25.4 mm.

The thermal tests were conducted in a vacuum system similar to systems used by other investigators[2-4] and the technique used was the same. The tests were done in a vacuum of $10^{-4}$ mm Hg. The system was maintained at a pressure of $10^{-4}$ mm Hg for four days prior to the actual testing so that all surfaces were thoroughly de-gassed. The heat flux based upon the apparent contact area was about 5000 BTU/hr.sq.ft. The load on the interface ranged from about 150 to 1500 psi. The pseudotemperature drop across the interface ranged from a low value of 0.5°F corresponding to the high loads up to a maximum value of about 6°F corresponding to the light loads. Temperature readings were taken every 45 minutes and the calculated thermal conductance ranged from about 800 to 12,000 BTU/hr.sq.ft. °F. Fouché conducted hardness tests and found the yield pressure of magnesium-zirconium to be 44,000 psi.

To facilitate the comparison of this theory with the test data of Bardon, Eq. (39) has been written in the following dimensionless form:

$$(\delta h_c/k_m)^{-1} = \pi^{-1}(k_m/k_1)[\ln(2\gamma/P^*) - 1]$$
$$\times [1+(k/k_2)/(1-2\alpha/\pi)]+C^{-1}(\sigma/\delta)(1/P^*)^{0.985} \quad (40)$$

For the interfaces investigated by Bardon, Eq. (40) becomes

$$(\delta h_c/k_m)^{-1} = 0.68[\ln(2\gamma/P^*) - 1]$$
$$+ (2.78 \times 10^{-5}/\delta)(1/P^*)^{0.985} \quad (41)$$

The geometric and physical characteristics of the contacting surfaces used in the test program are listed in Table 1.

Table 1 Geometric and physical characteristics of the surfaces

| Stainless steel | Magnesium-zirconium |
|---|---|
| $k_1$ = 10 BTU/hr ft °F | $k_2$ = 59 BTU/hr ft °F |
| $\sigma \doteq 0$ | $\sigma \doteq 10^{-6}$ ft |
| $\alpha \doteq 0$ | $\alpha \doteq 30°$ |
| $P_m$ = 360,000 psi | $P_m$ = 44,000 psi |
| | $\delta_1$ = 1/610 ft |
| | $\delta_2$ = 1/1220 ft |
| | $\delta_3$ = 1/2440 ft |

A comparison between experimental results and the predicted values calculated by means of Eq. (34) with $\gamma$ = 1 and by means of Eq. (41) with $\gamma$ = 1 are presented in (Figs. 6-8). Equation (34) is based upon the assumption that there are no surface roughness effects, while Eq. (41) takes into consideration roughness effects. The first interface tested $\delta_1$ (Fig. 6) had a total spiral length of about 3.3 ft and could be approximated

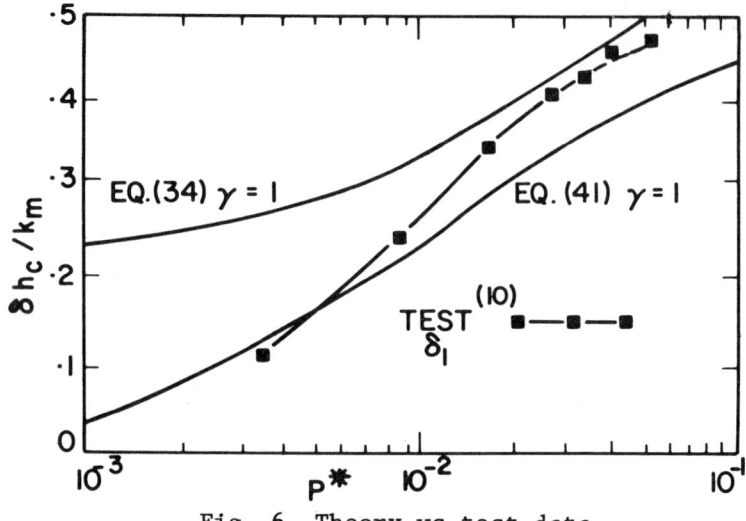

Fig. 6 Theory vs test data.

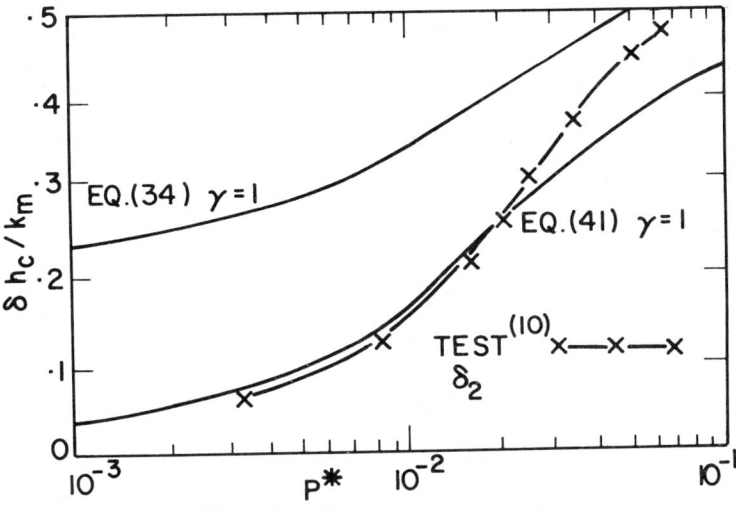

Fig. 7 Theory vs test data.

by 25 heat channels connected in parallel. There is good agreement between Eq. (34) and test data in the load range $P^* = 1.2 \times 10^{-2}$ to $5.2 \times 10^{-2}$, whereas in the load range $P^* = 3 \times 10^{-3}$ to $1.2 \times 10^{-2}$ there is good agreement between Eq. (41) and the data. The second interface tested $\delta_2$ (Fig. 7) had a total spiral length of about 6.6 ft and could be approximated by 50 heat channels in parallel. There is excellent agreement between Eq. (41) and the data from $P^* = 3.4 \times 10^{-3}$ up to $P^* = 3.5 \times 10^{-2}$, and then the test data falls on the values predicted by Eq. (34). The last interface tested $\delta_3$ (Fig. 8) had a total spiral length of about 13.2 ft and could be approximated by 100 heat channels in parallel. For this interface there

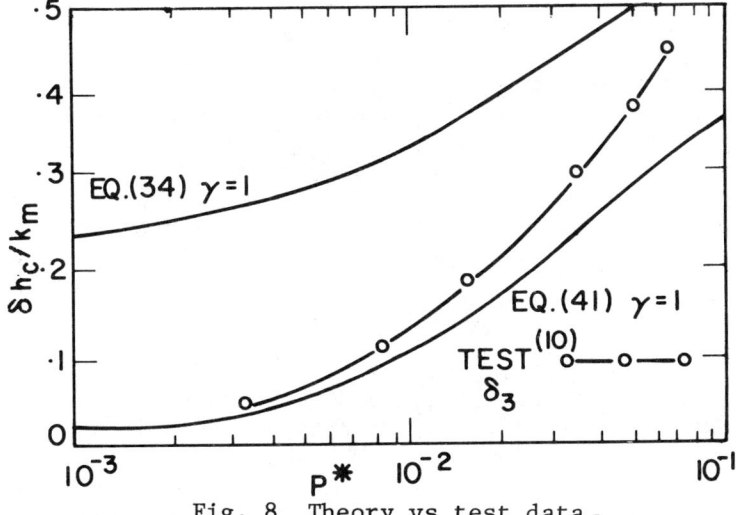

Fig. 8 Theory vs test data.

is very good agreement between Eq. (41) and test data in the load range $P^* = 3.3 \times 10^{-3}$ to $3 \times 10^{-2}$, then the test data tends towards those values predicted by Eq. (34). At the highest load there is less than 10% difference between the data and Eq. (34). From these Figures it would seem that at light loads the contact area behaves like individual contact spots lined up along a spiral while at high loads the contact area appears to behave as a quasi-continuous spiral. Equation (41) adequately predicts the thermal conductance at the light and moderate loads but fails at the high loads. Equation (34), on the other hand, can accurately predict the thermal conductance at the high loads.

## Conclusions

A theory has been presented which results in two expressions which give the upper and lower bounds for the thermal contact conductance at interfaces formed by smooth and turned surfaces. The assumptions upon which the theory is based appear to be valid for the available test data. The contact area at light loads was modelled as discrete contact spots so arranged on the apparent area of contact that they appear to form a quasi-continuous spiral. For this arrangement the overall conductance will depend upon both the contact spot conductance and the spiral conductance. This model is a good approximation if dimensionless load pressures are less than $3 \times 10^{-2}$. At high pressures the contact spots are very close together and many have merged, and they collectively behave as a continuous spiral. For this arrangement, surface roughness effect is negligible, and the expression based upon a continuous spiral adequately predicts the conductance. This model appears to be valid for dimensionless load pressures exceeding $5 \times 10^{-2}$.

Future work will compare the theory against more test data to be obtained by this investigator. The expression developed by Cooper et al.[5] will be extended or modified to include various shaped contact spots situated close together. It is hoped that an expression can be developed which better agrees with the test data in the intermediate load range from $3 \times 10^{-2}$ to $5 \times 10^{-2}$.

## References

1. Centinkale, T.N. and Fishenden, M., "Thermal Conductance of Metal Surfaces on Contact", International Conference of Heat Transfer, Inst. of Mechanical Engineering, London, pp 271-275.

2. Clausing, A.M., "Some Influences of Macroscopic Constric-

tions on the Thermal Contact Resistance", Rept. ME-TN-2422, 1965, NASA.

3. Mikic, B.B. and Rohsenow, W.M., "Thermal Contact Resistance", Rept. 4542-41, 1966, MIT, Cambridge, Mass.

4. Yovanovich, M.M. and Fenech, H., "Thermal Contact Conductance of Nominally Flat, Rough Surfaces in a Vacuum Environment", AIAA Progress in Astronautics and Aeronautics: Thermophysics and Temperature Control of Spacecraft and Entry Vehicles, Vol. 18, edited by G.B. Heller, Academic Press, New York, 166, pp 773-794.

5. Cooper, M.G., Mikic, B.B. and Yovanovich, M.M., "Thermal Contact Resistance", International Journal of Heat and Mass Transfer, Vol. 12, 1969, pp 279-300.

6. Tien, C.L., "A Correlation for Thermal Contact Conductance of Nominally Flat Surfaces in a Vacuum, Thermal Conductivity," Proceedings of the Seventh Conf. National Bureau of Standards, edited by D.R. Flynn and B.A. Peavy, Jr., 1968, pp 755-759.

7. Veziroglu, T.N. and Chandra, S., "Thermal Conductance of Two-Dimensional Constriction", AIAA Progress in Astronautics and Aeronatics: Thermal Design Principles of Spacecraft and Entry Bodies, Vol. 21, edited by J.T. Bevans, Academic Press, New York, pp 591-617.

8. Yovanovich, M.M. "Overall Constriction Resistance Between Contacting Rough, Wavy Surfaces", International Journal of Heat and Mass Transfer, Vol. 12, 1969, pp 1517-1520.

9. Roiron, G., Thesis (Docteur Ingenieur), 1964, University of Poitiers, France.

10. Bardon, J.P., Thesis (Docteur es Sciences), 1965, University of Poitiers, France.

11. Fouche, F., Thesis (Docteur es Sciences), 1966, University of Poitiers, France.

12. Dyachenko, P.E., Tolkacheva, N.N., Andreev, G.A. and Karpova, T.M., The Actual Contact Area between Touching Surfaces, Consultants Bureau, New York, 1964.

# THERMAL CONDUCTANCE OF A ROW OF CYLINDERS CONTACTING TWO PLANES

M. Michael Yovanovich[*]

University of Waterloo, Ontario, Canada

## Abstract

Analyses are presented for predicting the overall heat transfer coefficient for steady heat transfer between two smooth planes separated by a single row of a) uniformly spaced and b) nonuniformly spaced circular cylinders under vacuum conditions and negligible radiation heat transfer. The analyses incorporate the Hertzian theory as well as the thermal constriction theories of Veziroglu, valid for two-dimensional, rectangular heat channels, and of Yovanovich-Coutanceau, valid for circular cylinders, to determine the total thermal constriction resistance of a typical element. General expressions are developed relating the conductance to the thermal conductivities, modulii of elasticity, Poisson's ratios, cylinder diameter, spacing and apparent contact pressure. There is good agreement between the theory and some experimental data.

## Nomenclature

| | | |
|---|---|---|
| $a$ | = | half-width of contact area, Eq. (5) |
| $A_a$ | = | apparent contact area |
| $b$ | = | half-width of heat channel |
| $D$ | = | cylinder diameter |
| $e$ | = | modulus of elasticity |
| $f$ | = | force per cylinder per unit length |
| $F$ | = | total force on the joint |
| $h_j$ | = | thermal conductance of the joint, Eq. (1) |
| $k$ | = | physical property parameter, Eq. (6) |
| $L$ | = | length of cylinders and apparent area |
| $N$ | = | number of cylinders |
| $P_a$ | = | apparent contact pressure |

---

Presented as Paper 71-436 at the AIAA 6th Thermophysics Conference, Tullahoma, Tenn., April 26-28, 1971. The author acknowledges the financial support of the National Research Council of Canada

[*]Associate Professor of Mechanical Engineering

| | | |
|---|---|---|
| $P^*$ | = | dimensionless contact pressure, Eq. (15) |
| $Q$ | = | total heat flow rate |
| $R$ | = | thermal resistance |
| $S$ | = | pitch of cylinders, $S = \alpha D$ |
| $T$ | = | temperature |
| $\Delta T_j$ | = | pseudo-temperature drop across joint |
| $W$ | = | width of apparent area |
| $\alpha$ | = | dimensionless spacing, $\alpha = S/D$ |
| $\lambda$ | = | thermal conductivity |
| $\alpha^*$ | = | dimensionless spacing ratio, $\alpha^* = \alpha_1/\alpha_2$ |
| $\theta$ | = | contact angle |
| $\nu$ | = | Poisson's ratio |

Subscripts

| | | |
|---|---|---|
| 1,2,3 | = | planes 1 and 2, and cylinder, respectively |
| a | = | apparent |
| j | = | joint |

## Introduction

Heat transfer across a joint consisting of a single row of long smooth circular cylinders in elastic contact with two smooth planes (Fig. 1) is currently of interest not only to aerospace engineers but also to cryogenic engineers. The aerospace engineer often is called upon to predict the total thermal resistance present in the available heat path from the heat source (usually some precision electromechanical instrument) to the heat sink. Quite often these precision instruments are mounted on gimbaled platforms, and, therefore, heat must be conducted across roller bearings which can be modeled as a row of solid cylinders contacting two planes. The design engineer must be able to calculate that fraction of the total resistance which is due to the bearings. Once he knows the parameters (geometric, physical, and thermal) which determine this bearing resistance, he can then make a proper engineering decision to alleviate it. On the other hand the cryogenic engineer is concerned about heat leakage into the cryogens. The path available for the heat leakage often includes stand-offs which can be modeled as Fig. 1.

A survey of the open literature shows that this problem has not been solved. It will be assumed that the cylinders have a length-to-diameter ratio of at least two, but preferably larger. Under vacuum conditions this will permit us to neglect end effects and so model the cylinders as being very long. The cylinder surface as well as the surfaces of the two planes are taken to be smooth so that the Hertzian theory can be used to predict the size of contact. This is not a severe

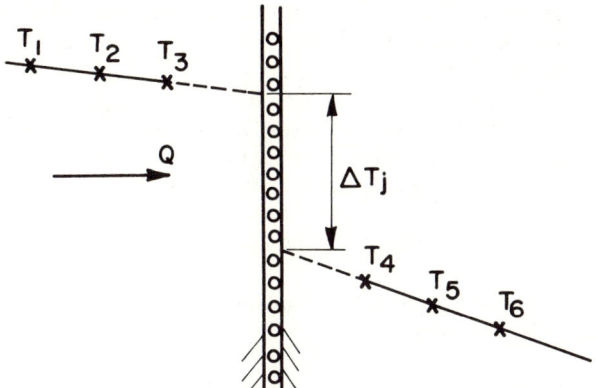

Fig. 1 Temperature field.

restriction, provided the actual surface conditions are relatively smooth and the load is relatively large.

The surfaces are assumed to be clean, free of oxides or lubricants. Neglecting oxides does not negate the results of this analysis. Provided the oxide layer does not alter the contact characteristics and its thermal properties are known, the effect of an oxide layer always can be added to the final results of this analysis. The presence of a lubricant changes completely the nature of the problem. Cylinders surrounded by a lubricant must be treated differently. This paper will be limited to the study of thermal constriction resistances only.

## Statement of the Problem

Identical circular cylinders are placed between two large plane solids. The cylinders are parallel to each other, and the pitch is uniform. When the smooth plane solids make elastic contact with the cylinders, the contact areas are very narrow strips located diametrically opposite each other. When the planes have different elastic properties, the contact areas will differ in their widths. As the contact load increases, the contact areas increase in width, but the width will always remain much smaller than the diameter of the cylinders.

When the joint described above is placed in a vacuum and there is steady linear heat conduction Q in both solids (Fig.1) a pseudo-temperature drop $\Delta T_j$ will be observed across the joint. This temperature drop is a direct measure of the thermal resistance of the joint. This resistance is the result of the convergence and divergence of the heat flow lines in the neighborhood of the contact areas. For joints placed in a vacuum, and if there is negligible radiation heat transfer across the no contact regions, there is only one path available for heat transfer across the joint - that is, conduction through the contact areas.

The objective, then, is to develop a general expression for predicting the joint thermal resistance or its reciprocal, the thermal conductance, which is defined as

$$h_j = (Q/Aa)/\Delta T_j \qquad (1)$$

where $(Q/Aa)$ is the heat flux based upon the apparent or nominal area of the joint.

The thermal problem will be treated as N typical heat channels (Fig. 2) thermally connected in parallel. Each heat channel will be subdivided into subelements: a cylinder and two rectangular heat channels (Fig. 3). Each rectangular heat channel will have a constriction resistance associated with it, and the cylinder will have two constrictive resistances because it has two different contact areas. These four constriction resistances will be thermally connected in series because all the heat entering a typical heat channel must flow through each subelement. The total resistance of a channel is, therefore, the sum of the four constriction resistances.

Total Resistance of a Typical Heat Channel

Figure 3 shows schematically the subelements used to determine the total resistance of a typical heat channel. Only one rectangular heat channel is shown because they are geometrically similar except for the contact width. Veziroglu[1] has shown that the symmetric constriction resistance of a rectangular channel of half-width b thermal conductivity $\lambda_1$, and half-width

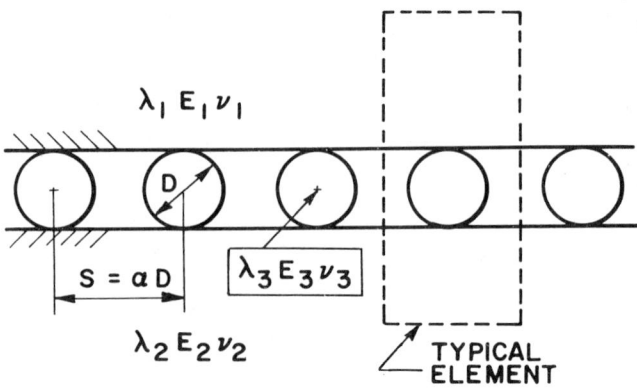

Fig. 2 Typical element for thermal analysis 1.

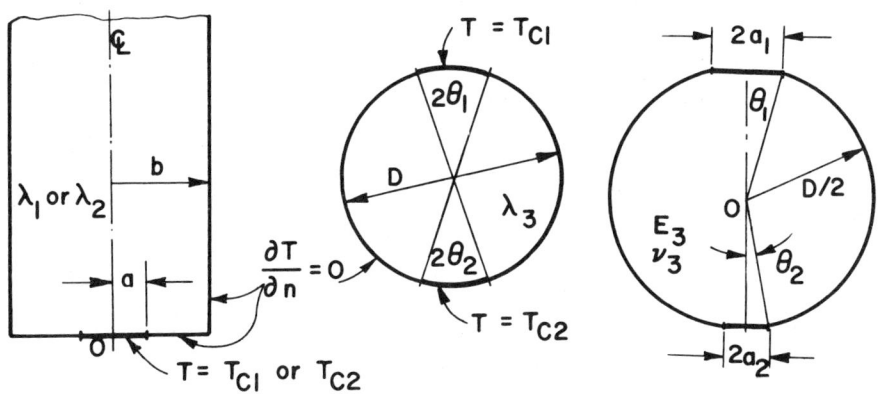

Fig. 3   Schematic of subelements.

of contact $a_1$, is given by

$$R_1 = (\pi\lambda_1)^{-1} \ln[1/\sin(\pi a_1/2b)] \tag{2}$$

per unit length of heat channel. A similar expression can be written for the resistance in the second rectangular channel where the half-width of contact is $a_2$ and the thermal conductivity is $\lambda_2$. Yovanovich and Coutanceau[2] recently determined the thermal resistance of a circular cylinder having identical sources and sinks placed diametrically opposite each other. The remainder of the boundary of the cylinder is perfectly insulated. The total resistance consists of two identical resistances: the resistance from the source to the midplane of the cylinder and the resistance from the midplane to the sink. The results of this analysis can be used to obtain the following expression:

$$R_3 = (\pi\lambda_3)^{-1} \ln[2(1+\cos\theta_1)/\sin\theta_1] + (\pi\lambda_3)^{-1} \ln[2(1+\cos\theta_2)/\sin\theta_2] \tag{3}$$

for the total constriction resistance across the cylinder when the heat enters through the contact area subtending an angle $\theta_1$ at the center of the cylinder and leaves through a second contact area subtending an angle $\theta_2$ (Fig. 3). The remainder of the cylinder boundary is impervious to heat flow.

The total constriction resistance of a typical heat channel is, therefore,

$$R = R_1 + R_2 + R_3 \tag{4}$$

### Elastic Contact Between Cylinder and Plane

Elasticity theory[3] shows that the half-width of the contact area between the plane solid 1 and the cylinder is given by

$$a_1 = \sqrt{2fD(k_1 + k_3)} \qquad (5)$$

$k_1$ and $k_3$ are physical parameters given by

$$k_i = (1 - \nu_i^2)/\pi E_i \quad (i = 1,2,3) \qquad (6)$$

A similar expression can be written for the half-width of the contact area between the cylinder and plane solid 2.

### Evaluation of Geometric Parameters

Consider the geometric parameter $(a/b)$ that appears in Eq. (2). This parameter can be related to the pitch $S = \alpha D$ and the cylinder diameter in the following manner:

$$a_1/b = a_1/(S/2) = 2a_1/\alpha D \qquad (7)$$

By means of Eq. (6) the ratio in Eq. (7) becomes

$$a_1/b = (2/\alpha)\sqrt{2f(k_1 + k_3)/D} \qquad (8)$$

The geometric parameter $(a/b)$ is always much less than unity – that is, the half-width of the contact areas are always much smaller than the half-width of the heat channels even when $\alpha = 1$. Therefore, for small values of $(a/b)$ $\sin(\pi a/2b)$ can be approximated by $(\pi a/2b)$ with very little error. This result will be used in the next section.

Since the half-width of the contact areas are much smaller than the cylinder diameters (Fig. 3) the subtended angles are very small and therefore the geometric parameter appearing in Eq. (3) can be approximated in the following manner:

$$(1 + \cos\theta_1)/\sin\theta_1 \doteq (D/a_1)[1 - (a_1/D)^2/2] \doteq D/a_1 \qquad (9)$$

### Total Joint Resistance and Conductance

The results of the preceding section can be used to evaluate the total joint resistance in terms of the physical and geometric properties of the joint. The total constriction resistance of a typical element, Eq. (4), can now be written as

$$R = (\pi\lambda_1)^{-1}\ln[\alpha/\pi\sqrt{2f(k_1+k_3)/D}] + (\pi\lambda_2)^{-1}\ln[\alpha/\pi\sqrt{2f(k_2+k_3)/D}] +$$

$$(\pi\lambda_3)^{-1}\ln[2/\sqrt{2f(k_1+k_3)/D}] + (\pi\lambda_3)^{-1}\ln[2\sqrt{2f(k_2+k_3)/D}] \quad (10)$$

per unit length of cylinder. Since the typical heat channels are thermally connected in parallel, the total joint resistance is $R_j = R/LN$ and the number of cylinders forming the joint is $N = W/S = W/\alpha D$. It can be shown that the following relationship exists between the joint resistance and the typical heat channel resistance:

$$R_j = \alpha DR/Aa \quad (11)$$

Since, $Q = h_j\, Aa\, \Delta T_j = \Delta T_j/R_j$, therefore,

$$\alpha D\, h_j = L/R \quad (12)$$

where R is given by Eq. (10).

Equation (10) is not in its simplest form. It can be reduced to a more convenient form if we note that the force on each cylinder per unit length can be written as

$$f = P_a\, Aa/NL = \alpha D\, P_a \quad (13)$$

Therefore, we can replace $2f/D$ by $2\alpha\, P_a$ and write Eq. (10) as

$$R = (2\pi)^{-1}[1/\lambda_1 + 1/\lambda_2 - 2/\lambda_3][\ln(\alpha/2)] -$$
$$\pi^{-1}[1/\lambda_1 + 1/\lambda_2]\ln\pi + (2\pi)^{-1}[1/\lambda_1 + 1/\lambda_3]\ln(1/P^*_{13}) +$$
$$(2\pi)^{-1}[1/\lambda_2 + 1/\lambda_3]\ln(1/P^*_{23}) \quad (14)$$

where $P^*_{13} \equiv P_a(k_1+k_3)$ and $P^*_{23} \equiv P_a(k_2+k_3) \quad (15)$

Equation (14) clearly shows the relationship between the total constriction resistance of a typical heat channel and the physical and geometric properties of the joint. For the special case of a joint consisting of identical materials ($\nu_1 = \nu_2 = \nu_3 = \nu$; $E_1 = E_2 = E_3 = E$; $\lambda_1 = \lambda_2 = \lambda_3 = \lambda$), Eq. (14) reduces to

$$R = (2/\pi\lambda)\ln[E/2\, P_a(1-\nu^2)] \quad (16)$$

and the thermal conductance, according to Eq. (12), can be written in the following dimensionless form:

$$(\alpha\, D\, h_j/\lambda) = (\pi/2)/\ln[E/2\, P_a(1-\nu^2)] \quad (17)$$

## Effect of Eccentricity

An occasion may arise where the cylinders are not placed upon a uniform pitch (Fig. 4). The pitch between adjacent cylinders may differ, but the two pitches will be repeated alternately. For this arrangement a typical heat channel can be taken as shown in Fig. 4. It will be noted that this heat channel exists for every cylinder. Furthermore, it will be

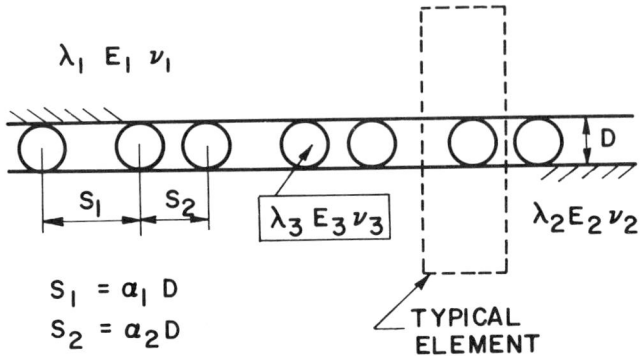

Fig. 4 Typical element for thermal analysis II.

noted that the center of the cylinder does not coincide with the centerline of the rectangular heat channel. The cylinder is displaced a distance e from the centerline of the heat channel (Fig. 5). The eccentricity will augment the constriction resistance within the rectangular heat channels but will not influence the constriction resistance within the cylinders. This effect of eccentricity has been considered by Veziroglu[1]. He found that the expression for the symmetric constriction, Eq. (2), should be altered to the following form to take into consideration the displacement from the symmetric position:

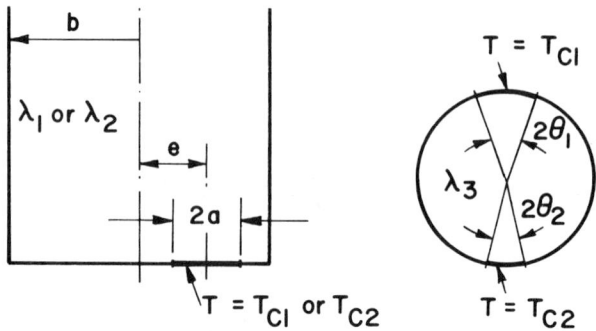

Fig. 5 Schematic of subelements.

$$R_1 = (\pi\lambda_1)^{-1} \ln[1/\cos(\pi e/2b) \sin(\pi a_1/2b)] \qquad (18)$$

It can be easily proven by means of Fig. 4 that

$$(\pi e/2b) = (\pi/2)(\alpha_1 - \alpha_2)/(\alpha_1 + \alpha_2) \qquad (19)$$

where $\alpha_1 > a_2$.

Without going into all the details of this analysis, which follows closely the work done in the previous sections, it can be stated that, for the case of eccentricity, the total constriction resistance of a typical heat channel can be expressed as

$$\pi R = (1/\lambda_1 + 1/\lambda_2)\{\ln(\sec[(\pi/2(\alpha_1 - \alpha_2)/(\alpha_1 + \alpha_2)]/\pi)\} +$$
$$2^{-1}(1/\lambda_1 + 1/\lambda_2 - 2/\lambda_3)\ln[(\alpha_1 + \alpha_2)/4] + \qquad (20)$$
$$2^{-1}(1/\lambda_1 + 1/\lambda_3)\ln(1/P^*_{13}) + 2^{-1}(1/\lambda_2 + 1/\lambda_3)\ln(1/P^*_{23})$$

Equation (20) is now the general equation for the total resistance of a typical heat channel. It can be shown that it reduces to Eq. (14) when there is no eccentricity ($\alpha_1 = \alpha_2 = \alpha$).

The thermal conductance of the joint when eccentricity is present is related to the total resistance in the following manner:

$$(\alpha_1 + \alpha_2)D\, h_j = 2/R \qquad (21)$$

Equation (21) reduces to Eq. (12) when there is no eccentricity.

Another interesting special case of Eq. (20) results when the materials are identical:

$$R = (2/\pi\lambda)\ln\{\sec[(\pi/2)(\alpha_1 - \alpha_2)/(\alpha_1 + \alpha_2)]/\pi P^*\} \qquad (22)$$

where, for this special case;

$$\pi P^* = 2P_a(1 - \nu^2)/E \qquad (23)$$

For identical materials the joint conductance with eccentricity, $h_j(\alpha_1 > \alpha_2)$, relative to the joint conductance without eccentricity, $h_j(\alpha_1 = \alpha_2)$, can be related to the total heat channel resistances, with and without eccentricity, in the following manner:

$$h_j(\alpha_1 > a_2)/h_j(\alpha_1 = \alpha_2) = [R(\alpha_1 > \alpha_2)/R(\alpha_1 = \alpha_2)]^{-1} \qquad (24)$$

because $2\alpha = \alpha_1 + \alpha_2$. The ratio of the resistance can be written as

$$R(\alpha_1 > \alpha_2)/R(\alpha_1 = \alpha_2)$$
$$= 1 + \ln\{\sec[(\pi/2)(\alpha^* - 1)/(\alpha^* + 1)]\}/\ln(1/\pi P^*) \qquad (25)$$

where $\alpha^* = \alpha_1/\alpha_2$. The maximum effect of the eccentricity for a particular contact pressure will occur when

$$(\alpha^*_{max} - 1)/(\alpha^*_{max} + 1) = (\alpha_1 + \alpha_2 - 2)/(\alpha_1 + \alpha_2) \qquad (26)$$

The effect of uneven spacing is a minimum at very light contact pressures and is a maximum at very high contact pressures where the contact area width is the largest.

## Test equipment and procedure

Tests were carried out in a conventional thermal contact resistance system.[4] The tests were done in a vacuum $10^{-4}$ mm Hg. The two large plane solids were 1 in. x 1 in. x 1 in. stainless steel (303) having the test surface optically flat. Four copper-constantan thermocouples were inserted to the centerline into both solids. The thermocouple nearest the test surface was located 1/8 in. from the surface, and the other three were located on 1/4 in. centers. The cylinders were 1/32 in. diameter of stainless steel (303) 1 in. long, relatively smooth and clean. Heat was supplied by a nichrome resistance wire cemented into machined grooves in the heater block. The heater operated on a.c. current with variable resistance controllers. The heat input was about 10 watts, and the heat flux based upon the apparent area was about 5000 Btu/ hr ft$^2$ °F. The apparent contact pressure ranged from 100 to 4000 psi. The pseudo-temperature drop across the joint ranged from a low value of about 8.5°F to a high of about 50°F. The low values corresponded to the highest loads and the smallest pitch ($\alpha = 1$), while the high values corresponded to the lowest loads and the largest pitch ($\alpha = 4$). Tests were conducted at pitches corresponding to $\alpha = 1, 2$, and 4. The error in the copper-constantan thermocouples used in the tests did not exceed 3/4% with an average joint temperature ranging from 150 to 280°F. The thermocouple millivolt output was monitored through a digital recorder having negligible error.

## Results and discussions

The tests results agree well with the theory, Eq. (17). The maximum discrepancies between data and theory occur at the lightest contact pressure. At this load the data for $\alpha = 1$

fall 5% below the theoretical values. As $\alpha$ is increased, this error decreases to about 1-2% for $\alpha = 4$. The difference between data and theory decreases with increasing contact pressure for all values of $\alpha$. At the highest pressures all the data agree to within 1-2% of the theoretical values. The best agreement over the entire load range occurs when the spacing is a maximum.

It is believed that oxides or perhaps the surface roughness of the cylinders may be responsible for the observed lower values at the low loads when the spacing is a minimum.

Future work will need to compare the theory with data obtained for different materials in contact and to test for the effect of eccentricity. The theory should be extended to include the effects of oxides and surface roughness, especially at light contact pressures. An analysis incorporating the effects of lubricants would be desirable, although very difficult to achieve.

## Conclusion

A theory has been presented that yields two expressions for the thermal conductance of uniformly spaced and for nonuniformly spaced cylinders. The theory agrees with a limited amount of test data, in spite of the fact that the cylinders were neither absolutely smooth nor absolutely clean.

## References

1. Veziroglu, T.N., "Thermal Conductance of Two-Dimensional Constrictions", NASA Grant NGR 10-007-010 SUB 3, Mech. Eng. Dept., University of Miami, January 1968.

2. Yovanovich, M.M. and Coutanceau, J., "Sur la détermination de la résistance thermique transversale d'un cylindre de révolution homogène isotrope avec des conditions aux limites mixtes," Comptes Rendus de l'Academie Sciences, Paris, Vol. 268, March 1969, pp. 821-823.

3. Seely, F. B. and Smith, J.O., Advanced Mechanics of Materials, John Wiley & Sons, Inc., 1967, Chap. 11.

4. Yovanovich, M.M. and Fenech, H., "Thermal Contact Conductance of Nominally Flat, Rough Surfaces in a Vacuum Environment," Progress Astronautics and Aeronautics: Thermophysics and Temperature Control of Spacecraft Entry Vehicles, edited by G.B. Heller, Vol. 18, Academic Press, New York, 1966, pp. 773-794.

RE-ENTRY THERMAL ANALYSIS OF VARIABLE
THICKNESS SPHERICAL VEHICLES

J. C. Dunn[*] and R. E. Nickell[†]

Bell Telephone Laboratories, Incorporated, Whippany, N. J.

Abstract

Spherical vehicles, with variable wall thickness, for the NIKE-X Re-entry Measurements Program were thermally analyzed to determine their capability to absorb the heat flux of re-entry without surface ablation and subsequent wake contamination. The analysis was carried out by using the finite-element computer program HTCON, which solves, approximately, the transient axisymmetric heat conduction problem, taking into account such factors as precise geometry, inhomogeneous and nonlinear material properties, time-dependent heat flux from laminar and turbulent boundary-layer flow, and radiative emission. The results indicate that peak stagnation point temperatures for 15-in.-diam beryllium spheres with tungsten ballast are slightly above the melting temperature of beryllium and that some wake contaminants should have been expected. Airborne optical data obtained after completion of the analysis are in agreement with the computational predictions.

Nomenclature

$h_s$ = stagnation enthalpy of external flow, Btu/lbm

$h_w$ = enthalpy of external flow at wall condition, Btu/lbm

$q_s$ = re-entry heat flux at stagnation point, Btu/ft$^2$ sec

$q_\theta$ = re-entry surface heat flux at θ when the boundary-layer is laminar, Btu/ft$^2$ sec

$q_T$ = re-entry surface heat flux for a turbulent boundary-layer, Btu/ft$^2$ sec

$R_N$ = vehicle nose radius, ft

---

Presented as Paper 71-424 at the AIAA 6th Thermophysics Conference, Tullahoma, Tenn., April 26-28, 1971.
 *Member of Technical Staff, SAFEGUARD Test Targets Group.
 †Supervisor, Solid Mechanics Group.

$Re_\delta$ = Reynolds number based on laminar boundary-layer momentum thickness, dimensionless.

$T_w$ = vehicle surface temperature, °R

$V_\infty$ = vehicle re-entry velocity, fps

$X$ = distance along body surface = $r_o \theta$, ft

$\varepsilon$ = hemispherical emissivity of vehicle surface, dimensionless

$\rho_o$ = standard air density at sea level, slugs/ft$^3$

$\rho_\infty$ = freestream air density, slugs/ft$^3$

$\sigma$ = Boltzmann constant, Btu/ft$^2$ sec, °R$^4$

Introduction

Hollow beryllium spheres, 15 in. in diam, were flown in phase B of the Re-entry Measurements Program (RMP-B). The spheres contained on-board experiments designed to measure or alter, in a controlled manner, observable wake phenomena. Wake contamination due to ablation was to be avoided. Thus, accurate analysis was necessary to predict vehicle thermal response during re-entry.

During the preliminary design of the 15-in.-diam spheres, the Raytheon Company[1] used a one-dimensional finite-difference solution to obtain re-entry shell temperatures. Their results show a maximum surface temperature of only 1875°F which is well below melting. However, it appears that values computed by Raytheon for "cold wall" heat flux at the stagnation point are about 25% low. This could easily be attributed to the use of preliminary trajectory information since heat flux is very sensitive to vehicle velocity.

In order to obtain more detailed and exact temperature information during re-entry, a finite-element computer code, HTCON, was used to solve the transient axisymmetrical heat conduction problem in the forward hemisphere. During the initial adaptation of this code, temperature distribution in a 15-in.-diam copper sphere, being considered for ICBM flight, was calculated.[2] More recently, re-entry temperatures were computed for the 15-in.-diam RMP-B spheres including exact geometry, temperature-dependent material properties, and laminar and turbulent boundary-layer flow. The computer code can be easily used for solution of temperature distributions in other re-entering spheres. Trajectory altitude and velocity data, material properties, and geometry information are required inputs. HTCON is presently being used for shell design and temperature prediction of re-entering spheres to be flown in the SAFEGUARD System Test Target Program.

## Sphere Design

The geometry of RMP-B spheres is shown in Fig. 1. The shell was beryllium with an outside radius $r_o$ = 7.5 in. and inner radius $r_i$ = 6.675 in. Inner and outer spherical surfaces had centers offset by e = 0.3 in. The eccentricity of these surfaces combined with the addition of tungsten ballast resulted in a center of gravity location approximately 1.5 in. forward of the outer surface's geometric center. This static margin produced the desired stabilization during re-entry.

Beryllium was selected as the shell material because of its superior heat-sink properties; the most outstanding of these is the specific heat, which is an increasing function of temperature as shown in Fig. 2. Thermal conductivity is also shown in this figure as a decreasing function of temperature, but is still high enough to allow sufficient thermal diffusion. These thermal properties, combined with a fairly high melting point of 2350°F and high strength-to-density ratio, make beryllium one of the best re-entry materials of the heat-sink variety. Several other materials (copper,[2] graphite, stainless steel, and titanium[3]), considered under the same re-entry constraints, were found to be inferior.

## Re-entry Environment

RMP-B spheres re-entered the Earth's atmosphere with initial conditions typical of ICBM trajectories. Slowdown within the atmosphere was determined by

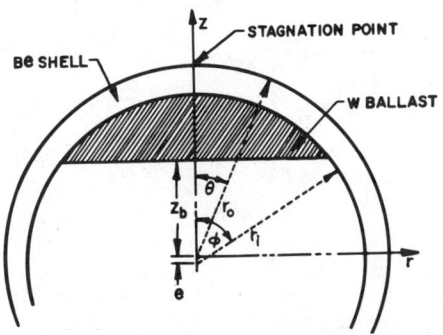

Fig. 1  Schematic beryllium-tungsten sphere.

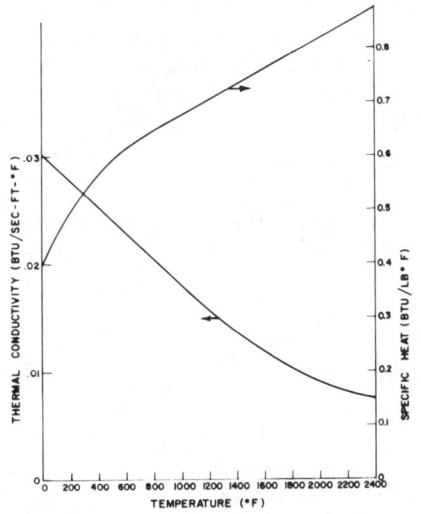

Fig. 2  Thermal properties of beryllium.

vehicle weight and geometry and is shown in Fig. 3.

Re-entry dynamics were somewhat complex due to center-of-gravity offset, induced tip off rates, and spin stabilization. An analysis by R. Pringle[4] shows that a nominal flight with spin stabilization would have angle-of-attack variation of less than 5° during periods of peak aerodynamic heating. Because of this small variation and for simplification of the thermal analysis, a zero angle of attack was assumed for the entire re-entry. This assumption is conservative.

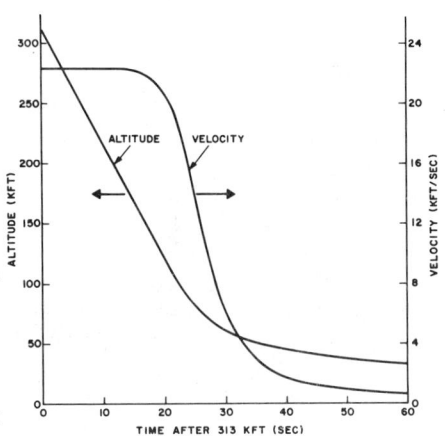

Fig. 3 RMP-B sphere re-entry trajectory.

Re-entry heat flux can, in general, be described as a function of flow parameters (freestream density, enthalpy, and velocity) and body parameters (diameter, temperature, and body angle). The functional relation has two forms; one is valid for laminar boundary-layer flow and the other for turbulent flow.

For laminar flow, a simplified Fay-Riddell[5] relation, as reported by Detra and Hidalgo,[6] was used to describe stagnation point heat transfer. Heat flux variation with body angle, $\theta$, was obtained from the work of Kemp et al.[7] using an assumed Newtonian pressure distribution. A least squares approximation to calculated $\theta$ dependence is given in Ref. 2. Thus, for laminar flow, the following heat-transfer relations apply:

$$q_s = 865/\sqrt{R_N} \left(V_\infty/10^4\right)^{3.15} \left(\rho_\infty/\rho_o\right)^{0.5} \left(1-h_w/h_s\right) - \sigma \varepsilon T_w^4 \qquad (1)$$

$$q_\theta = 865/\sqrt{R_N} \left(V_\infty/10^4\right)^{3.15} \left(\rho_\infty/\rho_o\right)^{0.5} \left(1-h_w/h_s\right)\left(0.0908+1.207 \cos^2\theta - 0.307 \cos^4\theta\right) - \sigma \varepsilon T_w^4 \qquad (2)$$

where the notation used is defined in the Nomenclature.

Boundary-layer transition to turbulent flow was based on a critical Raynolds number calculated using the laminar boundary-layer momentum thickness. This Reynolds number was evaluated by assuming a Newtonian pressure distribution and by taking a least squares fit of body angle dependence given in Ref. 6.

$$Re_\delta = \sqrt{12\,R_N}\,\left(\rho_\infty/\rho_o\right)^{0.5076}\left(V_\infty/10^3\right)^{0.477}$$
$$\cdot \left[275 - 180\cos^2\theta + 91\cos^4\theta - 136\cos^6\theta\right] \quad (3)$$

Following transition data reported by Libby,[8] the boundary layer was assumed turbulent whenever $Re_\delta$ exceeded 200. This procedure indicates that, for the front hemisphere, boundary-layer turbulence occurs first at the 90° station, then works its way forward, toward the stagnation point, as altitude decreases.

Turbulent boundary-layer heat flux was calculated using a procedure presented in Ref. 6. Assuming a maximum possible contribution of dissociation energy, the turbulent heat-transfer rate can be expressed by

$$\dot{q}_T = 1.575\times 10^4 (X)^{-0.2}\left(\rho_\infty/\rho_o\right)^{0.8}\left(V_\infty/10^4\right)^{3.18}$$
$$\cdot \cos^{1.367}\theta\left[1 - \cos^{1/3}\theta\right]^{0.4} \quad (4)$$

Body angle dependence is included in this relation with the maximum value occurring near the sonic point.

Equations (1), (2), and (4) were used to compute re-entry heat flux boundary conditions for RMP-B spheres. Equation (4) was used at a particular body station whenever $Re_\delta$, found from Eq. (3), exceeded 200. Time dependence is introduced in these relations through freestream velocity and density. Velocity vs altitude (or time) data was taken from the trajectory code used in the construction of Fig. 3. Density ratio, $\rho_\infty/\rho_o$, was computed using an exponential fit to density data given in the U.S. Standard Atmosphere, 1962.[9]

Net heat-transfer rates calculated for the stagnation and sonic points are given in Fig. 4 for RMP-B sphere re-entries. Heat flux variation with body angle is shown in Fig. 5 at times both before and after boundary-layer transition.

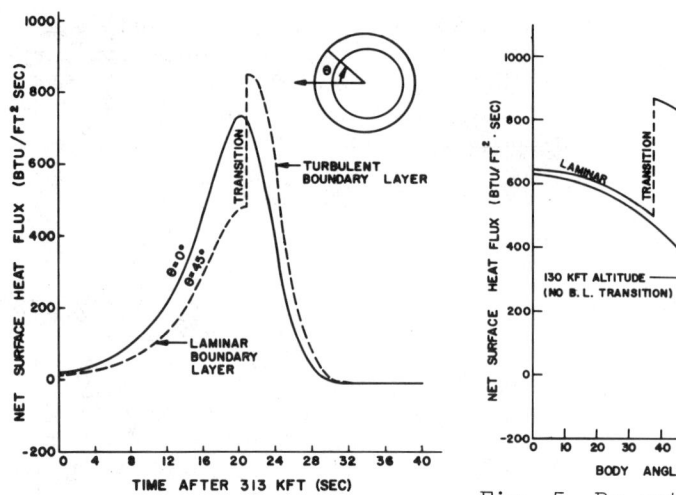

Fig. 4  Re-entry heat flux.

Fig. 5  Re-entry heat flux as a function of body angle.

### Finite Element Analysis

The two-dimensional finite element computer code HTCON was used to analyze, approximately, the transient temperature distribution in the sphere. This code is based on a variational principle derived by Gurtin[10] and adapted to finite element techniques by Wilson and Nickell.[11] Limited documentation of the code is available.[12] HTCON can analyze plane or axisymmetric solids composed of many different materials, each of which can have thermal properties (thermal conductivity and specific heat) that are specified to be temperature-dependent. The current version treats up to 25 different materials. Boundary conditions on heat flux and temperature can be specified to be time dependent; constant (in time) convective heat-transfer coefficients for boundary surfaces are also included.

The transient solution is obtained by a step-forward integration procedure - identical to the first order to the implicit Crank-Nicholson operator. The integration scheme can be shown to be unconditionally stable with respect to the choice of time increment. Variable time increments can be used.

Nonlinear thermal properties and nonlinear boundary conditions - such as radiative heat flux - are treated through a quasi-linearization procedure; i.e., the material properties and the radiative flux are determined, for the current time interval, on the basis of the temperatures computed at the end of the previous interval. In actual practice, however, it is

seldom economical to change the material properties at each
interval, since this involves the recomputation of the coefficient matrices (i.e., the thermal conductivity matrix and
the specific heat matrix), modifying these new matrices in
order to satisfy geometric (temperature) boundary conditions,
and retriangularizing the effective coefficient matrix. In
most cases, it suffices to change the material data every five
to ten time steps, thereby making a substantial saving in computation costs with little loss in accuracy. There is, of
course, no difficulty or real cost associated with changing
the radiative flux at every interval. If the choice of solution interval size is made with these considerations in mind,
and large temperature excursions between material or boundary
data changes are avoided, the quasi-linearization procedure
should be adequate without resort to iteration.

The temperature dependence of the thermal conductivity and
volumetric specific heats of beryllium (Fig. 2) indicated that
an assessment of the quasi-linearization was necessary in
order to develop confidence in the results. Therefore, a
study of solution accuracy as a function of interval size and
number of intervals before changing material data was undertaken. A one-dimensional axisymmetric model of the sphere -
at the stagnation point - was chosen as the vehicle for assessment. This model had an advantage, in addition to simplicity, in that it provided a check against previous finite-
difference calculations,[1] where a one-dimensional plane model
was used. Two effects were thus being considered: a) nonlinear material properties and b) curvilinear geometry.

First, a one-dimensional plane problem was analyzed in an
effort to assess these previous results. Fifteen elements
were used in each of the two materials (for a total of thirty
quadrilaterial elements and 62 nodal points). The elements
were insulated along lateral boundaries in order to insure
normal heat flow through the wall. A sketch of the mesh is
shown in Fig. 6a. Nodal points 1 and 32 (the inside of the
tungsten ballast) were insulated and the stagnation point re-
entry heat flux was applied to the surface between nodes 31 and
62. This heat flux, for the trajectory of interest, is shown
in Fig. 4. A time step size of 0.1 sec was used in the analysis and the material properties were adjusted at every time
step in order to optimize the accuracy of the quasi-linearization procedure. Comparison of the temperature profile through
the wall, at the time of peak stagnation point temperature,
from this analysis with a similar profile from Ref. 1 (for a
slightly different geometry) shows substantial difference.
The lower stagnation temperature calculated by Raytheon is

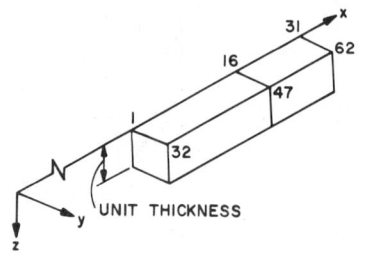

Fig. 6a  One-dimensional plane model.

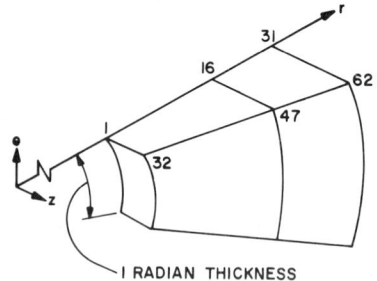

Fig. 6b  One-dimensional axisymmetric model.

primarily due to the use of lower stagnation heat flux, as mentioned previously.

Following this analysis, a one-dimensional axisymmetric problem was solved in order to determine the effects of curvature in the geometry. The mesh was chosen to be identical to that used in the plane problem (see Fig. 6b). Stagnation point re-entry heat flux was applied once again to the surface area between nodes 31 and 62 and the same time step size and quasi-linearization scheme was used. A comparison of the stagnation point surface temperature, as a function of time, for the one-dimensional plane and axisymmetric geometries is shown in Fig. 7. The differences between the two analyses can be traced to the ratio of surface area of re-entry heat flux application to the volume of material absorbing heat. The magnitude of the difference is about 6% at peak surface temperature and continues to increase with time as the surface area/volume error plays a more important role. Another comparison - the temperature profiles through the wall at the time of peak stagnation point temperature - is shown in Fig. 8. The uniformity of temperature difference between the two profiles, due to the surface area/volume ratio, is readily discernible. These two comparisons indicate that plane models for re-entry heating of curved bodies can be grossly unconservative unless radius to thickness ratios are larger than that used in this analysis.

The next step in the study was concerned with a fully two-dimensional axisymmetric analysis of the front half of the 15-in. sphere. The mesh layout for the computations consisted of 440 nodal points and 405 finite elements, with the bulk of the elements concentrated near the stagnation region (Appendix). The same time increment size (0.1 sec) was used as before but, for reasons of economy, the nonlinear material behavior was taken into account by adjusting the properties every second rather than at every time step. A comparison of the stagnation point surface temperatures vs time for the three cases is shown in Fig. 7. The effect of changing the material properties less often has a similar unconservative effect to that of neglecting curvature, leading to an error in peak temperature of about 3%. In all three cases, the peak temperatures are above the melting temperature of beryllium. Adjustment of the surface heat flux due to ablation was not considered in these analyses; hence, temperatures were allowed to fictitiously exceed the beryllium melting point. A wall temperature profile at the angle $\theta = 0°$ is shown in Fig. 8 for comparison with the one-dimensional examples.

In addition, the surface temperatures versus time for three body angles ($\theta = 0°$, $45°$, $60°$) are plotted in Fig. 9 for the two-dimensional axisymmetric case. The effect of laminar-to-turbulent flow transition can be seen in the sudden temperature changes. In Fig. 10, the surface temperature is shown as a function of body angle at the time of peak stagnation temperature. The extent to which turbulent flow has affected the temperature distribution is easily discernible.

## Conclusions

From the foregoing analyses an estimate of the presence of beryllium and/or its oxidation products as contaminants in the re-entry wake can be made. If all beryllium above the melting temperature (2350°F) is assumed to leave the surface and enter the wake, an estimate of 30 g is obtained. This figure will

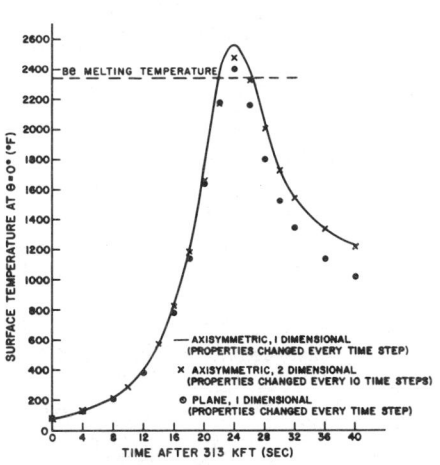

Fig. 7 Re-entry surface temperature; $\theta = 0°$.

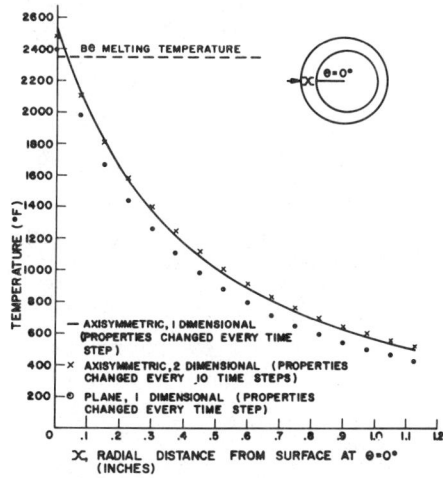

Fig. 8 Shell temperature; $\theta = 0°$, time = 24 sec.

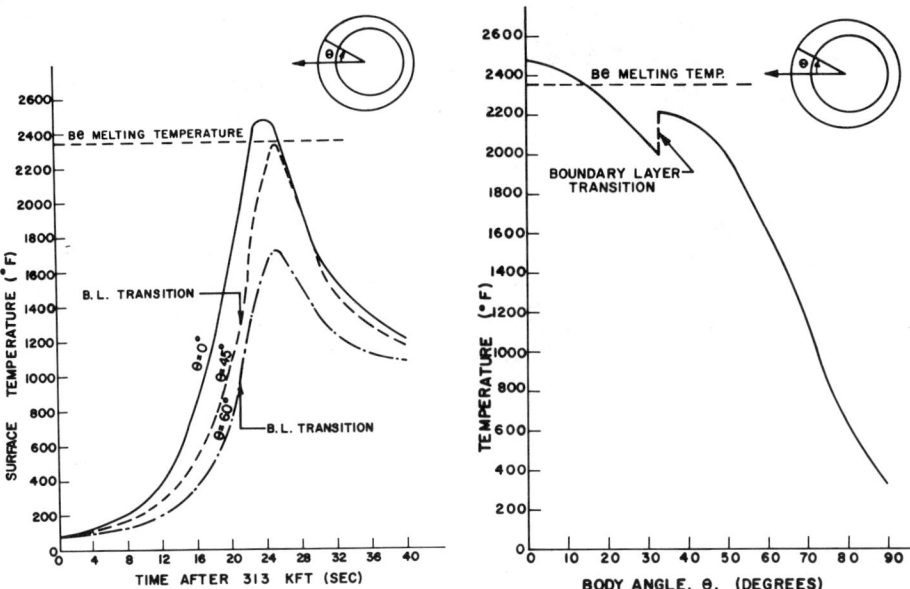

Fig. 9 Re-entry surface temperatures.

Fig. 10 Surface temperature vs. θ; time = 24 sec.

likely be slightly erroneous due to three factors: a) the sphere's angle of attack variation during re-entry will distribute the heat flux more evenly over the area and lower peak temperatures to some degree; b) ablation is an endothermic process; therefore, change-of-phase of beryllium at the surface will lower temperatures at interior points; c) ablation eliminates some heat storage capacity and exposes new material to the high surface heat-transfer rates; therefore, temperatures at interior points are raised. In any case, the analysis does show that a small amount of beryllium contaminants should enter the re-entry wake. By referring to the maximum surface temperature curve, Fig. 7, one would expect to see surface melting or wake contamination begin at 22 sec. or 104,000-ft altitude, and end about 26 sec, or 76,000 ft. Peak contamination would be predicted at 88,000-ft altitude.

The presence of beryllium oxide in the re-entry wake was, in fact, detected by optical instruments on several occasions; but, at the time, was thought to have been caused by localized high surface temperatures, as in the vicinity of experiment ports in the shell, or by mission anomalies such as poor re-entry orientation. After reviewing the optical data reports, one flight can be singled out as most ideal for investigation of the surface melting problem. Several factors contributed to this selection. The sphere had a continuous front hemisphere, flight trajectory and orientation were normal, and a complete set of optical data was obtained during re-entry. The optical data show good evidence of surface melting in the altitude region predicted by analysis. As shown in Fig. 11, spectral cameras detected beryllium oxide in the wake from 104,000-ft altitude to 75,000-ft altitude, in excellent agreement with predictions obtained from Fig. 7. During this same altitude interval, visible radiation intensity increased markedly and the body, along with a sizable wake, were visible in the cine camera data. Figure 12 shows the relative visible irradiance during re-entry as obtained by spot densitometer traces of the cine camera data. Peak intensity occurred at 87,000-ft altitude. Below 75,000-ft altitude radiation intensity had decreased to the point where the wake was no longer visible.

While the amount of melting would be difficult to predict from the optical data, melting time agreement with analysis suggests a high level of confidence can be attributed to the analytical simulation of the re-entry environment and heat-transfer processes.

An additional set of conclusions concerns the adequacy of mathematical simulation of transient heat-transfer during re-entry, especially for materials whose thermal properties are markedly nonlinear with temperature. The quasi-linearization procedure adopted here seems to be adequate, although unconservative, and could presumably be optimized. For instance, inspection of Fig. 7 reveals that noticeable deviation between the one- and two-dimensional axisymmetric analyses is confined to the area of most rapid temperature change with time - 18 to 26 sec. An optimal path might be to change properties every sec from the beginning of re-entry until t = 18 sec; then, change properties every 0.5 sec from t = 18 until t = 21;

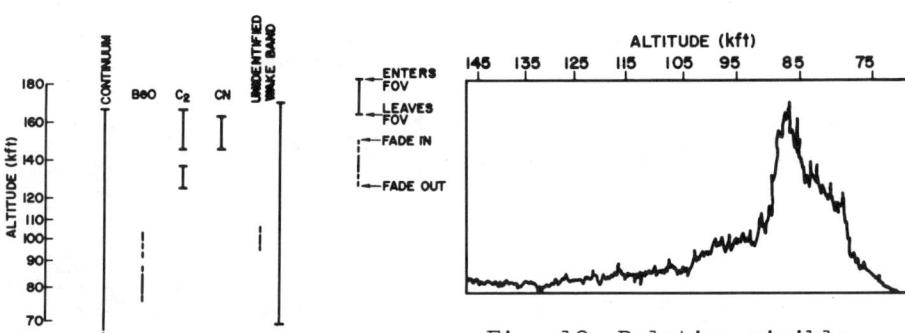

Fig. 11  Spectral emission.

Fig. 12  Relative visible irradiance.

change properties every time step (0.1 sec) from t = 21 until
t = 23; change properties every 0.5 sec from t = 23 until
t = 26; and finally, change properties every sec during the
remainder of re-entry. Such an analysis would be about twice
as costly as the current analysis but would be essentially as
accurate as the one-dimensional axisymmetric calculations.

Another conclusion drawn here concerns the use of plane
models for simulating heat transfer in curved geometries. The
effect is again unconservative, since re-entry heat flux is
applied externally to a disproportionately small surface area
compared to the volume of material absorbing heat. The combined effects of curvature and material nonlinearity could
seemingly lead to unconservative results sufficient to jeopardize the mission.

Appendix

The grid layout for the two-dimensional axisymmetric analysis is shown in Fig. 13. There are 440 nodal points and 405 elements in the grid. Typical element shapes are shown by the shaded areas in the figure. Note that the bulk of the elements are concentrated in the stagnation-point region.

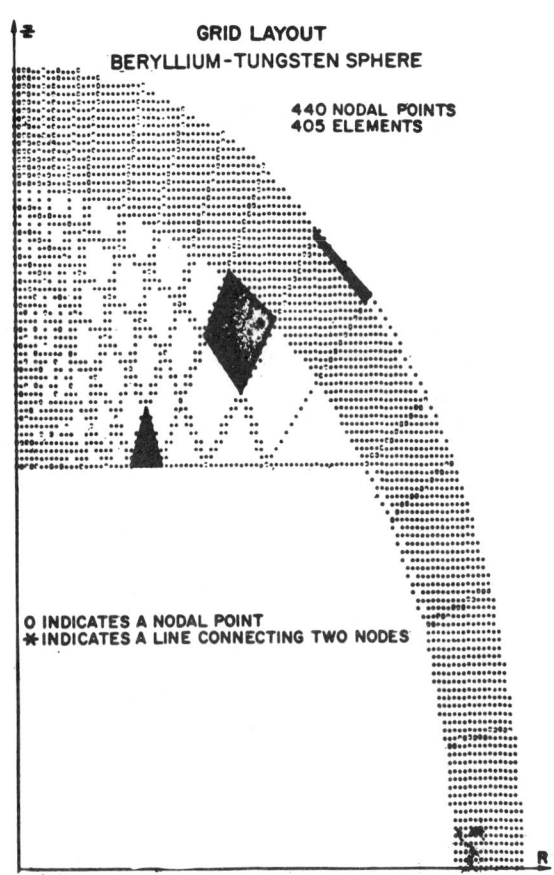

Fig. 13 Grid layout beryllium-tungsten sphere.

## References

[1] Raytheon Co., "Reentry Sphere - External Thermal Analysis," Final Report to Bell Telephone Labs., 1967.

[2] Williams, J. C., Bell Telephone Laboratories, private communication, Oct. 1969.

[3] "Study of Reentry Spheres," Preliminary Report, Contract Number 601742, April 10, 1967, Raytheon Co.

[4] Pringle, R., Bell Telephone Laboratories, private communication, Oct. 1968.

[5] Fay, J. A. and Riddell, F. R., "Theory of Stagnation-Point Heat Transfer in Dissociated Air," Journal of the Aerospace Sciences, Vol. 25, No. 2, Feb. 1958, pp. 73-85.

[6] Detra, R. W. and Hidalgo, H., "Generalized Heat Transfer Formulas and Graphs for Nose Cone Re-entry into the Atmosphere," ARS Journal, March 1961, pp. 318-321.

[7] Kemp, N. H., Rose, P. H., and Detra, R. W., "Laminar Heat Transfer Around Blunt Bodies in Dissociated Air," Journal of the Aerospace Sciences, Vol. 26, No. 7, July 1959, pp. 421-430.

[8] Libby, P. A., "A Survey of Heat Transfer Problems Connected with Space Vehicle Reentry," Paper presented at the Seminar in Astronautical Propulsion, Sept. 8-12, 1960, Varenna, Italy.

[9] U.S. Standard Atmosphere, 1962, U.S. Government Printing Office, Washington, D.C.

[10] Gurtin, M. E., "Variational Principles for Linear Initial-Value Problems," Quarterly Applied Mathematics, Vol. 22, No. 3, Oct. 1964, pp. 252-256.

[11] Wilson, E. L. and Nickell, R. E., "Application of the Finite Element Method to Heat Conduction Analysis," Nuclear Engineering and Design, Vol. 4, No. 3, Oct. 1966, pp. 276-286.

[12] Chaloupka, A. B., "A Computer Program for the Analysis of Two Dimensional Heat Conduction Using the Finite Element Technique," M.S. thesis, June 1969, U.S. Naval Postgraduate School, AD69 0450.

# COUPLING OF SHAPE CHANGE, HEATING DISTRIBUTION AND INTERNAL CONDUCTION FOR ABLATING BODIES

Jin H. Chin*

Lockheed Missiles & Space Company, Sunnyvale, Calif.

## Abstract

A computerized procedure is developed to determine the thermal response of axisymmetric ablating nosetips along their trajectories. The procedure includes generation of finite-difference nodal points for a rectangular grid system, pressure and heating distributions based upon the instantaneous surface contour, two-dimensional conduction by an alternating-direction-explicit scheme, and shape change by considering the movement of individual surface nodal points. Anisotropic and temperature-dependent properties, substrates and interior cavity boundary are permitted. Application of this procedure to nosetip analyses demonstrates that coupled calculations are required for accurate prediction of nosetip response.

## Nomenclature

| | | |
|---|---|---|
| $B'$ | = | $\dot{m}/\overline{C}_H$ |
| $c$ | = | specific heat |
| $C_H$ | = | nonblowing heat transfer coefficient including entropy swallowing |
| $C_{H,o}$ | = | value of $C_H$ at stagnation point excluding vorticity effects |
| $\overline{C}_H$ | = | blowing heat transfer coefficient |
| $F_{CH}$ | = | $C_H/C_{H,o}$ |
| $F_p$ | = | particle removal factor to obtain total recession rate $F_p \dot{m}$ |

---

Presented as Paper 71-413 at the AIAA 6th Thermophysics Conference, Tullahoma, Tenn., April 26-28, 1971. This work was performed under Air Force Contract No. F04701-69-C-0237 and F04701-68-C-0299.

*Staff Engineer, Aero-Thermodynamics Department, Engineering Technology.

| | | |
|---|---|---|
| $h_C$ | = | enthalpy of solid graphite at wall temperature |
| $H$ | = | total enthalpy of gaseous species |
| $k$ | = | thermal conductivity |
| $k_{11}$ | = | $k_\xi \cos^2\beta + k_\eta \sin^2\beta$ |
| $k_{12}$ | = | $k_{21} = (k_\xi - k_\eta) \cos\beta \sin\beta$ |
| $k_{22}$ | = | $k_\xi \sin^2\beta + k_\eta \cos^2\beta$ |
| $K_r$ | = | roughness correction factor for heat transfer |
| $\dot{m}$ | = | thermochemical mass ablation rate |
| $M$ | = | Mach number |
| $p$ | = | pressure |
| $q$ | = | heat flux |
| $r$ | = | radial distance |
| $R_N$ | = | nose radius |
| $Re_\theta$ | = | momentum-thickness Reynolds number = $\rho_e u_e \theta / \mu_e$ |
| $St$ | = | Stanton number = $C_H / \rho_e u_e$ |
| $t$ | = | time |
| $T$ | = | absolute temperature |
| $u$ | = | temperature for forward ADE traverse; velocity |
| $v$ | = | temperature for backward ADE traverse |
| $z$ | = | axial distance |
| $\alpha$ | = | constant for transitional heating equation |
| $\beta$ | = | angle between $\xi$- and z-axes |
| $\epsilon$ | = | surface emissivity |
| $\theta$ | = | momentum thickness |
| $\mu$ | = | viscosity |
| $\rho$ | = | density |
| $\sigma$ | = | Stefan-Boltzmann's constant |

Subscripts and Superscripts

| | | |
|---|---|---|
| cr | = | critical |
| e | = | edge of boundary layer |
| n | = | at time $t_n$ |
| T | = | turbulent boundary layer |

w        = at wall condition

η        = conductivity principal axis, generally considered as a-direction

ξ        = conductivity principal axis, generally considered as c-direction

0        = for Point 0

## I. Introduction

High performance hypersonic vehicles are subjected to severe aerothermal environment along their trajectories. In particular, the ablative nosetips of the vehicles will be subject to shape changes caused by thermo-mechanical erosion and possible internal fracture from thermo-mechanical stresses. For accurate design calculations, the pressure and heating distributions must be based upon the instantaneous shapes, particularly when transition from laminar to turbulent flow occurs. The instantaneous shapes are governed by the ablation history which is coupled to internal conduction and the material thermo-chemical ablation properties.

Previous works related to nosetip shape change predictions during atmospheric reentry were reviewed recently by Thyson et al.[1] They presented a shape-change analysis and compared the predicted results with experimental shape change data for paradichlorobenzene and 3-D quartz phenolic models. Internal conduction was not considered by using effective heats of ablation, $Q^*$. Popper, Toong and Sutton[2] considered the shape change and internal conduction for sphere cones. But the calculation was carried out in the laminar flow regime without a large change of the characteristic shape.

The thermophysical properties differ widely for different ablative materials. Charring ablator materials such as carbon phenolic have relatively low thermal conductivities. For these materials, severe temperature gradients are developed within a thin layer below the nosetip surface. The in-depth temperature response is of relatively minor importance. Consequently, the quasi-steady ablation model or the one-dimensional ablation conduction method of Gallagher et al.[3] gives adequate results. On the other hand, graphitic materials have relatively high thermal conductivities. Multi-dimensional heat conduction methods must be used to obtain the transient interior temperatures which are required for nosetip stress calculations. The formulation should permit a wide range of nosetip shapes including sphere cones, blunt laminar shapes, biconic turbulent shapes, triconics, and negative body-slope regions.

This paper presents the approaches employed in the graphite nosetip analysis codes.[4,5] Computed results for a relatively slender and a relatively blunt nosetip are given to demonstrate the importance of the coupled calculations.

## II. Analysis

### a) Multimaterial, Anisotropic, and Axisymmetric Conduction

The heat balance over an elementary volume of solid yields

$$\rho c \frac{\partial T}{\partial t} = \frac{\partial}{\partial z}\left[k_{11}\left(\frac{\partial T}{\partial z}\right)_r + k_{12}\left(\frac{\partial T}{\partial r}\right)_z\right]_r$$

$$+ \frac{1}{r}\frac{\partial}{\partial r}\left[rk_{21}\left(\frac{\partial T}{\partial z}\right)_r + rk_{22}\left(\frac{\partial T}{\partial r}\right)_z\right]_z \qquad (1)$$

In the numerical solution, a rectangular grid system as shown in Fig. 1 is used. Nonuniform grid spacings are allowed. The grids are drawn so that between two adjacent grid intersections there is no more than one material boundary. Automatic nodal point generation is provided with minimum input geometrical data. The boundary nodes are assumed to have zero heat capacitance. As shown in Fig. 1, a typical interior mode 0 is surrounded by eight nodes designated by numerals 1 to 8 and zero-capacitance nodes, if any, identified by alphabets.

An unconditionally stable, Alternating Direction Explicit (ADE) finite-difference method[6,7] is used to integrate Eq. (1). The computation between $t_n$ and $t_n + \Delta t_n$ is carried out in two directions. In the forward direction, traverse is made from left to right along each row and from the bottom to the top row. In the backward direction, traverse is made from top to bottom along each column and from the right to the left column. For Node 0, Eq. (1) is replaced by two finite-difference expressions for the forward and backward ADE directions to enable explicit calculations of $u_0^{n+1}$ and $v_0^{n+1}$, respectively. (Details of the numerical procedure and treatments of the surface and inner boundary points are given in Ref. 5.) The temperature at the next time step $t_{n+1}$ is then obtained by the average: $T_0^{n+1} = \left(u_0^{n+1} + v_0^{n+1}\right)/2$.

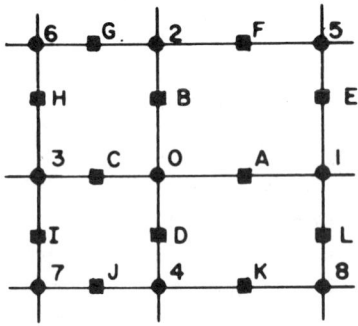

Fig. 1 Nodal point system.

b) Surface Energy Balance

The net conductive heat flux to the nosetip at a surface node is given by the surface energy balance

$$q_w = \overline{C}_H (H_e - H_w) - \dot{m}(H_w - h_C) - \epsilon \sigma T_w^4 \qquad (2)$$

The surface emissivity is assumed constant and $h_C$ is a function of $T_w$. With a diffusion-rate-controlled, equilibrium-sublimation model[8,11] of graphite ablation, the mass transfer parameter B' is a function of $p_e$ and $T_w$, and $H_w$ is a function of B' and $T_w$. With a rate-controlled-sublimation graphite ablation model,[9] $C_H$ becomes an additional parameter for B' and $H_w$. A considerable uncertainty exists in the thermodynamic properties of carbon vapor species. This uncertainty reflects a large variation of B' for the different models, as shown in Fig. 2.

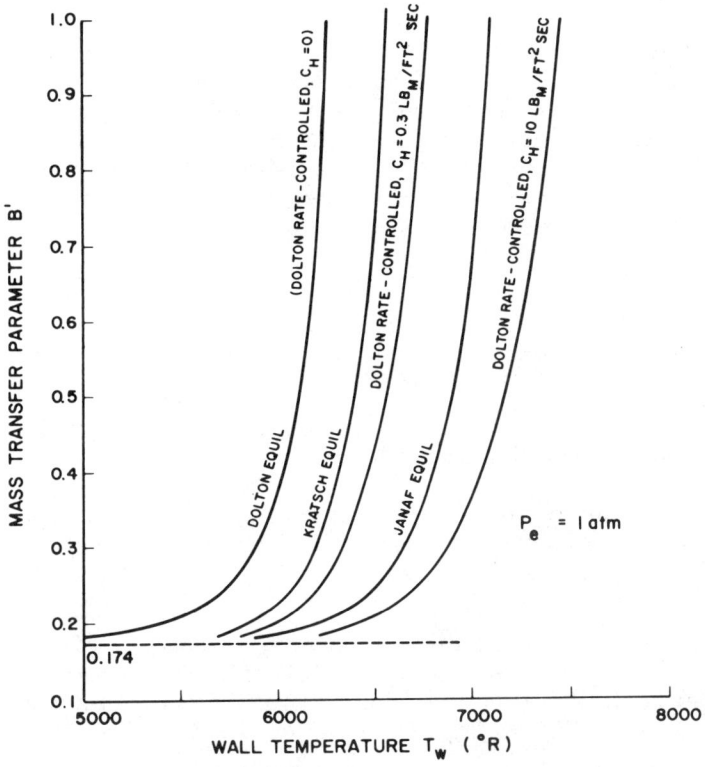

Fig. 2 Mass transfer parameter for different graphite ablation chemistry models.

Given $p_e$, $H_e$, $\epsilon$, $C_H$ together with an ablation chemistry model and a blowing-effectiveness correlation, Eq. (2) is linearized with respect to $T_w$ for use in conjunction with the ADE finite-difference energy balance expressions.

c) Heating and Pressure Distributions

Zero-blowing heating and pressure distributions are calculated first and then the heat transfer coefficient is reduced according to an empirical blowing effectiveness correlation.

For pressure distribution calculations, the nosetip shoulder is first determined. Forward of the shoulder, the ratio of the local to stagnation pressure depends solely upon the local body angle, according to a correlation of a series of blunt-body, method-of-characteristics solutions for spheres. Immediately aft of the shoulder, an overexpansion region exists. The pressure downstream is adopted from correlation of spherecone results.

The laminar heat transfer distribution is computed according to the Cohen methods.[12,13] The turbulent heat transfer relation is derived by Reynolds analogy and compressibility transformation from the skin-friction equation for incompressible flow over flat plates. The Persh method[3,4,14,15] for transitional boundary layers is used to account for the finite rate of increase of the convective heating downstream of the onset of transition:

$$St = St_T - \alpha/Re_\theta^2 \qquad (3)$$

The value of the constant $\alpha$ is determined by requiring a continuity of $C_H$ and $Re_\theta$ at the onset of transition. The value of $Re_\theta$ is computed by integration of the momentum integral equation invoking local similarity with entropy layer swallowing effects.

An approximate model[4,16] is used to account for the surface roughness effects on convective heating. It is assumed that surface roughness has negligible effects on laminar heating and that it increases turbulent heating by a multiplication factor $K_r$. The value of this roughness correction factor is estimated by means of the results of Fenter,[17] with an imposed upperlimit value 3.0.[18,19] For transitional heating, Eq. (3) is assumed valid for rough surfaces after $St_T$ is replaced by $K_r St_T$. This approximate model yields a reasonable prediction[16] of the heat transfer rates on smooth and roughened hemispheres and cones presented by Thyson et al.[1]

d) Nosetip Shape Change

The initial nosetip is a sphere cone or has an arbitrary surface contour (e.g., biconic, triconic). As ablation proceeds, the movement of the surface nodal points is considered individually. The radius of curvature at the stagnation point is determined by a three-point or least-square conic fit, dependent upon the number of surface nodes within the subsonic region. The inclination angle at the midpoint between two surface nodes is calculated by a central difference formula; it is assumed to vary linearly with the wetted distance between two midpoints.

At each surface node, the linear recession velocity normal to the surface is given by $F_p \dot{m}/\rho$.

e) Time Step and Frequency of Computation

Although the ADE scheme is unconditionally stable, the value of $\Delta t$ must not be so large that the truncation error may become excessive. A multiple (say 5) of $\Delta t_{stab}$ (a time step satisfying the stability criterion of a simple, explicit method and for all nodes without surrounding zero-capacitance boundary points) is used as one criterion for time-step selection. Other criteria include the maximum allowed temperature rise at monitored points, the maximum surface recession per time step, and the crossing of the grid intersections by the surface points.

The frequency of detailed heating distribution calculations is determined by the magnitude of shape change. Between detailed environment calculations, the normalized pressure and heating distributions are assumed frozen; the values at the surface points are obtained by interpolation with respect to the wetted distance and correcting for the changes at the stagnation point.

## III. Results and Conclusions

Calculations were performed for two typical reentry vehicles with widely different nosetip radii. The ATJ-S graphite nosetips are assumed to have a uniform initial temperature of 560°R at 250 kft altitude.

The equilibrium-sublimation graphite ablation chemistry model based upon the calculation of Dolton et al.[9] (see Fig. 2, Dolton Equilibrium) was employed to compute the thermo-chemical ablation rates. The effects of particle removal were computed by an empirical formula (with a threshold at 55 atm pressure and increasing nearly linearly from unity to 2 at 100 atm) which was obtained by correlation of CAL Wave Super Heater sphere-cone recession data.

Because applicable data characterizing the surface roughness of ablating graphite nosetips were not available, the initial roughness was assumed to be 0.002 in. (equivalent sand grain). A body station 1.25 in. from the initial stagnation point was selected to monitor the surface roughness. After the transition point passed over this monitor location, the surface roughness increased, as ablation continued, from the initial value to a maximum allowable sand-grain roughness of 0.010 in. Thereafter, the surface roughness was assumed to remain at the maximum value.

The transition of boundary layers depends upon a multitude of parameters including $M_e$ and surface roughness effects.[1,20] Because of the narrow range of $M_e$ for the nosetip region of interest and the lack of an accurate knowledge of the coupled effects of roughness formation and boundary layer transition, a constant $Re_{\theta,cr} = 210$ was used as the criterion for transition.

Figure 3 shows the initial configuration of a relatively slender nosetip ($R_N$ = 0.25 in.). To demonstrate the numerical features of the calculations, the substrates and the grid system assumed are also given. The instantaneous nose shapes and the normalized heat transfer coefficient $F_{CH}$ are shown in Figs. 4 and 5, respectively. The vorticity effects cause the nonunity value of $F_{CH}$ at the stagnation point. At high altitudes, the laminar heating distribution causes a slow nose blunting. As the vehicle descends, the transition point

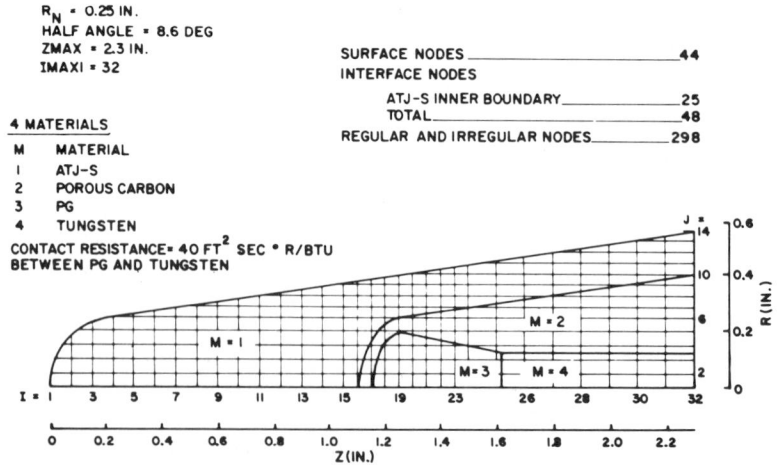

Fig. 3 Initial configuration and grid system—slender nosetip.

passes over the monitor location at 51.7 kft. The monitor surface roughness reaches the maximum allowed value at 46.5 kft. The upperlimit value 3.0 of the roughness correction factor $K_r$ is invoked

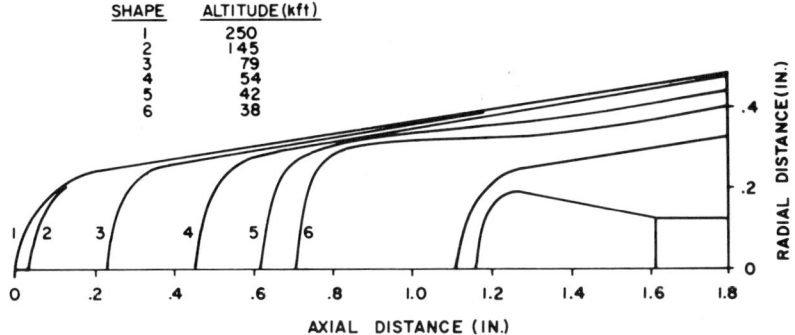

Fig. 4 Instantaneous shapes-slender nosetip.

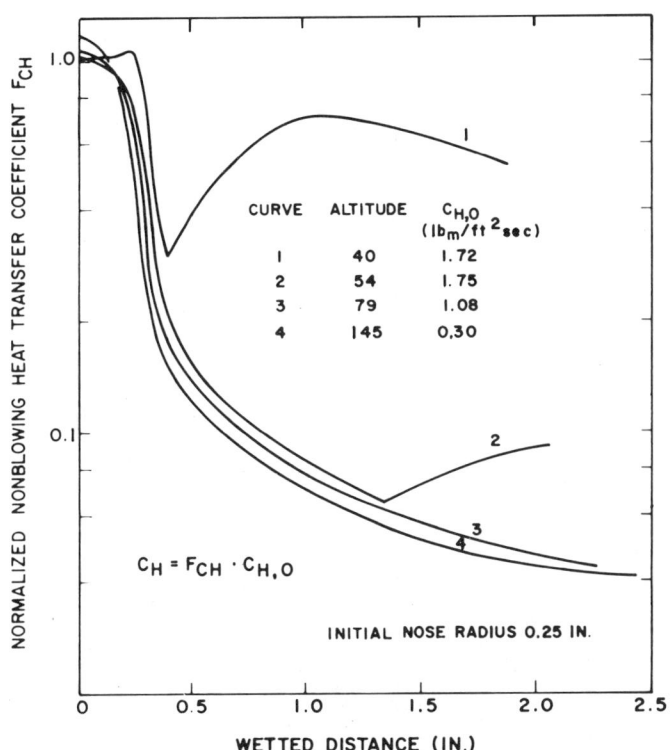

Fig. 5 Heat transfer coefficient distribution-slender nosetip.

at 47.5 kft. The calculation stops at 37.5 kft because the nosetip side-wall thickness will become less than the local radial grid spacing. At this time, the nose remains blunt because the transition point has not moved forward to the subsonic region.

The isotherms for the ATJ-S graphite region are given in Fig. 6. At high altitudes, the conduction is primarily one-dimensional in the axial direction, as shown in Fig. 6a. At intermediate altitudes, the

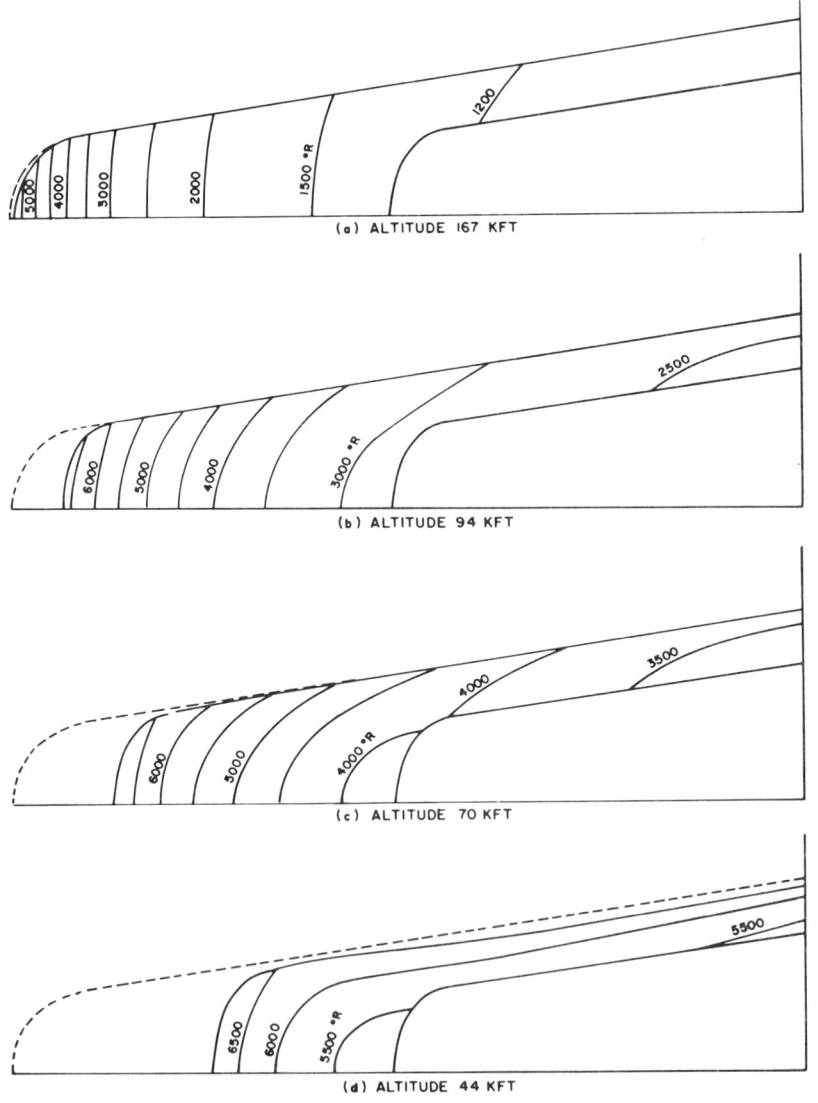

Fig. 6 Isotherm plots-slender nosetip.

side-wall heating becomes significant and two-dimensional conduction is evident (Figs. 6b and c). At 70 kft as shown in Fig. 6c, a hot spot forms near the shoulder of the graphite inner boundary, indicating a possible stress concentration region for potential mechanical fracture. As the vehicle descends further, the whole graphite region is heated to high temperatures, as indicated in Fig. 6d. The substrates also encounter a significant temperature rise.[5]

As an example of relatively blunt vehicles, a sphere cone with initial $R_N$ = 1.5 in. and cone half-angle 8.6 deg. is considered. Shown in Fig. 7 is the initial configuration as well as the instantaneous shapes. The heating distribution is given in Fig. 8. The upper-limit roughness correction factor is invoked at 66 kft. The calculation stops at 42 kft because of the small side-wall thickness again. As for the slender tip, the laminar heating distribution causes nose blunting at high altitudes. As the vehicle descends, the transition point moves forward to the nose front face; the extent of the blunt laminar region decreases with gradual transformation toward a biconic shape.

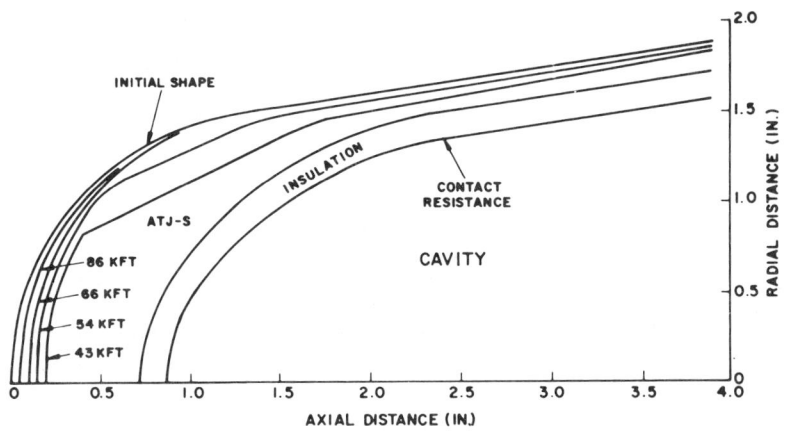

Fig. 7 Initial configuration and instantaneous shapes-blunt nosetip.

The foregoing results have demonstrated the interdependence of nosetip shape change, heating distribution, and internal conduction. The effects of the initial configuration (including size and location of the graphite inner boundary) and trajectory parameters are evident or implicated. It also is apparent that the computed results depend, to a varying degree, on the mathematical models and input parameters used, including: graphite ablation chemistry model, particle removal correction, transition criterion, and surface roughness effects.

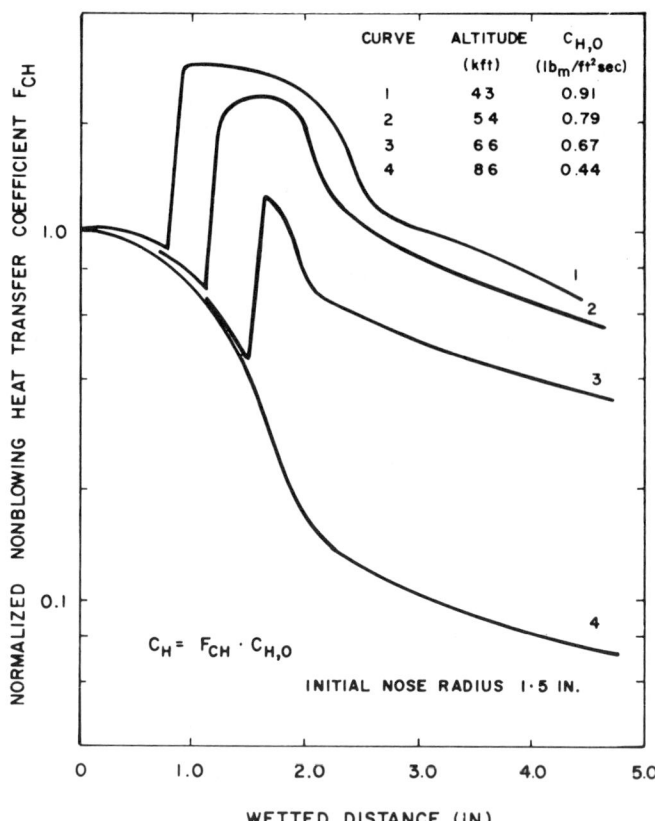

Fig. 8 Heat transfer coefficient distribution-blunt nosetip.

As indicated in the results for the slender nosetip, the upperlimit value 3.0 of the roughness correction factor is invoked before the surface reaches 0.010 in. equivalent sandgrain roughness. The upperlimit value would be smaller if the formula of Thyson et al.[1] or Nestler[21] was used. On the other hand, in view of the accuracy of the rough surface heat transfer results for lack of a more complete experimental data base, the upperlimit value is within the range of the roughness correction factor employed by Denman and Minges[20] and based on extensive AEDC-NOL empirical heat transfer data obtained under Air Force sponsorship on spherical and biconic shapes. The approximate method also is found to yield reasonable predictions of the NOL data presented by Thyson et al.[1]

The results indicated that rough surface turbulent heating greatly increases the side-wall recession. As the vehicle descends, accurate knowledge of the movement of the transition point along the nosetip surface and of the growth rate of surface roughnesses is therefore important.

The large variation of the mass transfer parameter for the different graphite ablation chemistry models, as shown in Fig. 2, will yield variations of shape change predictions. This is demonstrated in Fig. 9 by the results for the slender nosetip. To isolate the effects of the chemistry models, the particle removal correction is not included in the new predictions. The Dolton rate-controlled sublimation model[9] is seen to yield lower recessions than the Dolton equilibrium model. Particle removal is effective for the stagnation region as shown by comparison of the results for the equilibrium model.

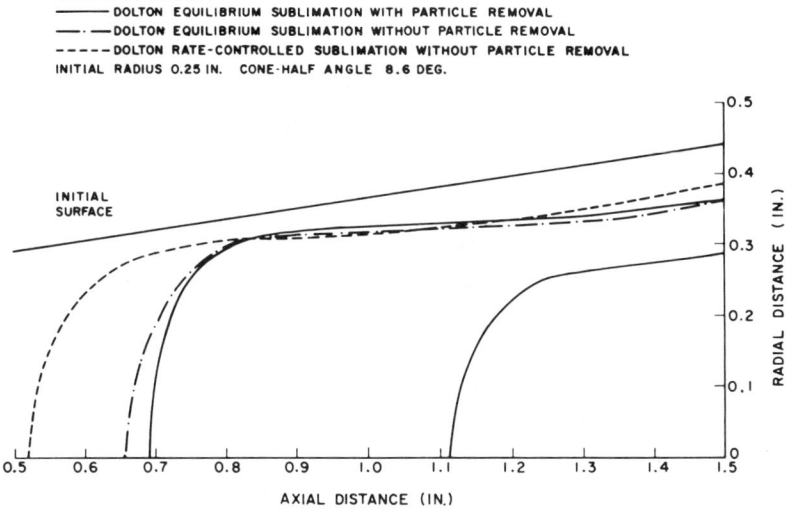

Fig. 9 Effects of graphite ablation chemistry model on shape change-slender nosetip.

It is concluded from this study that accurate prediction of nosetip response requires coupled calculation of shape change, heating distribution and internal conduction. Further theoretical and experimental results also are needed to reduce the uncertainty of computation models including graphite ablation chemistry, particle removal effects, and roughed surface heating for ablating bodies.

## References

[1] Thyson, N., Neuringer, J., Pallone, A., and Chen, K.K., "Nose Tip Shape Change Predictions During Atmospheric Re-entry", AIAA Paper 70-827, Los Angeles, Calif., 1970.

[2] Popper, L.A., Toong, T.Y. and Sutton, G.W., "Three-Dimensional Ablation Considering Shape Change and Internal Heat Conduction", AIAA Journal, Vol. 8, No. 11, Nov. 1970, pp. 2071-2074.

[3] Gallagher, L. W., Coleman, W. D. and Butler, W. R., "Advanced Composities II (RESEP II) CO No. P001 Vol. I, Computer User's Manual, Response of Charring Ablator Nosetips - Version 2", LMSC N-16-69-6, SAMSO-TR-70-16, Vol. I, Sept. 6, 1969, Lockheed Missiles & Space Co., Sunnyvale, Calif.

[4] Chin, Jin H., "Advanced Composites II (RESEP II) CO No. P001 Vol. II. Computer User's Manual, Prediction of the Response of Graphite Nosetips - Version 2", LMSC N-16-69-5, SAMSO-TR-70-16, Vol. II, Sept. 6, 1969, Lockheed Missiles & Space Co., Sunnyvale, Calif.

[5] Chin, Jin H. and Butler, W., "Nosetip Design Analysis and Test Program (NDAT) Vol. II. Computer User's Manual for the Graphite Nosetip Thermal and Structural Response (GRANTS)", LMSC D032066, SAMSO-TR-71-11, Vol. II, Dec. 15, 1970, Lockheed Missiles & Space Co., Sunnyvale, Calif.

[6] Larkins, B. K., "Some Finite Difference Methods for Problems in Transient Heat Flow", Heat Transfer, Cleveland, Chemical Engineering Progress Symposium Series, Vol. 61, No. 59, 1965, pp. 1-11.

[7] Barakat, H. E. and Clark, J. A., "On the Solution of the Diffusion Equations by Numerical Methods," Journal of Heat Transfer, Vol. 88C, No. 4, Nov. 1966, pp. 421-427.

[8] Kratsch, K. M., Hearne, L. F., and McChesney, H. R., "Theory for the Thermophysical Performance of Charring Organic Heat Shield Composites," LMSC 803099, 2-60-63-7, Oct. 18, 1963, Lockheed Missiles & Space Co., Sunnyvale, Calif.

[9] Dolton, T. A., Goldstein, H. E. and Maurer, R. E., "Thermodynamic Performance of Carbon on Hyperthermal Environments," AIAA Progress in Astronautics and Aeronautics: Thermal Design Principles of Spacecraft and Entry Bodies, Vol. 21, edited by Jerry T. Bevans, Academic Press, New York, 1969, pp. 169-201.

[10] Maurer, R. E., "Correlation Constants for Graphite Ablation Chemistry Models," private communication, Sept. 15, 1970, Lockheed Missiles & Space Co., Sunnyvale, Calif.

[11] Kratsch, K. M., Martinez, M. R., Clayton, F. I., Greene, R. B., and Wuerer, J. E., "Graphite Ablation in High-Pressure Environments," AIAA Paper 68-1153, Williamsburg, Va., 1968.

[12] Cohen, H. B., "Boundary-Layer Similar Solutions and Correlation Equations for Laminar Heat Transfer Distribution in Equilibrium Air at Velocities up to 41,000 Feet Per Second," TR-118, 1961, NASA.

[13] Hearne, L. F., Chin, J. H., and Woodruff, L. W., "Final Report, Study of Aerothermodynamic Phenomena Associated with Reentry of Manned Spacecraft," LMSC Y-78-66-1, May 1966, Lockheed Missiles & Space Co., Sunnyvale, Calif.

[14] Persh, J., "A Procedure for Calculating the Boundary Layer Development in the Region of Transition from Laminar to Turbulent Flow," NAVORD Report 4438, March 1957, Naval Ordnance Lab., White Oak, Md.

[15] Cresci, R. J., MacKenzie, D. A., and Libby, P. A., "An Investigation of Laminar, Transitional, and Turbulent Heat Transfer on Blunt-Nosed Bodies in Hypersonic Flow," Journal of the Aerospace Sciences, Vol. 27, No. 6, June 1960, pp. 401-414.

[16] Chin, Jin H., "Effects of Surface Roughness on Heat Transfer to Ablating Bodies," Journal of Spacecraft and Rockets, Vol. 8, No. 7, July 1971, pp. 804-806.

[17] Fenter, F. W., "The Turbulent Boundary Layer on Uniformly Rough Surfaces at Supersonic Speeds," Report DRL-437, Jan. 1960, Defense Research Lab., The University of Texas, Austin, Texas.

[18] Dipprey, D. F. and Sabersky, R. H., "Heat and Momentum Transfer in Smooth and Rough Tubes at Various Prandtl Numbers," International Journal of Heat and Mass Transfer, Vol. 6, No. 5, May 1963, pp. 329-353.

[19] Welsh, W. E., Jr., "Shape and Surface Roughness Effects on Nosetip Ablation," AIAA Journal, Vol. 8, No. 11, Nov. 1970, pp. 1983-1989.

[20] Denman, G. L. and Minges, M. L., "High Pressure Ablation of Plastic Composites and Graphites in the 50 Megawatt Arc," AIAA Paper 70-770, Los Angeles, Calif., 1970.

[21] Nestler, D. E., "Compressible Turbulent Boundary Layer Heat Transfer to Rough Surfaces," AIAA Journal, Vol. 9, No. 9, Sept. 1971, pp. 1799-1803.

# TEMPERATURE UNCERTAINTIES ASSOCIATED WITH SPACECRAFT THERMAL ANALYSES

R. G. Goble*

Martin Marietta Corporation, Denver, Colo.

## Abstract

The primary purpose of this study is to provide a method for determining those steady-state temperature uncertainties that result because the values of the physical parameters used in the thermal analysis of a system are not known exactly. Influence coefficients, obtained by partial differentiation of the steady-state heat balance equation, are used to statistically establish the temperature uncertainties for a typical orbiting spacecraft. The results indicate that the temperatures of those vehicle surfaces that receive direct solar radiation are most influenced by variations in the vehicle surface optical properties while those surfaces that receive no direct solar radiation are most affected by variations in the emissive power of the earth.

## Nomenclature

$a_{ij}$ = linear conductance (=KA/L or hA), BTU/hr-°F
$A$ = linear conduction area, ft$^2$
$A_i$ = nodal area of ith node, ft$^2$
ALB = albedo constant
$b_{ij}$ = radiation interchange coefficient (=$\sigma \mathcal{F}_{ij} A_i$), BTU/hr-°F$^4$
B = an N x N coefficient matrix with general element

$$B_{ij} = \delta_{ij} \sum_{m=1}^{NT} \left(1-\delta_{im}\right)\left(a_{im}+4b_{im}T_i^3\right) - \left(1-\delta_{ij}\right)\left(a_{ij}+4b_{ij}T_j^3\right)$$

$B\alpha$ = an NS x NS coefficient matrix with general element

$$B_{\alpha_{ij}} = \delta_{ij}\left(1 - F_{ii}\rho_{\alpha_i}\right) - \left(1 - \delta_{ij}\right)F_{ij}\rho_{\alpha_j}$$

---

Presented as Paper 71-430 at the AIAA 6th Thermophysics Conference, Tullahoma, Tenn., April 26-28, 1971. This work was performed under Contract No. NAS8-24000, sponsored by the National Aeronautics and Space Administration.
*Senior Engineer, Thermophysics Section.

$B_\epsilon$ = an NS x NS coefficient matrix with general element
$$B_{\epsilon_{ij}} = \delta_{ij}\left(1 - F_{ii}\rho_{\epsilon_i}\right) - \left(1 - \delta_{ij}\right) F_{ij}\rho_{\epsilon_j}$$
$F_{ij}$ = view factor from ith node to jth node
$\mathcal{F}_{ij}$ = radiation interchange factor from ith node to jth node
$h$ = film coefficient for convective heat transfer, BTU/hr-ft$^2$-°F
$K$ = thermal conductivity, BTU/hr-ft-°F
$L$ = linear conduction length, ft
$n$ = number of influencing parameters considered in an analysis
$N$ = number of nodes with variable temperatures
NS = number of nodes in model (excluding space)
NT = number of nodes in model (including space)
$P_m$ = general nomenclature for influencing parameters
$P_i$ = internal power generated at ith node, BTU/hr
$Q_i$ = total heat input at ith node, BTU/hr
$QIA_i$ = incident albedo heat flux at ith node, BTU/hr-ft$^2$
$QIP_i$ = incident planetary heat flux at ith node, BTU/hr-ft$^2$
$QIS_i$ = incident solar heat flux at ith node, BTU/hr-ft$^2$
QS = solar constant, BTU/hr-ft$^2$
QP = planetary emission, BTU/hr-ft$^2$
$T_i$ = temperature of ith node, °R
w = watts
$\alpha_i$ = short wavelength absorptivity of ith node
$\beta\alpha_{ij}$ = short wavelength radiation absorption factor (portion of energy leaving node i that is finally absorbed by node j)
$\beta\epsilon_{ij}$ = long wavelength radiation absorption factor (portion of energy leaving node i that is finally absorbed by node j)
$\Delta T_i$ = temperature uncertainty of ith node, °F
$\delta_{ij}$ = Kronecker delta ($\delta_{ij}=1$, i=j; $\delta_{ij}=0$, i≠j)
$\epsilon_i$ = long wavelength emissivity and absorptivity of ith node
$\rho\alpha_i$ = short wavelength reflectivity of ith node (=1-$\alpha_i$)
$\rho\epsilon_i$ = long wavelength reflectivity of ith node (=1-$\epsilon_i$)
$\sigma$ = Stefan-Boltzmann constant, 1.714x10$^{-9}$ BTU/hr-ft$^2$-°R$^4$
$\partial T_i/\partial p_m$ = influence coefficient relating changes in temperature of the ith node to variations in the influencing parameter, $p_m$

Abbreviations

AM = airlock module
ATM = Apollo telescope mount
CBRM = charger/battery/regulator/module
CSM = command and service modules

FAS  = fixed airlock shroud
MDA  = multiple docking adapter
OA   = orbital assembly
OWS  = orbital workshop

### Symbols

$[\ ]$  = square matrix
$\{\ \}$ = column matrix

### Introduction

Temperatures obtained by the thermal analysis of a system are not exact, but are subject to many sources of uncertainty which can be divided into the following general categories: 1) analyst errors, 2) failure of the analytical model to accurately simulate the system, 3) the use of numerical approximations in the computations, and 4) inaccuracies in the physical parameters used in the analysis.

In general, it is not possible to assess the temperature uncertainties that result from analyst errors of from the failure of the model to simulate the actual system. These uncertainties must be minimized by the careful application of standard modeling techniques within the constraints imposed by computer program size limitations and by budgetary considerations. Temperature uncertainties resulting from the use of numerical approximations can be estimated by analyzing the assumptions and techniques used in the various computer programs involved in the analysis. This paper, however, is concerned only with those temperature uncertainties that can be attributed to inaccuracies in the physical parameters.

### Method of Analysis

Due to the complexity of the analysis, this paper is concerned with the steady-state case only. To determine uncertainties in the predicted temperatures, it becomes necessary to obtain "influence coefficients" that relate variations in each of the influencing parameters to each of the nodal temperatures. This is accomplished by partial differentiation of the steady-state heat balance equation following the general method presented by Ishimoto and Bevans.[1] Table 1 shows the matrix equations obtained by this procedure for each of the influencing parameters considered.

The following assumptions and ground rules were used in the development and application of the equations presented in Table 1: 1) radiating surfaces of the system to be analyzed

Table 1 Matrix equations used in determining influence coefficients

| Influencing parameter | Matrix equation |
|---|---|
| QS | $\left\{\dfrac{\partial T_i}{\partial QS}\right\} = \left[B_{ij}\right]^{-1} \left\{\alpha_i A_i \dfrac{QIS_i + QIA_i}{QS} + \sum\limits_{m=1}^{NS} \rho_\alpha \beta_\alpha \alpha_m A_m \dfrac{QIS_m + QIA_m}{QS}\right\}$   $i=1,2,\cdots,N$ |
| ALB | $\left\{\dfrac{\partial T_i}{\partial ALB}\right\} = \left[B_{ij}\right]^{-1} \left\{\alpha_i A_i \dfrac{QIA_i}{ALB} + \sum\limits_{m=1}^{NS} \rho_\alpha \beta_\alpha \alpha_m A_m \dfrac{QIA_m}{ALB}\right\}$   $i=1,2,\cdots,N$ |
| QP | $\left\{\dfrac{\partial T_i}{\partial QP}\right\} = \left[B_{ij}\right]^{-1} \left\{\varepsilon_i A_i \dfrac{QIP_i}{QP} + \sum\limits_{m=1}^{NS} \rho_\varepsilon \beta_\varepsilon \varepsilon_m A_m \dfrac{QIP_m}{QP}\right\}$   $i=1,2,\cdots,N$ |
| $P_k$ | $\left\{\dfrac{\partial T_i}{\partial P_k}\right\} = \left[B_{ij}\right]^{-1} \left\{\delta_{ik}\right\}$   $i=1,2,\cdots,N$   $k=1,2,\cdots,N$ |
| $A_k$ | $\left\{\dfrac{\partial T_i}{\partial A_k}\right\} = \left[B_{ij}\right]^{-1} \left\{\left(\alpha_i \delta_{ik} + \rho_\alpha \beta_\alpha \alpha_{ki}\right)(QIS_k + QIA_k) + \left(\varepsilon_i \delta_{ik} + \rho_\varepsilon \beta_\varepsilon \varepsilon_{ki}\right) QIP_k + \delta_{ik} \sigma \varepsilon_i \sum\limits_{m=1}^{NT} \beta \varepsilon_{im} \left(T_m^4 - T_i^4\right)\right\}$   $i=1,2,\cdots,N$   $k=1,2,\cdots,NS$ |
| $T_k$ | $\left\{\dfrac{\partial T_i}{\partial T_k}\right\} = \left[B_{ij}\right]^{-1} \left\{a_{ik} + 4b_{ik} T_k^3\right\}$   $i=1,2,\cdots,N$   $k=N+1, N+2, \cdots, NT$ |
| $\alpha_k$ | $\left\{\dfrac{\partial T_i}{\partial \alpha_k}\right\} = \left[B_{ij}\right]^{-1} \left\{A_k(QIS_k + QIA_k)\left(\delta_{ik} - \beta_{\alpha_{ki}}\right) + \sum\limits_{m=1}^{NS} \rho_\alpha A_m (QIS_m + QIA_m) \dfrac{\partial \beta_{\alpha_{mi}}}{\partial \alpha_k}\right\}$   $i=1,2,\cdots,N$   $k=1,2,\cdots,NS$ |

where

$$\dfrac{\partial \beta_{\alpha_{ir}}}{\partial \alpha_k} = \left[B_{\alpha_{ij}}\right]^{-1} \left\{\left(\delta_{kr} - \beta_{\alpha_{kr}}\right) F_{ik}\right\} \quad i=1,2,\cdots,N \quad k=1,2,\cdots,NS \quad r \text{ successively fixed at } 1,2,\cdots,NS$$

# SPACECRAFT TEMPERATURE UNCERTAINTIES

$$\left\{\frac{\partial T_i}{\partial \varepsilon_k}\right\} = \left[B_{ij}\right]^{-1}\left\{A_k \, QIP_k\left(\delta_{ik} - \beta_{\varepsilon_{ki}}\right) + \sum_{m=1}^{NS}\rho_{\varepsilon_m} A_m QIP_m \frac{\partial \beta_{\varepsilon_{mi}}}{\partial \varepsilon_k} + \sigma A_i \sum_{m=1}^{NT}\left(\beta_{\varepsilon_{im}}\delta_{ik} + \varepsilon_i \frac{\partial \beta_{\varepsilon_{im}}}{\partial \varepsilon_k}\right)\left(T_m^4 - T_i^4\right)\right\}$$

$$i=1,2,\cdots,N \quad k=1,2,\cdots,NS$$

where

$$\left\{\frac{\partial \beta_{\varepsilon_{ir}}}{\partial \varepsilon_k}\right\} = \left[B_{\varepsilon_{ij}}\right]^{-1}\left\{\left(\delta_{kr} - \beta_{\varepsilon_{kr}}\right)F_{ik}\right\} \quad i=1,2,\cdots,N \quad k=1,2,\cdots,NS$$

$r$ successively fixed at $1,2,\cdots,NS$

$$\left\{\frac{\partial T_i}{\partial a_{kq}}\right\} = \left[B_{ij}\right]^{-1}\left\{\left(T_q - T_k\right)\left(\delta_{ik} - \delta_{iq}\right)\right\} \quad i=1,2,\cdots,N \quad k=1,2,\cdots,NS \quad q=k,k+1,\cdots,NS$$

$$\left\{\frac{\partial T_i}{\partial F_{kq}}\right\} = \left[B_{ij}\right]^{-1}\left\{\sum_{m=1}^{NS}\rho_{\varepsilon_m}\left(QIS_m + QIA\right)A_m \frac{\partial \beta_{\alpha_{mi}}}{\partial F_{kq}} + \sum_{m=1}^{NS}\varepsilon_m QIP_m A_m \frac{\partial \beta_{\varepsilon_{mi}}}{\partial F_{kq}} + \sigma \varepsilon_i A_i \sum_{m=1}^{NT}\left(T_m^4 - T_i^4\right)\frac{\partial \beta_{\varepsilon_{im}}}{\partial F_{kq}}\right\}$$

$$i=1,2,\cdots,N \quad k=1,2,\cdots,NS \quad q=k,k+1,\cdots,NT$$

where

$$\left\{\frac{\partial \beta_{\alpha_{ir}}}{\partial F_{kq}}\right\} = \left[B_{\alpha_{ij}}\right]^{-1}\left\{\delta_{ik}\left(\alpha_r \delta_{rq} + \rho_{\alpha_q}\beta_{\alpha_{qr}}\right) + \delta_{iq}\left(1-\delta_{kq}\right)\left(\alpha_r \delta_{rk} + \rho_{\alpha_k}\beta_{\alpha_{kr}}\right)A_k/A_q\right\}; \quad i=1,2,\cdots,NS$$

$k=1,2,\cdots,NS \quad q=k,k+1,\cdots,NT$

$r$ successively fixed at $1,2,\cdots,NS$

and

$$\left\{\frac{\partial \beta_{\varepsilon_{ir}}}{\partial F_{kq}}\right\} = \left[B_{\varepsilon_{ij}}\right]^{-1}\left\{\delta_{ik}\left(\varepsilon_r \delta_{rq} + \rho_{\varepsilon_q}\beta_{\varepsilon_{qr}}\right) + \delta_{iq}\left(1-\delta_{kq}\right)\left(\varepsilon_r \delta_{rk} + \rho_{\varepsilon_k}\beta_{\varepsilon_{kr}}\right)A_k/A_q\right\} \quad i=1,2,\cdots,NS$$

$k=1,2,\cdots,NS \quad q=k,k+1,\cdots,NT$

$r$ successively fixed at $1,2,\cdots,NS$

353

are considered to be semigray (accounts for absorption and reflection but no emission in the short wavelength region; and accounts for absorption and reflection as well as emission in the long wavelength region), 2) all system surfaces emit and reflect diffusely (emitted or reflected radiation has the same flux density in all directions), 3) influencing parameters are considered to be independent and normally distributed, 4) the influencing parameters whose effects are to be investigated are the solar constant, planetary albedo, planetary emission, internal power to be dissipated, nodal areas, boundary temperatures, surface optical properties ($\alpha$ and $\epsilon$), linear conductors, and view factors.

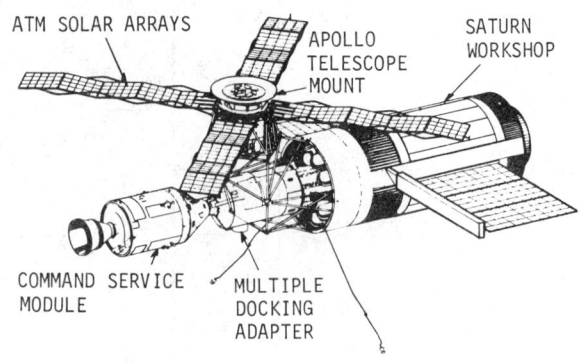

Fig. 1  Skylab orbital assembly.

A computer program called SANDEA (Sensitivity AND Error Analysis) has been written for use in conjunction with a thermal radiation analyzer program such as MTRAP[2] (which calculates view factors, incident environmental heat fluxes, radiation interchange factors, and orbital average heat rates) and a thermal analyzer program such as MITAS[3] (which calculates orbital average nodal temperatures). SANDEA solves the matrix equations given in Table 1, thereby determining the influence coefficients for each of the influencing parameters, and calculates the steady-state temperature uncertainties by means of

$$\Delta T_i = \left[ \sum_{m=1}^{n} \left( \frac{\partial T_i}{\partial p_m} \Delta p_m \right)^2 \right]^{1/2} \qquad i=1,2,\cdots,N \qquad (1)$$

Although Eq. (1) can be rigorously applied only for normally distributed variables, it also yields accurate results for variables with frequency distributions other than normal.[4]

### Application of the Method

The procedure described here was applied for a typical mission of the operational Skylab OA which is formed by the Saturn OWS, the MDA, the CSM, and the ATM with deployed solar panel arrays arranged as shown in Fig. 1.

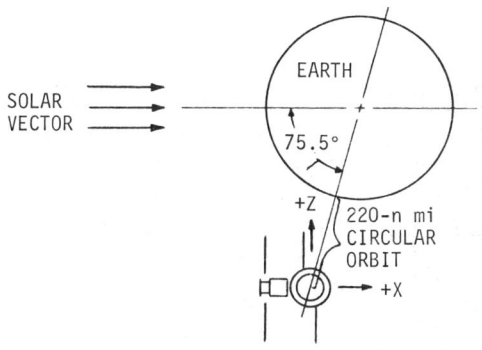

Fig. 2 Vehicle orientation and orbital parameters.

Hot case environmental constants were used in the analysis with the OA in a 220-nautical mile (n mi) circular orbit. The orbital plane was inclined at an angle of 75.5° to the earth-sun line as shown in Fig. 2. The vehicle was solar oriented with the ATM toward the sun and the OWS trailing at the subsolar point of the orbit as illustrated in Fig. 2.

Table 2 shows the influencing parameters considered, their expected values, and/or the uncertainties applied to them.

Table 2  Influencing parameter values and/or uncertainties

| Influencing parameter | Expected value | Uncertainty |
|---|---|---|
| [a]Solar constant, BTU/hr-ft$^2$ | 444.0 | ±13.0 |
| [a]Albedo constant | 0.30 | ± 0.12 |
| [a]Earth emission, BTU/hr-ft$^2$ | 77.7 | ±13.4 |
| Internal power, w | --- | ±10% |
| Nodal areas, ft$^2$ | --- | ± 5% |
| Boundary temperatures, °R | --- | ±10.0 |
| Surface optical properties-- | | |
| White paint- | | |
| $\alpha$ | 0.5 | ±20% |
| $\epsilon$ | 0.9 | +11% −20% |
| Black paint- | | |
| $\alpha$ | 0.9 | +11% −20% |
| $\epsilon$ | 0.9 | +11% −20% |
| Linear conductors, BTU/hr-°F | --- | ±20% |
| View factors | --- | ±10% |

[a] The expected values and the uncertainties used for the solar constant and the albedo constant were those presented in NASA TM X-53457.[5] The expected value for the solar constant includes a 3.4% seasonal bias. The value for earth emission was obtained from the relationship, QP=QS(1-ALB)/4, for thermal equilibrium of the earth.

## Description of Model

The external surfaces of the Skylab OA were modeled in a manner compatible with the MTRAP and SANDEA programs, thereby forming an analytical model which described the vehicle configuration and specified the optical properties of the vehicle surfaces. Figures 3 and 4 show the model configuration and the nodal breakdown used in the analysis.

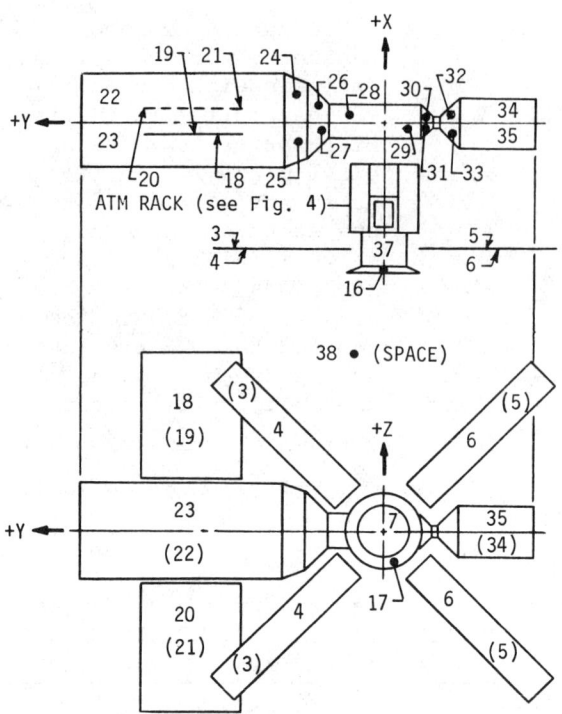

Fig. 3  Analytical model nodal assignments.

Of particular interest in this study were the hot case temperature uncertainties of the CBRMs and the ATM rack which is an octagonal structure that supports the experiment canister and various auxiliary equipment such as the CBRMs. Considering this interest in hot case temperature uncertainties, node 37 (the ATM experiment canister radiator) was treated as a boundary node by maintaining its temperature at its highest expected average value of 510 °R.

Steady-state (orbital average) temperatures were obtained for all other nodes in the model excluding the space node (38) by use of the MITAS program to impress on the model the orbital average heat rates calculated by MTRAP and the internal power generated by the various

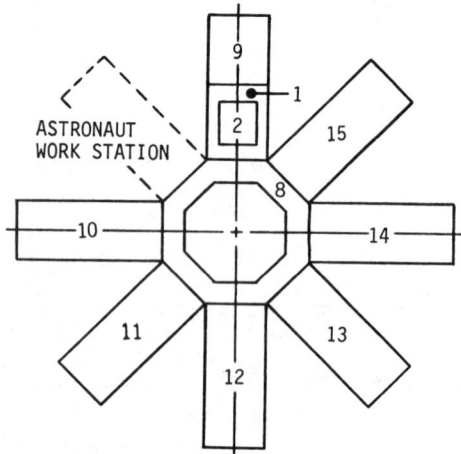

Fig. 4  ATM rack nodal assignments.

ATM rack-mounted components. Table 3 shows the internal power that must be dissipated and indicates the nodes to which this power was applied.

Table 3  Internal power to be dissipated.

| Node | Power, w |
|------|----------|
| 2    | 429.0    |
| 9    | 47.3     |
| 10   | 355.6    |
| 11   | 309.6    |
| 12   | 476.3    |
| 13   | 249.5    |
| 14   | 476.3    |
| 15   | 251.5    |

### Results

Steady-state (orbital average) temperature uncertainties are presented in Table 4 for the hot case Skylab mission described. The contribution of the individual influencing parameter variations to the uncertainty of each of the nodal temperatures is also given.

A study of the results indicates that, in general, the temperatures of those nodes that receive direct solar radiation are most influenced by variations in the vehicle surface optical properties while those nodes that receive no direct solar radiation are most affected by variations in the emissive power of the earth. The results also indicate that infrared radiation dominated nodes are significantly affected by uncertainties in the system view factors.

### Conclusion

The method described herein for use in the determination of steady-state temperature uncertainties is readily adaptable to the thermal analysis of any system for which the "nodal" technique can be applied in the construction of a representative analytical model.

It should be pointed out that a fundamental problem exists in the ability to establish the uncertainties associated with some of the influencing parameters. Data is available in the literature for parameters such as the solar constant, planetary albedo, planetary emission, and surface optical properties, but establishment of uncertainties in calculated parameters such as view factors is an ambitious study in itself. In analyzing the effects of such parameters, the utilitarian value of the method presented here lies, not so much in the ability to determine the associated temperature uncertainties, but more in the ability to indicate to which of these parameters the nodal temperatures are most sensitive. This knowledge then points out where greater care must be exercised in the construction of the model and in the determination of input parameter values.

Table 4  Skylab steady-state temperature uncertainties, hot case

Contributions of individual influencing parameter variations to nodal temperature uncertainties, $\pm °F$

| Node | Surface | $\Delta T_i,$ $\pm °F$ | QS | ALB | QP | $P_k$ | $A_k$ | $T_k$ | k | k | $a_{kq}$ | $F_{kq}$ |
|---|---|---|---|---|---|---|---|---|---|---|---|---|
| 1 | CBRM mounting panel | 8.53 | 0.87 | 2.03 | 6.89 | 1.63 | 0.74 | 0.04 | 0.88 | 2.13 | 1.37 | 3.15 |
| 2 | CBRMs | 11.52 | 0.69 | 1.90 | 7.17 | 4.46 | 2.30 | 0.02 | 1.06 | 4.80 | 0.62 | 5.25 |
| 3 | ATM solar panel (+X,+Y) | 22.66 | 3.65 | 0.69 | 2.57 | 0.00 | 2.81 | 0.03 | 18.03 | 8.87 | 9.04 | 1.17 |
| 4 | ATM solar panel (-X,+Y) | 30.34 | 4.62 | 0.70 | 2.42 | 0.00 | 2.69 | 0.02 | 27.11 | 10.79 | 5.82 | 0.35 |
| 5 | ATM solar panel (+X,-Y) | 26.62 | 3.26 | 0.67 | 2.78 | 0.01 | 3.61 | 0.00 | 20.54 | 9.92 | 12.51 | 0.26 |
| 6 | ATM solar panel (-X,-Y) | 33.26 | 4.51 | 0.71 | 2.64 | 0.01 | 3.59 | 0.01 | 29.63 | 10.90 | 8.31 | 0.05 |
| 7 | Sun end of ATM canister | 23.18 | 4.25 | 0.48 | 2.25 | 0.00 | 0.02 | 0.00 | 16.04 | 16.01 | 0.00 | 0.08 |
| 8 | Sun end of ATM rack | 8.22 | 0.89 | 1.53 | 6.52 | 0.59 | 0.92 | 0.13 | 0.63 | 2.76 | 3.54 | 0.34 |
| 9 | Rack side (+Z) | 8.21 | 1.07 | 2.19 | 6.99 | 0.79 | 0.77 | 0.01 | 1.82 | 2.46 | 1.29 | 0.64 |
| 10 | Rack side (+Y) | 7.73 | 1.62 | 1.18 | 5.30 | 2.21 | 0.86 | 0.01 | 1.49 | 2.40 | 3.68 | 0.75 |
| 11 | Rack side (+Y,-Z) | 7.77 | 1.17 | 1.01 | 6.59 | 2.36 | 0.81 | 0.01 | 0.86 | 2.23 | 1.58 | 0.29 |
| 12 | Rack side (-Z) | 9.08 | 0.46 | 0.58 | 6.08 | 3.39 | 1.86 | 0.01 | 0.36 | 5.05 | 2.07 | 0.08 |
| 13 | Rack side (-Y,-Z) | 8.62 | 0.70 | 0.90 | 7.25 | 2.13 | 1.27 | 0.01 | 0.81 | 3.59 | 0.81 | 0.14 |
| 14 | Rack side (-Y) | 9.01 | 0.89 | 1.07 | 5.61 | 3.07 | 1.80 | 0.01 | 1.51 | 4.72 | 3.21 | 0.41 |
| 15 | Rack side (-Y,+Z) | 8.54 | 1.07 | 1.78 | 6.48 | 1.93 | 0.98 | 0.01 | 2.88 | 2.93 | 2.40 | 0.52 |
| 16 | Sun shield (+X) | 6.84 | 2.12 | 1.06 | 5.36 | 0.03 | 0.97 | 0.89 | 1.97 | 2.10 | 0.69 | 1.38 |
| 17 | Sun shield (-X) | 22.85 | 4.18 | 0.46 | 2.20 | 0.00 | 0.01 | 0.00 | 15.82 | 15.79 | 0.00 | 0.04 |

SPACECRAFT TEMPERATURE UNCERTAINTIES

| Node | Description | | | | | | | | | | | |
|---|---|---|---|---|---|---|---|---|---|---|---|---|
| 18 | OWS solar panel (-X,+Z) | 21.63 | 4.65 | 0.78 | 2.71 | 0.00 | 3.17 | 0.01 | 15.96 | 10.18 | 8.32 | 0.80 |
| 19 | OWS solar panel (+X,+Z) | 15.06 | 2.73 | 0.70 | 3.07 | 0.00 | 3.52 | 0.00 | 9.28 | 9.35 | 4.83 | 0.54 |
| 20 | OWS solar panel (-X,-Z) | 23.29 | 4.81 | 0.62 | 2.55 | 0.00 | 3.36 | 0.00 | 17.29 | 11.30 | 8.59 | 0.54 |
| 21 | OWS solar panel (+X,-Z) | 15.83 | 2.77 | 0.56 | 2.91 | 0.00 | 3.80 | 0.00 | 9.89 | 9.87 | 4.91 | 0.55 |
| 22 | OWS (+X) | 12.04 | 0.75 | 3.40 | 11.31 | 0.00 | 0.45 | 0.00 | 1.53 | 1.34 | 0.63 | 0.43 |
| 23 | OWS (-X) | 23.51 | 4.55 | 0.78 | 2.15 | 0.00 | 0.34 | 0.00 | 16.22 | 16.20 | 0.31 | 0.80 |
| 24 | FAS (+X) | 13.99 | 0.40 | 4.06 | 13.28 | 0.00 | 0.06 | 0.00 | 1.16 | 1.11 | 0.13 | 0.16 |
| 25 | FAS (-X) | 23.22 | 4.45 | 0.77 | 2.24 | 0.01 | 0.15 | 0.00 | 16.02 | 15.99 | 0.42 | 1.09 |
| 26 | AM (+X) | 14.28 | 0.41 | 4.51 | 13.41 | 0.00 | 0.07 | 0.00 | 1.30 | 1.22 | 0.23 | 0.31 |
| 27 | AM (-X) | 21.67 | 4.21 | 0.95 | 2.59 | 0.04 | 0.17 | 0.01 | 14.90 | 14.87 | 0.38 | 1.15 |
| 28 | MDA cylinder (+X) | 14.19 | 0.25 | 2.42 | 13.95 | 0.00 | 0.04 | 0.00 | 0.69 | 0.63 | 0.12 | 0.21 |
| 29 | MDA cylinder (-X) | 11.84 | 2.56 | 1.25 | 5.78 | 0.02 | 0.29 | 0.01 | 7.02 | 6.80 | 0.78 | 1.59 |
| 30 | MDA cone (+X) | 14.29 | 0.43 | 4.65 | 13.21 | 0.00 | 1.06 | 0.00 | 1.31 | 1.22 | 0.25 | 1.87 |
| 31 | MDA cone (-X) | 13.43 | 2.67 | 2.11 | 5.43 | 0.02 | 4.63 | 0.00 | 7.53 | 7.15 | 1.57 | 2.79 |
| 32 | Command module (+X) | 14.08 | 0.40 | 4.11 | 13.35 | 0.00 | 0.06 | 0.00 | 1.11 | 1.04 | 0.18 | 0.82 |
| 33 | Command module (-X) | 21.76 | 4.25 | 0.90 | 2.56 | 0.03 | 0.19 | 0.01 | 14.93 | 14.89 | 0.45 | 1.83 |
| 34 | Service module (+X) | 14.28 | 0.23 | 2.39 | 14.05 | 0.00 | 0.05 | 0.00 | 0.73 | 0.65 | 0.18 | 0.04 |
| 35 | Service module (-X) | 19.53 | 3.70 | 0.64 | 3.24 | 0.01 | 0.16 | 0.00 | 13.37 | 13.32 | 0.56 | 0.70 |

*The temperature uncertainty for any given node is equal to the root-sum-square of the uncertainty contributions of the individual influencing parameters.

## References

[1] Ishimoto, T. and Bevans, J. T., "Temperature Variance in Spacecraft Thermal Analysis," *Journal of Spacecraft and Rockets*, Vol. 5, No. 11, Nov. 1968, pp. 1372-1376.

[2] Holmstead, G. M., "Martin Thermal Radiation Analyzer Program, Vol. I - User's Manual," TM-0478-70-05, July 1970, Martin Marietta Corp., Denver, Colo.

[3] Kannady, R. E., et al., "Martin Interactive Thermal Analysis System," MDS-SPLPD-71-FD238 (Rev 2), July 1971, Martin Marietta Corp., Denver, Colo.

[4] Kline, S. J. and McClintock, F. A., "Describing Uncertainties in Single-Sample Experiments," *Mechanical Engineering*, Vol. 75, Jan. 1953, pp. 3-8.

[5] Weidner, D. K., Editor, "Space Environmental Criteria Guidelines for Use in Space Vehicle Development," NASA TM X-53457, 1969 Rev., Oct. 1969, George C. Marshall Space Flight Center, Huntsville, Ala.

SPACECRAFT THERMAL DESIGN
VERIFICATION THROUGH MODELING

R. K. MacGregor*

The Boeing Company, Seattle, Wash.

## Abstract

The results of a program to examine the interaction of numerical modeling with scale modeling for the verification of a spacecraft thermal design are presented. Both numerical and experimental modeling (half-scale) studies were conducted and thermal performance compared with that of a full-scale thermal model. The numerical model was upgraded through the use of scale model experimental data. The scale model results were corrected for known violations of the scaling criteria through use of the numerical model. It is concluded that the combined modeling approach adequately simulates thermal model performance and provides a numerical model for performance predictions beyond the range of chamber test conditions.

## Introduction

Spacecraft temperature control is a critical technical problem demanding careful design. The development of a spacecraft thermal control system involves a combination of analysis and test. Thermal tests are most often utilized in the system development program for establishing or verifying the design concept, or for verifying the analytical model. As spacecraft have grown in size and complexity, larger and more complex simulation facilities have been required to accomplish the necessary testing. It has been postulated that the use of small scale models for thermal tests would allow the use of smaller and less expensive test facilities.

---

Presented as Paper 71-439 at the AIAA 6th Thermophysics Conference, Tullahoma, Tenn., April 26-28, 1971. Portions of this work were performed for the NASA Marshall Space Flight Center under Contract NAS8-21422.
*Specialist Engineer.

Thermal scale modeling, then, is an alternative to full-scale thermal testing that becomes attractive when large spacecraft are involved. The basic similitude criteria for the radiation-conduction system of an unmanned spacecraft can be developed either from dimensional analysis or from the differential equations which describe the behavior of the system. Extensive studies [1-6] into the derivation of the scale modeling criteria and the application of these criteria to spacecraft have been presented. General scaling criteria and the equations governing the "temperature preservation technique" are summarized in Table 1.

The objective of this study is an examination of the interactions of numerical modeling and scale modeling techniques for the verification of thermal designs without the necessity of full-size thermal testing.

Experimental Investigation

Model Configuration

The vehicle geometric configuration, as shown in Fig. 1, was selected with ease of manufacturing being a primary consideration. Although the final design does not resemble an actual spacecraft, it does incorporate the following characteristics typical of flight hardware, namely, lightweight exterior skin panels, a relatively heavy structural frame, an over-extended

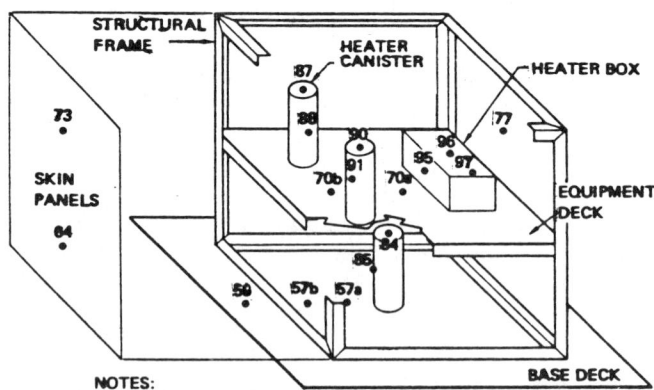

NOTES:
(1) ● THERMOCOUPLE LOCATIONS
(2) THERMOCOUPLES 91 AND 97 ARE HIDDEN FROM DIRECT VIEW
(3) CLOSURE DECK AND FRONT SKIN PANEL ARE NOT SHOWN
(4) THERMOCOUPLES 79 AND 81 ARE ON THE CLOSURE DECK
(5) THERMOCOUPLE 83 IS INTERNAL TO THE BASE DECK HEATER

Fig. 1  Schematic of the experimental configuration.

Table 1  Thermal Scale Modeling Criteria[a]

| | General criteria | | Temperature preservation criteria | |
|---|---|---|---|---|
| | Cartesian coordinates, 3-dimensional | Cartesian coordinates, geometric distortion | Cartesian coordinates, 3-dimensional | Cartesian coordinates, geometric distortion |
| | $k^* = L^* T^{*3}$ | $k^* = L^{*2} T^{*3}/d^*$ | $k^* = L^* = R$ | $k^* = L^{*2}/d^*$ |
| | $t^* = L^*(\rho c_p)^*/T^{*3}$ | $t^* = d^*(\rho c_p)^*/T^{*3}$ | $t^* = L^*(\rho c_p)^*$ | $t^* = d^*(\rho c_p)^*$ |
| | $q^* = k^* T^*/L^{*2}$ | $q^* = k^* T^*/L^{*2}$ | $q^* = 1/L^* = 1/R$ | $q^* = 1/d^*$ |
| | $S^* = k^* T^*/L^*$ | $S^* = d^* k^* T^*/L^{*2}$ | $S^* = 1$ | $S^* = 1$ |
| | $h^* = k^*/L^*$ | $h^* = k^*/L^*$ | $h^* = 1$ | $h^* = L^*/d^*$ |

[a]NOTE: a) * Denotes model/prototype ratio (i.e. $L^* = L_m/L_p$); b) assumes radiative properties preserved; c) assumes proportional temperature dependance for thermal conductivity and capacity. d = thickness (ft), h = joint conductance (Btu/hr-ft$^2$-°R), k = conductivity (Btu/hr-ft-°F), L = major dimension (ft), q = energy dissipation (Btu/hr-ft$^3$), R = scale ratio ($L_m/L_p$), S = energy flux (Btu/hr-ft$^2$), t = time (hr), T = temperature (°R), $(\rho c)_p$ = thermal capacity (Btu/ft$^3$-°R).

base deck which results in solar reflections back onto exterior surfaces, and energy sources interior to the spacecraft in discrete compartments to simulate electronics components.

The diameter of the simulated solar beam in the test chamber provided an upper limit on the prototype major dimensions of approximately 42 in. The final configuration took the form of a 20 in. cube on a 30-in. square plate, with the basic material being 1/16-in. 6061-T6 sheet stock. The pertinent dimensions and component design details are shown in Ref. 7. Nominal thermal properties are presented in Table 2.

The scale model was designed in accordance with the requirements of the temperature preservation technique of modeling at a nominal temperature of 535°R.

The two-dimensional "geometric distortion" scaling criteria (Table 1) was used to calculate the gages required for a wide range of materials for a nominal half-scale model. These results, for the half-scale model, indicated that 7075-T6 could be substituted for 6061-T6 at a scale ratio of 0.499 and 2024-0 could be substituted for 7075-T6 at a scale ratio of 0.496. This results in an absolute temperature error on the order of -0.6% at 535°R.

The "angle iron" frame was welded together and the skin panels and equipment deck were bolted to the frame. The bolts were sized and spaced such that the joint conductance was large compared to the conduction paths in the skin panels. A conducting silicone grease film along the joints also promoted conduction contact. As three heater elements were attached to the equipment deck it was made of a heavier gage to force stronger conductive coupling to the structural frame.

Table 2   Material thermal properties at 535°R

| Material | Density ($LB_M/FT^3$) | Specific Heat ($BTU/LB_M - °R$) | Thermal Conductivity ($BTU/HR-FT-°R$) |
|---|---|---|---|
| Aluminum silicate | 143.42 | 0.20 | 1.21 |
| 6061-T6 Aluminum | 169.34 | 0.215 | 96.6 |
| 7075-T6 Aluminum | 174.53 | 0.207 | 72.6 |
| 2024-0 Aluminum | 172.8 | 0.21 | 111.46 |

The heaters consisted of #28 gage (#36 gage on the half-scale model) nichrome wire in a helical wrap around the aluminum silicate heater core. Both ends of the nichrome wire were staked to the aluminum silicate where power leads (#24 gage wire) and voltage taps (#30 gage wire) were connected. The heater canister and box assemblies were also bolted together in such a manner as to minimize the resistance across the joints.

The surface finishes used on the vehicles consisted of one of two thermal control coatings or bare chem-cleaned aluminum. All exterior surfaces were coated with a 10-mil thickness of the Boeing developed B-1060 white thermal control coating. The single exception to this was the outer surface of the closure deck which was left as chem-cleaned aluminum. All interior surfaces were coated with a Sherwin-Williams flat-black thermal control coating. The one exception to this was the single side of the heater box which faced across the equipment deck enclosure to the two heater canisters. This surface was also left as bare aluminum. The radiative properties of these surfaces are presented in Table 3.

Instrumentation

Four regulated D. C. power supplies were connected to the spacecraft heaters. A decade box with fixed resistance was wired in series with each of the heaters. Voltage taps across both heater and decade resistances lead through a switch box to a Fluke digital voltmeter. The voltage taps across the heaters remove the energy dissipation in the lead wires from the calculation.

Twenty chromel-constantan thermocouples (paired wire with a double layer of stranded fiberglass insulation) were installed in each of the vehicles. Number 32 gage thermocouple wire was used in the half-scale model. All of wires were taken from the same spool. The locations of the thermocouples are shown in Fig. 1.

Two additional thermocouples were staked to small aluminum disks and used to monitor the chamber test conditions. One disk was tied into the chamber adjacent to and facing the shrouded cryo-wall. The other disk was supported in the vicinity of the model and faced into the solar beam.

The twenty-two thermocouples were calibrated relative to each other over a 250°R temperature range. Over this range the maximum deviation between the twenty-two thermocouples was ±0.002 millivolts. This corresponds to a temperature range of

Table 3    Radiative properties[a]

| Surface | | $\alpha_s$ | $\rho_s^s$ | $\rho_s^d$ | $\varepsilon_{IR}$ |
|---|---|---|---|---|---|
| Flat black on | 2024-0 | 0.962 | 0. | 0.038 | 0.841 |
| | 6061-T6 | 0.967 | 0. | 0.033 | 0.843 |
| | 7075-T6 | 0.967 | 0. | 0.033 | 0.875 |
| B-1060 on | 2024-0 | 0.184 | 0.014 | 0.802 | 0.894 |
| | 6061-T6 | 0.184 | 0.015 | 0.801 | 0.892 |
| | 7075-T6 | 0.181 | 0.012 | 0.807 | 0.905 |
| Aluminum | 2024-0 | 0.308 | 0.220 | 0.472 | 0.055 |
| | 6061-T6 | 0.241 | 0.212 | 0.547 | 0.052 |
| | 7075-T6 | 0.229 | 0.431 | 0.340 | 0.040 |

[a] $\alpha_s$ = solar absorptivity, $\rho_s^s$ = specular component of solar reflectivity, $\rho_s^d$ = diffuse component of solar reflectivity, $\varepsilon_{IR}$ = infrared emissivity.

±0.05°R. An absolute calibration was conducted on samples of wire from the same spool at the Boeing Metrology Laboratory. This calibration has NBS traceability and claims an accuracy of ±0.04°R for any particular calibration point.

The digital readout system utilized to obtain the test data, over the range of the test temperatures, has a digital least count of 0.6°R. As the digital least count far overshadowed the absolute calibration error and the spread of the thermocouple readings observed in the relative calibration, the standard NBS thermocouple calibration curve was used to reduce the data.

Space Environment Simulator

The tests were conducted in Chamber B at the Boeing Space Environment Simulation Laboratory utilizing a 4-ft solar simulator. A description is presented in Ref. 8. A brief description of the B chamber and the 4-ft solar simulator system follows.

The chamber is a vertical cylinder 10 ft in diameter and 18-ft high (Fig. 2). The top head contains the ion and sublimation pumping systems and supports the top and cylindrical portions of the cold-wall shroud, and the 47-in. by 56-in. off-axis parabolic collimating mirror. Test specimens are mounted on the bottom head and raised into the test zone by a hydraulic lift.

Vacuum pumping systems connected to the space simulator allow a variety of environmental conditions to be established from launch pressure profiles to long-term ultra-high vacuum as low as $10^{-11}$ torr. Top, sidewall and bottom cold-wall zones are cooled by 80 psia, subcooled, single-phase liquid nitrogen.

The solar beam is circular in cross section, measuring 42 in. in diameter. The height of the work zone is 96 in. The beam uniformity is $\pm 5\%$ at the base and upper planes and $\pm 4\%$ at the mid-plane position. Change in uniformity through the test volume depth is $\pm 1\%$. The apparent sun, as viewed from the test zone, subtends an average half angle of 1.8°.

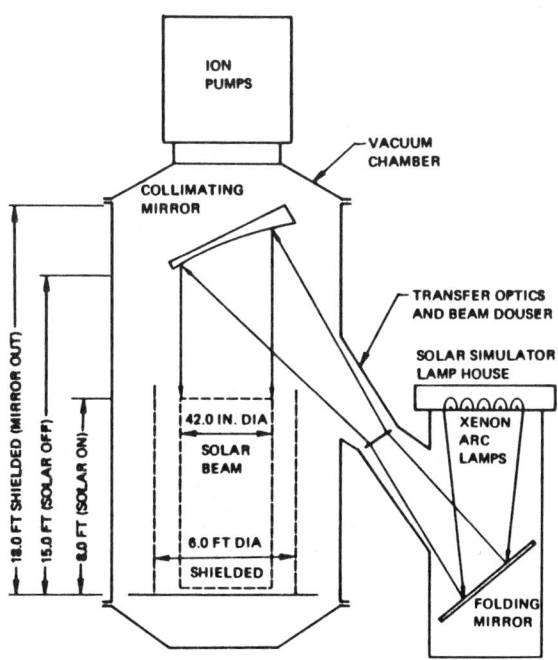

Fig. 2 Schematic of space environment chamber with solar simulator.

## Test Conditions

The test sequence and heater power levels are shown in Table 4. The test sequence covered a range of cases for both simulated solar and internal heater loads. The individual tests were conducted to allow a detailed examination of the behavior of the experimental vehicle as compared to the response of the numerical experiment.

Comparing the numerical and experimental results for test number one allows an evaluation of the interaction of the vehicle with the cold space environment and direct solar irradiation. Comparing the numerical and experimental results for

Table 4   Test sequence and heater power

|  | Prototype vehicle | Half-scale vehicle |
|---|---|---|
| Test #1 – Solar | On[a] | On |
| Power | Off | Off |
| Test #2 – Solar | On | On |
| Heater #1 | 119.62[b] | 30.14 |
| Test #3 – Solar | Off | Off |
| Heater #1 | 119.84 | 30.10 |
| Test #4 – Solar | On | On |
| Heater #2 | 60.01 | 15.11 |
| H   r #3 | 80.56 | 20.13 |
| Heater #4 | 119.59 | 30.11 |
| Test #5 – Solar | Off | Off |
| Heater #2 | 60.25 | 15.04 |
| Heater #3 | 80.81 | 20.15 |
| Heater #4 | 119.84 | 30.13 |

[a] Solar loads: prototype = 701.96 Btu/hr; half scale = 167.51 Btu/hr.
[b] Heater power levels. Nominal prototype power: heater #1 = 120, heater #2 = 60, heater #3 = 80, and heater #4 = 120 Btu/hr.

test number three allows an evaluation of the conduction-radiation dissipation from a heater canister and an evaluation of conduction paths internal to the vehicle. Comparing the numerical and experimental results for test number five allows an evaluation of the conduction-radiation interchange in geometrically complex vehicle enclosures. Comparing the experimental results of tests numbers two and four allows an evaluation of thermal scale modeling of spacecraft under complex external and internal environments.

The power dissipated by the spacecraft heaters during the tests is tabulated in Table 4. The small deviations from the nominal test condition are a result of inability to adjust the power supplies with adequate resolution and the temperature-dependent nature of the thermal resistance which allowed a drift in power dissipation over the transient portion of the test.

The nonuniformity of the solar beam results in a variation in actual power input to each of the external surfaces from that of an idealized one solar constant beam. Isointensity plots of the solar beam were taken prior to each test. The isointensity plots with a plan view of the spacecraft model superimposed are shown in Figs. 3 and 4. Figure 3 shows the prototype test condition. Figure 4 shows the test condition for the half-scale model. The isointensity plots were integrated over each external surface node and the absorbed solar load at each node was calculated using the surface area and solar absorptivity. The resulting solar heating rates for each exterior node are summarized in Table 4.

### Numerical Analysis

Substantial numerical analysis of these vehicle configurations was conducted for several reasons. First, a comparison of analysis and experiment allows a better understanding of the accuracy and limitations of numerical techniques. Second, the numerical model can be used to correct experimentally obtained temperatures for known compromises of the scaling criteria. Finally, a verified numerical model can be quickly "tested" for many environmental situations while actual chamber testing is expensive, time consuming, and often can not be made to achieve the desired external environmental heat loads.

### Thermal Analysis

The Boeing Thermal Analyzer program was used to reduce the thermal analysis network[9] and its time varying boundary condi-

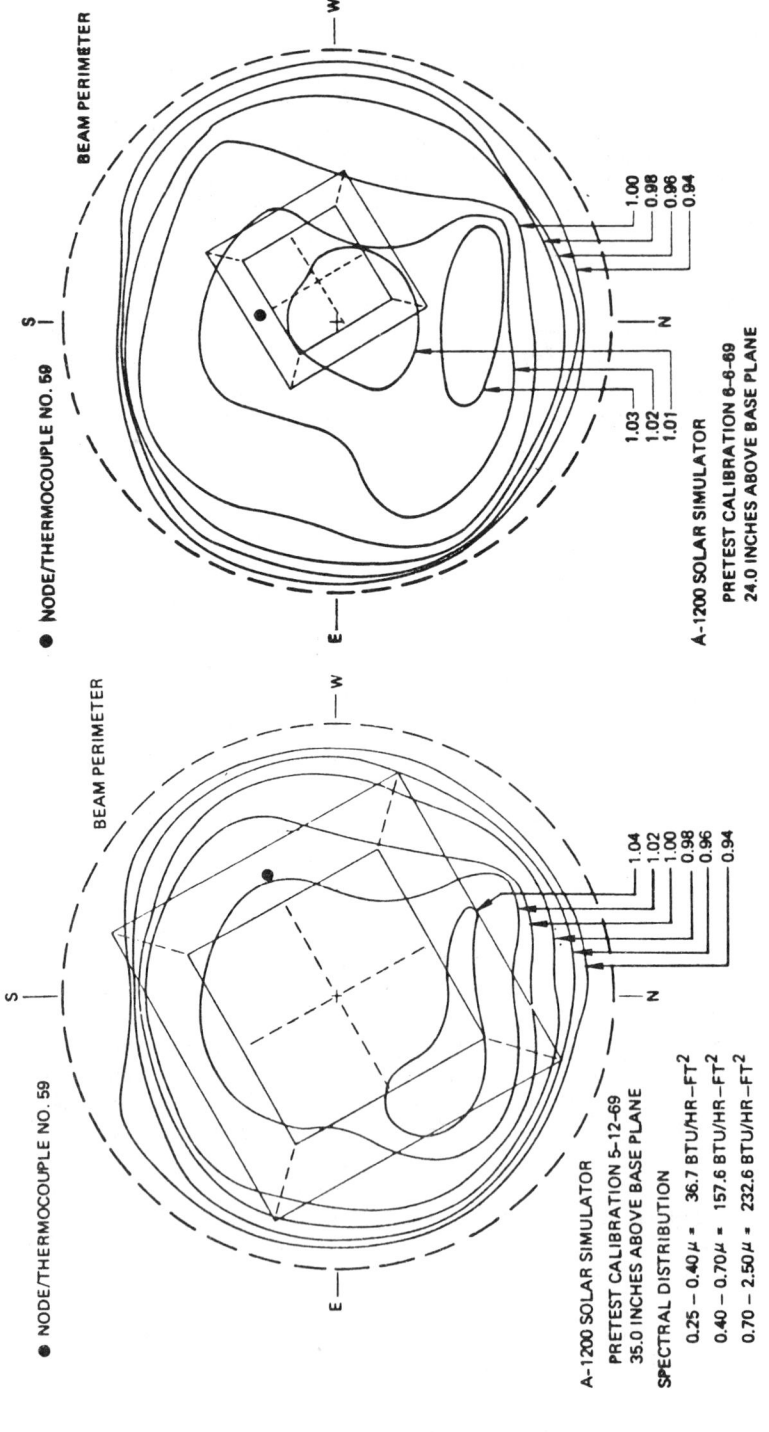

Fig. 3   Solar simulator isointensity plot overlaid by prototype plan view.

Fig. 4   Solar simulator isointensity plot overlaid by model plan view.

to a node by node temperature history of the model.
ive Interchange Factors (script $\mathcal{F}$'s) are computed using
e-Carlo based computer program.$^{10}$ This program utilizes
e emittance and specular-diffuse reflectance models to
terize the radiative properties of the surfaces.

### dal Model

rical modeling of the experiment requires modeling not
f the vehicle but modeling of the space simulation cham-
well. The finite size of the chamber and its nonblack
ter allow interactions which are not present under space
ions. Solar beam reflections and reflections of vehicle
d energy back onto the vehicle force a consideration of
amber in the numerical model.

ctuality, two nodal models were used. The detailed nodal
was used for thermal analysis while a simplified model
ed to determine the coefficients for radiation inter-
 between surfaces. As conductivity is not a considera-
n these latter calculations, only those nodes which have
icant surface areas or high temperatures are considered.
ural elements of high conductance but minimal surface
e.g., the structural frame, are omitted from the radia-
odal network with a negligible effect on the final ther-
alysis.

two nodal models utilized in this study are detailed in
. The radiative interchange model utilized 50 nodal
es while the thermal analyzer model utilized 102 nodes
33 conduction and 696 radiation connectors between nodes.

### Discussion of Results

he completion of this study, numerical and experimental
or both the prototype and half-scale vehicles were avail-
or comparison. In the following sections, these results
esented and compared. The discussion centers on three
: a comparison of numerical with experimental results
ch of the two vehicles, a comparison of the experimental
s, and a comparison of numerically adjusted half-scale
mental results with the prototype experimental data.

### cal and Experimental Results

bulated comparison of the numerical and experimental
s for the half-scale and prototype test sequences are

presented in Tables 5 and 6, respectively. The ratios of experimental to analytic temperatures (absolute) have been summarized in Figs. 5 and 6.

Examination of Figs. 5 and 6 show that the analyses for both the prototype and half-scale model have a 5-1/2% systematic error for test number three. Test number three was prolonged cold soak with only a single heater operating. The mean steady-state temperature was on the order of 270°R in both cases. A re-examination of the emissivity of the B-1060 white thermal control coating indicated a strong temperature dependency at low temperatures. The emissivity was subsequently measured as 0.7, at this low temperature. Incorporating the temperature-dependant emissivity of B-1060 into the analysis would eliminate this systematic error.

Examination of the temperature distributions across the base deck (Tables 5 and 6) indicates temperature differences as great as 45°R in a region assumed isothermal and represented by a single node in the numerical analysis. The noding in the base deck should have been more detailed.

Examination of the temperature distributions on the heater canisters and heater box when those heaters are operating (Tables 5 and 6) indicates that the heater temperatures are being poorly predicted. The heater core is noded as a single lumped node as shown in Fig. 7. The problem is one of obtaining the correct balance of radiation and conduction transport between the heater core and the heater shell. The present

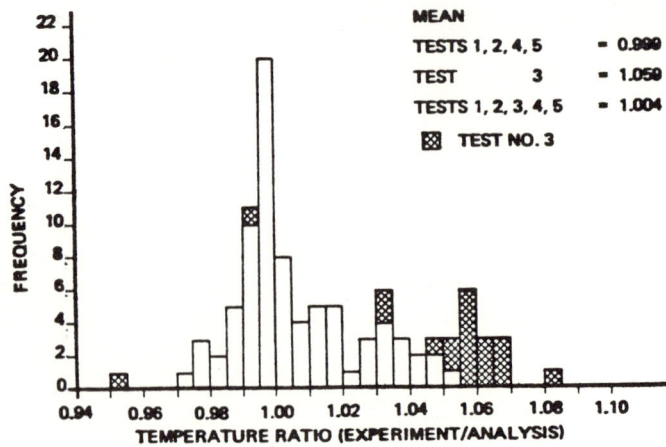

Fig. 5 Distribution of deviations between half-scale experiment and analysis.

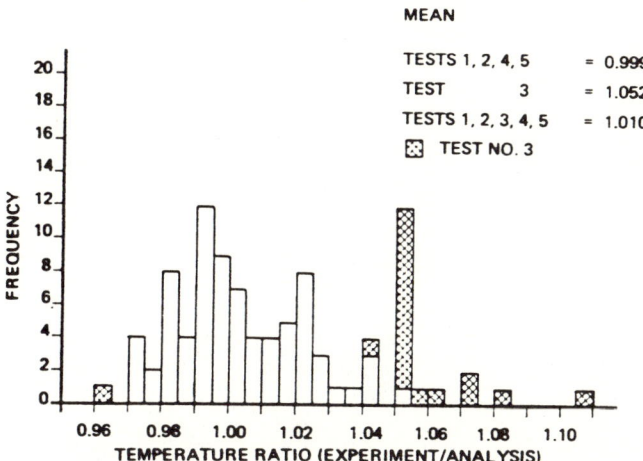

Fig. 6   Distribution of deviations between prototype experiment and analysis.

nodal model operates at a lower heater core temperature than the core surface which radiates predominately to the cylindrical portion of the surrounding shell. This nodal model thus forces more energy around the shell (nodal path 83-84-85-57) than occurs in the actual configuration. The proposed nodal model shown would correct this problem.

An examination of the results for nodes 79 and 81 on the external skin (on the sun facing upper closure) as shown in Tables 5 and 6 indicates relatively high temperatures predicted by analysis over the experimentally determined results. This is due to inaccuracies in the experimentally determined value of solar absorptivity/emissivity for the bare aluminum surface.

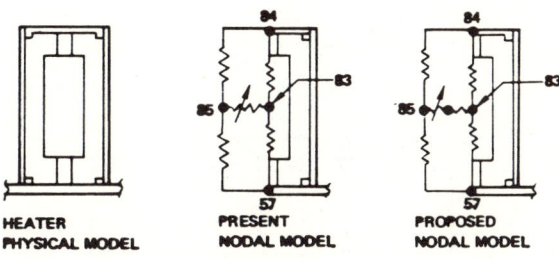

Fig. 7   Schematic of heater noding.

Table 5  Steady-state temperatures (half-scale model)

| Node | Test 1 Experiment (°R) On | Test 1 Analysis (°R) | Test 2 Experiment (°R) On | Test 2 Analysis (°R) | Test 3 Experiment (°R) Off | Test 3 Analysis (°R) | Test 4 Experiment (°R) On | Test 4 Analysis (°R) | Test 5 Experiment (°R) Off | Test 5 Analysis (°R) |
|---|---|---|---|---|---|---|---|---|---|---|
| Solar load | On | | On | | Off | | On | | Off | |
| Base deck | | | | | | | | | | |
| 57a | 381.5 | 365.4 | 428.1 | 408.6 | 318.6 | 304.0 | 408.0 | 396.0 | 310.6 | 299.3 |
| 57b | 384.3 | 365.4 | 410.0 | 408.6 | 288.7 | 304.0 | 412.7 | 396.0 | 313.0 | 299.3 |
| 59 | 391.4 | 400.7 | 407.3 | 413.1 | 275.0 | 253.9 | 414.7 | 423.6 | 308.2 | 297.8 |
| Heater one | | | | | | | | | | |
| 83 | 375.1 | 366.7 | 709.9[a] | 719.1 | 686.2[a] | 664.7 | 401.3 | 398.7 | 305.7 | 300.4 |
| 84 | 380.1 | 367.7 | 468.8 | 482.7 | 385.8 | 388.7 | 408.7 | 400.6 | 309.0 | 301.3 |
| 85 | 380.1 | 367.5 | 483.9 | 469.4 | 384.3 | 371.9 | 408.0 | 400.2 | 309.8 | 301.1 |
| Equipment deck | | | | | | | | | | |
| 70a | 398.5 | 398.8 | 410.7 | 414.3 | 268.0 | 254.6 | 458.6 | 459.4 | 358.9 | 355.3 |
| 70b | 398.5 | 398.8 | 410.7 | 414.3 | 268.0 | 254.6 | 455.5 | 459.4 | 354.4 | 355.3 |
| Heater two | | | | | | | | | | |
| 87 | 397.8 | 399.0 | 411.3 | 412.6 | 267.2 | 252.3 | 477.1[a] | 483.9 | 379.4[a] | 380.7 |
| 88 | 397.8 | 398.6 | 411.3 | 412.2 | 267.2 | 252.3 | 472.0 | 475.9 | 373.7 | 370.7 |
| Heater three | | | | | | | | | | |
| 90 | 398.5 | 398.8 | 411.3 | 412.4 | 267.2 | 252.4 | 488.8[a] | 493.6 | 393.6[a] | 392.3 |
| 91 | 397.8 | 398.4 | 412.7 | 412.0 | 268.0 | 252.4 | 482.1 | 483.7 | 385.8 | 379.8 |
| Heater four | | | | | | | | | | |
| 95 | 398.5 | 398.8 | 411.3 | 413.0 | 267.2 | 252.9 | 478.4[a] | 471.9 | 380.1[a] | 367.2 |
| 96 | 399.9 | 400.4 | 412.0 | 414.2 | 267.2 | 252.5 | 488.8 | 481.2 | 391.4 | 377.0 |
| 97 | 397.1 | 398.7 | 409.3 | 412.4 | 267.2 | 252.4 | 472.0 | 471.7 | 373.7 | 366.8 |
| External skin | | | | | | | | | | |
| 64 | 389.3 | 387.1 | 403.3 | 401.8 | 271.5 | 254.6 | 422.7 | 425.1 | 321.7 | 320.4 |
| 77 | 408.0 | 402.6 | 417.4 | 413.1 | 264.6 | 247.6 | 445.3 | 443.3 | 326.2 | 327.7 |
| 73 | 406.8 | 401.3 | 416.7 | 412.1 | 264.6 | 248.1 | 444.0 | 443.5 | 328.4 | 329.2 |
| 79 | 453.5 | 456.7 | 461.8 | 466.3 | 263.7 | 248.0 | 487.0 | 495.7 | 327.6 | 330.9 |
| 81 | 451.6 | 457.4 | 459.3 | 466.9 | 263.7 | 247.8 | 484.5 | 495.8 | 326.9 | 330.4 |

[a] Denotes indicated heater on during test.

Table 6   Steady state temperatures (prototype vehicle)

| Node | Test 1 Experiment (°R) | Test 1 Analysis (°R) | Test 2 Experiment (°R) | Test 2 Analysis (°R) | Test 3 Experiment (°R) | Test 3 Analysis (°R) | Test 4 Experiment (°R) | Test 4 Analysis (°R) | Test 5 Experiment (°R) | Test 5 Analysis (°R) |
|---|---|---|---|---|---|---|---|---|---|---|
| Solar load | On | | On | | Off | | On | | Off | |
| **Base deck** | | | | | | | | | | |
| 57a | 382.5 | 371.2 | 429.7 | 411.0 | 321.2 | 296.2 | 406.2 | 400.8 | 310.0 | 303.1 |
| 57b | 387.4 | 371.2 | 408.9 | 411.0 | 285.6 | 296.6 | 410.2 | 400.8 | 310.0 | 303.1 |
| 59 | 398.1 | 405.2 | 410.9 | 417.5 | 276.9 | 250.3 | 416.3 | 427.8 | 307.6 | 300.9 |
| **Heater one** | | | | | | | | | | |
| 83 | b | 372.4 | a | 779.9 | a | 742.2 | b | 403.6 | b | 304.0 |
| 84 | 382.5 | 373.2 | 497.6 | 484.4 | 402.9 | 376.2 | 409.6 | 405.2 | 312.5 | 304.7 |
| 85 | 382.5 | 373.1 | 487.8 | 475.8 | 392.4 | 365.6 | 407.6 | 404.8 | 310.8 | 304.6 |
| **Equipment deck** | | | | | | | | | | |
| 70a | 400.2 | 403.0 | 411.6 | 418.1 | 266.6 | 255.0 | 459.5 | 466.1 | 361.2 | 359.4 |
| 70b | 400.8 | 403.0 | 410.9 | 418.1 | 268.3 | 255.0 | 456.3 | 466.1 | 358.2 | 359.4 |
| **Heater two** | | | | | | | | | | |
| 87 | 404.2 | 402.6 | 413.6 | 415.9 | 266.6 | 253.5 | 482.9 | 481.9 | a 386.0 | 371.1 |
| 88 | 401.5 | 404.2 | 412.3 | 415.4 | 266.6 | 253.5 | 478.6 | 476.3 | 383.2 | 364.4 |
| **Heater three** | | | | | | | | | | |
| 90 | 403.5 | 402.7 | 413.6 | 415.9 | 267.5 | 253.4 | 490.2 | 491.4 | a 394.5 | 381.5 |
| 91 | 401.5 | 402.2 | 413.6 | 415.5 | 266.6 | 253.4 | 485.4 | 484.5 | 388.8 | 373.4 |
| **Heater four** | | | | | | | | | | |
| 95 | 402.9 | 403.6 | 412.9 | 417.2 | 267.5 | 253.8 | 480.5 | 493.9 | a 382.5 | 388.5 |
| 96 | 402.9 | 405.4 | 413.6 | 418.8 | 266.6 | 253.6 | 490.2 | 505.3 | 393.8 | 402.1 |
| 97 | 400.8 | 403.4 | 410.9 | 416.7 | 266.6 | 253.5 | 476.7 | 491.3 | 378.2 | 386.0 |
| **External skin** | | | | | | | | | | |
| 64 | 390.3 | 392.5 | 403.5 | 406.5 | 271.8 | 255.1 | 422.3 | 428.1 | 324.2 | 319.0 |
| 77 | 412.3 | 403.8 | 419.6 | 414.2 | 263.1 | 250.0 | 447.3 | 443.7 | 331.8 | 326.7 |
| 73 | 410.9 | 403.0 | 419.0 | 413.6 | 264.0 | 250.2 | 446.8 | 443.8 | 332.5 | 327.4 |
| 79 | 467.1 | 467.9 | 472.9 | 477.4 | 263.1 | 250.2 | 496.9 | 506.3 | 330.3 | 329.5 |
| 81 | 468.5 | 468.5 | 473.5 | 478.0 | 263.1 | 250.1 | 497.6 | 506.9 | 329.5 | 329.5 |

[a]Denotes indicated heater on during test.  [b]Thermocouple out.

It is important to note that the problem areas located in the numerical model were as evident by comparison with the scale-model as with the prototype vehicle. In general, the problems located in the numerical analysis can be rectified. Modifications in the nodal analysis and more accurately determined radiative property values would improve the results. However, the present results (Figs. 5 and 6) show a standard deviation of the temperature ratio on the order of 2%. Certainly an acceptable range of error for the level of detail to which the models were noded.

Experimental Results

A tabulation of the experimental results for the half-scale and prototype vehicles is presented in Tables 5 and 6. The ratios of the half-scale/prototype temperatures have been summarized in Fig. 8. From the figure it is evident that the correlation between the scale model and prototype test results is better than the correlation between numerical and experimental results. This is mainly due to the inability of the numerical models to represent the physical reality of the experimental models which includes: 1) temperature dependence of the radiative properties of the thermal control coatings, and 2) an equivalent continuous model rather than the discrete nodal model.

All of the half-scale data points predict the prototype temperatures within 2% except for the two nodes (numbers 79 and 81) on the bare aluminum sun facing upper closure surface. Here, temperature differences on the order of 2 to 4% result from a lack of preservation of the radiative properties of the materials.[7] Differences between the models as a result of variable temperature dependence of thermal conductivity, and the use of aluminum silicate for the heater cores in both models appeared to have had a minor effect on the results.

Numerical Adjustment of Experimental Data

After the numerical analysis had been completed for the half-scale model (as it had been built) in an effort to duplicate the experimental results, the nodal network was adjusted to predict the half-scale model results as it should have been built. That is, the following changes in the numerical analysis were made: 1) The thermal conductivity of the half-scale model materials were given the same relative temperature dependence as the prototype materials. 2) The heater power dissipated in each of the heaters was scaled to that dissipated in the prototype tests. 3) The solar heating loads on each of the external nodes were scaled to those loads incident on the prototype. 4) The radiative properties of the prototype were substituted to comply with the scaling requirement for preservation of radiative properties.

The results of this study are shown in Fig. 9. While this correction scheme has shifted the distribution, only a slight improvement has been effected in the correlation. This is due to the coarse noding of the base deck and heater elements which force larger errors in the analysis than existing in the scaled experimental data. This correction technique would have shown more promise if the numerical analysis had been redone to account for the previously noted deficiencies and a smaller scale model (on the order of 1/4 - 1/6) had been tested with its resulting larger compromise induced deviations from the prototype results.

Summary

Table 7 has been prepared as a summary of the results. Each of the comparisons previously discussed is entered in the table which indicates its systematic error, its range of error, and its standard deviation. In cases of experiment and numerical analysis, the analysis is compared to the experiment. In the experimental comparisons the half-scale tests are compared to the prototype results. Systematic error, range of error, and standard deviation percentages are based on the distributions of temperature ratio as shown in Figs. 5, 6, 8, and 9.

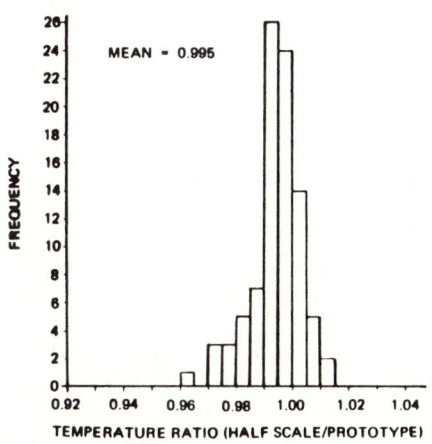

Fig. 8 Distribution of deviations between half-scale and prototype experimental data.

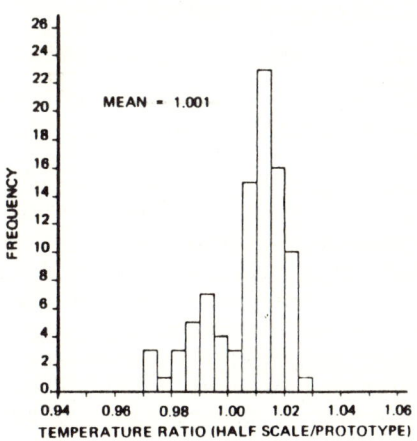

Fig. 9 Distribution of deviations between corrected half-scale and prototype experimental data.

Table 7    Summary of errors

| Comparison | Systematic[a] error (%) | Range of[b] error (%) | Standard[c] deviation (%) |
|---|---|---|---|
| Tests 1, 2, 4, 5 Experiment and analysis | | | |
| Half-scale | − 0.1 | + 5.3 − 2.8 | ± 1.5 |
| Prototype | − 0.1 | + 5.3 − 2.9 | ± 2.2 |
| Test 3 Experiment and analysis | | | |
| Half-scale | + 5.9 | + 2.4 −10.9 | ± 1.0 |
| Prototype | + 5.2 | + 5.4 − 8.9 | ± 0.7 |
| Tests 1-5 | | | |
| Experiment | − 0.5 | + 1.8 − 3.1 | ± 0.7 |
| Adjusted experiment | + 0.1 | + 1.5 − 4.0 | ± 1.0 |

[a] Displacement of mean from 1.0
[b] Total range of deviations about mean
[c] Range of deviations about mean which encompass 68% of the data points shown.

A comparison of experiment and analysis for tests numbers 1, 2, 4 and 5 shows a negligible (0.1%) systematic error and a standard deviation of approximately 2%. It may seem surprising that the analysis predicted the half-scale experiment with a smaller standard deviation than the prototype experiment. This is due to the measure of error selected as the basis of comparison. The distributions obtained are not ideally gaussian and other measures of error (i.e. a two sigma distribution) would show nearly identical results for the two vehicles. Note that the total range of error is 8.1% for the half-scale and 8.2% for the prototype.

A comparison of test and analysis for test number three shows the 5.5% systematic error which was discussed previously as a result of the temperature dependency of the B-1060 thermal control coating. The standard deviation on the order of 1% is typical of the results of a single test as opposed to the group of tests reported previously.

A comparison of the two experiments, before and after correction of the half-scale experimental data, shows an interesting effect. The raw data indicated a 0.5% systematic error and a 0.7% standard deviation. The adjusted experimental comparison shows a 0.1% systematic error and a 1.0% standard deviation. The reduction in systematic error is due to accounting for the nonsimilar temperature dependencies of the thermal conductivities of the materials. As both the standard deviation and the range of error increased slightly it is probable that utilizing the numerical model and its inaccuracies worsened the experimental comparison.

Good agreement has been shown between analysis and experiment and comparison of the experimental results themselves. Numerical correction of the experimental data has also assisted in reducing the systematic error in the half-scale experiment.

## Conclusions

This study has shown that numerical analysis can predict the performance of a complex spacecraft. The major difficulty with a numerical model is that it requires a complete understanding of the spacecraft, its material properties and their temperature dependence. The major limitations on numerical analysis are related to the completeness of our understanding of the physical problem and the ability of our computer programs to simulate the problem.

This study has shown that thermal scale modeling can successfully verify the thermal design of a spacecraft with complex conductive/radiative interchange. Scale modeling is limited, as are all experimental approaches, by the difficulty of producing complex external environments in existing vacuum chamber facilities.

A combination of the experimental and numerical methods, which utilizes the advantages of each method, appears to be the best approach to thermal design verification. A small scale model could be tested and used to aid the development of an accurate mathematical model. The upgraded mathematical model could then be used to predict the performance of a full-size vehicle over a much larger range of simulated flight conditions than could be achieved in the space simulator.

## References

[1] Vickers, J. M. F., "Thermal Scale Modeling: Basic Considerations," Jet Propulsion Lab Space Programs Summary IV, 37-18, Dec. 1962, pp. 80-83.

[2] Katzoff, S., "Similitude in Thermal Models of Spacecraft," TN D-1631, April 1963, NASA.

[3] Chao, B. T. and Wedekind, G. L., "Similarity Criteria for Thermal Modeling of Spacecraft," Journal of Spacecraft and Rockets, Vol. 2, No. 2, March-April 1965, pp. 146-152.

[4] Gabron, F., Johnson, R. W., Vickers, J. M. F., and Lucas, J. W., "Thermal Scale Modeling of the Mariner IV Spacecraft," AIAA Progress in Astronautics and Aeronautics: Thermophysics and Temperature Control of Spacecraft and Entry Vehicles, edited by G. B. Heller, Vol. 18, Academic Press, New York, 1966, pp. 675-695.

[5] Jones, B. P., "Theory of Thermal Similitude with Applications to Spacecraft - A Survey," Astronautics Acta, Vol. 12, No. 4, July - Aug. 1966, pp. 258-271.

[6] Doenecke, Jochen, "Thermal Scale Modeling Without Similitude," International Journal of Heat and Mass Transfer, Vol. 10, July-Dec. 1967, pp. 1894-1899.

[7] MacGregor, R. K., "Limitations in Thermal Similitude," Document D2-121352-1, Dec. 1969, The Boeing Co.

[8] Harmon, H. Neil, "Four-Foot Solar Simulation System," Institute of Environmental Sciences 1966 Annual Technical Meeting Proceedings, Institute of Environmental Sciences, Mt. Prospect, Ill., 1966, pp. 531-536.

[9] Oppenheim, A. K., "Radiation Analysis by the Network Method," Transactions ASME, Vol. 78, 1956, pp. 725-735.

[10] Corlett, R. C., "Direct Monte Carlo Calculation of Thermal Radiation in Vacuum," Transactions ASME, Journal of Heat Transfer, Vol. 88, Series C, pp. 376-382, 1966.

# III  Heat Pipes

# EFFECTS OF FRICTION ON THE SONIC VELOCITY LIMIT IN SODIUM HEAT PIPES

E. K. Levy[*]

Lehigh University, Bethlehem, Pa.

## Abstract

The sonic velocity limitation on the performance of heat pipes with metallic working fluids operating at low-vapor pressures has been the subject of several earlier analytical and experimental studies. In this paper, analytical results are presented which demonstrate the effect which the wall shear stress acting in the vapor flow passage has on the behavior of a sodium heat pipe in the sonic limit regime. It is shown that because of the wall shear stress in the adiabatic region, gas-dynamic choking will occur at the exit plane of the adiabatic region rather than at the evaporator exit. In addition, at a given value of operating temperature, the shear stress will reduce the maximum rate of heat transfer based on the sonic limit to a value lower than is expected from the frictionless analysis. This reduction in heat transfer capability can be quite large at the lower values of operating temperature.

## Nomenclature

$A$ = cross-sectional area $A = \pi R_\omega^2$
$C_{p1}$ = constant pressure specific heat for monatomic species
$D$ = diameter of vapor passage
$f$ = friction factor
$h$ = mixture enthalpy per unit mass

---

Presented as Paper 71-407 at the AIAA 6th Thermophysics Conference, Tullahoma, Tenn., April 26-28, 1971. This investigation is sponsored by the United States Atomic Energy Commission under Contract AT(30-1)-4095. The calculations were performed at the Lehigh University Computing Center.

[*]Assistant Professor, Department of Mechanical Engineering and Mechanics.

$h_E$ = enthalpy per unit mass of injected fluid
$h_L$ = enthalpy per unit mass of liquid in wick
$h_{fg}$ = enthalpy of evaporation ($h_{fg} = h_g - h_f$)
$\hat{h}$ = convective heat-transfer coefficient
$K_p$ = equilibrium constant
$K_T$ = thermal conductivity
$\ell_D$ = enthalpy of dimerization $\ell_D = h_2 - h_1$
$L_A$ = length of adiabatic region
$L_c$ = length of condenser
$L_E$ = length of evaporator
$\dot{m}$ = mass flow rate
$\dot{m}_1$ = mass flow rate of monatomic species
$\dot{m}_E$ = flow rate of injected mass per unit length
$M$ = Mach number
$p$ = pressure of mixture
$P_s$, $P_{sat}$ = saturation pressure
$P_r$ = Prandtl number
$Q_E$ = rate of heat transfer to evaporator per unit length
$Q_{conv}$ = rate of convective heat transfer from liquid-vapor interface to vapor per unit length
$Q_{choke}$ = maximum rate of heat transfer
$R_2$ = gas constant of diatomic species
$r_w$, $R_w$ = radius of vapor passage
$R_{eD}$ = Reynold's number ($\rho VD/\mu$)
$T$ = temperature of vapor
$T_E$ = temperature of liquid vapor interface
$T_s$, $T_{sat}$ = saturation temperature
$\Delta T_i$ = interfacial temperature drop
$V$ = vapor velocity
$V_E$ = radial velocity of injected mass
$V_L$ = velocity of liquid in wick
$x$ = axial distance
$\alpha$ = degree of dissociation

$\alpha_E$ = degree of dissociation of injected mass
$\rho$ = density
$\sigma$ = condensation coefficient
$\tau_\omega$ = shear stress
$\mu$ = absolute viscosity

## Introduction

The heat pipe, a heat-transfer device with a high effective thermal conductivity, operates through the continuous evaporation and condensation of its working fluid. One possible configuration of the heat pipe is represented schematically in Fig. 1. This illustration shows a cylindrically shaped, constant cross-sectional area vapor flow passage surrounded by an annular wick. Heat is transferred to the device at the evaporator section and rejected at the condenser end. In most applications the evaporator and condenser regions are separated by an insulated or adiabatic region. Vapor, formed within the evaporator at the liquid-vapor interface, flows longitudinally through the vapor flow passage to the condenser region where it condenses giving up its latent heat of evaporation. After condensing, the working fluid returns through the wick to the heat source where it is re-evaporated. The capillary force developed at the liquid-vapor-wick interface is used to supply the pumping power required to overcome the longitudinal static pressure losses which occur within both the vapor and liquid flow passages.

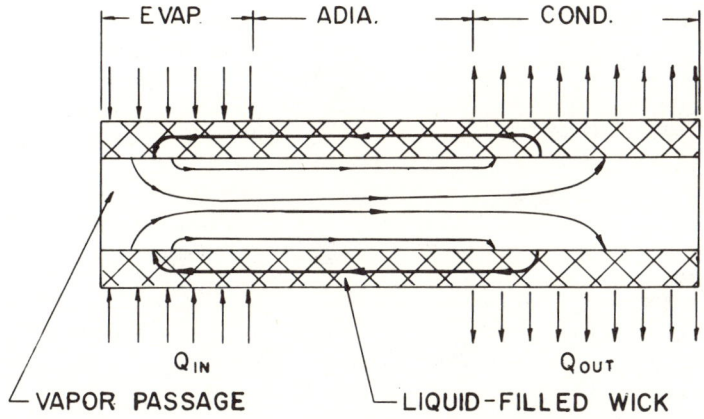

Fig. 1 A typical heat pipe.

The axial flow of vapor through the vapor passage of a heat pipe of the type described in Fig. 1 can be considered analogous to the flow of a gas through a convergent-divergent nozzle. The flow in the evaporator region, with the mass injection (evaporation) from the liquid-vapor interface into the vapor passage, is similar in many respects to a gas flow through a convergent nozzle; and the vapor flow in the condenser region, with the mass suction (condensation) at the liquid-vapor interface, behaves as a gas flow through a divergent passage. In most heat pipe situations, the axial components of vapor velocity are sufficiently small when compared to the velocity of sound in the fluid that the vapor can be considered incompressible. In these cases, the widely used analysis of Cotter (an incompressible vapor analysis) for the maximum rate of heat transfer in a heat pipe which is wick limited is generally valid.[1,2] However, in the case of heat pipes with metallic working fluids operating at the low end of the vapor pressure curve, the vapor densities are extremely small. Thus even for relatively low values of heat-transfer rates, axial components of vapor velocity can be generated which are of the same order of magnitude as the sonic velocity. In these instances, vapor compressibility effects become important. As in the flow of a perfect gas through a convergent nozzle, in the evaporator region the injected mass causes the vapor Mach number to increase in the direction of the flow toward a value of unity. In addition, the wall shear stress (the shear stress acting on the vapor at the liquid-vapor-wick interface) causes such a subsonic flow to accelerate toward a Mach number of unity. It is readily shown from the theory of compressible fluid mechanics[3] that the maximum Mach number which one can attain in the evaporator section is unity. It can also be shown that the maximum mass flow of vapor which can be generated in the evaporator corresponds to a value of $M = 1$ at the downstream end of the evaporator section. This is the condition of gasdynamic choking. In those cases where an adiabatic region separates the evaporator and condenser sections, the wall shear stress will cause the choking point to occur at the exit plane of the adiabatic region rather than at the end of the evaporator. In both instances, the choking point is analogous to the geometric throat of a convergent-divergent nozzle.

In the first analytical investigations of the effect of gasdynamic choking on the maximum rate of heat transfer in sodium heat pipes, the wall shear stress was ignored; and the sodium vapor was treated as a pure monatomic perfect gas.[2,4] Dzakowic et al.[5] pointed out the possibility that the dissociation-recombination reaction

$$2N_a \rightleftarrows N_{a2}$$

could influence heat pipe performance. Levy$^2$ also analyzed the problem by treating the vapor as a saturated two phase mixture in liquid-vapor phase equilibrium. Wall shear stress, dissociation and recombination effects, and the effects of vapor supersaturation were not accounted for in Levy's analysis. It is likely that a combination of all these phenomena is important in a sonic velocity limited heat pipe.

The purpose of this paper is to point out the relative importance of the wall shear stress and the extent to which the shear stress in the adiabatic region can influence the sonic limit. To accomplish this, the vapor is treated as a two-component mixture ($N_a$ and $N_{a2}$) of perfect gases which is in chemical equilibrium and is frozen with respect to phase equilibrium. Thus though the vapor dissociation and recombination reaction is assumed to proceed sufficiently rapidly for the local equilibrium chemical composition to occur everywhere in the vapor passage, it is assumed that there is no tendency for liquid droplets to be formed within the vapor as the vapor temperature drops below the local value of saturation temperature.

In a later paper, the influences on the sonic limit of local chemical and phase non-equilibrium will be examined. The kinetics of the sodium dissociation-recombination reaction and of the liquid droplet formation and growth processes will be included in the analysis at that time.

Chemical Equilibrium Flow Analysis - Evaporator Region

Consider a stationary cylindrical control volume of width dx (Fig. 2). The cylindrical surface of this control volume is drawn on the vapor side of the liquid-vapor-wick interface. Though in the real flow, the axial component of velocity V and the vapor temperature T may vary radially from the centerline of the vapor passage to the liquid-vapor interface, the vapor flow is assumed here to be one dimensional. That is, the mass, momentum, and enthalpy fluxes are represented by the mean values of velocity and temperature. In addition, any transverse variations in static pressure p are ignored. The vapor being generated by evaporation at the liquid-vapor interface crosses the cylindrical surface of the vapor control volume with a mass flow rate per unit length $\dot{m}_E$. It is assumed that as it enters the vapor control volume with a velocity $V_E$ the injected mass $\dot{m}_E$ has a zero component of velocity in the axial direction. The liquid-vapor interface is at a temperature $T_E$ and the injected mass has a static enthalpy $h_E$.

The vapor is taken to be a mixture of the two species $N_a$ and $N_{a2}$, each of which behaves as a perfect gas. At a given axial position, the analysis may require that the vapor be at an average temperature which is less than the local vapor saturation temperature. It is assumed that any calculated vapor supersaturation can be maintained with no tendency for the homogeneous nucleation and growth of liquid droplets. Thus though the two-component vapor mixture is assumed to be in local equilibrium with respect to the chemical reaction $2N_a \rightleftarrows N_{a2}$, the bulk of the vapor is taken to be frozen with respect to phase change. The injected fluid $\dot{m}_E$ is superheated vapor with an equilibrium composition $\alpha_E$ corresponding to the local static pressure p and the local interface temperature $T_E$. Finally it is assumed that the flow is steady.

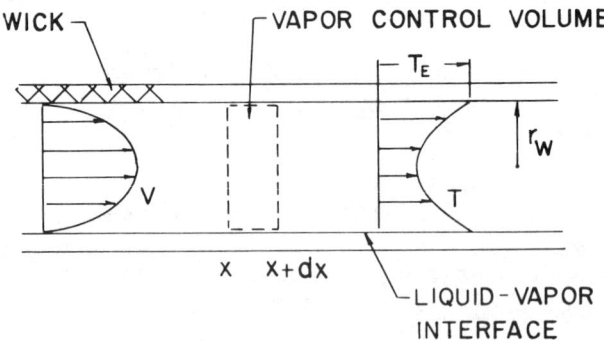

Fig. 2 Vapor control volume with typical velocity and temperature profiles.

Conservation of mass requires that

$$A d(\rho V)/dx = \dot{m}_E \qquad (1)$$

and conservation of momentum can be written

$$-\frac{dp}{dx} - \frac{\dot{m}}{A}\frac{dV}{dx} = \frac{V}{A}\dot{m}_E + \tau_\omega \frac{2\pi R_w}{A} \qquad (2)$$

Since in a real heat pipe the vapor temperature might vary radially, there must be a convective heat-transfer term from the liquid-vapor interface to the vapor in the energy equation. The rate of heat transfer per unit length is denoted by $Q_{conv}$.

Viscous dissipation has been neglected. Conservation of energy requires that

$$\frac{d}{dx}[h + \frac{V^2}{2}] = \frac{\dot{m}_E}{\dot{m}}[h_E - h + \frac{V_E^2 - V^2}{2}] + \frac{Q_{conv}}{\dot{m}} \quad (3)$$

Treating the $N_a$ (subscript 1) and $N_{a2}$ (subscript 2) species as perfect gases, $dh/dx$ becomes

$$\frac{dh}{dx} = [\alpha C_{p1} + (1-\alpha)C_{p2} - \ell_D(\frac{\partial \alpha}{\partial T})_p]\frac{dT}{dx} + [-\ell_D(\frac{\partial \alpha}{\partial p})_T]\frac{dp}{dx} \quad (4)$$

where $\alpha$ = degree of dissociation and $\ell_D$ = enthalpy of dimerization.

The equation of state of a 2 component mixture of perfect gases is

$$P = (1+\alpha)\rho R_2 T \quad (5)$$

which after some manipulation can be written as

$$\frac{d\rho}{dx} = [\frac{\rho}{p} - (\frac{\rho}{1+\alpha})(\frac{\partial \alpha}{\partial p})_T]\frac{dp}{dx} - [\frac{\rho}{T} + (\frac{\rho}{1+\alpha})(\frac{\partial \alpha}{\partial T})_p]\frac{dT}{dx} \quad (6)$$

In the case of an equilibrium mixture $\alpha$ is a function only of the pressure and temperature

$$\alpha = 1/\sqrt{1+4pK_p} \quad (7)$$

[Equations (4-7) are derived in detail in Ref. 6.] Equations (1-4, 6 and 7) can be rearranged and solved for the quantities $dT/dx$, $dV/dx$, and $dP/dx$. These equations are of the form

$$dT/dx = f_1(T, P, V, \dot{m}_E, \tau_\omega, Q_{conv}) \quad (8)$$

$$dP/dx = f_2(T, P, V, \dot{m}_E, \tau_\omega, Q_{conv}) \quad (9)$$

$$dV/dx = f_3(T, P, V, \dot{m}_E, \tau_\omega, Q_{conv}) \quad (10)$$

and constitute a system of three simultaneous first-order differential equations which can be solved numerically provided $\dot{m}_E$, $\tau_\omega$, and $Q_{conv}$ are specified.

Considering a thin control volume drawn around the liquid-vapor interface (Fig. 3), neglecting the kinetic energy terms, and using the approximation $(h_E - h_L) \approx h_{fg}$, one obtains a relationship between $Q_E$, $Q_{conv}$, and $\dot{m}_E$.

$$Q_E = Q_{conv} + \dot{m}_E h_{fg} \quad (11)$$

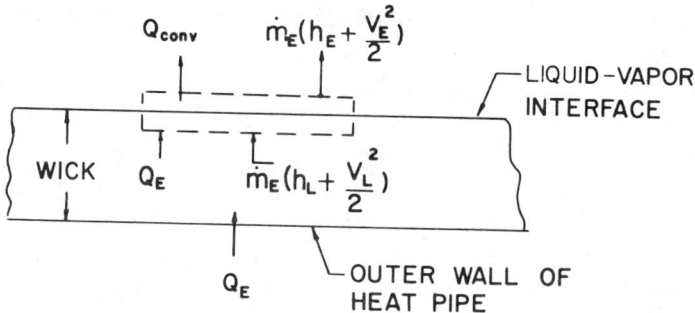

Fig. 3 Interfacial control volume.

### State of the Injected Vapor $\dot{m}_E$

The injected vapor, formed during the evaporation process, must be slightly superheated with respect to the static pressure of the bulk vapor flow. The relative order of magnitude of the interfacial superheat can be obtained from Schrage's analysis of the interfacial mass transfer resistance.[7] This yields

$$\Delta T_i = T_E - T_{sat} = \left(\frac{2-\sigma}{2\sigma}\right)\left(\frac{\dot{m}_E}{2\pi r_\omega}\right)\frac{(2\pi R)^{1/2} T_{sat}^{5/2} R}{P h_{fg}} \quad (12)$$

where $\sigma$ = condensation coefficient, $r_\omega$ = radius of vapor passage and $R$ = gas constant.

Hence in the evaporator region, the temperature of the liquid-vapor interface is

$$T_E = T_{sat}(P) + \Delta T_i \quad (13)$$

The data of Refs. 8 and 9 suggest that $\sigma \simeq 1$. In the temperature range and heat flux range of interest, $\Delta T_i \sim 1°C$.

### Wall Shear Stress and Convective Heat Transfer

The magnitude of the evaporator wall shear stress depends on whether the vapor is in a laminar or turbulent flow regime. Busse[10] analyzed the laminar, steady flow of an incompressible fluid in a circular duct with wall injection and found that in the evaporator region, with sufficiently large injection flow rates, the vapor velocity profile at a given axial location approaches a cosine curve.

$$u(r,x)/V = (\pi/2)\cos[(\pi/2)(r^2/r_\omega^2)] \quad (14)$$

where u = axial velocity component at a given radial position r and axial position x, and v = average axial velocity at x. From Eq. (14), the friction factor f becomes

$$f \equiv \tau_\omega/\rho V^2/2 = 2\pi^2/\rho V 2 r_\omega/\mu = 2\pi^2/R_{eD} \qquad (15)$$

The friction factor for fully developed laminar tube flow with no injection is $f = 16/R_{eD}$, where $R_{eD}$ = diameter Reynolds number. Equation (15) was used for all laminar shear stress calculations in the evaporator. The transition between laminar and turbulent flow was assumed to occur at $R_{eD} = 2000$. For turbulent flow in the evaporator, the equation

$$f = 0.079/(R_{eD})^{1/4} \qquad (16)$$

was used for f (Ref. 11). Kinney[12] and Yuan and Finkelstein[13] investigated the effect of mass injection on the heat transfer to a laminar incompressible flow in a circular duct and found that the effect of the wall injection is to decrease the heat-transfer coefficient to a value below that of the zero injection case. An upper bound on the heat-transfer term can be obtained by using the expression for the rate of heat transfer in a tube flow with no wall injection. The value

$$\hat{h} = 3.66 \, k_T/2r_\omega \qquad (17)$$

was used for the laminar pipe flow convective heat transfer coefficient and the value

$$\hat{h} = 0.023(k_T/2r_\omega)(R_{eD})^{0.8}(P_r)^{0.3} \qquad (18)$$

was used for the turbulent flow case.[14] A value of Prandtl number of $P_r = 0.85$ was used in all the turbulent heat transfer calculations.

Chemical Equilibrium Flow Analysis - Adiabatic Region

Whereas in the evaporator region, the total heat-transfer rate per unit length to the outside wall of the heat pipe is $Q_E$, this quantity is essentially zero in the adiabatic region. Of course, in a real device, axial heat conduction along the container wall and through the wick and the heat losses to the surroundings could tend to make $Q_E$ either slightly positive or negative. In this analysis, the adiabatic section was modeled mathematically by requiring that $\dot{m}_E = 0$ and $V_E = 0$; however, the convective heat-transfer term $Q_{conv}$ was allowed to be nonzero. (The effects of this approximation on the numerical results were checked and were found to be negligible).

In the evaporator region, the liquid-vapor interface is at a temperature $T_E$ which is slightly larger than the local saturation temperature. Everywhere in the adiabatic region, the interface must be at the local saturation temperature.

As an upper bound, the wall shear stress was determined from Eq. (15) for laminar flow and from Eq. (16) in the case of turbulent flow. Equations (17) and (18) were used to determine the convective heat-transfer coefficient.

Method of Solution and Results of Numerical Analysis

Equations (8, 9 and 10) were solved by digital computer by means of a Runge Kutta integration scheme. All solutions were made for the case of uniform heat addition in the evaporator region. At the upstream end of the evaporator ($x \to 0$), the vapor Mach number is sufficiently small that the incompressible equations of motion can be used to generate the initial conditions needed for the first step in the integration. Thus for a given upstream operating pressure p, the vapor at $x = \Delta x$ is assumed to have a temperature T given by $T_E$ in Eq. (13) and a mass flow rate given by the relation $\dot{m} = \dot{m}_E \Delta x$. These quantities are sufficient for an initiation of the Runge Kutta integration process.

For purposes of comparison all calculations were performed for a heat pipe with a 1.846-cm-diam vapor passage, a 17.8-cm-long evaporator, and a 5.1-cm-long adiabatic region. These dimensions are the same as those of the experimental sodium heat pipe of Dzakowic, Tang, and Arcella of the Westinghouse Astronuclear Laboratory.[5]

Figure 4 shows the computed variations of the vapor Mach number, composition $\alpha$, and static temperature T for the Westinghouse heat pipe operating at an upstream vapor temperature of 425°C and uniform evaporator heat addition rate of 2.84 watts/cm. Over the 17.8-cm-long evaporator, this corresponds to a heat-transfer rate of 50.7. Since at this power input rate and upstream vapor temperature, a vapor Mach number of unity was achieved at the exit plane of the adiabatic region, the quantity Q = 50.7 corresponds to the gasdynamic choking limit. The infinite gradients in Mach number, composition $\alpha$, and vapor static temperature occurring at the adiabatic exit plane are characteristic of the gasdynamic choking phenomenon.[3] It is clear from Fig. 4 that rather than occurring at the evaporator exit plane, choking will occur at the exit to the adiabatic region. The acceleration of the fluid occurring within the adiabatic region is caused primarily by the wall shear stress $\tau_\omega$. The convective heat-transfer term $Q_{conv}$ is relatively

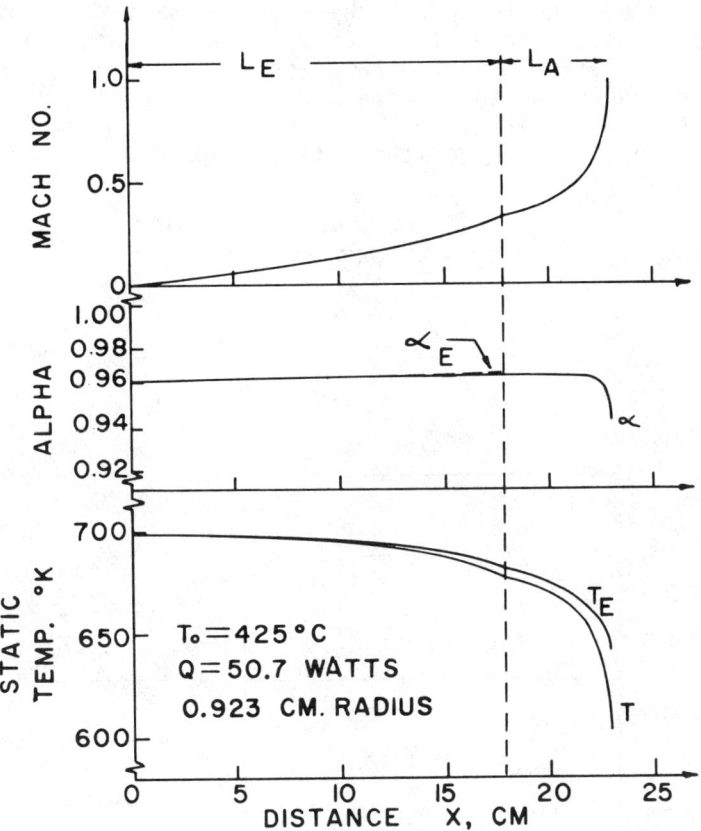

Fig. 4 Computed axial variation of vapor temperature, composition, and Mach number - 425°C.

small in this case and contributes little towards promoting sonic velocities. The quantities $T_E$ and $\alpha_E$ are the axial variations of the liquid-vapor interface temperature and the chemical composition of the vapor as it is formed by evaporation at the liquid-vapor interface. At any axial position x, the difference between $T_E$ and $T$ is approximately equal to the degree of supersaturation of the sodium vapor. Figures 5 and 6 show calculations similar to those on Fig. 4 at upstream vapor temperatures of 500 and 600°C. In each of these instances, choking occurred at 22.1 cm.; the heat-transfer rates shown in the figures correspond to the choking limits at these temperatures. Note that the effect of increasing the upstream vapor temperature is to increase the Mach number and the degree of supersaturation of the vapor at the evaporator exit plane. That is, at higher operating temperatures, the larger evaporator heat fluxes tend to reduce the relative importance of the wall shear stress. Note also that whereas at 450°C, the degree of dissociation α increased slightly over the entire evaporator length and then decreased abruptly near the exit plane of the adiabatic region, at 500 and 600°C, α decreased steadily with distance over the entire evaporator and adiabatic regions.

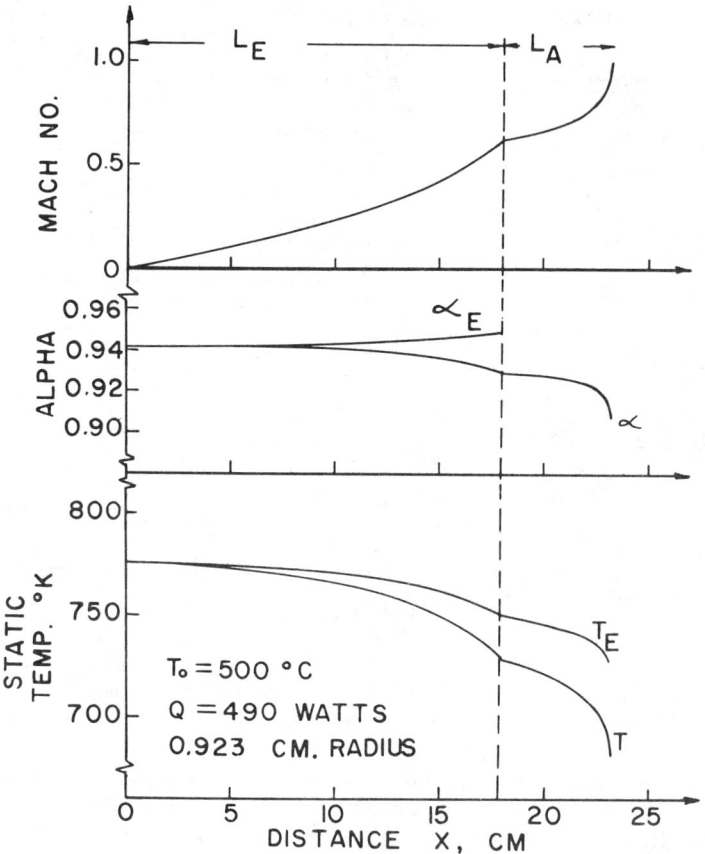

Fig. 5 Computed axial variation of vapor temperature, composition, and Mach number - 500°C.

Similar calculations were carried out for the Westinghouse heat pipe at various points in the interval from 425-600°C. These results are summarized in Figs. 7-9. Curve D of Fig. 7 shows the variation of the choking limited heat transfer rate with the temperature of the vapor at the upstream end of the evaporator. In the vicinity of x = 0, this temperature also corresponds to the temperature of the liquid-vapor interface. The data points shown were taken from Ref. 5. The agreement

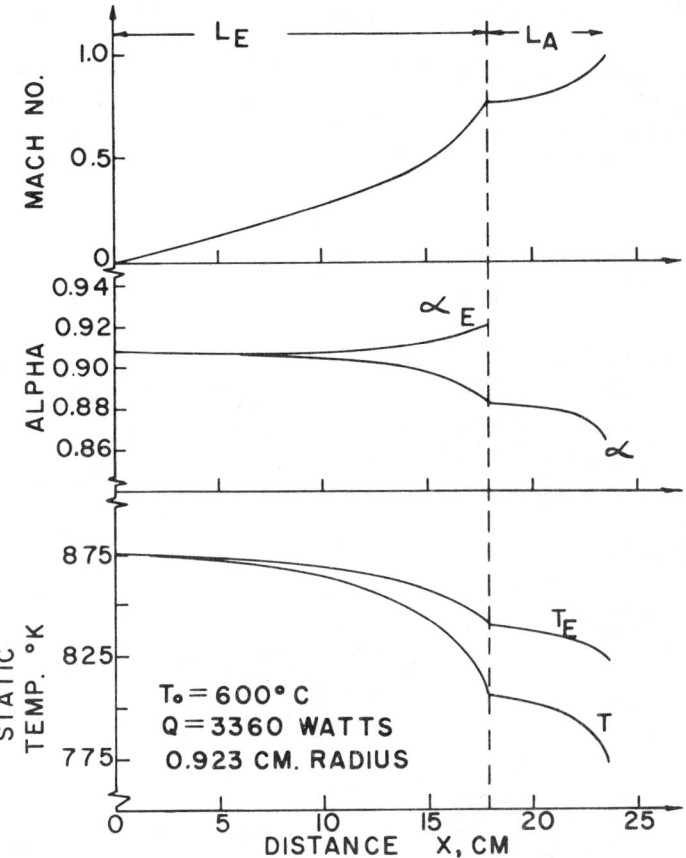

Fig. 6 Computed axial variation of vapor temperature, composition, and Mach number - 600°C.

between the data and the chemical equilibrium flow analysis seems to be excellent except at the higher operating temperatures where a discrepancy of approximately 30% occurred. The sudden dropping off of the experimental data at 560°C suggests that at temperatures greater than this, the maximum rate of heat transfer was being limited by a phenomenon other than the sonic limit. Curve D of Fig. 8 shows a comparison between the analytically and experimentally obtained axial variations in interfacial temperature. The ordinate $\Delta T_E$ is the difference in value between the liquid-vapor interface temperature measured at the upstream end of the evaporator and that at the choking point. Correcting for radial conduction losses through the heat pipe wall and the wick, $\Delta T_E$ is the axial temperature drop which one would observe along the outside surface of the heat pipe. Finally the two curves labeled "D" on Fig. 9 show the variation in axial drop in vapor static temperature and the variation in the degree of supersaturation of the vapor occurring at the choking point. The axial drop in vapor static temperature was taken to be the difference between the vapor static temperature at x = 0 and that at the exit plane of the adiabatic region.

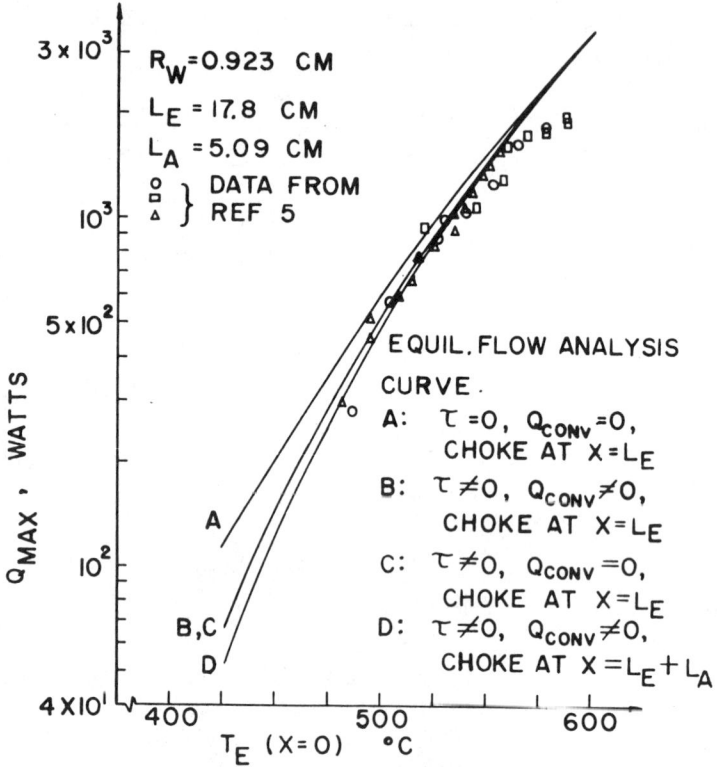

Fig. 7 Maximum heat transfer based on sonic limit.

To study the effect of shear stress $\tau_\omega$ and convective heat transfer $Q_{conv}$ on the temperature variations and on the choking limit $Q_{choke}$, a group of calculations was made for the Westinghouse heat pipe with $\tau_\omega$ and $Q_{conv}$ alternately set equal to zero. These results are summarized with curves A, B, and C on Figs. 7-9. The curves labeled "A" show the behavior of the system when it is assumed that $\tau_\omega = Q_{conv} = 0$. In this case, choking must occur at the evaporator exit plane. The B curves show the system characteristics with nonzero shear stress and convective heat-transfer terms but with choking occurring at the evaporator exit plane. The B and D curves can be compared to determine the relative importance of the 5.09-cm-long adiabatic section. Finally the C curves show the effect which the convective heat-transfer rate has on the system behavior. In general, the influences of the convective heat-transfer, shear stress, and adiabatic section are most prominent at the lower operating temperatures where the mass injection per unit length $\dot{m}_E$ has its smallest values. At a given value of operating temperature, the wall shear stress $\tau_\omega$ restricts the choking heat-transfer rates $Q_{choke}$, increases the axial drop in static pressure and thus increases the axial drop in interface

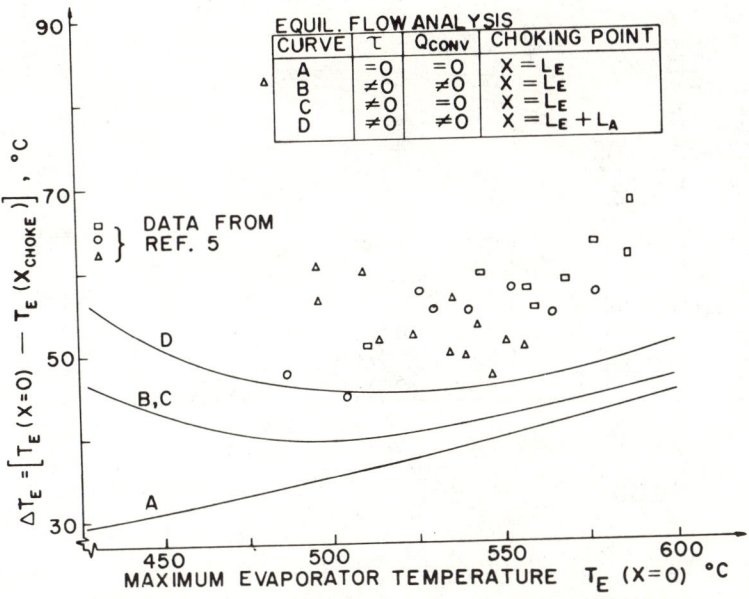

Fig. 8 Axial drop in interfacial temperature.

temperature (Fig. 8), increases the axial drop in vapor static temperature, and decreases the degree of vapor supersaturation at the choking point. The convective heat-transfer term $Q_{conv}$ has a negligible effect on the choking heat-transfer rate and on the axial interfacial temperature drop, but it does tend to decrease the axial drop in vapor static temperature and to decrease the vapor supersaturation at the choking point.

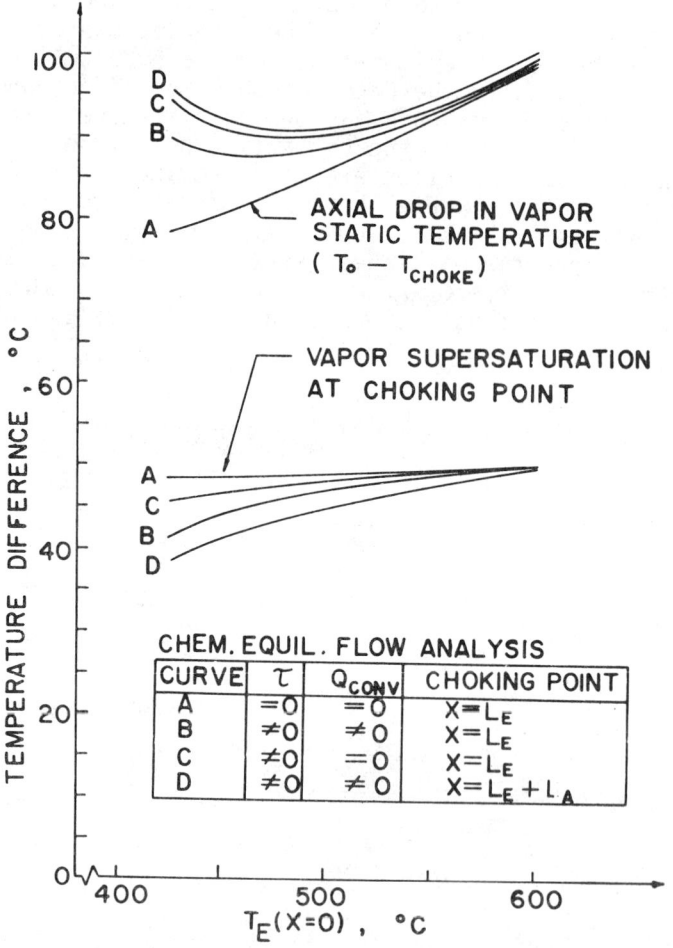

Fig. 9 Axial drop in vapor static temperature and vapor supersaturation at choking point.

### Summary and Concluding Remarks

A theoretical investigation of the effects of the wall shear stress on the sonic limit in a sodium heat pipe has been described. Assuming the vapor (a mixture of monatomic and diatomic sodium) to be in chemical equilibrium but frozen with respect to phase change, solutions were presented over a wide range of operating temperatures and were compared to some recent experimental results. Generally good agreement was obtained between the analysis and the data on the maximum rate of heat transfer and on the axial variations of heat pipe temperature. The calculations indicate that the wall shear stress in the adiabatic region will cause gasdynamic choking to occur at the end of the adiabatic region rather than at the evaporator exit. Further, by neglecting the shear stress, one will overestimate the maximum rate of heat transfer based on the sonic limit. This error is appreciable at the lower values of operating temperature.

Further work is now in progress to determine the effects of chemical nonequilibrium and of the vapor supersaturation and liquid droplet nucleation phenomena on the performance of alkali metal heat pipes operating in the sonic limit regime. The results of this study will be reported in a later paper.

## References

[1] Cotter, T. P., "Theory of Heat Pipes," LA-3246-MS, Los Alamos Scientific Laboratory, Los Alamos, New Mexico, 1965.

[2] Levy, E. K., "Theoretical Investigation of Heat Pipes Operating at Low Vapor Pressures," Transactions of the ASME Journal for Industry, Vol. 90, Series B, No. 4, Nov. 1968, pp. 547-552.

[3] Shapiro, A., The Dynamics and Thermodynamics of Compressible Fluid Flow, Vol. 1, Ronald Press, New York, 1953, pp. 219-260.

[4] Kemme, J., "Ultimate Heat Pipe Performance," IEEE Transactions on Electron Devices, Vol. Ed-16, No. 8, Aug. 1969, p. 717.

[5] Dzakowic, G., Tang, Y., and Arcella, F., "Experimental Study of Vapor Velocity Limit in a Sodium Heat Pipe," ASME Paper 69-HT-21, Minneapolis, Minnesota.

[6] Levy, E. K., and Chou, S. F., "Vapor Compressibility Effects in Heat Pipes - Progress Report 1," Rept. NYO-4095-1, May 1970, Dept. of Mechanical Engineering, Lehigh Univ., Bethlehem, Pa.

[7] Schrage, R., A Theoretical Study of Interphase Mass Transfer, Columbia University Press, New York, 1953.

[8] Wilcox, S., and Rohsenow, W., "Film Condensation of Potassium Using Copper Condensing Block for Precise Wall Temperature Measurement," ASME Paper 69-WA/HT-29, Los Angeles, Calif.

[9] Barry, R., Sartar, A., and Balzhiser, R., "Condensing Heat Transfer Considerations Relevant to Rubidium and Other Alkali Metals," AIChE Preprint 5, Aug. 3, 1969, Minneapolis, Minnesota.

[10] Busse, C., "Pressure Drop in the Vapor Phase of Long Heat Pipes," *Proceedings of the 1967 IEEE Thermionic Conversion Specialist Conference*, Nov. 1967, pp. 391-398.

[11] Schlichting, H., *Boundary Layer Theory*, 4th ed., McGraw-Hill, New York, 1960, p. 503.

[12] Kinney, R., "Fully Developed Frictional and Heat Transfer Characteristics of Laminar Flow in Porous Tubes," *International Journal of Heat and Mass Transfer*, Vol. 11, 1968, pp. 1393-1401.

[13] Yuan, S., and Finkelstein, A., "Heat Transfer of a Laminar Pipe Flow with Coolant Injection," *Proceedings of the Heat Transfer and Fluid Mechanics Institute*, Vol. 9, June 1956, pp. 79-88.

[14] Rohsenow, W., and Choi, H., *Heat, Mass, and Momentum Transfer*, Prentice-Hall, Englewood Cliffs, N.J., 1961, p. 192.

POSSIBLE APPLICATION OF ELECTRO-OSMOTIC
FLOW PUMPING IN HEAT PIPES

M.M. Abu-Romia*
Polytechnic Institute of Brooklyn, Brooklyn, N.Y.

Abstract

s paper employs electro-osmosis for flow pumping or pres-
generation for the possibility of either increasing the
um capability of the heat pipe or overcoming the vapor
in the evaporator section of the pipe, respectively. The
e of electro-osmosis utilizes the presence of a naturally
ring potential at the interface of a fluid and capillary
which in turn creates a redistribution of charged ions
nt within the channel. By applying a potential across
orous electrodes embedded within the wet portion of the
an electro-osmotic force is generated and utilized to
he capillary force of the heat pipe. This results in a
r liquid velocity in the wick, and consequently an in-
e in the maximum capability of the heat pipe. In this
a review of electrokinetic relations in a capillary,
with an analysis for predicting the effect of electro-
ic flow pumping on the heat capability of the heat pipe
resented. Representative results for a supposed heat
utilizing glass beads as a wick material and different
e water solutions as working fluids, indicate that an in-
e in heat pipe capability of several orders of magnitude
be achieved by employing this proposed scheme.

Nomenclature

= cross-sectional area of wick

= dielectric constant of fluid

= particle diameter

= electronic charge

= applied potential difference

---

sented as Paper 71-423 at the AIAA 6th Thermophysics
rence, Tullahoma, Tenn., April 26-28, 1971.
sociate Professor of Mechanical Engineering.

| | | |
|---|---|---|
| $F_{eo}$ | = | local electro-osmotic force |
| $\bar{F}_{eo}$ | = | net electro-osmotic force |
| $F(\theta,t)$ | = | elliptic integral of the first kind |
| $g_r$ | = | acceleration due to gravity near sea level |
| $g_c$ | = | gravitational constant |
| $G_f$ | = | mass flux of fluid in wick |
| $G_g$ | = | mass flux of vapor normal to inner wick surface |
| $h$ | = | half-channel width |
| $h_f$ | = | enthalpy of fluid |
| $h_g$ | = | enthalpy of vapor |
| $h_{fg}$ | = | enthalpy of evaporation |
| $H_r$ | = | capillary equilibrium height |
| $k$ | = | Boltzmann constant |
| $K$ | = | wick permeability |
| $L$ | = | length |
| $L_p$ | = | distance of + ve electrode from end of adiabatic section |
| $n$ | = | ion concentration |
| $Q$ | = | maximum capability of heat pipe |
| $r_h$ | = | hydraulic radius |
| $R$ | = | radius of curvature of liquid-vapor interface |
| $t$ | = | temperature |
| $T$ | = | absolute temperature |
| $x$ | = | axial coordinate |
| $y$ | = | lateral coordinate measured from center of channel |
| $Y$ | = | potential gradient, $\Delta E/L$ |
| $z$ | = | valence of positive and negative ions |
| $\alpha$ | = | ratio of surface ionic to thermal energy, $ez\psi_o/kT$ |
| $\varepsilon$ | = | wick porosity |
| $\psi, \psi_o, \psi_c$ | = | potential as defined by Eq. (4), at the wall and in the center, respectively |
| $\phi$ | = | local potential in capillary |
| $\rho$ | = | fluid density |

| | | |
|---|---|---|
| $\rho_e$ | = | local charge density |
| $\bar{\rho}_e$ | = | net charge density |
| $\mu$ | = | fluid viscosity |
| $\omega$ | = | inverse Debye length, $= (8\pi n_o e z \alpha / D \psi_o)^{1/2}$ |
| $\Omega$ | = | electrokinetic radius |
| $\sigma$ | = | surface tension |

Subscripts

| | | |
|---|---|---|
| a | = | adiabatic section |
| c | = | condenser section |
| e | = | evaporator section |
| f | = | fluid |
| g | = | vapor |

## I. Introduction

The heat pipe, as a device capable of transporting large quantities of heat with very small temperature drops, has been employed in many aspects of thermal engineering.[1-4] In the laboratory, it has been used as a detector for precision measurement of radiant emissivity of surfaces, as a thermostat for measurement of vapor pressure of liquid metals, and as a thermal conductor and heat flux transformer in demonstration models of thermoionic convertors. It is used a great deal in electronic circuits and in space applications requiring thermal control; space and weight of course being the major influencing factors in choosing the heat pipe against conventional devices. Many other applications connected with high temperature and direct energy conversion are currently under consideration.

Providing sufficient heat-transfer areas for both the condenser and evaporator sections, the heat pipe can transport a wide range of heat-transfer rates. A heat-transfer rate is eventually reached at which the heat pipe starts to suffer from a rise in temperature at the evaporator section of the pipe, and as a result, the heat pipe ceases to be practically operable. The maximum capability of the heat pipe is usually characterized by this limiting heat-transfer rate. Recent heat pipe studies have indicated that the most important factors which limit the maximum heat capability of the pipe are the capacity of its capillary pump and/or the presence of vapor lock in the evaporator section of the pipe.[5-7] It is therefore natural to look for ways of aiding the capillary pumping of the heat pipe and overcoming any pressure rise in the evaporator section. Of the different ways to achieve this goal, we propose the scheme of electro-osmosis for both flow pumping and pressure generation. Electro-osmotic flow pumping would aid the capillary forces, increasing the liquid velocity within the wick, and as a result, the maximum capability of the heat pipe would increase. In the case of any vapor pressure rise in the evaporator section of the pipe which would block the liquid flow in the wick, sufficient electro-osmotic pressure could be generated to overcome this vapor rise, and this would restore the operation of the heat pipe. In this investigation, a heat pipe configuration incorporating two porous electrodes across which a potential is applied, is considered. An analysis is carried out to predict the effect of electro-osmotic flow pumping on the increase in the maximum heat capability of the heat pipe. Numerical calculations for a supposed heat pipe utilizing glass beads as a wick material and operating with different dilute water solutions are presented. The present investigation indicates that electro-osmotic flow pumping could be very effective in boosting up the heat capability of the heat pipe. However, quantitatively, the results will be subjected to subsequent experimental investigation to confirm the degree of their accuracy. No detailed evaluation of the effect of electro-osmotic pressure generation on overcoming the vapor lock in the pipe is presented here because of the lack of available data on such mechanism.

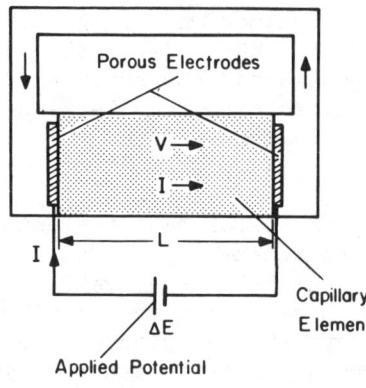

Fig. 1 Electro-osmotic flow pumping system

Electrokinetic phenomena, which encompass electro-osmotic flow pumping and pressure generation, depends on the presence of a naturally occurring potential or a charge accumulation at the interface of the fluid and capillary wall. The mechanism by which this potential is produced is generally attributed to preferential surface adsorption of ions.[8-9] The surface ionic potential causes a redistribution of the charged ions that are present in the fluid. When a fluid is pumped through a capillary, a net charge accompanies the movement of the bulk of the fluid, and as a result an axial streaming potential gradient is generated. Conversely, as shown in Fig. 1, when an axial potential gradient is applied to a capillary element, an electro-osmotic flow is generated which is proportional to the potential gradient. The external current in such a case would be the sum of the conduction and convection components. It is interesting to mention that the phenomena of electrokinetics, discovered by Reuss in 1809[10] and studied by Helmholtz in 1879,[11] were utilized only recently in engineering applications of energy conversion and fluid pumping fields.[12-15]

## Analysis

### Review of Electrokinetic Relations in a Capillary Channel

The capillary channel, taken here as a plane slit of length L and width 2h, contains a fluid which is considered to be a binary electrolyte with each ion possessing z electric charge. When the capillary wall is insulated the net lateral current is zero. An equilibrium balance is therefore established between the ionic diffusion and conduction currents in the lateral direction, and this necessitates a Boltzmann distribution for ionic concentrations $n_{\pm}$ given by

$$n_{\pm} = n_o \exp(\mp ze\psi/kT) \qquad (1)$$

$\psi$ is the local lateral component of the potential in the fluid, and its value at the wall $\psi_o$ is taken as a constant for a specific fluid and capillary material combination. The local charge density $\rho_e$ is the excess of positive over negative charge, or

$$\rho_e = ez(n_+ - n_-) = -2n_o ez \sinh(\alpha\psi/\psi_o) \qquad (2)$$

where $\alpha = ez\psi_o/kT$, is defined as the surface ionic energy ratio. The charge density and the potential are also related through the one-dimensional Poisson equation

$$d^2\psi/dy^2 = 4\pi\rho_e/D \qquad (3)$$

where y is the lateral distance measured from the center of the channel, and D the dielectric constant of the fluid. The total potential $\phi$ at any location in the channel is influenced by both the applied potential $\Delta E$ and the surface ionic potential $\psi_o$. It is given by

$$\phi = E(x) + \psi(y) \qquad (4)$$

where x is the axial distance along the capillary channel. The fact that the charge density is independent of the axial direction indicates that the axial potential E has the distribution

$$E(x) = \Delta E \, (L-x)/L \tag{5}$$

where L is the length of the capillary channel. Following Ref. 13, Eqs. (2) and (3) are combined in the form

$$d^2(\alpha\psi/\psi_o)/d(\omega y)^2 = \sinh(\alpha\psi/\psi_o) \tag{6}$$

where the inverse Debye length is defined by

$$\omega = [8\pi n_o e z \alpha / D \psi_o]^{1/2} \tag{7}$$

Equation (6) is subject to the following boundary conditions

$$d(\alpha\psi/\psi_o)/d(\omega y) = 0 \text{ and } \psi = \psi_c \text{ at } y = 0 \tag{8}$$

Integrating Eq. (6) once and utilizing the boundary conditions of Eq. (8), the result is

$$\frac{d(\alpha\psi/\psi_o)}{d(\omega y)} = 2\left[\cosh^2\left(\frac{\alpha\psi}{2\psi_o}\right) - \cosh^2\left(\frac{\alpha\psi_c}{2\psi_o}\right)\right]^{1/2} \tag{9}$$

Further integration of Eq. (9) yields the result

$$\omega y = F(\theta_y, t)/\cosh(\alpha\psi_c/2\psi_o) \tag{10}$$

where $F(\theta_y, t)$ is the elliptic integral of the first kind

$$\theta_y = \sin^{-1}\left\{ \frac{[\cosh^2(\alpha\psi/2\psi_o) - \cosh^2(\alpha\psi_c/2\psi_o)]^{1/2}}{\sinh(\alpha\psi/2\psi_o)} \right\} \tag{11}$$

$$t = [\cosh(\alpha\psi_c/2\psi_o)]^{-1}$$

Equation (10) specifies the value of the lateral potential $\psi$ at any distance y from the center of the slit. The center potential $\psi_c$ can be obtained from Eq. (10) as

$$\Omega/2 = \omega h = F(\theta_h, t)/\cosh(\alpha\psi_c/2\psi_o) \tag{12}$$

where $\Omega$ is the electrokinetic radius defined as twice the product of the channel area and the inverse Debye length divided by the wetted perimeter (or twice the ratio of hydraulic radius to Debye length). The ratio of the center potential to the surface potential can be determined according to Eq. (12) for different values of $\alpha$ and electrokinetic radius. It is shown in Ref. 13 that, for electrokinetic radii greater than ten, this ratio is essentially zero.

The local electro-osmotic force $F_{eo}$ per unit volume in the axial direction is given by

$$F_{eo} = -\rho_e \, \partial\phi/\partial x = \rho_e \, \Delta E/L = \rho_e \, Y \qquad (13)$$

where Y is the potential gradient and $\rho_e$ the charge density. Integrating Eq. (13) on the channel cross-sectional area gives the net electro-osmotic force in the capillary

$$\bar{F}_{eo} = (1/2h) \int_{-h}^{h} F_{eo} \, dy = Y(1/2h) \int_{-h}^{h} \rho_e \, dy = Y \bar{\rho}_e \qquad (14)$$

where $\bar{\rho}_e$ is the net charge density. It can be evaluated analytically by use of Eqs. (2), (6) and (9) as

$$\bar{\rho}_e = (8n_o ez/2\omega h)[\cosh^2(\alpha/2) - \cosh^2(\alpha\psi_c/2\psi_o)]^{1/2} \qquad (15)$$

The variation of the net charge density with electrokinetic radius is plotted in Fig. 2 for different values of $\alpha$. For moderate or large values of electrokinetic radii, Eq. (15) degenerates to

$$\bar{\rho}_e = (8n_o ez/2\omega h) \sinh(\alpha/2) \qquad (16)$$

## Capillary and Electro-Osmotic Pumped Heat Pipe

Electro-osmotic pumping could be simply achieved by placing two porous electrodes, across which a potential $\Delta E$ is applied, in the heat pipe wick as shown in Fig. 3. The rest of the heat pipe components are the same as those of a conventional heat pipe. The placement of the electrodes in the wet section of the wick should be in such a way as to maximize the electro-osmotic force. Such procedure has not been followed here, and simply the negative electrode is located between the adiabatic and evaporator sections and the positive electrode is placed at a distance $L_p$ from the end of adiabatic section in the condenser region. To determine the increase in the maximum heat capability of the heat pipe due to electro-osmotic pumping, we utilize the same mathematical model as of Refs. 2 and 5. In addition to the assumptions used by these two references, the analysis of electro-osmotic pumping through the wick is based on the following: 1) The wick consists of identical number of parallel capillary channels, each with an electrokinetic radius of $2\omega r_h$ replacing that of a slit of width 2h. 2) Electrical heating has a negligible effect on the heat capacity of the working solution. 3) The potential gradient is

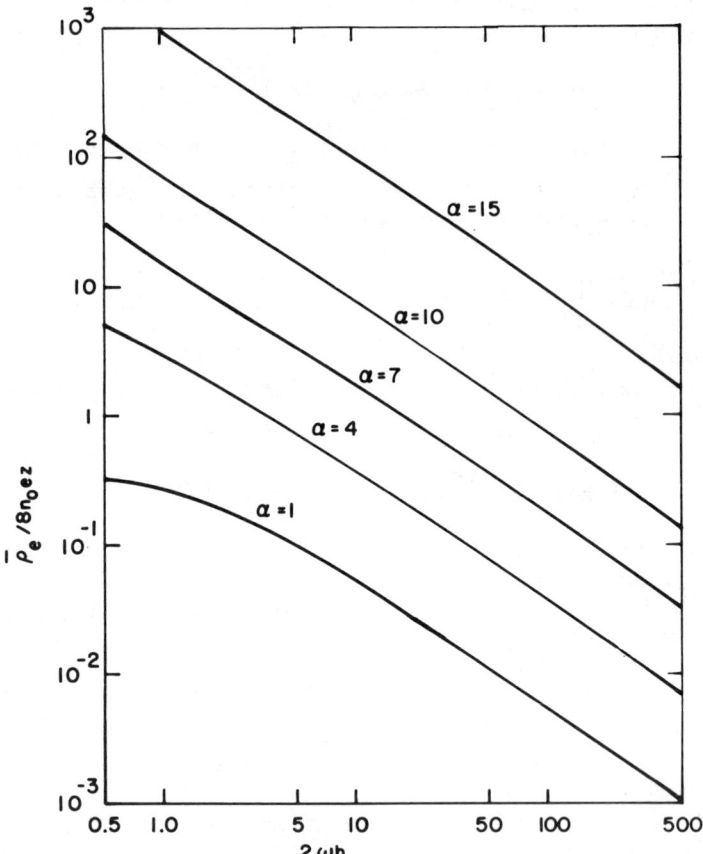

Fig. 2 Variation of net charge density with electrokinetic radius.

constant and is equal to the applied potential divided by the distance between the electrodes. 4) No variation in the ion concentration is taking place along the wick length or due to liquid evaporation. 5) The electric field intensity within the wick is below the breakdown value. 6) The variation in the capillary force due to the applied potential is negligible.

Taking into account the previous assumptions, and confining our analysis to the cylindrical heat pipe of Fig. 3, the governing mass and energy equations take the form

$$d/dx\ (G_f(x)) = (\pi d/\varepsilon A)\ G_g \qquad (17)$$

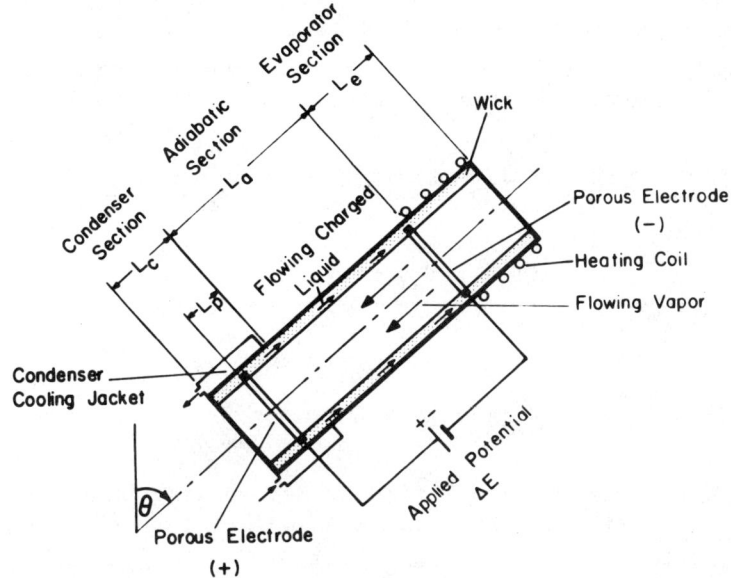

Fig. 3 Heat pipe configuration with electro-osmotic flow pumping.

$$h_f \cdot d/dx \, (G_f(x)) + Q/\varepsilon AL_c = (\pi d/\varepsilon A) \, h_g G_g \qquad (18)$$

and the momentum equation for fluid flow in the wick is

$$2\sigma \cdot d/dx(1/R(x)) + \bar{\rho}_e(x) \, Y - (g/g_c)\rho_f \cos\theta - (\mu_f/K\rho_f)G_f(x)$$
$$= (1/g_c\rho_f)d/dx(G_f^2(x)) \qquad (19)$$

where the second term in the left-hand side of Eq. (19) represents the electro-osmotic force, and the friction force is accounted for by using Darcy's law.

Following Ref. 5, Eqs. (17) and (18) can be combined to give

$$d/dx \, (G_f(x)) = Q/\varepsilon AL_c h_{fg} \quad 0 \le x \le L_c \qquad (20)$$

which when integrated gives the mass velocity in the condenser and adiabatic sections as

$$G_f(x) = (Q/\varepsilon A h_{fg})(x/L_c) \quad 0 \le x \le L_c$$

and
$$G_f(x) = Q/\varepsilon A h_{fg} \quad L_c \le x \le L_c + L_a \qquad (21)$$

If the temperatures of the condenser and adiabatic sections of the heat pipe are represented by average values $T_c$ and $T_a$, respectively, Eq. (19) can be integrated over the combined length of the two sections. This leads to the result

$$\frac{1}{g_c \rho_f(T_c)} \left[\frac{Q}{\varepsilon A h_{fg}}\right]^2 + (1/2K)\left[\frac{\mu_f(T_c)}{\rho_f(T_c)} L_c + \frac{2\mu_f(T_a)}{\rho_f(T_a)} L_a\right][Q/\varepsilon A h_{fg}]$$

$$= 2\left[\frac{\sigma(T_a)}{R(L_c+L_a)} + \frac{\sigma(T_c)-\sigma(T_a)}{R(L_c)} - \frac{\sigma(T_c)}{R(0)}\right] - (g/g_c)\cos\theta\,[\rho_f(T_c)L_c$$

$$+ \rho_f(T_a)L_a] + Y\,[\bar{\rho}_e(T_c) L_p + \bar{\rho}_e(T_a) L_a] \qquad (22)$$

Equation (22) can be simplified by recognizing that the first term of the left-hand side is much smaller than the second term, and therefore could be neglected.[2] Further simplification can be achieved by noting that the heat pipe is assumed to be flooded at both the condenser and adiabatic sections, i.e., $R(0) = R(L_c) = \infty$. With these simplifications, Eq. (22) reduces to

$$Q = [2K\varepsilon A h_{fg}]\,\{2\sigma(T_a)/R(L_c+L_a) - (g/g_c)\cos\theta\,[\rho_f(T_c)L_c$$

$$+ \rho_f(T_a)L_a] + Y\,[\bar{\rho}_e(T_c) L_p + \bar{\rho}_e(T_a)L_a]\}/[\mu_f(T_c)L_c/\rho_f(T_c)$$

$$+ 2\mu_f(T_a)L_a/\rho_f(T_a)\,] \qquad (23)$$

Equation (23) gives the maximum capability of the heat pipe limited by the capacity of its capillary and electro-osmotic pump.

Numerical Results

To illustrate the effect of electro-osmotic flow pumping on the increase in the maximum heat capability of the heat pipe, numerical results are obtained for a heat pipe with a configuration similar to that of Fig. 3. The heat pipe considered is of the same dimensions as of heat pipe "A" utilized by Cosgrove et al. in their experimental study.[5] This enabled us to utilize the temperature distributions reported by these authors in calculating the electro-osmotic pumping force. The working fluid in the pipe is taken as one of four dilute water solutions; distilled water in equilibrium with atmospheric carbon dioxide, $10^{-5}$ N-HCl, $10^{-5}$ N-KCl, and $10^{-5}$ N-KOH. The wick material is chosen to be glass beads based on the knowledge of surface properties of glass in contact with the water solutions.[8,15] Table 1 contains the physical and computed electrokinetic parameters for the four solutions.

Table 1 Physical constants and electrokinetic parameters of solutions

| Solution | Concentration, moles/liter | $\psi_o$, Volts | $\alpha(T_o)$ | $\omega(T_o) \times 10^{-4}$ cm$^{-1}$ |
|---|---|---|---|---|
| Dist.$H_2O$ in equil. with $CO_2$ | $4.38 \times 10^{-6}$ ($HCO_3^-$) | -0.160 | 6.4 | 6.82 |
| HCl | $10^{-5}$ | -0.080 | 3.2 | 10.3 |
| KCl | $10^{-5}$ | -0.140 | 5.6 | 10.3 |
| KOH | $10^{-5}$ | -0.155 | 6.2 | 10.3 |

In evaluating the contribution of electro-osmotic pumping to the maximum capability of the heat pipe, as given by Eq. (23), the temperatures of the adiabatic and condenser sections are taken at their zero-potential values. The net charge density at any temperature is evaluated for different values of the electrokinetic radius $2\omega r_h$ and surface ionic energy ratio $\alpha$ according to Eq. (15). The hydraulic radius $r_h$ is determined from the bead size $D_p$ and porosity of the wick $\epsilon$ by the relation[16]

$$r_h = \epsilon D_p / 6(1-\epsilon) \qquad (24)$$

Values of the radius of curvature R in Eq. (23) are calculated from the capillary equilibrium data as

$$2\sigma(T_a)/R(L_c + L_a) = (g_r/g_c)[\sigma(T_a)/\sigma(T_r)] \rho_f(T_r) H_r \qquad (25)$$

where the capillary equilibrium height $H_r$ is determined from the reported measurements of Ref. 5.

$$H_r = 2.493 \times 10^{-4} D_p^{-1.135}, \text{ ft} \qquad (26)$$

In addition, the permeability K of the wick material is calculated using the Blake-Kozeny equation

$$K = D_p^2 \epsilon^2 / 150 (1-\epsilon)^2 \qquad (27)$$

such relation has been confirmed by the measurements of Ref. 5.

The maximum capability of the heat pipe, utilizing a wick of glass beads with a diameter size of $0.346 \times 10^{-3}$ ft and with an applied potential of 20v, is shown in Fig. 4 as a function of the cosine of the angle of inclination. Also shown in the figure for comparison, are the results of Cosgrove et al.[5] which correspond to the zero-potential solution of Eq. (23). As can be noted, the maximum capability of the heat pipe increases by three times its zero-potential value for the $10^{-5}$ N-KOH solution. The addition of an acid by the same concentration to the distilled water will have the effect of decreasing the surface ionic potential, and therefore would result in a lower increase in the maximum heat capability of the pipe. Fig. 5 shows the increase in the maximum heat capability of the heat pipe with the applied potential for a specific inclination of the pipe. Although the results of this figure indicate that such relation is linear, in reality, however, lower rate of increase in the heat capability will take place due to the rise in the temperature of the condenser and adiabatic sections of the heat pipe together with the influence of electrical heating. These effects are expected to be significant at large values of the applied potential.

The application of electro-osmotic flow pumping in heat pipes could have the additional advantage of utilizing the pipe in cases where the capillary force cannot overcome the gravitational force. Such situations will occur when the wick is made of a material with large pores or the capillary force is reduced due to an increase in the heat pipe temperature. With electro-osmotic pumping, the heat pipe can operate with larger pore sizes as shown in Fig. 6. In this figure, the maximum capability of the heat pipe is plotted as a function of the particular diameter according to Eqs. (23-27). As to be noted, optimum design of the heat pipe to achieve maximum heat capability will depend on the applied potential and values of the electrokinetic parameters in addition to those of the capillary height and permeability.

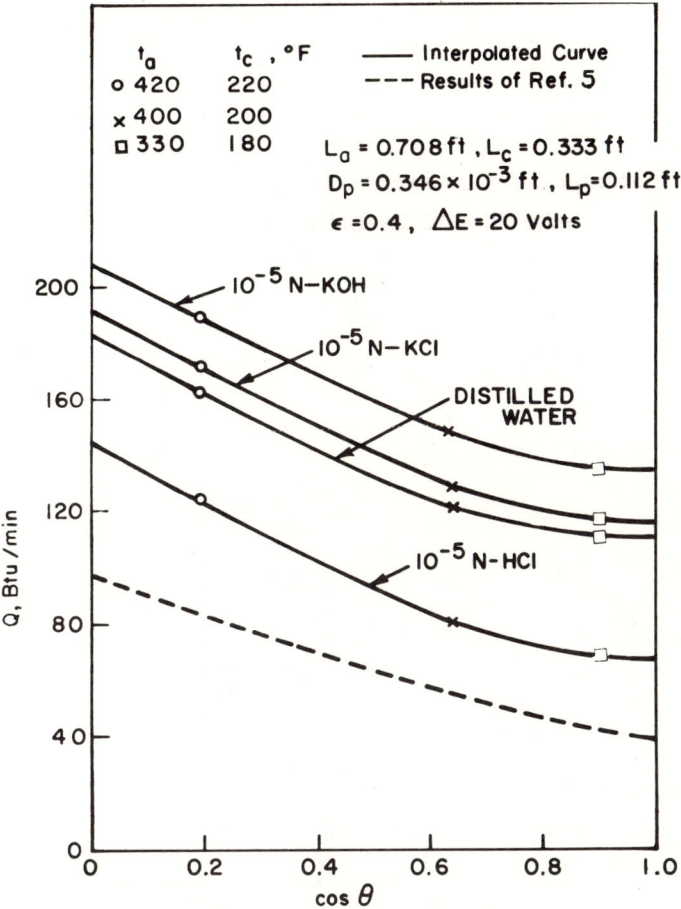

Fig. 4 Maximum capability of heat pipe "Q" as a function of cos θ for several dilute water solutions.

### Conclusion

The scheme of utilizing electro-osmotic flow pumping in heat pipes has been investigated to determine its effect on the increase in the maximum heat capability of the heat pipe. The results of the investigation indicate that an increase in the maximum heat capability of several orders of magnitude could be achieved by applying moderate values of electric potential. Electro-osmotic pumping may also be used in starting-up the pipe and therefore reduce the transient period. Several limitations on the working fluids and wick material, such as high electrical resistance materials, are required. Further

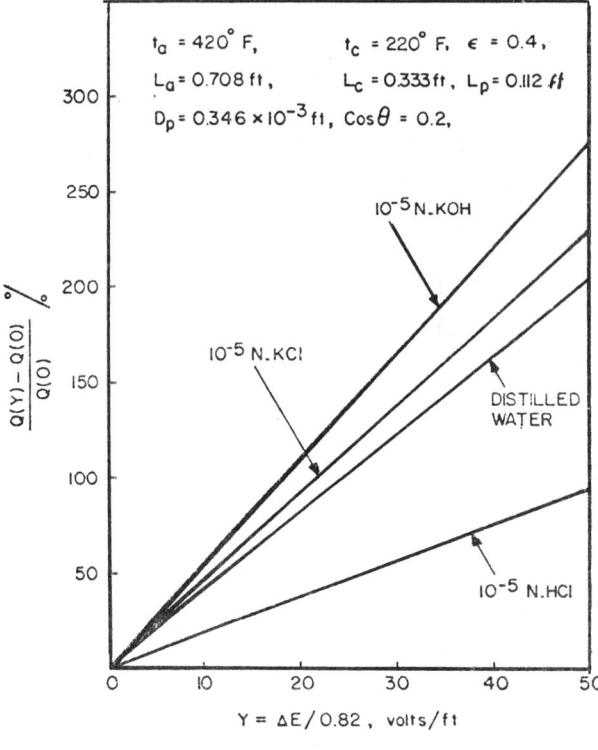

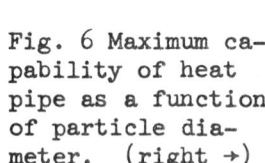

Fig. 5 Increase in maximum capability of heat pipe with applied potential. (← left)

Fig. 6 Maximum capability of heat pipe as a function of particle diameter. (right →)

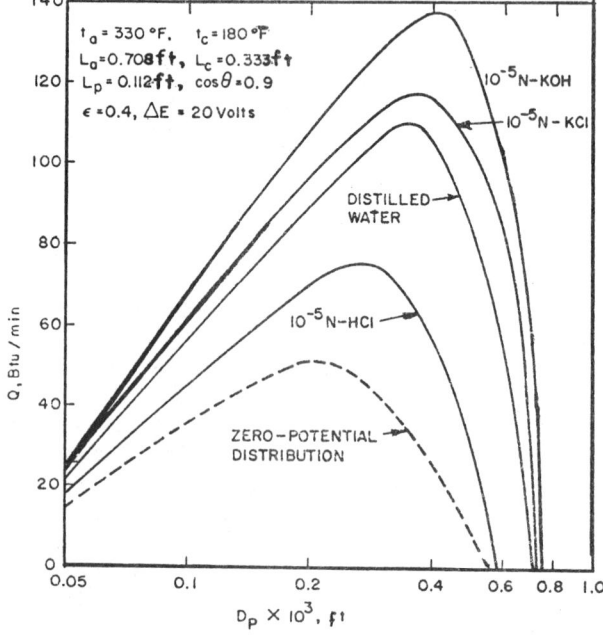

theoretical and experimental research in the area will provide the optimum design of the heat pipe in connection with electro-osmotic flow pumping.

References

[1] Proceedings of the Joint AEC/Sandia Laboratories Heat Pipe Conference, Rept SC-M-66-623, Oct. 1966, Sandia Lab., Albuquerque, N. Mex.

[2] Kunz, H.R., Wyde, S.S., Nashick, G.H. and Barnes, J.F., "Vapor-Chamber Fin Studies; Operating Characteristics of Fin Models," CR-1139, Aug. 1968, NASA.

[3] Katzoff, S., "Heat Pipes and Vapor Chambers for Thermal Control of Spacecraft," AIAA Progress in Astronautics and Aeronautics: Thermophysics of Spacecraft and Planetary Bodies, Vol. 20, edited by G.B. Heller, Academic Press, New York, 1967, pp. 761-819.

[4] Abu-Romia, M.M. and Bhatia, B., "Measurement of Stagnation-Point Heat Transfer from Plasma Torch by Using Heat Pipe Calrimetry," AIAA Paper 71-81, New York, 1971.

[5] Cosgrove, J.H., Ferrell, J.K. and Carnesale, A., "Operating Characteristics of Capillarity-Limited Heat Pipes," Journal of Nuclear Energy, Vol. 21, No. 7, Jan. 1967, pp. 547-558.

[6] Ferrell, J.K. and Johnson, H.R., "The Mechanism of Heat Transfer in the Evaporator Zone of a Heat Pipe," ASME Paper 70-HT/SPT-12, Space Technology and Heat Transfer Conf., June 1970, Los Angeles, Calif.

[7] Moss, R.A. and Kelly, A.J., "Neutron Radiographic Study of Limiting Planar Heat Pipe Performance," International Journal of Heat and Mass Transfer, Vol. 13, No. 3, Mar. 1970, pp. 491-502.

[8] Davies, J.T. and Rideal, E.K., Interfacial Phenomena, Academic Press, New York, 1961.

[9] Bikerman, J.J., Surface Chemistry, Academic Press, New York, 1958.

[10] Reuss, F.F., *Memoires de la Societe Imperiale de Naturalistes de Noscou*, Vol. 2, 1809.

[11] Helmholtz, H.F.H., "Studien uber electrische Grenzschicten," *Annalen der Physik und Chemie, von G. Wiedemann*, Vol. 7 1879.

[12] Dresner, L., "Electrokinetic Phenomena in Charged Microcapillaries," *Journal of Physical Chemistry*, Vol. 67, Aug. 1963, pp. 1635-1641.

[13] Burgreen, D. and Nakache, F.R., "Electrokinetic Flow in Ultrafine Capillary Slits," *Journal of Physical Chemistry*, Vol. 68, No. 5, May 1964, pp. 1084-1091.

[14] Morrison, F.A., Jr. and Osterle, J.F., "Electrokinetic Energy Conversion in Ultrafine Capillaries," *Journal of Chemical Physics*, Vol. 43, No. 6, Sept. 1965, pp. 2111-2115.

[15] Burgreen, D. and Hildreth, D. "Electrokinetic Power Conversion," presented at 5th Inter-society Energy Conversion Engineering Conference, Sept. 1970, Las Vegas, Nev.

[16] Bird, R.B., Stewart, W.E. and Lightfoot, E.N., *Transport Phenomena*, Wiley, New York, 1962.

EXPERIMENTAL PERFORMANCE OF GROOVED HEAT PIPES
AT MODERATE TEMPERATURES

N. Kosowski[*] and R. Kosson[†]

Grumman Aerospace Corporation, Bethpage, N. Y.

Abstract

Heat transfer characteristics and heat transport capacity values are presented for two essentially similar grooved heat pipes. The pipes were made of 1/2-in. o.d. aluminum, and had 30 internal longitudinal grooves. Tests were conducted with Freon-21, Freon-113, and ammonia working fluids, at various temperatures, charge levels, and tilt conditions. The grooved heat pipes are sensitive to gravity effects, but offer attractive performance coupled with a very simple and reliable construction.

I. Introduction

Grooved heat pipes have a relatively simple geometry which lends itself to analysis (Refs. 1-3) and appears to offer attractive performance for many applications. Experimental data reported in the literature (Refs. 4, 5) has been relatively meager, however.

The objective of the experimental program discussed in this paper was to study the characteristics and performance of grooved heat pipes in the moderate temperature range suitable to space vehicle applications. The grooved heat pipe was selected for study because it offered the potential of relatively high heat-transfer rates in conjunction with the low-conductiv-

---

Presented as Paper 71-409 at the AIAA 6th Thermophysics Conference, Tullahoma, Tenn., April 26-28, 1971.

This work was sponsored by the Grumman Advanced Development Program. The authors wish to acknowledge the efforts of J. Westell, J. Jackson, K. Hill, and B. McGrory in obtaining some of the experimental data.

[*]Principal Investigator, Thermophysics Group; presently Project Engineer, Universal Desalting Corp.

[†]Group Leader, Thermophysics Group.

ity working fluids required for the moderate temperature range, and because its inherent simplicity promised greater reliability and durability than other more complex heat pipe constructions. These same reasons led to selection of grooved heat pipes for a flight test experiment on the OAO spacecraft. The paper presents a considerable amount of data not previously available in the literature.

## II. Experimental Configuration

Test results and performance data presented in this paper were obtained for two essentially similar experimental grooved heat pipes. The heat pipes were made of the same internally grooved tubing, nominally 0.5-in. o.d. with 30 grooves cut on the inside tube wall. The average groove dimensions were 0.030-in. wide and 0.035-in. deep. The cross section of the heat pipes is shown in Fig. 1. The individual groove cross-sectional area, as determined from planimeter measurements of enlarged photographs of eight sections of the sample, was $1.042 \times 10^{-3}$ in.$^2$ $\pm$ 3.5%. This was used as the liquid flow area in computing the pipe transport capacity. A somewhat larger groove cross-sectional area, $1.17 \times 10^{-3}$ in.$^2$ was determined by weighing sections of pipe empty and filled with water, and computing the weight of water required to fill the grooves. This allows for small holes and irregularities in the surfaces, and was used in computing the amount of working fluid required to level fill the grooves.

Fig. 1 Pipe cross section ($\frac{1}{2}$ in. o.d.)

The evaporator section of the heat pipe is covered with a nichrome ribbon heater. The condenser section is wrapped with two counterflow coils of 1/8-in. diameter copper tubing, potted in cerrobend to improve thermal contact between the cooling coils and the condenser. A small heat pipe transport section joins the evaporator with the condenser. The characteristic dimensions for each heat pipe are given in Table 1.

Table 1  Pipe Lengths

| Heat pipe | Heat pipe length, in. | Evaporator, in. | Condenser, in. | Transport section, in. |
|---|---|---|---|---|
| 1 | 34.125 | 12.25 | 19.25 | 2.625 |
| 2 | 35.0 | 12.75 | 20.25 | 2.0 |

A total of 56 copper constantan thermocouples were installed on top of the outside tube walls and six additional thermocouples were attached at two axial locations to the bottom and sides of the pipe wall - one such ring being on the condenser, the other on the evaporator. A vapor thermocouple probe was inserted into the pipe from the condenser end.

All thermocouples on the heat pipe walls and the vapor probe were connected to Bristol recorders for continuous monitoring of the temperature distribution on the heat pipe. More accurate test readings were taken periodically of selected thermocouples connected through a switching system to a Leeds and Northrup K-4 system. Thermopiles installed in the coolant loop plenum chambers were read out on a Leeds and Northrup Microvolt Amplifier.

The heat pipe is held in three micarta brackets each supported by two bolts from a flat levelling plate. The levelling plate is hinged to a two-column frame support at the evaporation side, and flexibly attached to a threaded column at the condenser side. The latter can be moved up or down to vary the pipe tilt by means of a high resolution crankshaft. The three support columns are anchored to a baseplate. During installation the micarta brackets were carefully aligned using a laser beam. A machinist's level, coupled with dial gages, was placed on the levelling table and used to obtain the desired pipe level.

Both pipes were well insulated prior to test.

### III. Test Procedure

For convenience, a reference 100% working fluid quantity for the heat pipe was defined as the sum of the weights of the liquid which would fill all grooves, assuming flat menisci in the entire pipe, and of the vapor filling the remaining volume, all at 70°F. The heat pipes were tested with several amounts of working fluid and at various vapor temperatures. The weight of working fluid in the heat pipe was determined with an accuracy of $\pm$ 0.5% of the total.

Based on the measured resistance of the nichrome ribbon heaters, a range of electric heat inputs was applied to the evaporator. Either a-c line current was used with a variac voltage regulator or a d-c source was included in the circuit. No significant differences in thermocouple readings were observed with either method of applying heat to the evaporator. Temperature-controlled FC-75 coolant was circulated in the two counterflow loops of the condenser. FC-75 entering the loops was maintained at the desired temperature by means of an air-to-liquid heat exchanger, with the air either heated or cooled in a Wiley unit with liquid nitrogen. The amount of FC-75 flowing through the coils was regulated and measured by two flowmeters. Thermopiles in the coolant loop plenums recorded the temperature rise of the coolant. It was thus possible to control and measure the electrical heat input into the evaporator, and the heat carried away from the condenser. To obtain a complete heat balance, it was necessary to account for leakages through the insulation to and from the heat pipe. In initial tare tests, just enough heat was removed or applied to keep the heat pipe nearly isothermal at several selected temperatures. Knowing the heat pipe and ambient temperatures and the corresponding heat input or removal required, an overall insulation conductance factor was determined for a range of temperatures for the entire length of the heat pipe. In later tests, where the evaporator and condenser were at different temperatures, the loss or gain through the insulation was determined by assuming that the insulation conductance was linearly uniform, and corrective factors $l_{ev}/l_{hp}$ and $l_{cond}/l_{hp}$, respectively, were applied to the conductance in each case. This method of accounting for insulation losses, together with evaporator input and condenser output measurements resulted in heat balances with differences which were less than 1 watt in all cases. In tests where the vapor temperatures were close to ambient temperatures, which accounts for the bulk of the program, the effect of losses or gains through the insulation were insignificant.

Each test setting was monitored continuously on Bristol temperature recorders. Periodic readings were made of thermocouples on the K-4 system, of thermopiles, of input current and voltage, and of coolant flow rates. Equilibrium was considered achieved when no temperatures as measured on the K-4 changed over a period of one hour. The data obtained in millivolts was converted and average temperatures were calculated for the evaporator and condenser, respectively. The end thermocouples were not used in computing averages to eliminate edge effects. Continuity checks on all thermocouples were performed before each test series, and on some during the tests, when called for. Thermocouples which became detached from the heat pipe were not included in computation of averages. Out of approximately 60 thermocouples, 3 to 4 were eliminated for these reasons.

### IV. Visual Observation Grooved Heat Pipe

A 34-in. long, 1/2 in. o.d. aluminum heat pipe with 30 longitudinal grooves (slightly different from those of pipe No. 1) was made with glass viewing ports in the end walls to permit visual observation of the liquid distribution within the pipe. With Freon-21 at 100% charge and the pipe adiabatic (no heat flow), it was noted that the upper grooves were partially drained, and a puddle formed along the bottom of the pipe. When the heaters were turned on, the puddle decreased in size and the upper grooves became more nearly filled. Some liquid was observed flowing down the window at the condenser end. At high values of heat flux the puddle nearly disappeared, but some liquid was observed flowing down the condenser end wall and circumferentially within the condenser. The upper grooves tended to be nearly full at high values of heat load, but were the first to dry out as heat load was increased to dryout.

With ammonia working fluid, the effects of gravity were qualitatively similar but much smaller in magnitude, as might be anticipated from the lower surface tension to liquid density ratio, $\sigma/\rho_\ell$. All grooves were full at low heat loads.

V. Test Results - Normal Heat Pipe Operation

Freon-21 Working Fluid

The first tests were run with Freon-21 working fluid to determine effects of temperature, amount of charge, and tilt on heat pipe performance. Results for a heat pipe with 24.5 gms of Freon-21, corresponding to approximately 89% of the amount required to level fill the grooves, are shown in Fig. 2 for temperatures ranging from 10°F to 70°F with the pipe level. Figure 2 shows over-all temperature difference (mean outside wall temperature of top of evaporator minus mean outside wall temperature of top of condenser) as a function of heat flux. The data has a little scatter ($\pm 1/2$°F), but no trend with temperature. Over-all pipe conductance is very nearly constant at about 5.3 w/°F for heat flows from 10 to 50 w.

Figure 3, for charge levels of 99 and 108%, uses the vapor thermocouple reading to establish an evaporator to vapor $\Delta T$ and a vapor to condenser $\Delta T$. Results are plotted as a function of heat per ft of length $Q/\ell$ (of evaporator or condenser) and may be converted to film coefficient, h, using the relation

$$h = \frac{1}{\pi D} \frac{(Q/\ell)}{\Delta T} \qquad (1)$$

Based on an inside diameter D of 0.452-in., the evaporator film coefficients are approximately 200 Btu/hr-ft²°F. Condenser coefficients are difficult to evaluate because of the small temperature differences involved, but are in the range of 300 to

400 Btu/hr-ft$^2$°F. For reference purposes, dashed lines corresponding to h=200 and h=300 are shown along with the evaporator and condenser results.

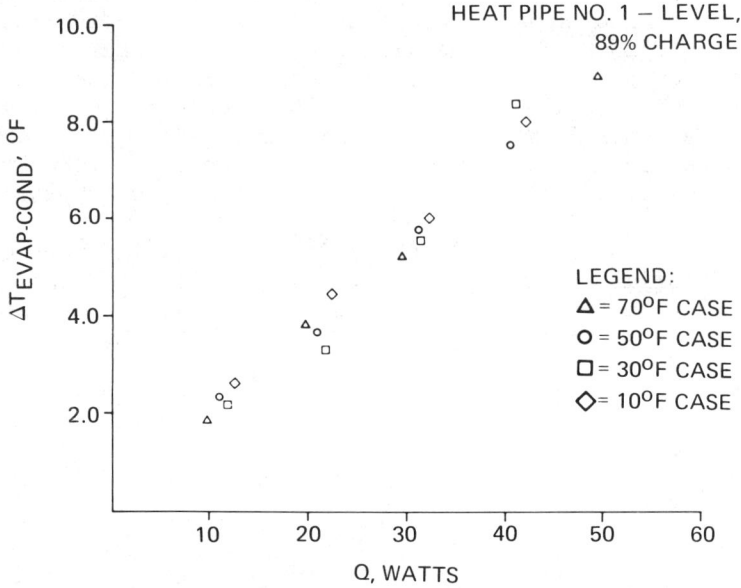

Fig. 2 Freon-21, over-all temperature drop vs heat flow.

The results shown have been corrected for heat leakage through the insulation, as noted under procedure.

Evaporator and condenser temperature differences are shown in Figs. 3-A and 3-B, both level and at slight tilt (0.05 in.), for a pipe temperature of 70°F. The level condenser temperature differences in Fig. 3-B are marginally higher at 108% charge than those obtained at lower charge levels, reflecting the thermal resistance of a larger puddle at the bottom of the pipe. In the evaporator, the effects of the larger puddle might be offset by having slightly more liquid in some of the upper grooves, as a result of the increased top-side condensation.

At highest heat flux levels, the effect of tilt was to increase the evaporator temperature difference by approximately 1°F at 99.3% charge and about 1/2°F at 108% charge. Tilt increases the partial draining of the upper grooves at the condenser end wall (see notes on visual observation), but also

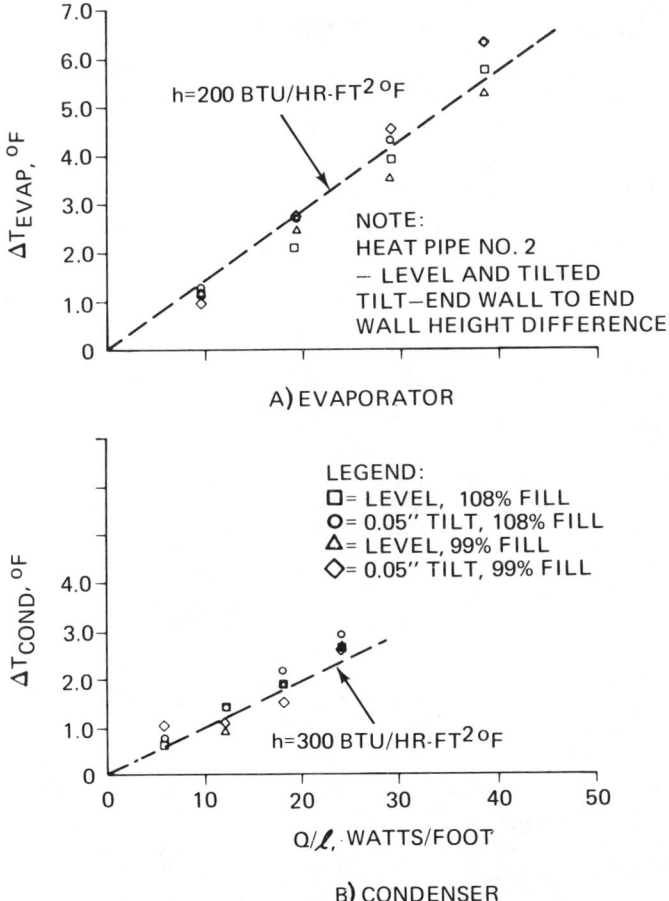

Fig. 3 Freon-21, 99 and 108% charge, temperature drop vs heat flux.

eliminates the puddle in the evaporator, which probably accounts for the very small increase at 108% charge. In the condenser, at 108% charge, there seems to be a small increase in temperature drop, but it is of the same order as the scatter in the data.

The most significant effect of tilt is, of course, on maximum heat transport capacity, and the results for Freon-21 are shown in Fig. 4. In each case a range is shown bracketing the maximum heat flux. At the lower value, no dryout was observed at the evaporator end wall, while at the higher value, dryout (as evidenced by a rise in wall temperature) was well advanced. The amount of tilt is taken as the end-wall to end-wall height difference.

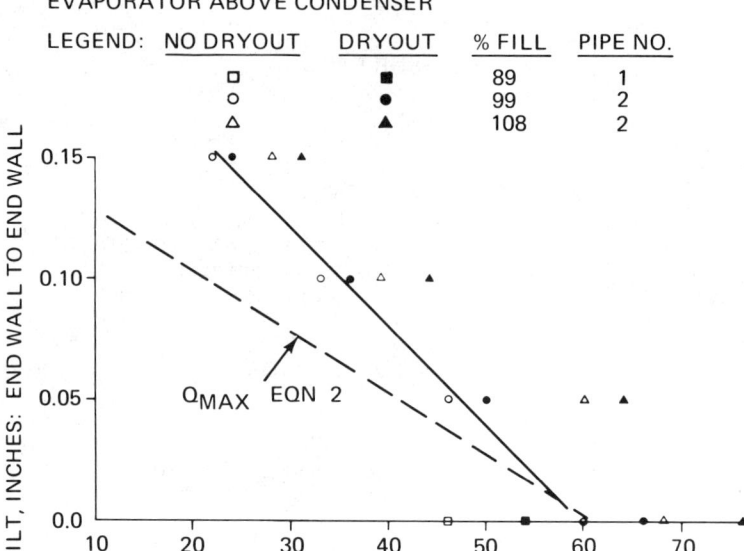

Fig. 4  Freon-21, heat pipe tilt vs maximum heat transport.

As shown, maximum transport capacity increases with increasing amount of charge, with the effect being somewhat more pronounced at low values of tilt, where the puddle at the bottom of the pipe can contribute directly as an auxiliary wick. At high values of tilt the puddle is at the condenser end of the pipe and only contributes indirectly to the capacity of the bottom grooves by reducing the length and height against which the grooves must pump.

Neglecting effects such as groove liquid loss at the condenser end (which might be increased by tilt) and vapor pressure drop, the theoretical pumping capacity, $Q_{max}$, in the absence of puddle contributions is given by:

$$Q_{max} = \left[\frac{\sigma}{r_c} - \frac{\rho_\ell g h}{g_c}\right] \frac{g_c \rho_\ell \lambda A_\ell D_\ell^2}{32 \mu_\ell L_e} \qquad (2)$$

where  $\sigma$ = surface tension, $r_c$ = minimum two-dimensional meniscus radius of curvature, $\rho_\ell$ = liquid density, g = gravity acceleration = $417 \times 10^6$ ft/hr$^2$, $g_c$ = conversion constant = $417 \times 10^6$ lbm ft/lbf hr$^2$, $\lambda$ = latent heat, $\mu_\ell$ = liquid viscosity, $A_\ell$ = liquid flow cross-sectional area, $D_\ell$ = hydraulic diameter of an individual groove, and $L_e$ = effective pumping length (midpoint of evaporator to midpoint of condenser). (Equation 2 is derived using Cotter's equations (Ref. 6), equating the capillary pressure rise to the sum of the liquid viscous and gravity pressure drops.)

From this, $Q_{max}$ decreases linearly with increasing h, as shown by the dashed curve in Fig. 4. The experimental data for $Q_{max}$ shown in Fig. 4 clearly falls off with increasing tilt much less than would be indicated by Eq. 2. Implicit in Eq. 2, are the assumptions that condensation occurs uniformly along the length of the condenser, and that the meniscus is flat (vapor pressure equal to liquid pressure) at the end of the condenser. These, plus the puddle and groove end wall draining may account for the discrepancy.

Extrapolating the tilt data to zero tilt probably provides a better estimate of zero "g" performance than direct use of the level data. Such an extrapolation is indicated on Fig. 4 for the 100% fill, giving a value approximately equal to the theoretical value of 61 w.

## Ammonia Working Fluid

Tests with ammonia working fluid were run at approximately 70°F with charges of 9.4, 11.4, 12.7 and 14.0 grams, corresponding to 74, 90, 100 and 110% of the amount required to level fill the grooves.

Over-all temperature difference with the pipe level is shown in Fig. 5 for the four fill conditions. All the data plots very nearly on a straight line corresponding to a conductance of 9.4 w/°F. Also shown on Fig. 5 are lines representing data taken at 0.095 in. tilt for the 74 and 90% fill conditions, which gave very nearly constant conductance values of 8.0 and 8.7 w/°F, respectively.

The evaporator and condenser temperature differences as a function of heat flux corresponded to film coefficients (based on Eq. 1 with D = 0.452 in.) of 400 and 500 Btu/hr-ft$^2$°F, respectively.

Maximum heat transport capacity as a function of tilt is illustrated in Fig. 6 for the four fill conditions along with the theoretical curve computed using Eq. 2. Extrapolating the high tilt data for the nominal 100% charge gives a dryout flux of approximately 300 w, compared with a theoretical value from Eq. 2 of only 221 w, based on a meniscus contact angle of 30°. This may reflect the fact that the nominal 100% charge is based on a flat meniscus. At maximum heat flux, the meniscus depression of a 30° contact angle can cause a liquid cross-sectional area reduction of as much as 15% at the end of the evaporator. The average drop in liquid cross-sectional area over the entire pipe would be less than half this, however. This makes the nominal 90% fill case more representative for zero "g" predictions, and extrapolating the high tilt data to zero tilt gives a value of $Q_{max}$ about equal to the theoretical value. Dryout could not be reached at zero tilt for the three higher fill conditions, because of heater limitations (approximately 300 w).

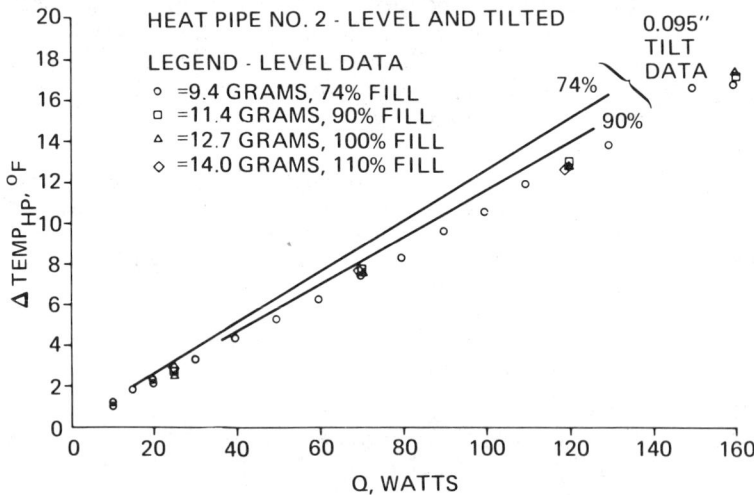

Fig. 5  Ammonia, over-all temperature drop vs heat flow.

## Freon-113 Working Fluid

A limited number of tests were conducted with Freon-113 at 70°F at a nominal 100% charge based on level full grooves. The evaporator and condenser temperature differences corresponded to film coefficients (based on Eq. 1 with D = 0.452 in.) of 115 and 200 Btu/hr-ft²°F, respectively.

Maximum heat transport capability as a function of tilt is shown in Fig. 7 along with a theoretical curve using Eq. 2. Extrapolating the tilt data linearly to zero tilt gives a burnout flux of 24 w, compared with a theoretical value of 22 w.

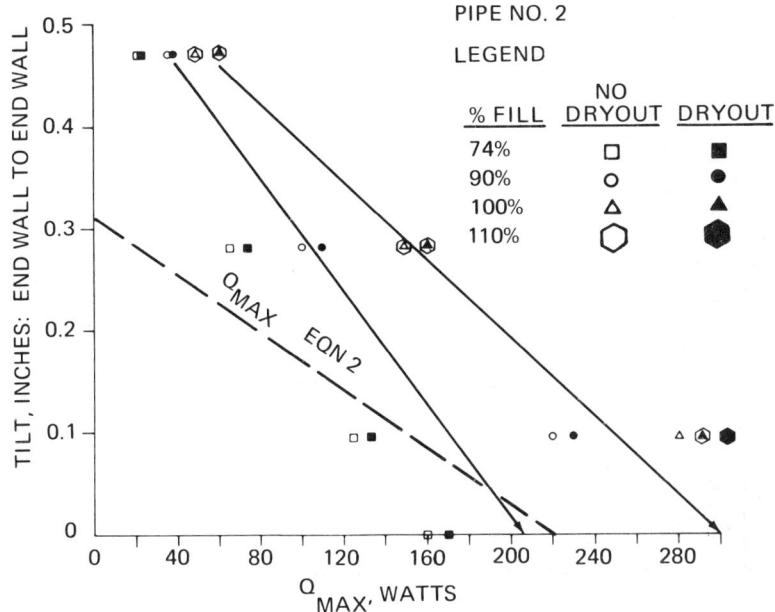

Fig. 6  Ammonia, heat pipe tilt vs maximum heat transport.

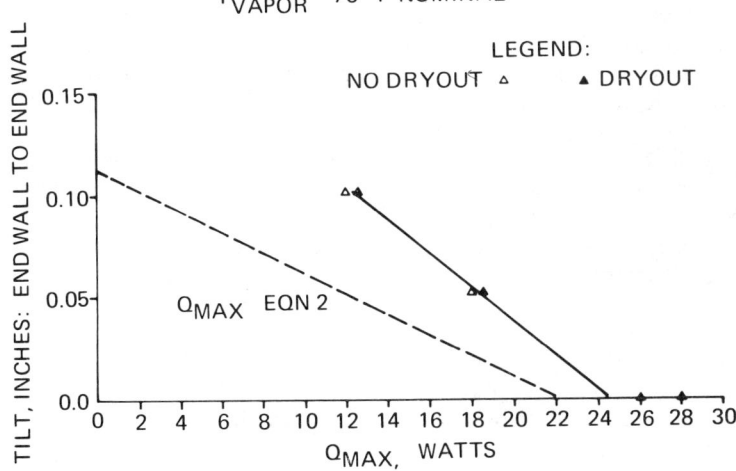

Fig. 7  Freon-113, heat pipe tilt vs maximum heat transport.

## Data Comparisons

A summary of the film coefficient values obtained and some pertinent fluid properties for the three fluids tested is given below:

Table 2   Performance and Property Data

| Fluid | $h_{EV}$ Btu/hr-ft$^2$ °F | $h_{COND}$ | Liquid conductivity Btu/hr-ft °F | Theo. cap. rise, in. ($\theta = 30°$) | Apparent cap. rise in. |
|---|---|---|---|---|---|
| NH$_3$ | 400 | 500 | 0.300 | 0.31 | 0.57 |
| F-21 | 200 | 300 | 0.064 | 0.12 | 0.20 |
| F-113 | 115 | 200 | 0.038 | 0.11 | 0.18 |

The film coefficient values shown reflect the higher heat flux, level pipe data, and are believed representative of what would be obtained under zero "g" conditions. The variations at lower heat flux values and with tilt are to a large extent explainable in terms of gravity affecting the liquid distribution. The changes in liquid distribution are dependent on pipe length and tilt as well as heat flow and fluid properties. Because of this, attempts to correlate the low and high heat flux data in terms of nondimensional parameters would involve pipe geometry more strongly than the conventional Nusselt, Prandtl and Reynolds numbers, and should involve tests on a variety of pipe cross-sectional shapes.

Comparing evaporator data for the three fluids, it may be noted that film coefficients for the two Freons are roughly proportional to the liquid thermal conductivity. The ammonia evaporator film coefficient, while higher, does not differ as much from Freon-21 as might be expected, both from the conductivity values and from the visual observation that the evaporator grooves were more completely filled. Visual observation indicates that vaporization takes place from the liquid surface, with no discernible nucleate boiling. Since the fins have very high conductivity relative to the liquid, most of the heat transfer and vaporization takes place in the vicinity of the meniscus attachment line, where contact angle should greatly affect the results. From the data, ammonia would appear to have a larger contact angle than the Freons.

The condenser data comparisons are similar to those for the evaporator. Film coefficients for the two Freons are roughly proportional to their thermal conductivity; the ammonia film coefficient, while higher than for the Freons, is not higher in proportion to its thermal conductivity. The condensation is believed to occur primarily on the fin ends (lands), where a higher contact angle for the ammonia could result in a slightly thicker liquid layer (of order 0.001 in.), offsetting some of the benefits of the higher conductivity.

The tilt data for all three fluids showed less fall off in $Q_{max}$ with increasing tilt than would be indicated by Eq. 2 (assuming $\theta = 30°$ in all cases). Multiplying the h value in Eq. 2 by 0.6 would give a much better approximation to the data, though the reason for this is not apparent. In each case, a linear extrapolation of the high tilt data to zero tilt, while crude because of the limited number of data points, indicates a maximum heat capacity close to that predicted for zero "g". The level data, particularly for overcharged pipes, gives considerably higher values.

## VI. Conclusions

Gravity (tilt) has a strong effect on maximum transport capacity, but only a minor effect on temperature differences. Variation in temperature level has only a minor effect on temperature differences. For pipes with approximately 100% charge, there is good agreement between the predicted value for zero "g" maximum transport capacity and a value obtained by extrapolating high tilt data to zero tilt. Level test data can over estimate the zero "g" capacity.

## References

1. Frank, S., "Optimization of a Grooved Heat Pipe", Intersociety Energy Conversion Engineering Conference Proceedings, pp 833-845, August 1967.

2. Bahr, A., Burck, E., Hufschmidt, W., "Liquid-Vapor Interaction and Evaporation in Heat Pipes", Second International Conference on Thermionic Electrical Power Generation, May 1968, Stresa, Italy, pp 543-556.

3. Green, M., "An Improved Method of Calculating Frictional Pressure Losses Along Grooved Heat Pipes", Grumman Advanced Development Report ADN 04-02-69.1, June 1969.

4. Kemme, J. E., "High Performance Heat Pipes", Thermionic Conversion Specialist Conference Proceedings, Oct. 1967, Palo Alto, Cal., pp 355-358.

5. Bilenas, J. A., and Harwell, W., "Orbiting Astronomical Observatory Heat Pipes - Design, Analysis and Testing," ASME Paper 70-HT/SpT-9, 1970.

6. Cotter, T. P., "Theory of Heat Pipes", Los Alamos Scientific Laboratory Report LA-3245-MS, March 1965.

OPERATING CHARACTERISTICS AND
LONG LIFE CAPABILITIES OF ORGANIC FLUID HEAT PIPES

A. Basiulis[*] and M. Filler[†]

Hughes Aircraft Company, Torrance, Calif.

Abstract

A comprehensive test program on heat pipes using organic working fluids has demonstrated good performance for most cooling applications. These heat pipes have been effectively utilized to fill operating temperature gaps existing in the low- and medium-temperature ranges. The low thermal conductivity of the organic fluids had negligible effects on the evaporator temperature gradient between the heated surface and the saturated vapor. Additionally, life test compatibility results show very high reliability. This was further verified by a post test analysis of a 20,000 hr heat pipe which showed negligible internal mass transport, corrosion and fluid property changes.

Nomenclature

$A$ = area, $cm^2$

$f_m$ = figure of merit, $Cal/cm^2$ - sec

$\Delta H$ = latent heat of vaporization, $Cal/gram$

$g$ = acceleration due to gravity, $cm/sec^2$

$Q$ = heat load, w

$T_S$ = temperature evaporator surface, $°C$

---

Presented as Paper 71-408 at the AIAA 6th Thermophysics Conference, Tullahoma, Tenn., April 26-28, 1971.

[*]Head, Mechanical Engineering, Manufacturing Engineering Department, Electron Dynamics Division.
[†]Member of the Technical Staff, Applied Mechanics and Thermal Devices, Mechanical Engineering Section, Manufacturing Engineering Department, Electron Dynamics Division.

$T_{SAT}$ = temperature saturated vapor °C
$U$ = over-all heat transfer coefficient, w/cm² °C
$\gamma$ = surface tension, dynes/cm
$\rho_\ell$ = density, liquid, g/cm³
$\rho_v$ = density, vapor, g/cm³
$\nu$ = kinematic viscosity, cs

## Introduction

The use of organic fluids as working fluids for heat pipe applications has often been ignored by most investigators, primarily for two reasons. First, organic fluids generally have poor physical properties, such as low heat of vaporization, low surface tension, and low-thermal conductivity, as compared to such inorganic fluids as water, ammonia or the liquid metals. Second, there is a lack of information concerning the long term compatibility of organic fluids with potential envelope and wick materials. However, some of the organic fluids offer unique advantages important enough to merit further serious investigation. For instance, some of the organic fluids are particularly suitable for electronic component cooling because of their low boiling points and good dielectric properties. In addition, certain organic fluid heat pipes can be used to fill the important operating temperature gap existing between water and mercury heat pipes and between ammonia and water heat pipes.

Studies and tests with organic fluid heat pipes have indicated that in most applications for electronic component cooling there is sufficient capacity to dissipate the required heat load while maintaining the components at desired temperatures.[1] It was shown that, in many cases, existing conventional liquid cooling systems could be replaced with much smaller and lighter air cooled heat pipes.

The performance of organic fluid heat pipes, and the potential incompatibilities of certain heat pipe materials with the resultant three basic types of failure modes was evaluated. The first potential failure mode, which usually occurs during initial testing, is a breakdown of the working fluid. It is most often caused by either direct reaction between the fluid and the envelope or the fluid and the noncondensable gas. The second potential failure mode, due to heat pipe material incompatibility, is a decomposition of the wick material. This can be caused by chemical corrosion, physical deterioration due to dynamic action of the fluid, and electrolytic action caused by dissimilar metals. The third potential failure mode is the decomposition of the material of the inner wall of the heat pipe envelope. The causes here can be either chemical corrosion or electrolytic action. The last two failure modes are slow processes and require extended life tests to determine the long range effects. The post test analysis of heat pipes using organic fluids has shown small amounts of deterioration of the wick similar to that observed in liquid metal heat pipes, including those using mercury.[2]

## Operating Characteristics

Organic fluids are so endowed by nature that they possess relatively low values of surface tension, heat of vaporization, and thermal conductivity. This, of course, is undesirable from a heat pipe performance standpoint. However, the variety of highly useful boiling points and good dielectric properties of many of these fluids makes them very attractive as potential heat pipe fluids.

The relatively low thermal conductivity of the organic fluids was the property of primary concern. It has been thought by many investigators that, in the evaporator zone of the heat pipe, heat is transferred by conduction through the wick to the wick inner surface where evaporation takes place. It can be noted that in experiments with water in flooded wicks reported by E. C. Phillips and J. P. Hinderman,[3] that at low flux densities the mechanism of heat transfer appears to be conduction as indicated by a large $\Delta T$ across the evaporator wick. However, when these same wicks were no longer flooded and permitted to operate in a heat pipe mode, operation where the fluid is returned from the condenser to the evaporator by capillary pumping without the aid of gravity, the $\Delta T$ decreased substantially indicating that the heat-transfer mechanism could not be conduction across the entire wick. Moreover, the authors and other investigators have recently shown that in actual heat pipes this $\Delta T$ is very small even at low heat fluxes and using working fluids with low thermal conductivity. One possible explanation of the mechanism of wick heat transfer is that there is nucleation within the wick itself. This assumption finds support in the observations of bubble formation in the wick by P. J. Marto and W. L. Mosteller.[4] These findings were contradicted by the work of J. K. Ferrell[5] and H. R. Johnson, where they concluded that there is no boiling in the wick. The authors have compiled results of heat pipe performance from their own experiments, from selected data of Phillips and Hinderman, and from selected data of B. D. Marcus.[6] Analysis of these results by the authors have shown that whatever the heat-transfer mechanism is, it is independent of wick thickness. This is shown in Fig. 1 where the conduction assumption error indicated by the ratio of the calculated $\Delta T$ to the measured $\Delta T$ is plotted against thermal heat flux. The calculated $\Delta T$ here is based on the assumption of conduction through a layer of fluid the same thickness as the wick. It can be seen from the figure that the error of this assumption is clearly greater in heat pipes with thicker wicks or those using fluids of lower thermal conductivity.

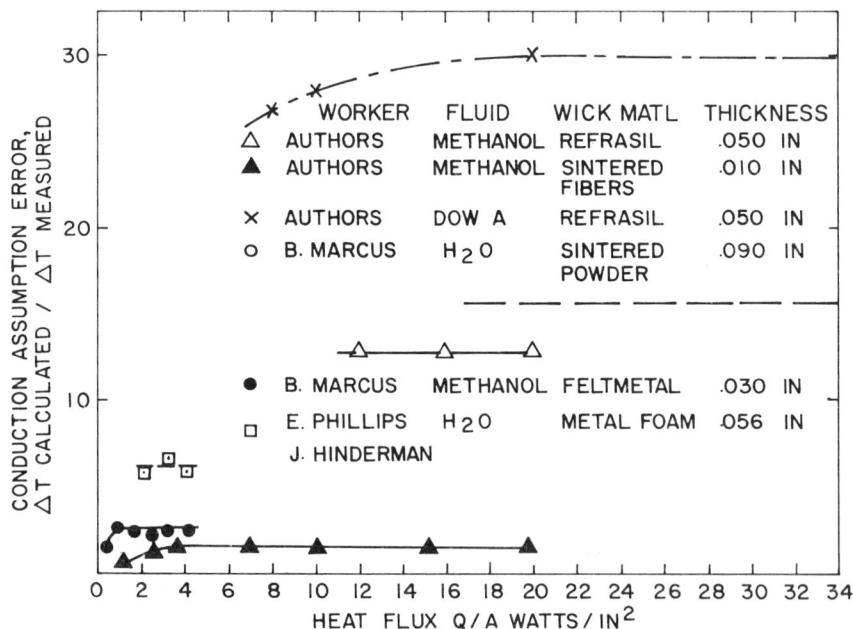

Fig. 1  Ratio of ($\Delta T$ calculated/$\Delta T$ measured) heat flux is based on the conductivity of a given fluid.

In order to obtain a better understanding of the heat-transfer mechanism in the evaporator zone of a heat pipe, an experimental heat pipe was constructed. The heat pipe, shown at the top of Fig. 2, was of the radial type configuration with a re-entrant type wick structure. Tests were performed using Dowtherm A® and then identical tests were repeated using methanol. The heat pipe, prior to the insertion of each fluid, was baked-out under vacuum conditions. The heat pipe was charged with working fluid and then pinched-off. Test results are shown in Fig. 3. As a reference, the theoretical heat-transfer values based on conduction through a layer of Dowtherm A approximately the same thickness as a single layer of Refrasil® wick are also plotted. The figure shows that the actual heat-transfer values are one order of magnitude greater than those based on pure conduction. This could be explained only if there was nucleate boiling in the wick, or the fluid level in the wick had receded to a thin film such that the conduction losses through it would be minimal [i.e., the equation: $Q/A = U(T_{surf} - T_{sat})$,[5] could be used]. During further experiments it was found that this temperature gradient did not change when the wick thickness in the evaporator zone was doubled.

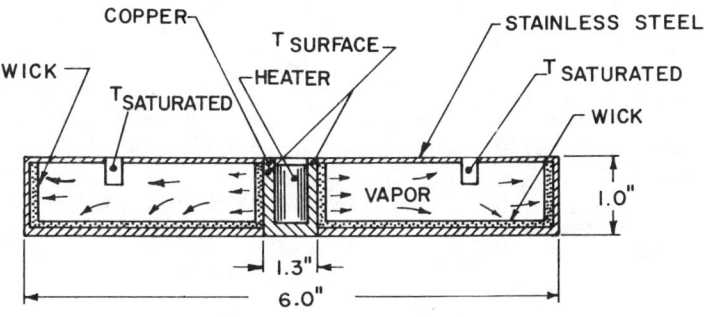

RADIAL FLOW HEAT PIPE WITH A REENTRANT WICK

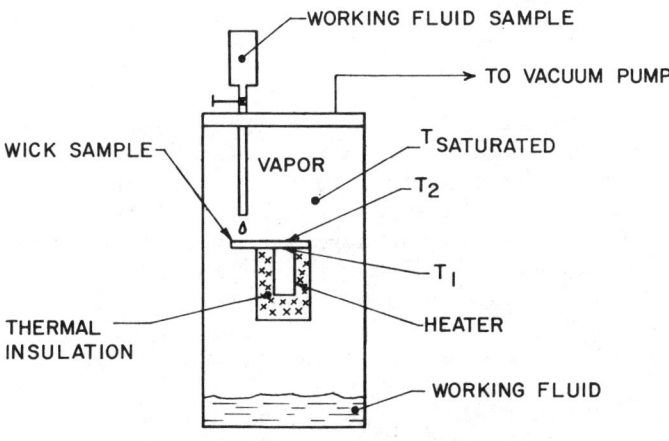

CRITICAL FLUX MEASUREMENT APPARATUS

Fig. 2  Test vehicles for evaporation experiments.

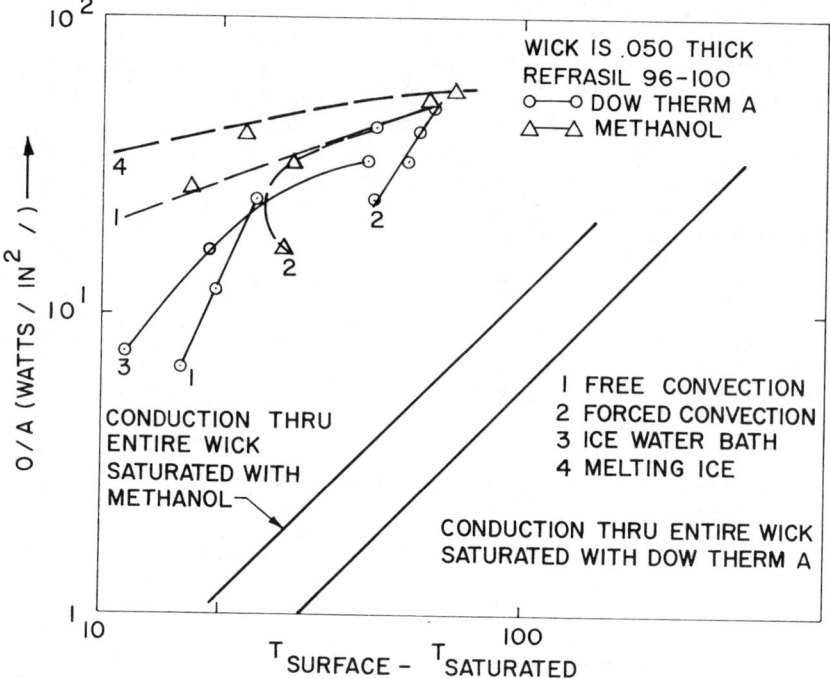

Fig. 3  Heat-transfer characteristics of wick saturated with organic fluids.

It was also observed that in all heat pipes tested using organic fluids dry out occurs because they are wick pumping limited and not because they reach critical heat flux. This was verified using the apparatus shown at the bottom of Fig. 2. Working fluid was supplied to the evaporator by a wick which was an extension of the evaporator wick. During the experiment, fluid column height was varied, thereby varying wick pumping, until dry-out occurred. The results are plotted in Fig. 4. The burn-out due to critical flux density was never obtained, since when column height was increased above 3/8 in. the wick became flooded and no longer represented actual heat pipe operation. These experiments verified that the temperature gradient between the heated heat pipe surface and interior saturated vapor is independent of wick thickness, and in many cases wicks up to 1/8-in.-thick can be used without increasing the temperature gradient.

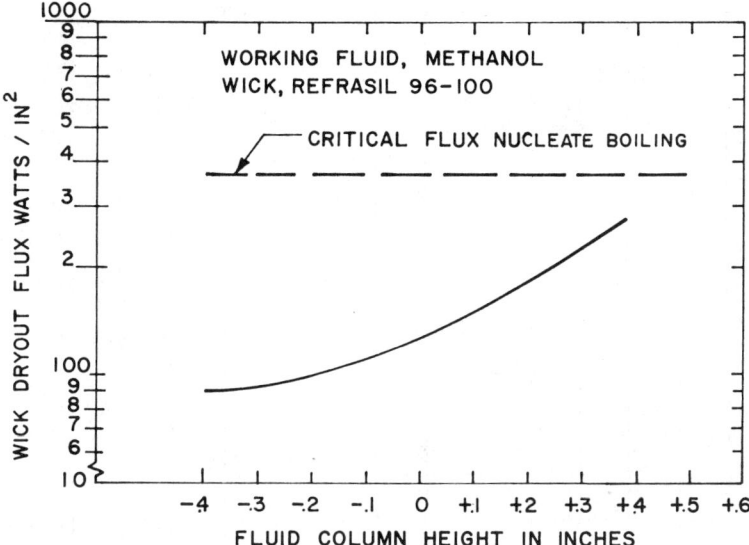

Fig. 4 Dry out heat flux for methanol.

## Selection of Working Fluids

Candidate organic working fluids were initially selected from the analysis of a literature search. Good fluids, in addition to being compatible with potential heat pipe materials, should have high latent heat of vaporization, high surface tension, low viscosity, and relatively high thermal conductivity. Since all these characteristics in any one fluid are quite improbable, the best compromise fluid for any given application must be chosen.

The fluids with relatively high figures of merit and high nucleate boiling critical flux values were selected for initial tests. These tests were started by processing the fluid in a closed test vehicle, while operating it in a refluxing mode, and observing thermocouple readings for deviation from isothermal operation. This provided a good indication as to whether noncondensable gases were forming or to whether the fluid was breaking down. This type of failure usually can be observed during the first 24 hr of operation. Working fluids that passed this initial test were processed into a heat pipe and put on life test. Physical characteristics of the selected fluids are listed in Table 1. The critical flux shown is calculated from the equation given by Kutateladze.[7] This equation was chosen because it gave results that agreed well with experimental results in nucleate boiling systems. The nucleate boiling flux was selected

Table 1 Working fluid properties[a]

| Fluid | Boiling point °F / °C | Melting point °F / °C | Density Liquid lb/ft³ / g/cm³ | Density Vapor lb/ft³ / g/cm³ | Viscosity Liquid cps / cs | Viscosity Vapor cps / cs | Surface tension dyne/cm | Latent heat BTU/lb / cal/g | Specific heat BTU/lb°F / cal/g°C | Heat of fusion BTU/lb / cal/g | Critical flux w/in² / w/cm² |
|---|---|---|---|---|---|---|---|---|---|---|---|
| Freon 12 $C_1Cl_2F_2$ | -21.6 / -29.8 | -252 / -158 | 92.6 / 1.48 | 0.395 / 0.00633 | 0.37 / 0.249 | 0.0127 / 2.01 | 16.6 | 71.04 / 39.47 | 0.21 / 0.21 | | 138 / 21.5 |
| Ethyl Chloride $CH_3CH_2Cl$ | 54 / 12.27 | -218 / -138.7 | 56.0 / 0.898 | 0.1872 / 0.0030 | 0.30 / 0.326 | 0.0093 / 3.1 | 20.39 | 163.0 / 90.55 | 0.36 / 0.36 | | 203 / 31.5 |
| Acetone $CH_3COCH_3$ | 133 / 56.6 | -139 / -95 | 48 / 0.77 | 0.135 / 0.0022 | 0.6 / 0.78 | 0.0085 / 3.88 | 17.4 | 224 / 124.5 | 0.55[b] / 0.55[b] | 42 / 23.4 | 218 / 33.8 |
| Methanol $CH_3OH$ | 148.3 / 64.65 | -144 / -97.8 | 49.8 / 0.796 | 0.0756 / 0.0012 | 0.34 / 0.43 | 0.0135 / 11.15 | 22.6 | 473 / 262.8 | 0.566 / 0.566 | 28.6 / 16.4 | 368 / 57 |
| Hexaflouro-benzene $C_6F_6$ | 176 / 80.2 | 39.2 / 3.8 | 100 / 1.62 | 0.402 / 0.0064 | | | 21.4 | 73 / 40.6 | 0.285 at 24°C / 0.285 | 26.8 / 14.88 | 131 / 20.4 |
| DC-200 - 65 CS | 212 / 100 | -90 / -67.7 | 47.5 / 0.763 | 0.075[b] / 0.001[b] | 0.45 65 at 25°C | 0.01[b] / 10.0[b] | 15.9 at 25°C | 95.8 / 52.4 | 0.504 / 0.504 | | 74 / 11.5 |
| Dowtherm-E | 352 / 177.78 | 0 / -18 | 70.8 / 1.127 | 0.269 / 0.0046 | 0.335 / 0.337 | 0.01093 / 2.38 | 37 at 25°C | 119.1 / 66.3 | 0.411 / 0.411 | 38 / 21.1 | 192 / 30 |
| Dowtherm-A | 495.8 / 257.7 | 53.6 / 12 | 53.28 / 0.854 | 0.2485 / 0.00398 | 0.27 / 0.316 | 0.0103 / 2.59 | 40 at 25°C | 128 / 71.3 | 0.524 / 0.524 | 42.2 / 23.4 | 172 / 26.7 |
| CP-9 | 566.6 / 297 | -85 / -65 | 50[b] / 0.81 | 0.25[b] / 0.004 | 0.25[b] / 0.32 | 0.01[b] / 2.5 | 41 at 20°C | 120.5 / 67 | | | 162[b] / 25 |
| Water $H_2O$ | 212 / 100 | 32 / 0 | 60 / 0.9584 | 0.0373 / 0.0006 | 0.2838 / 0.299 | 0.0125 / 20.9 | 72 at 25°C | 970 / 539.55 | 1.0 / 1.0 | 144 / 79.71 | 703 / 109 |
| Amonia $NH_3$ | -28.1 / -33.34 | -107.9 / -77.7 | 42.5 / 0.681 | 0.0552[b] / 0.000708 at 0°C | 0.25 / 0.367 | 0.0079 | 42.41 | 587.59 / 326.44 | 1.07 / 1.07 | 142.6 / 79.4 | 446 / 59.1 |

[a] All values given at the boiling point unless otherwise noted.
[b] Estimated or approximate value.

because it is independent of wick pumping capacity. The figure of merit is calculated using the expression given by Katzoff.[8] Figures of merit and critical heat fluxes of selected fluids of are shown in Fig. 5. The critical flux is defined as

$$Q/A = \pi/24 \, (\Delta H) \, \rho_v^{1/2} \left[ \gamma_g (\rho_\ell - \rho_v) \right]^{1/4}$$

### Selection of Test Vehicles

Test vehicle materials selection was based on the criteria of compatibility of envelope materials with the working fluid and the possible requirement for use of these materials in electronic component cooling. The selection was based on published data and on in-house experimental results. Compatibility of many typical heat pipe and wick materials combinations is given in Table 2. A standardized size heat pipe was selected for use in most of the life tests. It was a thin wall cylinder one inch in diameter and six inches long with flat closure end caps on both ends. The six-inch length

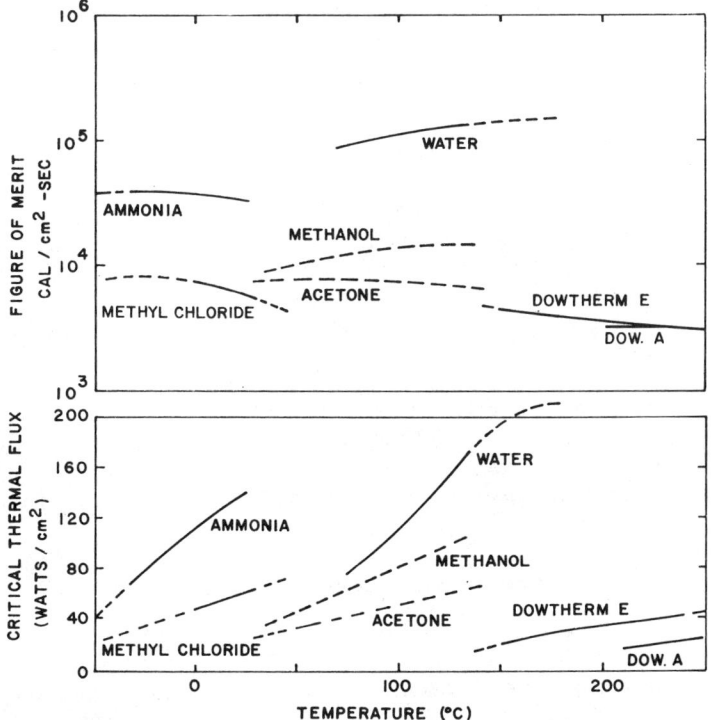

Fig. 5 Figure of merit and critical heat flux vs temperature.

was selected because nearly all of the selected organic fluids will operate against gravity to a column height of at least six inches. The operation against gravity is very important in compatibility life testing since deterioration of the wick can not be effectively observed when heat pipe operation is in a gravity aided reflux mode.

Table 2  Heat pipe materials compatibility recommendations

| Wick material | Working fluids[a] | | | | | |
|---|---|---|---|---|---|---|
| | Water | Acetone | Ammonia | Methanol | Dow-A | Dow-E |
| Copper | RU | RU | NR | RU | RU | RU |
| Aluminum | GNC | RL | RU | NR | UK | NR |
| Stainless Steel | GNT | PC | RU | GNT | RU | RU |
| Nickel | PC | PC | RU | RL | RU | RL |
| Refrasil | RU | RU | RU | RU | RU | RU |

[a] RU = recommended by past successful usage, RL = recommended by literature, PC = probably compatible, NR = not recommended, UK = unknown, GNC = generation of noncondensable gas at all temperatures, GNT = generation of noncondensable gas at elevated temperatures, when oxide is present.

Life Tests

Life testing was designed to detect heat pipe failure modes which depend on continuous relatively long-term operation, and to collect positive data showing the long-life reliability of heat pipes using organic working fluids. A current summary of this continuing test program is shown in Table 3. Most heat pipes are the standardized one inch diameter, six-inches-long geometry, however, some special geometries of particular interest have also been added to the life test program. These include a radial heat pipe designed to cool a traveling-wave tube vacuum barrel assembly, a variable conductance constant temperature heat pipe, and a flexible heat pipe. One heat pipe was built with electrical feed throughs into the vapor space so as to permit periodical checks on the dielectric strength of the working fluid.

The life test heat pipes are loaded with a constant heat input using external resistance heaters and the temperatures along the heat pipe length are recorded once a month after an initial test series. During the initial test series of 100 hr of operation the temperatures are recorded daily. To

CHARACTERISTICS OF ORGANIC HEAT PIPES 441

Table 3  Heat pipes on life test (March 1, 1971)

| HP | Type (Body material) | Wick | Fluid | Interface | Hr of operation |
|---|---|---|---|---|---|
| HP-1 | Axial (Cu) | SS Screen | DC-200 | Pressure | 23,000[a] |
| HP-3 | Axial (SS) | SS Screen | Dow E | Pressure | 29,078 |
| HP-5 | Radial(Cu/SS) | Cu Powder | Dow E | Sintered | 31,270 |
| HP-7 | Axial (Cu) | SS Screen | Dow E | Pressure | 30,814 |
| HP-9 | Axial (Cu) | Cu Powder | Dow 209 | Sintered | 30,344 |
| HP-10 | Axial (Cu) | Refrasil | Dow A | Pressure | 20,500[a] |
| HP-11 | Axial (Cu) | Refrasil | CP-9 | Pressure | 23,410 |
| HP-W-2 | Axial (Cu) | Refrasil | Dow A | Brazed | 16,594 |
| HP-W-3 | Axial (Cu) | Refrasil | Dow A | Brazed | 15,826 |
| HP-W-4 | Axial (SS) | Refrasil | Water | Pressure | 15,299 |
| HP-W-5 | Axial (SS) | Refrasil | Dow A | Brazed | 13,117 |
| HP-W-6 | Axial (Cu) | Refrasil | $C_6F_6$ | Pressure | 14,332 |
| HP-W-11 | Axial (Al) | Nickel | DC-200 | Pressure | 8,070 |
| HP-W-12 | Axial (Cu) | Refrasil | $C_2Cl_4$ | Pressure | 7,008 |
| HP-SS-10 | Axial (Cu) | Refrasil | Methanol | Pressure | 14,010 |
| HP-SS-11 | Axial (Cu) | Refrasil | Acetone | Pressure | 14,010 |
| HP-SS-12 | Axial (SS) | SS Screen | Ammonia | Pressure | 8,750 |
| HP-SS-13 | Axial (SS) | SS Screen | Ammonia | Pressure | 8,750 |
| HP-SS-14 | Axial (Al) | Al Felt | Ammonia | Pressure | 8,070 |
| HP-SS-15 | Axial (Al) | Ni Felt | Ammonia | Pressure | 8,070 |
| HP-F-1 | Axial (SS) | Refrasil | Water | Pressure | 3,744[a] |
| HP-W-14 | Axial (Cu) | Refrasil | $(CH_3)_2S$ | Pressure | 6,382 |
| HP-W-16 | Axial (Cu) | Cu Felt | CP-32 | Pressure | 1,128 |
| 1318-H-AL-A | Axial (Al) | SS Fiber | Acetone | Pressure | 1,008 |

[a]Test terminated.

date, life tests of three heat pipes have been terminated. The first containing DC-200 fluid, the second containing Dowtherm A fluid, and the last being a special flexible center section geometry containing water. Both DC-200 and Dowtherm A organic fluids are of particular interest since they are very good dielectrics and they have been used in heat pipes which are integral parts of electron tubes. The post test analysis of the DC-200 heat pipe, designated HP-1, and the preliminary analysis of the Dowtherm A heat pipe, designated HP-10, has been completed.

Post Test Analysis

The life test of heat pipe HP-1 was terminated after 23,000 hr. of successful operation. The heat pipe envelope was constructed of OHFC copper, the wick was a 200-mesh stainless steel screen, and the working fluid was DC 200 (viscosity of 1.0 C.S.). This heat pipe (as was done in the processing of all heat pipes prior to the filling operation) had been baked-out under vacuum conditions at a temperature at least 100°C higher than the intended operating temperature. The vacuum bake-out operation removes all traces of contaminants from the heat pipe envelope, and since the working fluid is vacuum distilled in the heat pipe prior to pinch-off, any contaminants found in the heat pipe during post test analysis must have been generated within the heat pipe during operation.

For the post test analysis of HP-1, the heat load was first removed and the heat pipe was allowed to cool to room temperature. Then the pinched-off processing port was opened and the fluid distilled into a sample container. Some of the fluid was returned to the Dow Corning Company for analysis and the remainder was used to test the fluid dielectric properties. The heat pipe internal envelope was filled with a special clear potting compound, split in half along its length, and then sectioned for metallographic analysis.

The dielectric properties of the removed life test working fluid were compared to a standard sample of fresh DC-200 fluid. It was found that the dielectric strength of the life test fluid had not measurably deteriorated during the 23,000 hr of operation. Analysis of the fluid sample by Dow Corning showed no change in fluid viscosity, and their IR spectrum tests indicated no measurable changes due to operation. A GLC analysis, however, did show some slight traces of impurities. Mr. R. Standford[9] of Dow Corning indicated that, in his opinion, considering the long time of continuous operation, the impurities were not significant and should not affect the performance of the fluid.

Preliminary observations of the potted sections of this heat pipe showed the envelope and the wick to be in relatively good condition. However, close examination of the photomicrographs showed that some corrosion and mass transfer had taken place, especially in the evaporator section. Both samples on the vapor side and envelope side of the condenser, showed similar results. Here, the deteriorated layer was uniform and measured $6 \times 10^{-5}$ to $7 \times 10^{-5}$-in-thick. The small amount of deterioration of the wick was caused either by corrosion or by the leaching out of the impurities from the stainless steel wires through the rinsing action of the fluid. The deposit on the wires in the evaporator zone indicates that some mass transfer has probably taken place. This build-up is similar in appearance to the build-up observed by S. E. Deverall[2] on stainless steel wires in mercury filled heat pipes. There was no evidence found of either corrosion or build-up on the walls of the heat pipe envelope. A preliminary post test analysis of heat pipe HP-10, which used Dowtherm A fluid and a Refrasil wick, showed no visible physical deterioration of the wick.

## Conclusions

The work to date with heat pipes using organic working fluids has shown that these heat pipes perform much better than originally expected, and can be effectively used to fill the operating temperature gap existing between water and mercury heat pipes and ammonia and water heat pipes. In many applications the organic fluid heat pipe is the only choice because of their good dielectric properties. Many organic fluid heat pipes have, in addition, shown excellent long life reliability.

It was observed that organic fluid heat pipes reach evaporator dry-out conditions because of insufficient wick pumping capacity, and nucleate boiling critical heat flux densities could not be reached in actual heat pipes. The temperature gradients between envelope surface and saturated vapor are independent of wick thickness when using Refrasil® (glass cloth), Feltmetal® (sintered fibers), or sintered metal powder. Moreover, it has been shown that the mechanism of heat transfer in the evaporator is not conduction through a layer of working fluid equal to the wick thickness. Therefore, the low thermal conductivity of organic fluids is not a serious detriment to effective usage in heat pipes.

The life test and compatibility results indicate that organic fluid heat pipes are very reliable even when used continuously for long periods of time. The small amount of internal mass transport observed in the post test analysis of the 23,000 hr heat pipe appears to have had a negligible effect on heat pipe operation. It can be assumed that this heat pipe could have continued successful operation for an order of magnitude increase in life time.

## References

1. Basiulis, A. and Filler, M., "Heat Pipes Meet Unique Requirements in Electronic Component Cooling," Paper S30-2, The Fifth Intersociety Energy Conversion Engineering Conference, Las Vegas, Nev., Sept. 1970.

2. Deveral, J. E., "Mercury As a Heat Pipe Fluid," ASME Paper 70-HT/SpT-8, Space Technology and Heat Transfer Conference, Los Angeles, Calif., June 1970.

3. Phillips, E. C. and Hinderman, I. D., "Determination of Properties of Capillary Media Useful in Heat Pipe Design," ASME Paper 69-HT-18, ASME-AICHE Heat Transfer Conference, Minneapolis, Minn., August 1969.

4. Marto, P. S. and Mosteller, W. L., "Effect of Nucleate Boiling on the Operation of Low Temperature Heat Pipes," ASME Paper 69-HT-24, ASME-AICHE Heat Transfer Conference, Minneapolis, Minn., August 1969.

5. Ferrell, J. K. and Johnson, H. R., "The Mechanism of Heat Transfer in the Evaporator Zone of a Heat Pipe," ASME Paper 70-HT/SpT-12, Space Technology and Heat Transfer Conference, Los Angeles, Calif., June 1970.

6. Marcus, B., Private Communication, February 17, 1971, TRW.

7. Kutateladze, S. S., "Hydrodynamic Theory of Changes in the Boiling Process Under Free Convection Condition," *Izzestia Akadamii Nauk S.S.R.*, Otdelenia Tekh. Nauk, 1951, pp. 529-536. Translation from Russian, AEC-TR-1441.

8. Katzoff, S., "Heat Pipes and Vapor Chambers for Thermal Control of Spacecraft," *AIAA Progress in Astronautics and Aeronautics: Thermophysics of Spacecraft and Planetary Bodies*, Vol. 20, edited by G. B. Heller, Academic Press, New York, 1967, pp. 761-819.

9. Standford, R. W., Private Communication, March 30, 1970, Dow Corning Corp.

# DESIGN AND PERFORMANCE OF NONCONDENSIBLE GAS CONTROLLED HEAT PIPES

J. D. Hinderman* and E. D. Waters†
McDonnell Douglas Astronautics Company, Richland, Wash.
and
R. V. Kaser‡
University of Oklahoma, Norman, Okla.

## Abstract

A sequence of guidelines are presented to facilitate the design of noncondensible gas-controlled heat pipe systems. Working fluid selection, reservoir sizing, reservoir wicks, and ambient conditions are among the factors considered. The performance and controllability of typical gas-controlled heat pipes is compared with theoretical predictions; agreement is generally good. Reservoir wicking increases the sensitivity of temperature control to reservoir temperature variations.

## Nomenclature

| | |
|---|---|
| $A_c$ | = cross-sectional area for tubing between condenser and reservoir |
| $A$ | = constant of proportionality (Eq. 5) |
| $h_{f_g}$ | = latent heat of vaporization |
| $\ell_c$ | = length of front motion region from minimum to maximum heat rejector |
| $n_A$ | = moles diffused at time t |
| $n_{A\infty}$ | = moles diffused at infinite time |
| $M$ | = molecular weight |
| $P_e$ | = vapor pressure corresponding to $T_e$ |

---

Presented as Paper 71-420 at the AIAA 6th Thermophysics Conference, Tullahoma, Tenn., April 26-28, 1971. The work described was conducted by McDonnell Douglas Astronautics Co. as part of an independent research and development program.

*Laboratory Scientist/Specialist, Donald W. Douglas Labs.
†Section Chief, Donald W. Douglas Labs.
‡Associate Professor, School of Aerospace, Mechanical and Nuclear Engineering.

| | |
|---|---|
| $P_c$ | = vapor pressure corresponding to $T_c$ |
| $P_x$ | = vapor pressure corresponding to $T_x$ |
| $P_r$ | = pressure ratio = $P_{e_{max}}/P_{e_{min}}$ |
| $R$ | = universal gas constant |
| $T$ | = temperature |
| $T_e$ | = temperature in evaporator region |
| $T_c$ | = temperature in machine condenser region |
| $T_x$ | = temperature in reservoir region |
| $V_c$ | = volume of condenser region swept out by front in moving from minimum to maximum position |
| $V_x$ | = volume of NCG excess volume reservoir |
| $\Delta H_v$ | = latent heat of vaporization on a molar basis |
| $\Delta T_{control}$ | = evaporator temperature variation available for controlling front position |

## Introduction

The heat pipe as a heat-transfer device has been investigated extensively during the past few years and much of this work has been reported in the literature. In addition to high thermal conductivity, the heat pipe has excellent passive thermal control characteristics. This aspect of heat pipe operations as well as inert gas control have been discussed by Marcus and Fleischman[1] and Bienert,[2] but there are specific areas which require further investigation in using a noncondensible gas (NCG) to achieve temperature control.

## Basic Operating Characteristics

The purpose of a NCG heat pipe system is to control the operating temperature of the component, device, or region to which it is connected against variations in component power dissipation and environmental thermal conditions. Varying heat levels can be transported by a heat pipe without a change in operating temperature. By coupling the heat pipe to a noncondensible gas reservoir (Fig. 1), thermal energy removed from the condenser can be controlled within very close limits. On start-up, working fluid and gas are separated because both are swept to the condenser zone but only the working fluid condenses and is pumped back to the evaporator zone. The noncondensible gas is trapped in the condenser end, producing an inactive condenser zone. This effectively limits the radiator area available for heat dissipation. When heat input

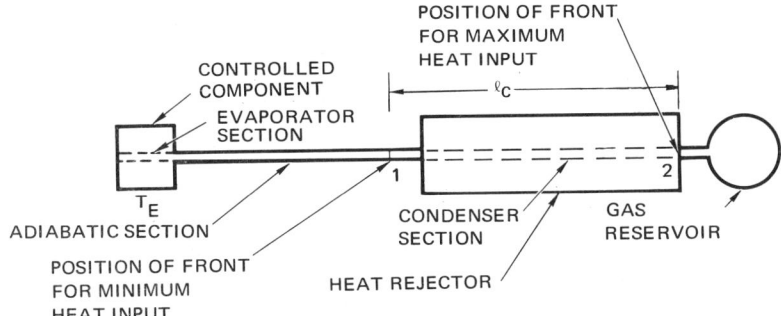

Fig. 1 NCG controlled heat pipe.

increases, a large movement of the inert-gas working-fluid front is accomplished with only a slight pressure increase in the inert gas. This amplification of front motion is accomplished by the noncondensible gas reservoir in the same way as the bulb in a conventional thermometer. The interface moves to increase condenser area for more heat rejection capability.

The position of the front is a function of temperature in the NCG reservoir region as well as pressure in the pipe. In addition, the front position is a function of the amount of noncondensible gas and working fluid vapor which are in the reservoir region. All of these effects must be considered in a proper design.

Effect of Reservoir Temperature Variations on Controllability

The effect of reservoir temperature variation on controllability was studied for two similar heat pipes; one had a wick in the reservoir, the other had no wick in the reservoir. The two heat pipes are made of 3/8-in. o.d. x 0.010-in. wall stainless-steel tubing. The pipes are 48 in. long with a 10-in.-long cylindrical reservoir (1.25-in. o.d. x 0.016-in. wall) welded to the end of each. A continuous wick is used from the beginning of the evaporator to the end of the reservoir in one pipe; the wick stops 9 in. from the reservoir leaving an unwicked separation region of 9 in. The pipes were loaded with ammonia working fluid and He NCG.

Both pipes have a 10-in. evaporator, 9-in. adiabatic section, and 20-in. condenser. Heater tape was wrapped around the pipe in the evaporator to create a uniform heat input zone; the condenser was formed from flattened copper tubing for good thermal contact with the heat pipe. Heater tape and coolant coils were used on the reservoirs to control reservoir temperature independent of the heat pipe temperature. The entire system was then insulated and tested. During this sequence of tests, a constant power was applied to the evaporator and varying power was applied to the reservoir to vary its temperature.

The front position (Figs. 2 and 3) was near the entrance to the reservoir, and the position remained relatively constant as reservoir temperature increased. Evaporator temperature is plotted as a function of reservoir temperature for both pipes in Fig. 4. The extreme sensitivity of evaporator temperature

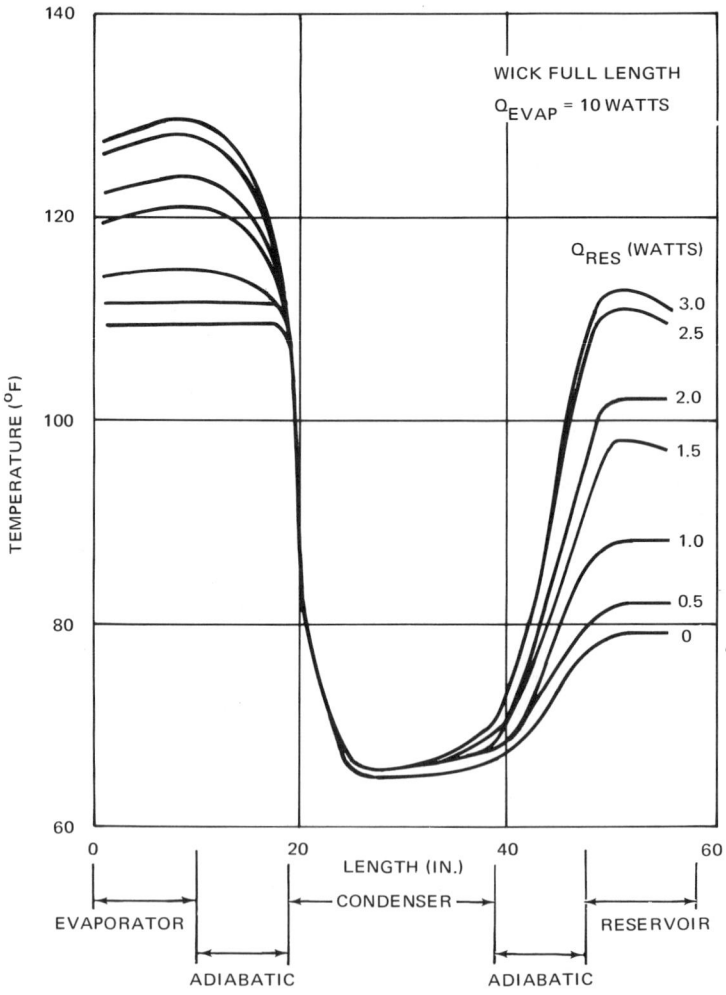

Fig. 2  Heat pipe temperature distribution with reservoir wicking.

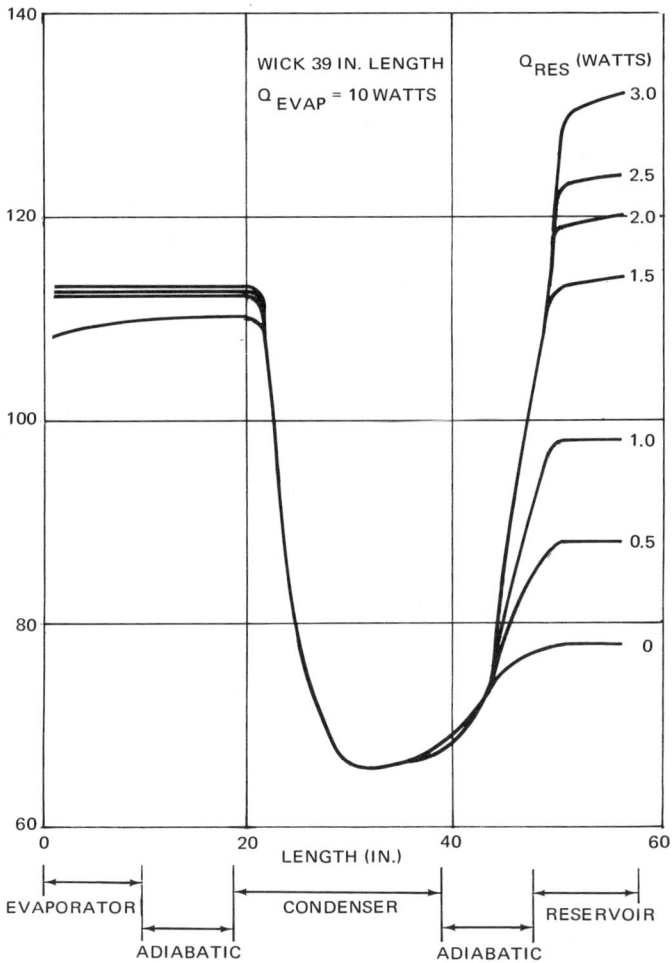

Fig. 3 Heat pipe temperature distribution with wickless reservoir.

to reservoir temperature variation for the wicked reservoir is apparent. The evaporator temperature of the heat pipe with no wick in the reservoir does change as a function of reservoir temperature, but the change is relatively insensitive compared to the wicked reservoir heat pipe.

A theoretical curve for these two cases is obtained by assuming that the heat pipe is divided into three isothermal regions: the evaporator and active portion of the condenser (characterized by temperature $T_e$ and corresponding vapor pressure ($P_e$), the inactive portion of the condenser ($T_c$ and

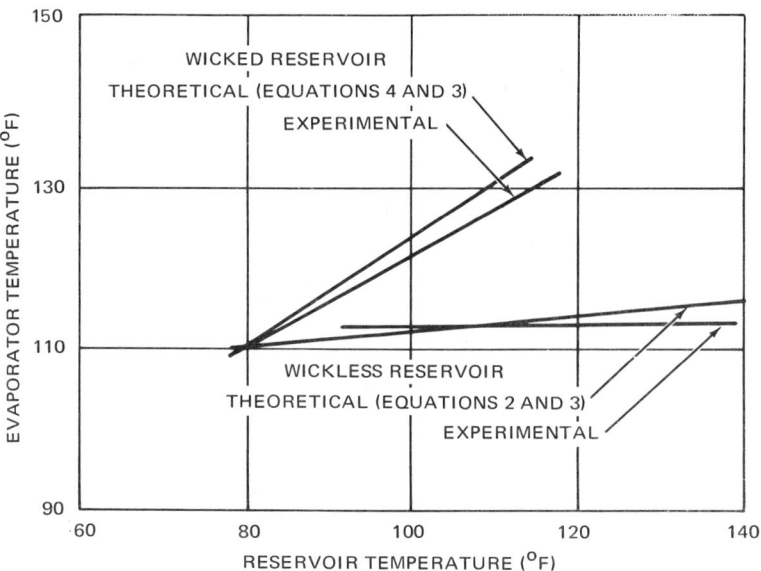

Fig. 4 Reservoir temperature effect on nominal control temperature.

$P_c$), and the inactive portion of the reservoir ($T_x$ and $P_x$). An equation expressing the conservation of NCG moles may be written; the subscript "o" designates a set of reference conditions. For the case where there is a wick in the condenser but not in the reservoir

$$\frac{P_e}{R}\left(\frac{V_x}{T_x} + \frac{V_c}{T_c}\right) - \frac{P_c V_c}{R\,T_c} = \frac{P_{e_o}}{R}\left(\frac{V_x}{V_{x_o}} + \frac{V_{c_o}}{T_{c_o}}\right) - \frac{P_{c_o} V_{c_o}}{R\,T_{c_o}} \qquad (1)$$

Solving for $P_e$ produces

$$P_e = \frac{P_{e_o}\left(\dfrac{V_x}{T_{x_o}} + \dfrac{V_{c_o}}{T_{c_o}}\right) + \left(\dfrac{V_c P_c}{T_c} - \dfrac{V_{c_o} P_{c_o}}{T_{c_o}}\right)}{(V_x/T_x) + (V_c/T_c)} \qquad (2)$$

The temperature corresponding to $P_e$ is obtained from the vapor pressure temperature relationship for ammonia

$$T_e = T_{e_{NH_3}}(P_e) \qquad (3)$$

Equations 2 and 3 are combined to obtain $T_e$ as a function of $T_x$ and plotted in Figure 4. The correlation is quite good.

The differences are likely attributable to the fact that the interfaces between regions have a finite thickness, and therefore no region can be exactly characterized by a single temperature.

A similar equation describes the wicked reservoir

$$P_e = \frac{P_{e_o}\left(\dfrac{V_{c_o}}{T_{c_o}} + \dfrac{V_x}{T_{x_o}}\right) + \left(\dfrac{P_c V_c}{T_c} - \dfrac{V_{c_o} P_{c_o}}{T_{c_o}}\right)}{(V_c/T_c) + (V_x/T_x)}$$

$$+ V_x \left(\frac{P_x}{T_x} - \frac{P_{x_o}}{T_{x_o}}\right) \bigg/ \left(\frac{V_c}{T_c} + \frac{V_x}{T_x}\right) \qquad (4)$$

Equations 3 and 4 are also plotted in Fig. 4 for comparison with the wicked reservoir data; as before, agreement is quite good.

Reservoir temperature can have an effect on the nominal control temperature of a NCG controllable heat pipe system. The presence of a wick in the reservoir increases the sensitivity of this variation. The reason for the significant effect with a wick in the reservoir is that reservoir pressure varies according to the vapor pressure temperature relationship, whereas for the wickless reservoir, pressure varies according to the perfect gas law. The value of dp/dt is generally much larger for a vapor pressure temperature relation than for the perfect gas law.

## Diffusion Effects on NCG Control

The controllability of a NCG heat pipe system can be described with two parameters: 1) the nominal control temperature and 2) the control range. For example, a component to be controlled at 50°F ±2°F has a nominal control temperature of 50°F and a control range of 4°F. The volume of the NCG reservoir determines the control range and the amount of NCG determines the nominal control temperature. Working fluid vapor which gets into the reservoir can act as a noncondensible gas if its partial pressure is below the equilibrium saturation pressure at the reservoir temperature. Therefore, changes in the amount of working fluid vapor in the reservoir can change the nominal control temperature of the heat pipe. This effect has been observed at DWDL and has been reported by Marcus and Fleischman.[1]

If the reservoir is assumed to be pure NCG at time (t) = 0, and there is a non-zero concentration of working fluid vapor at the entrance to the reservoir, the entire reservoir tends to approach the concentration at the entrance because of diffusion. The rate at which the concentration in the reservoir approaches the equilibrium amount corresponding to reservoir entrance conditions is a function of heat pipe geometry, working fluid and NCG (diffusion coefficients), and separation between the condenser and reservoir. Figure 5 shows the ratio $n_A/n_{A\infty}$ as a function of time for a hypothetical NCG heat pipe system where $n_A$ is equal to the number of moles of working fluid vapor diffused into the reservoir at time t assuming a constant concentration of vapor is applied at reservoir entrance at t = 0; and $n_{A\infty}$ is equal to the number of moles which diffuse into the reservoir after an infinite time with a constant vapor concentration (corresponding to some temperature) at the entrance.

The ordinate on the right shows the change in nominal control temperature for this particular system caused by such diffusion. These results are obtained analytically by a solution of the diffusion equation in a cylindrical system with two regions of different length and area to represent the separation tubing and reservoir.

This discussion has assumed that initially there was no vapor in the reservoir and observed the change in nominal control temperature. If, however, there is vapor in the reservoir at the desired nominal control temperature and that the amount of vapor corresponds to the vapor pressure at reservoir entrance conditions (i.e. $n_A = n_{A\infty}$), then the nominal control temperature will change if the temperature at the reservoir entrance changes.

To minimize the large change in nominal control temperature which occurs because of vapor diffusion into an initially vaporless reservoir, the initial load should have $n_{A\infty}$ moles of vapor in the reservoir corresponding to some average temperature at the reservoir entrance. As the reservoir entrance temperature changes, vapor will diffuse into or out of the reservoir depending on whether the temperature rises or falls.

## Design Guidelines

Performance characteristics as well as several other system characteristics can be summarized and compiled in a general set of design guidelines: 1) Definition of thermal control problem and environmental constraints. 2) Design of radiator for maximum and minimum heat rejection. 3) Selection of working fluid. 4) Selection of wick. 5) Determining size and configuration of NCG reservoir.

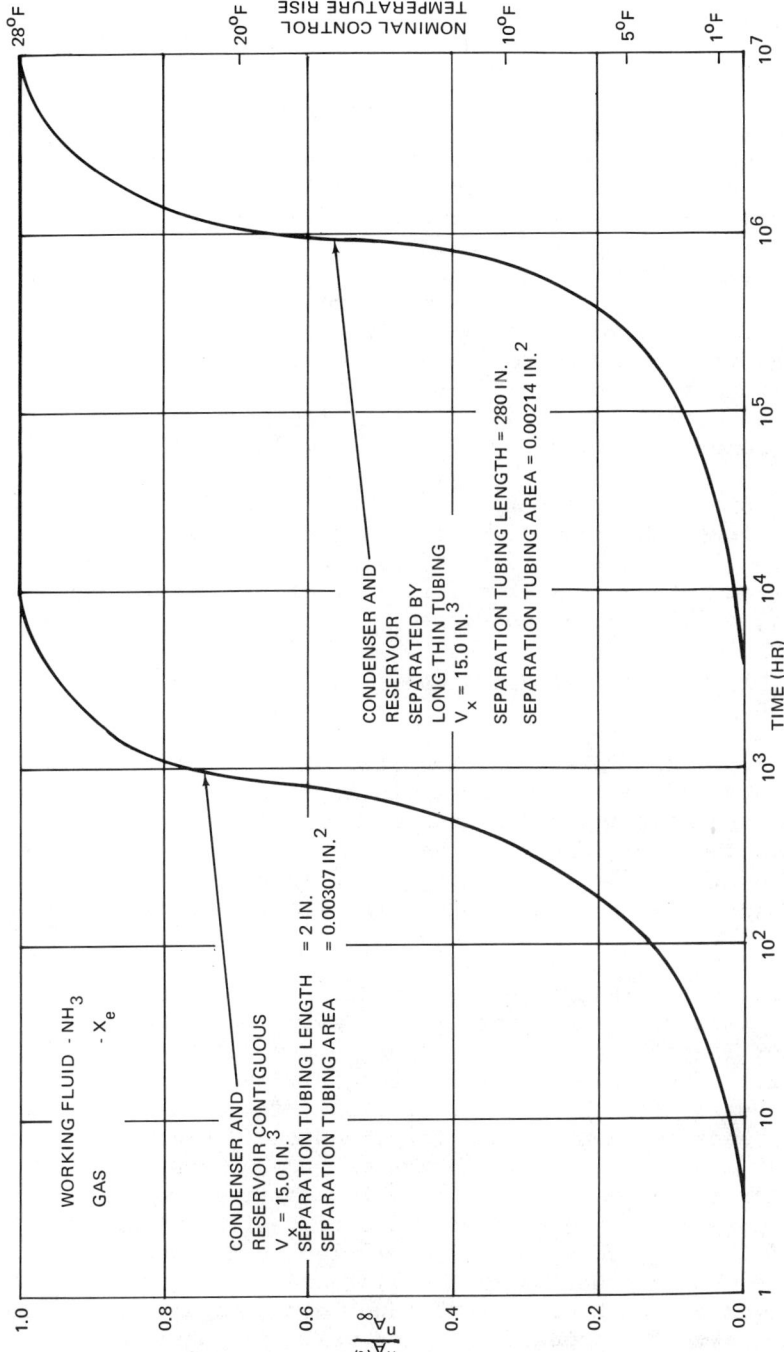

Fig. 5 Variation of reservoir vapor content and heat pipe control temperature with time.

## Definition of Thermal Control Problem and Environmental Constraints

In general, there are three important regions of a NCG heat pipe system which must be considered in adequately defining the problem; these are the evaporator, the heat rejector/condenser, and the noncondensible gas volume. The thermal environments of these three regions can all be different, and therefore, the temperatures could all be different in the absence of any control or interconnection between the regions. For example, the evaporator exterior surface temperature is to remain in the range of 60° to 65°F, while the component is dissipating between 1 and 50 w which is transferred to the evaporator; the environment of the condenser region dictates a temperature range from -150° to 100°F; the noncondensible gas volume temperature will vary in this range unless controlled within some other range.

In addition to the temperatures in the condenser and reservoir regions, the mode of heat transfer (i.e. convection, radiation, conduction) must be defined and the important parameters (heat-transfer coefficient, emissivity, shape factors, etc.) must be specified.

## Design of Heat Rejection for Maximum and Minimum Heat Rejection

The type, configuration, and size of the heat rejector is determined by the design constraints, especially the minimum and maximum heat rejection constraints. It is desirable to minimize the heat loss when the component is dissipating very little heat (the "off" condition) and to activate the entire heat rejector when the component is dissipating its maximum load.

*Minimum heat rejection.* It is desirable to obtain a very high thermal resistance between the evaporator and heat rejector fin to minimize heat loss in the "off" condition. There are three general methods to minimize longitudinal fin heat transfer for the "off" condition (Fig. 6). 1) Maximizing the distance between the NCG front and the beginning of heat rejector 1. However, as 1 increases, the volume of the NCG container increases because there is a larger front motion required, and therefore, there is a tradeoff. 2) Slotting the fins of the heat rejector perpendicular to the heat pipe, to minimize heat conduction in the fin parallel to the heat pipe. 3) Maximizing the thermal resistance of the heat pipe tubing by minimizing the conductivity and wall thickness.

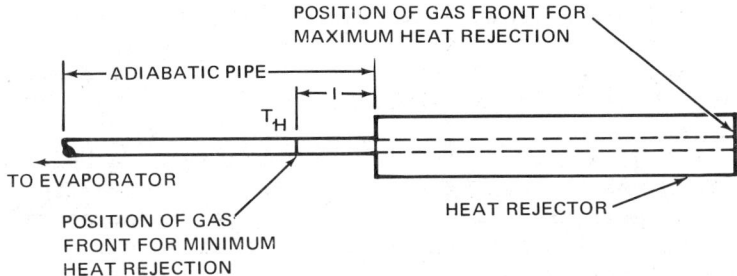

Fig. 6  Heat rejector design.

Maximum heat rejection. Once the basic configuration of the heat rejector system has been determined by the minimum heat rejection criterion, the size of the rejector must be determined by the maximum heat rejection criterion. The size can be determined by a heat balance on the entire heat pipe/heat rejector system. The heat input to the system is from the heat source and possibly solar energy, and the heat rejector exchanges heat with the surroundings. For a convective or conductive rejector located inside of a spacecraft, the solar term can be neglected.

Selection of Working Fluid

The criteria for fluid selection of an ordinary constant conductance heat pipe are adequately defined in the literature (e.g., Ref. 3). In addition, two other criteria can be very important; these are freezing point and slope of the vapor pressure temperature curve.

In a constant conductance heat pipe, the freezing point must be below the operating heat pipe temperature, but in a NCG variable conductance heat pipe the fluid must have a freezing point below all portions of the system, including the inactive portion.

If a working fluid is used which has a freezing point above the steady-state radiator temperature, a series of events can occur which will eventually lead to drying of the wick in the evaporator with no chance of recovery until the radiator increases in temperature to a value greater than the freezing point of the working fluid. The mechanism of failure is as follows: The working fluid diffuses through the NCG region and condenses on the cold surfaces of the pipe. Because of the low surface temperature, the working fluid freezes on condensation and does not return to the evaporator. There is then an accumulation of frozen working fluid in the cold condenser and a depletion of working fluid from the evaporator until eventually evaporator dryout occurs. The time for this to occur is a function of the rate of diffusion of the working fluid through the noncondensible gas.

Front motion and therefore controllability of a NCG heat pipe system, for a given reservoir volume, is a function of the internal pressure change with evaporator temperature change. The magnitude of the internal pressure change for a given control temperature range is a function of the working fluid. A plot of this pressure ratio as a function of $\Delta T_{control}$ is shown in Fig. 7 for several working fluids. The fluid which will be most effective for control will be the one which has the highest value of $P_r$ for a given $\Delta T_{control}$. A simple method of comparing various working fluids for controllability can be devised by approximating the vapor pressure in

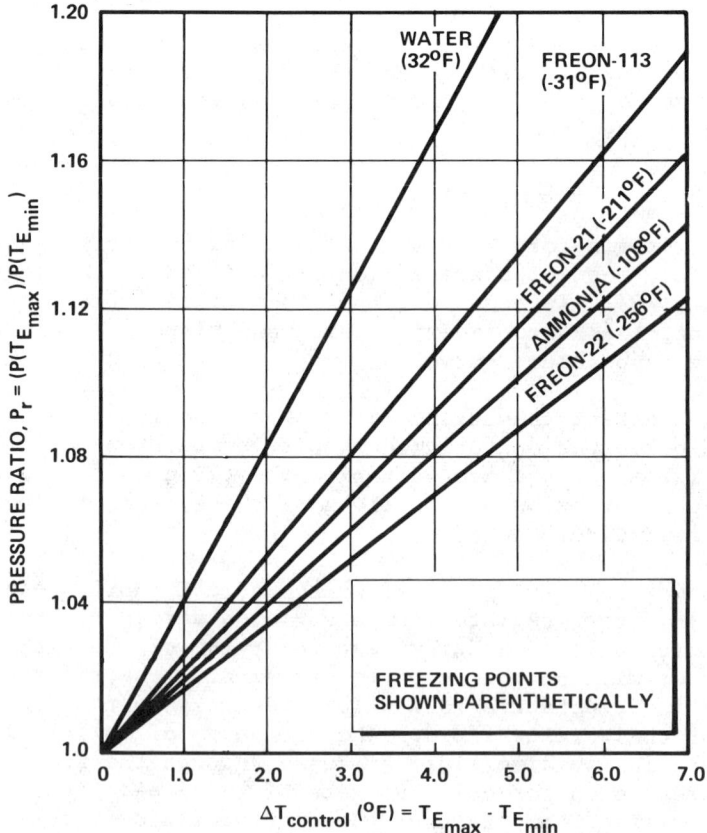

Fig. 7 Pressure ratio vs $\Delta T_{control}$ for various working fluids.

a small temperature range by the integrated value of the Clausius-Clapeyron equation.

$$P = A e - h_{fg}M/RT \quad (5)$$

where $A$ = a constant (lb/ft$^2$), $h_{fg}$ = latent heat of vaporization (Btu/lb$_m$), $M$ = molecular weight (lb$_m$/lb$_m$ mole), $R$ = universal gas constant (Btu/lb$_m$-°R), and $T$ = absolute temperature (°R).

This assumes that the vapor behaves as an ideal gas and that $h_{fg}$ is independent of temperature over the range of interest. Therefore, if $\Delta T_{control}$ is small ($\sim$10°F), the variation in $h_{fg}$ will generally be small. $P_r$ is given by

$$\frac{P_{max}}{P_{min}} = \frac{A e - (h_{fg}M/RT_{e_{max}})}{A e - (h_{fg}M/RT_{e_{min}})} = \exp\left[-\frac{h_{fg}M}{R}\left(\frac{1}{T_{e_{max}}} - \frac{1}{T_{e_{min}}}\right)\right] \quad (6)$$

For a given value of $T_{e_{max}}$ and $T_{e_{min}}$ (i.e. $\Delta T_{control}$) the working fluid which has the highest value of $P_r$ has the highest value of $\Delta H_v = h_{fg}M$. Therefore a fluid figure-of-merit for controllable heat pipe fluids is the latent heat on a molar basis; a tabulation of $\Delta H_v$ for various fluids is the only quantity required.

## Wick Selection

The selection of a wick is based on three factors all of which are a consideration in designing a constant conductance heat pipe: 1) condensate return; 2) minimum thermal resistance; and 3) compatibility of material with working fluid. These are all discussed in the literature (e.g., Ref. 3). The second criteria can be especially important in NCG variable-conductance heat pipe operation because any temperature drop required to get heat into the system reduces the temperature change available for pressure change. This is especially true if controlled power ranges from a value close to zero to a relatively large value.

## Determining NCG Reservoir Size and Configuration

When the working fluid has been selected, the degree of controllability is a function of the volume of the NCG reservoir. The volume required to perform this control can be determined by a mole balance of the type in Eqs. 2 and 4. These equations can then be solved for $V_x$. The reservoir volume $V_x$ is a function of: 1) the distance the front must be moved from minimum rejection to maximum rejection; 2) the temperature variation in the reservoir; and 3) the temperature control required at the evaporator ($P_r$ vs $\Delta T_{control}$ from Fig. 7). The volume required is also a function of whether there is a wick in the reservoir and whether the amount of working fluid vapor in the reservoir can change because of diffusion. Variations in reservoir temperature can be decreased by locating the reservoir near the controlled evaporator (Fig. 8).

The variation in nominal control temperature because of vapor diffusion into the reservoir (Fig. 5) can be mitigated in several ways: 1) using a long thin separation tubing between the condenser and the reservoir and 2) choosing a large value of $V_x/A_c$ which is the ratio of reservoir volume to cross-sectional diffusion area of the connecting tubing.

As previously mentioned, the two basic design equations are obtained from the mole balance on the NCG of the type given by Eqs. 2 and 4; Eq. 2 for no vapor in the reservoir, Eq. 4 where vapor exists in the reservoir. The measured controllability ($\Delta T_{control}$) of heat pipe systems is compared with the calculated values based on vapor in the reservoir or no vapor in the reservoir in Table 1.

During the Viking effort, heat pipes were designed to control a variable thermal load against variation in thermal conditions corresponding to a Mars day/night cycle. A controllable heat pipe was also built for a communications satellite to control a variable thermal load (1 to 65 w) against thermal variations imposed by a synchronous equatorial orbit including complete shadowing of the radiator for portions of the orbit and solar eclipse during portions of the orbit when the sun is shining on the radiator.

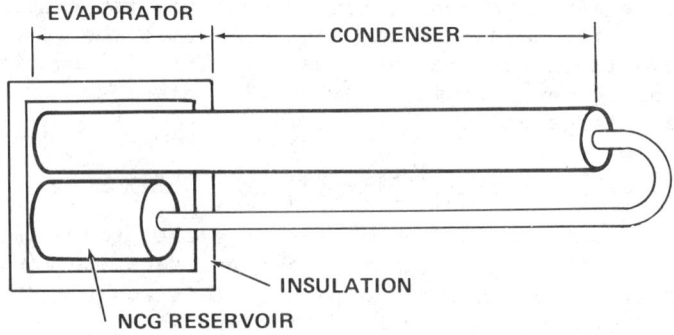

Fig. 8 Technique for controlling NCG reservoir temperature.

Table 1 Experimental and analytical performance comparison

| Experiment | Reservoir | $Q_{range}$ (w) | $V_x/A_c$ (in.) | No Vapor In Reservoir $\Delta T_{control}$ Calc 1 (°F) | $\Delta T_{control}$ Measured (°F) | Vapor In Reservoir $\Delta T_{control}$ Calc 2 (°F) |
|---|---|---|---|---|---|---|
| Viking | Wickless | 40–80 | 29.3 | 12.2 | 16 | 16.3 |
| Viking | Wickless | 30–285 | 20.5 | 32.1 | 64 | 63.8 |
| Viking | Wickless | 10–50 | 32.3 | 11.1 | 25 | 17.7 |
| Res temp | Wick | 10–30 | 129 | 2.3 | 3 | 3.0 |
| Res temp | Wick | 5–40 | 129 | 4.9 | 4 | 4.7 |
| Res temp | Wick | 10–40 | 129 | 4.8 | 2 | 4.0 |
| Res temp | Wick | 10–40 | 129 | 2.6 | 2 | 1.1 |
| Res temp | Wick | 10–30 | 129 | 2.5 | 1 | 1.3 |
| Res temp | Wickless | 1–30 | 129 | 14.1 | 10 | 17.7 |
| Res temp | Wickless | 20–60 | 129 | 5.0 | 6 | 2.0 |
| Res temp | Wickless | 20–60 | 129 | 5.4 | 6 | 7.6 |
| Res temp | Wickless | 10–60 | 129 | 7.3 | 3 | 3.6 |
| Experimental system | Wickless | 5–50 | 363 | 1.5 | 2 | 7.9 |
| Experimental system | Wickless | 5–50 | 733 | 0.4 | 1 | 1.9 |
| Comsat | Wickless | 1–65 | 6576 | 1.5 | 1.7 | 32.1 |

The "Reservoir" column in Table 1 indicates whether $\Delta T_{calc\ 1}$ or $\Delta T_{calc\ 2}$ should correlate with $\Delta T_c$ measured. Any heat pipe which has a wick in the reservoir should be correlated by $\Delta T_{calc\ 2}$. With no wick in the reservoir, measured values may be correlated by either $\Delta T_{calc\ 1}$ or $\Delta T_{calc\ 2}$ depending on whether the diffusion time constant is small or large. For example, in Fig. 5 the contiguous condenser and reservoir have a diffusional time constant of approximately $10^2$ hr, and the separated condenser and reservoir have a diffusional time constant of approximately $10^5$ hr. If the time constant is small (small $V_x/A_c$, no separation tubing between condenser and reservoir), a change in temperature at the end of the condenser is rapidly transmitted to a change in reservoir vapor concentration (i.e. diffusional equilibrium exists). If the diffusion time constant is large, however, the condenser and the reservoir are not in diffusional communication and a change in reservoir entrance conditions is not reflected in a change in reservoir vapor concentration. The effect of diffusion time constant as exhibited by $V_x/A_c$ is shown in Table 1. The performance of the wickless reservoir Viking heat pipes with a small value of $V_x/A_c$ is correlated by $\Delta T_{calc\ 2}$ (diffusional equilibrium); however, the performances of the wickless reservoir temperature, experimental system, and Comsat heat pipes are correlated by $\Delta T_{calc\ 1}$ because the diffusion time constant is large enough (large $V_x/A_c$) to prevent communication of the condenser end with the reservoir. The correlations in Table 1 (values enclosed with dashed lines) are reasonable on an intuitive physical basis.

## Summary and Conclusions

A number of NCG variable-conductance heat pipe systems have been built and tested and guidelines have been identified for designing NCG heat pipes. The following conclusions can be made as a result of this work: 1) Reservoir temperature variations affect the controllability of NCG heat pipe systems. 2) The sensitivity of the effect of reservoir temperature variations on controllability is increased by the presence of a wick in the reservoir. 3) Diffusion of working fluid vapor into or out of the reservoir can affect the nominal control temperature. This effect can be mitigated by increasing $V_x/A_c$ or using a long unwicked separation tubing between the condenser and the reservoir. 4) Heat loss in the "off" condition can be minimized by using thin wall, low conductivity tubing for the heat pipe, slotting the heat rejector fins perpendicular to the heat pipe, and locating the front position a certain distance away from the beginning of the condenser for the "off" condition. 5) A fluid figure-of-merit for NCG controllable heat pipe operation is given by the latent heat of vaporization on a molar basis. 6) The thermal resistance of an evaporator wick is especially important because it affects the amount of $\Delta T_{control}$ available for working fluid pressure change. 7) The controllability of a NCG system can be predicted by one of two equations describing operation for vapor in the reservoir or no vapor in the reservoir.

## References

[1] Marcus, B. D., and Fleischman, G. L. "Steady-State and Transient Performance of Hot Reservoir Gas-Controlled Heat Pipes" Paper 70-HT/SpT-11, ASME Space Technology and Heat Transfer Conference, Los Angeles, Calif., June 21-24, 1970.

[2] Bienert, W., "Heat Pipes for Temperature Control," Proceedings: Fourth Intersociety Energy Conversion Engineering Conference, Sept. 22-26, 1969, Washington, D.C., pp. 1033-1041.

[3] Phillips, E. C., "Low-Temperature Heat Pipe Research Program," CR-66792 NASA, June 1969.

# FEEDBACK CONTROLLED VARIABLE CONDUCTANCE HEAT PIPES

W. B. Bienert* and P. J. Brennan[†]

Dynatherm Corporation, Cockeysville, Md.

and

J. P. Kirkpatrick[‡]

NASA Ames Research Center, Moffett Field, Calif.

## Abstract

Electrical and mechanical methods of achieving feedback control in variable conductance heat pipes are evaluated. A steady-state analysis is used to determine desirable design characteristics along with the performance of feedback controlled heat pipes. Although both electrical and mechanical methods are technically feasible, in general, the electrical system offers a greater degree of temperature control and can be designed more practically than a mechanical bellows system. Test results obtained with a breadboard model of an electrical feedback system are presented.

## Nomenclature

A = area
h = heat transfer coefficient
k = spring constant
L = length
m = mass
p = pressure

---

Presented as Paper 71-421 at the AIAA 6th Thermophysics Conference, Tullahoma, Tenn., April 26-28, 1971. This paper is based on NASA CR-73475 for the NASA Ames Research Center under Contract NAS2-5722.
*Director of Engineering.
†Project Engineer.
‡Research Scientist.

Q = heat load
R = gas constant
$R_s$ = thermal impedance between the heat source and the vapor
T = temperature
V = volume
y = active condenser length
β = coefficient of volumetric expansion
η = bellows displacement
χ = coefficient of isothermal expansion
π = vapor pressure in inactive part of condenser or in storage volume

Subscripts

a = auxiliary fluid
c = condenser
e = evaporator
g = noncondensible gas
H = high power condition
ic = inactive condenser section
L = low power condition
o = sink condition
s = heat source
st = storage volume
sto = bellows at minimum extension
v = vapor

## I. Introduction

The principle of a variable conductance heat pipe has been known and understood for years.[1-5] Serious efforts are now being made to apply the principle to various spacecraft temperature control problems.[6,7] Ideally, a conventional variable conductance heat pipe is able to maintain its own temperature nearly constant while heat input or environmental conditions are changed.[8] If the thermal impedance between the heat source and the heat pipe is small, the source temperature will also remain nearly constant. However, frequently this impedance is non-negligible and the source temperature will vary with changing heat input even if the heat pipe temperature is maintained at a constant level. A second difficulty with a conventional variable conductance heat pipe is its sensitivity to fluctuations of the sink temperature. Variable sink conditions affect the temperature of the noncondensing gas therein causing it to expand or compress. Also, the temperature of saturated vapor in the inactive part of the condenser and gas reservoir, and therefore the partial volume available for gas storage, is dependent upon the sink temperature.

Except for the ideal case which requires an infinite gas storage volume, either of these effects can cause large excursions of a conventional system's vapor temperature and correspondingly of the heat source temperature. Finally, gas generation in the heat pipe during operation will change a conventional system's operating characteristics.

The variations in source temperature associated with the above problems can be reduced significantly by the use of a feedback system. The purpose of this paper is to describe the work completed toward the development of feedback controlled heat pipes. As in the case of conventional thermal control heat pipes, a noncondensing gas is employed to control the heat rejection area of the heat pipe, but now the storage volume is variable and linked to the heat source. In principle, the feedback mechanisms that are evaluated monitor the source temperature and adjust the gas storage volume. Both active (electrical) and passive (mechanical) methods of feedback control are considered. A detailed steady-state analysis is performed to evaluate the control that is afforded by feedback systems and also to establish the influence of various design parameters. Conceptual designs for both active and passive systems are presented, and the degrees of control afforded by equivalent active and passive systems are determined for a reference application.

## II. Principle of Operation

A feedback controlled variable conductance heat pipe is shown schematically in Fig. 1.

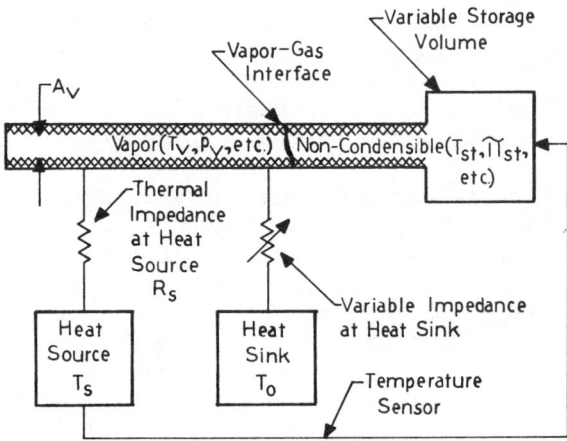

Fig. 1 Feedback controlled variable conductance heat pipe.

When a noncondensing gas is present in a heat pipe, it is entrained by vaporized working fluid and swept to the condenser end of the pipe where it is compressed to an extent dependent on the heat load, sink condition, and available storage volume. In the absence of axial conduction and diffusion effects which can be minimized, the part of the condenser which is blanketed by the noncondensible is not available for heat rejection. By controlling the axial location of the gas vapor interface, the heat rejection area and therefore the condenser's conductance, can be regulated to provide temperature control. As the heat load and/or sink temperature increase, the interface moves toward the storage reservoir. At the high power/high sink temperature condition, the condenser will be fully active and its maximum conductance will be realized. The opposite applies to reductions in heat load and/or sink temperature.

The location of the gas-vapor interface in a variable conductance heat pipe is determined from conservation of mass and energy requirements. In a conventional system, even though a variable gas reservoir may be employed, the interface is continually adjusted by the system's vapor pressure, and the best control that can be accomplished is to maintain the system's vapor temperature constant.[8] In a feedback-controlled system, the available storage volume is regulated by the temperature of the heat source. Consequently, the condenser's conductance is controlled essentially by the source temperature. As a result, the location of the interface or the active condenser length can be adjusted independently of the system's vapor pressure to give a vapor temperature which is required to establish the desired heat source temperature control. In other words, the system vapor temperature (or pressure) is the controlling factor in a conventional system, whereas the application of feedback permits the source temperature to be the controlling parameter.

In the passive feedback system, a variable storage volume is affected by physical displacement of a bellows system. A passive feedback controlled variable conductance heat pipe is shown schematically in Fig. 2. The control system consists of two bellows and a sensing bulb located near the heat source. The inner bellows contains an auxiliary fluid and is connected to the sensing bulb by a capillary tube. As will be demonstrated later, it is best to use an incompressible liquid as the auxiliary fluid. Variations in the source temperature cause a change in the temperature of the auxiliary fluid. The corresponding pressure change results in displacement of the

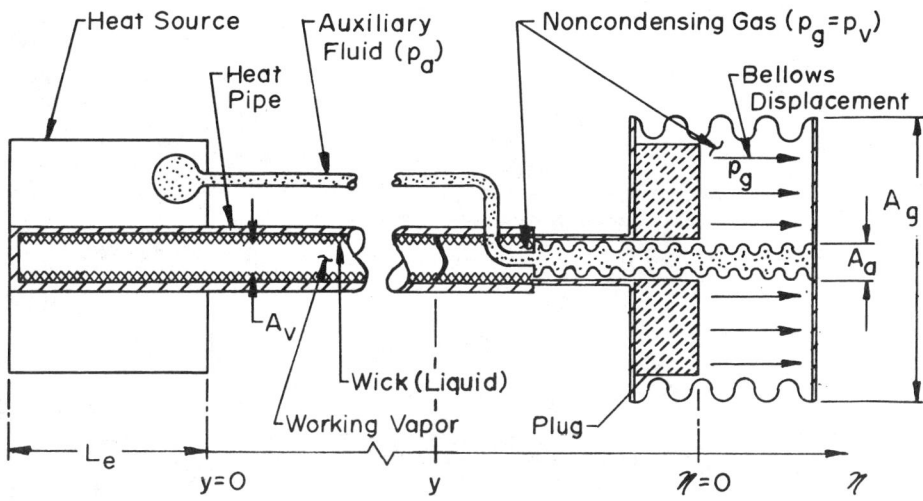

Fig. 2  Passive feedback controlled heat pipe system.

inner bellows and therefore of the outer bellows which is used as the storage volume for the noncondensible. This in turn causes a displacement of the gas-vapor interface or, more important, an adjustment of the variable conductance. Thus, by relating the variable conductance to the heat source, a feedback system which regulates the source temperature is effected.

Instead of a bellows system and an auxiliary fluid reservoir which is used in the passive system, a thermistor, an electronic controller, and a heated storage volume are used to provide active control. An active feedback controlled heat pipe system is shown schematically in Fig. 3. A continuous wick extends into the storage volume which is located beyond the condenser end of the heat pipe. As a result, saturated working fluid is always present in liquid form in the storage volume and the partial pressure of the vapor in equilibrium with this liquid is determined by the temperature of the storage reservoir. The total pressure which is determined by the vapor temperature in the active section of the heat pipe must be uniform throughout the system. Since the partial pressure of the vapor in equilibrium with the liquid in the storage volume is determined by the reservoir temperature, the partial pressure of the noncondensible, and therefore the partial volume it occupies in the reservoir, also are determined by the reservoir temperature. The vapor pressure of any fluid is sensitive to temperature, and consequently small

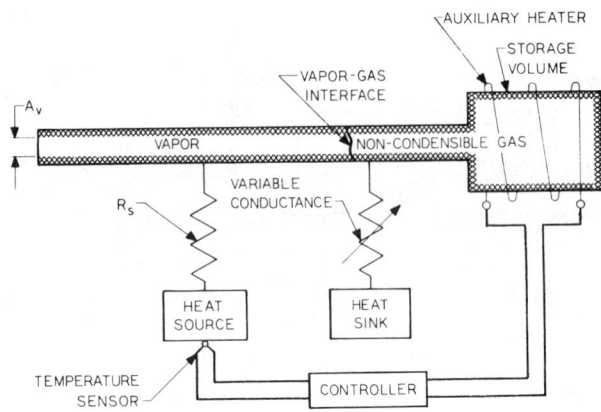

Fig. 3 Active feedback controlled heat pipe system.

variations of the reservoir temperature suffice to achieve large changes in effective reservoir volume occupied by the noncondensible. These changes result in movement of the gas out of the storage volume into the condenser (or vice versa), thereby changing the conductance of the heat pipe. By controlling the auxiliary heat input with a feedback temperature sensor attached to the heat source so that proper displacement of the noncondensible occurs, sharp control of the source temperature can be maintained.

### III. Analytic Evaluation

#### A. General Formulation

The analysis of a variable conductance heat pipe in Ref. 8 has been extended to include the effects of feedback control. The mathematical model describes in differential form the variation of the heat source temperature $T_s$ as affected by changes in the heat input $Q$, the heat sink temperature $T_o$, and other independent variables. The model is general in the sense that it is not limited to a specific configuration and treats both active and passive methods for obtaining thermal control in functional form. The only limitation imposed upon the system is that a noncondensing gas is used to control the available heat rejection areas. A sharp gas-vapor interface is assumed, and axial conduction is neglected.

A system of four partial differential equations is required to define the model. These differential equations are determined by taking the total derivative of the following general steady-state equations.

1. Conservation of energy:

$$F(T_v, y, Q, T_o) = 0 \qquad (1)$$

$$\text{e.g., } Q = h P y (T_v - T_o) \qquad (2)$$

2. Conservation of mass:

$$G(p_v, V_{ic}, V_{st}, T_o, \pi_o, T_{st}, \pi_{st}, m_g) = 0 \qquad (3)$$

e.g., assuming noncondensible obeys the Ideal Gas Law,

$$\frac{(p_v - \pi_o)}{R_g T_o} V_{ic} + \frac{(p_v - \pi_{st}) V_{st}}{R_g T_{st}} = m_g \qquad (4)$$

where $V_{ic} = A_v (L_c - y)$ \qquad (5)

and, for the passive system,

$$V_{st} = V_{sto} + A_{ga} \eta \qquad (6a)$$

with $A_{ga} = A_g - A_a$ \qquad (6b)

Equations (1) and (3) are common to both active and passive systems. Because of the presence of an auxiliary fluid, two additional equations are required for the passive system. They are

3. Force balance:

$$H(p_v, \eta, p_a) = 0 \qquad (7)$$

For the bellows configuration shown in Fig. 2 the following steady-state force balance applies:

$$p_a A_a + p_v A_{ga} = k(\eta - \eta_{fl}) \qquad (8)$$

4. Equation of state for auxiliary fluid (assuming $T_a = T_s$, i.e., no transient effects):

$$J(T_s, \eta, p_a, m_a) = 0 \qquad (9)$$

e.g., for an incompressible liquid

$$dp_a = \frac{\beta}{\chi} dT_s - \frac{A_a}{\chi V_a} d\eta \qquad (10)$$

In addition to Eq. (9) and (10), the following auxiliary equations apply:

$$T_s = T_v + R_s Q \qquad (11)$$

and, for a saturated fluid,

$$p_v = p_v(T_v), \quad \pi = \pi(T) \qquad (12)$$

By taking the total derivative of the functional equations (e.g., $dF = \frac{\partial F}{\partial T_v} dT_v + \frac{\partial F}{\partial y} dy + \frac{\partial F}{\partial Q} dQ + \frac{\partial F}{\partial T_o} dT_o = 0$), and making the appropriate substitutions, one can solve for the variation of the source temperature $(dT_s)$ as

$$dT_s = \left[ \left( \frac{\partial T_v}{\partial Q} R_s (1+S+S_1) \right) dQ + \left( \frac{\partial T_v}{\partial T_o} + S \psi_o \right) dT_o \right.$$
$$\left. S \psi_{st} dT_{st} \right] / 1 + S + S_1 + S_2 \qquad (13)$$

where $\psi \equiv \frac{1}{(p\alpha)_v} \left[ \pi \gamma \left(1 + \frac{\partial p_v}{\partial \pi}\right) + \frac{\partial p_g}{\partial T} \right] \qquad (14)$

with $\alpha_v = \frac{\partial \ln p_v}{\partial T_v}, \quad \gamma = \frac{\partial \ln \pi}{\partial T} \qquad (15)$

The parameters $S_i$ are associated with passive control and will be discussed in detail later. In deriving Eq. (13), control parameters such as either active or passive injection of the noncondensible gas $(dm_g)$ or of an auxiliary fluid $(dm_a)$ have been neglected. Equation (13) describes the change in source temperature as affected by changes in the heat load (Q), sink temperature $(T_o)$, and storage temperature $(T_{st})$. Variations in Q and $T_o$ are associated with the performance requirements of the system (duty cycle, orbit, etc.). Uncontrolled changes in the storage temperature could result because of changes in the heat sink environment. The effect of these variations must be minimized to prevent undesired variations in the source temperature. In a passive system with or without feedback, changes in $T_s$ resulting from variations in Q, $T_o$, etc., are attenuated by the parameters $S_i$. In the active system, the storage temperature is controlled by an electronic feedback unit. In this case, the controlled variation in $T_{st}$ is used essentially to negate those changes in $T_s$ that arise as a result of changes in heat load, etc.

Before considering feedback control, a brief evaluation of the factors that influence all variable conductance heat pipes which utilize a noncondensible gas to establish thermal control will be given.

$\partial T_v/\partial Q$ ... This is the inverse of the variable conductance of the condenser and is related to the effect of the heat dissipation requirement on the vapor temperature.

$R_s$ ... This is the thermal impedance between the heat source and the vapor and defines the interaction between $T_s$, $T_v$, and Q through Eq. (11).

$\partial T_v/\partial T_o$ ... This partial derivative reflects the effect of sink temperature on the vapor temperature to satisfy heat dissipation requirements.

$\psi$ ... The parameter $\psi$ is associated with establishing conservation of the mass of the noncondensible within the inactive part of the condenser and the storage reservoir. In particular it relates to the influence of sink and/or storage temperature on the displacement of the noncondensible gas and therefore on the location of the gas-vapor interface.

The displacement of the interface associated with $\psi_o$ which relates to the inactive condenser section, will be present in all variable conductance heat pipes employing a noncondensible gas. In those cases where a non-wicked reservoir is used (e.g., in "hot" reservoirs[5] or in a bellows system where a wick is just not practical) there should be no working fluid present in the reservoir, and therefore, neglecting diffusion, $\pi_{st} = 0$. Under these circumstances the only effect of changes in storage temperature will be to cause expansion or compression of the noncondensible (ideal) gas in the reservoir. For most applications this effect will be relatively small.

Unless special designs are employed, the reservoir temperature of a conventional thermal control heat pipe and also of a passive bellows system will be essentially equal to the sink temperature. In those applications where variations in the sink temperature as they affect the displacement of the gas in the reservoir prove troublesome, one possible solution is to utilize a "hot" reservoir which is contained within the active part of the heat pipe.[5] The reservoir temperature in such a design then will be the same as the vapor temperature, which can be kept nearly constant. One also could employ an auxiliary heater to keep the reservoir at a constant temperature through varying sink conditions. However, since this requires the use of feedback, it would be more beneficial to utilize an active feedback system that will maintain control of the source temperature rather than the reservoir temperature.

B. Passive System

The degree of control that is afforded by a passive system is determined by the magnitude of the parameters $S_i$ in Eq. (13). Each of these parameters has a physical interpretation. S describes the effect of using a noncondensing gas and a fixed storage volume, as in conventional thermal control heat pipes. $S_1$ applies when a variable storage volume, such as a bellows whose displacement is affected only by the system's vapor pressure, is used to contain the noncondensible. Finally, $S_2$ is associated with passive feedback control and is applicable when an auxiliary fluid which senses the source temperature is used to provide the driving force to vary the storage volume.

In order to define design parameters for the bellows system as well as select an auxiliary fluid it is necessary to study these parameters in more detail. Their general expressions are as follows:

$$S = \frac{\partial T_v}{\partial y} \frac{\partial V_{iv}}{\partial p_c} \left(\frac{p\alpha}{A}\right)_v \tag{16}$$

$$S_1 = \frac{\partial T_v}{\partial y} \frac{\partial V_{ic}}{\partial V_{st}} \frac{\partial \eta}{\partial p_v} A_g \frac{p\alpha}{A}\bigg|_v \Theta \qquad (17)$$

$$S_2 = \frac{\partial T_v}{\partial y} \frac{\partial V_{ic}}{\partial V_{st}} \frac{\partial \eta}{\partial p_a} \frac{A_g}{A_v} \omega_1 \Theta \qquad (18)$$

with

$$\Theta \equiv (1 - \frac{\partial \eta}{\partial p_a} \omega_2)^{-1} \qquad (19)$$

The parameters $\omega_1$ and $\omega_2$ are related to the type of auxiliary fluid which is used and are defined as:

|  | Incompressible liquid | Saturated vapor |  |
|---|---|---|---|
| $\omega_1 =$ | $\beta/\kappa$ | $(p\alpha)_a$ | (20) |
| $\omega_2 =$ | $-(A/\kappa V)_a$ | 0 | (21) |

Evaluation of these parameters for the bellows configuration shown in Fig. 2 gives

$$S = \frac{(T_v - T_o)}{y(p_v - \pi_o)} (\frac{p\alpha}{A})_v \left[ V_{ic} + V_{st}(\frac{T_o}{T_{st}}) \right] \qquad (22)$$

$$S_1 = \frac{(T_v - T_o)}{y} \frac{p_v}{(p_v - \pi_o)} \frac{T_o}{T_{st}} \frac{A_g^2}{k} (\frac{p\alpha}{A})_v \Theta \qquad (23)$$

$$S_2 = \frac{(T_v - T_o)}{y} \frac{p_v}{(p_v - \pi_o)} \frac{T_o}{T_{st}} \frac{A_g A_a}{k A_v} \omega_1 \Theta \qquad (24)$$

with

$$\theta = (1 - \frac{A_a}{k} \omega_2)^{-1} \qquad (25)$$

A detailed discussion of the effect of these parameters in attenuating variations in source temperature caused by changing heat load and/or sink conditions is given in Ref. 9. In general, the magnitude of S is always greater than zero. Furthermore, to have stable control using feedback, one must have $S_2$ greater than zero. This corresponds to having expansion of the storage volume with increasing source temperature. The parameter $S_1$ may be made less than, equal to, or greater than zero depending on the bellows configuration employed. The optimum configuration will depend on what variables (heat load, sink temperature, etc.) have the most marked effect on the source temperature. However, as indicated by Eq. (13), the magnitude of the feedback parameter $S_2$ should be as large as practical. In particular $S_2$ should be large, independently of S and $S_1$. The use of an auxiliary fluid that is very sensitive to temperature changes allows one to accomplish this. Both saturated vapors and incompressible liquids qualify. However, the pressure change associated with a given temperature change is several orders of magnitude greater for a liquid than that experienced by a saturated vapor. This is reflected in the relative magnitude of $\beta/\chi$ and $(p \propto )_a$ for the incompressible liquid and the saturated vapor, respectively. Consequently, larger values of $S_2$ relative to S and $S_1$ are realized when the auxiliary fluid is an incompressible liquid.

The steady-state performance of a mechanical feedback controlled heat pipe system has been determined for a reference application. Performance requirements are listed in Table 1. The design of the heat pipe and bellows system is summarized in Table 2. Standard off-the-shelf bellows were used in the design. Methanol was selected as a working fluid because it has a relatively low vapor pressure at the operating temperature, and therefore mass diffusion effects should be negligible. Methanol also was selected as the auxiliary liquid because it has a high ratio of volumetric expansion to isothermal compressibility.

The variation of heat source temperature with changes in the heat input has been determined by simultaneous solution of the appropriate steady-state equations and is compared to that for an ideal conventional thermal control heat pipe ($V_{st} \to \infty$) in Fig. 4. The corresponding variations in the

Table 1  Nominal performance requirements

| | |
|---|---|
| Nominal source temp. ($T_s$) | 18°C |
| Nominal sink temp. ($T_o$) | 200°K |
| Power range (Q) | 0-30w |
| Nominal heat source resistance ($R_s$) | 0.3°C/w |

Table 2  Reference passive feedback controlled heat pipe design

| | |
|---|---|
| Heat pipe geometry | 0.94 cm o.d. x 0.05 cm wall |
| Evaporator length | 46 cm |
| Transport section | 16.5 cm |
| Condenser length | 43 cm |
| Vapor area | 0.47 cm$^2$ |
| Working fluid | Methanol |

Bellows System

| Bellows parameters | Auxiliary bellows | Gas bellows |
|---|---|---|
| o.d. (cm) | 0.8 | 8 |
| Effective area (cm$^2$) | 0.32 | 32 |
| Stroke (cm) | 2 | 2 |
| Spring rate (newtons/m) | $1.06 \times 10^3$ | $3.4 \times 10^3$ |
| Auxiliary fluid | Methanol | |
| Auxiliary fluid storage vol. (cm$^3$) | 65 | |

vapor temperature of the working fluid also are shown. The value of feedback control becomes obvious from Fig. 4. By driving the variable storage volume with a fluid which senses the source temperature, the variable volume will be expanded to the extent that the vapor temperature will decrease with increasing heat load. In the conventional thermal control heat pipe $dT_v$ is greater than or equal to zero. Thus from Eq.(11), in the limit of an ideal conventional heat pipe system, the source temperature will vary directly as the heat load. However, since $dT_v$ can be made negative by using feedback, the variation of source temperature with changes in heat load can be made to go to zero. Although this example was not based on an optimum design, the comparison indicates a decrease in the variation of the source temperature of approximately 50% when compared to an ideal conventional system.

C. Active System

A steady-state analysis was performed to evaluate the control capability along with the storage volume requirements for an active feedback controlled heat pipe system. The analysis accounts for variable heat load and sink conditions and is based on satisfying Conservation of Mass (Eq. 4), taking into account the following principles of operation:

1) <u>At the high power, high sink condition</u> the entire condenser section will be active, and all of the noncondensible will be compressed within the storage volume. Also, to provide maximum storage at the high power condition, the storage reservoir should be at the lowest temperature it can achieve which for this condition is the maximum sink temperature. Thus

$$m_g = (p_v - \pi_{st})_H V_{st} / R_g T_{o_H} \qquad (26)$$

2) <u>At the low power, low sink condition</u> the entire condenser section will be inactive with the noncondensible expanded throughout the condenser length which will be at the minimum sink temperature. Ideally, the temperature of the storage reservoir will approach the vapor temperature of the working fluid at this condition in which case all of the gas will be in the condenser section. Thus

$$m_g = (p_v - \pi_o)_L V_c / R_g T_{o_L} \qquad (27)$$

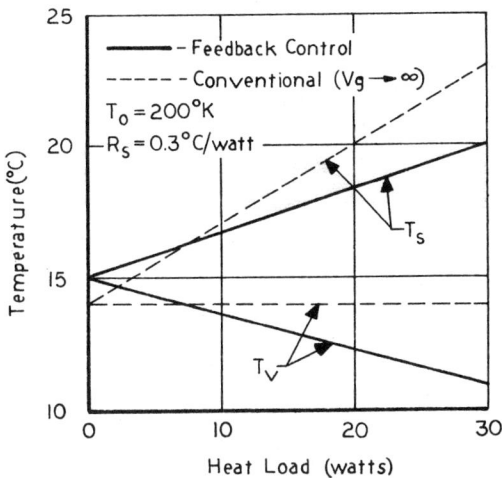

Fig. 4 Steady-state temperature control.

It has been assumed that the condenser section is totally inactive (i.e., $V_{ic} = V_c$) at this condition. At intermediate power and/or sink conditions the gas-vapor interface will be located at a point within the condenser which gives a vapor temperature which is consistent with the desired degree of temperature control. Simultaneous solution of Eqs. (26) and (27) gives

$$\frac{V_{st}}{V_c} = \frac{(p_v - \pi_o)_L}{(p_v - \pi_{st})_H} \frac{T_{o_H}}{T_{o_L}} \quad (28)$$

By use of Eq. (28) and Eq. (11) the storage volume requirements can be determined as a function of the heat load and sink condition for a specified degree of source temperature control. Results of the analysis based on absolute temperature control ($\Delta T_s = 0$) are shown in Fig. 5 for methanol and ammonia. The following parameters were specified in determining these results:

(1) $T_s = 291°K$

(2) $T_{o_L} = 200°K$

(3) $T_{v_L} = T_s$ (i.e., $R_s Q_L \sim 0$)

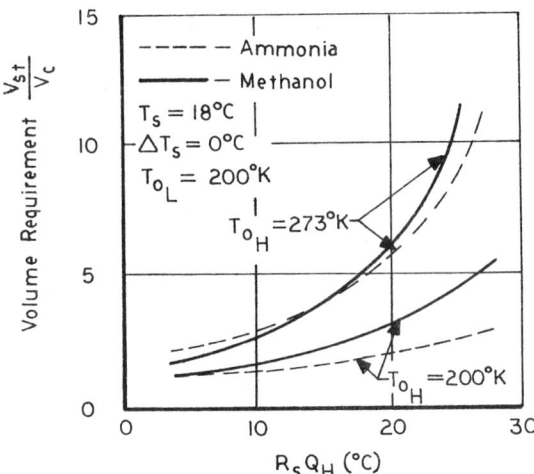

Fig. 5  Storage volume requirements for active system.

The following conclusions can be drawn from these results and Eq. (28):

1) Storage volume requirements are greater and increase more rapidly with increasing $R_s Q_{max}$ for a fluid with a low vapor pressure at the operating condition. This is because the percentage change in pressure is greater for a fluid with a low vapor pressure at the operating temperature. As a result, there is a larger change in the system's vapor pressure with changing heat load in order to maintain a specified degree of control. This results in a greater expansion of the noncondensible as it moves into the reservoir in going from the low to high power condition. Very simply,

$$\frac{(p_v - \pi_o)_L}{(p_v - \pi_{st})_H}$$

is greater for a low pressure fluid, hence the larger storage volume requirements.

2) The storage volume requirements increase as the maximum sink temperature increases. This is due partly to the fact that as the sink temperature increases the partial pressure of the working fluid in the reservoir increases therein reducing the partial volume available for the gas. In addition, the gas expands in the reservoir due to the increase in temperature.

## D. Comparison of Active and Passive Systems

In general, better temperature control will be obtained using an active feedback controlled heat pipe than with an equivalent passive system. Ideally, at the low power/low sink condition all of the gas will be in the condenser section in an active system. However, in a passive system, even though a plug is used as indicated in Fig. 2, there will be an appreciable amount of the noncondensible within the bellows when they are at their minimum extension. This excess gas must be accommodated when the bellows system expands to its maximum extension at the high power/high sink condition. Hence, storage requirements generally will be greater for a passive system which affords the same degree of temperature control as an active system, or, conversely, better temperature control can be obtained with an active system than with an equivalent passive system.

Also, a flexible wick which lines the bellows convolutions does not appear very practical. Consequently it is necessary to have a semipermeable plug between the condenser section and the bellows to prevent the working fluid from accumulating within the storage volume. The plug must be impervious to both the liquid and vapor but permeable to the noncondensible. This is just another complexity to an already complicated design. The design of an active system, on the other hand, is rather straightforward and makes this control method all the more attractive.

## IV. Experimental Electrical Feedback Controlled Heat Pipe

### A. Test System

An experimental model of an electrical feedback-controlled variable-conductance heat pipe has been tested to demonstrate its temperature control capability. A drawing of the model is shown in Fig. 6. The heat pipe contains an annular wick configuration. Several layers of 200-mesh screen are attached to the inner wall of storage volume. This screen is interconnected with the annular wick in a transition section between the condenser and the reservoir. Water is used as the working fluid and the noncondensible is argon. A heater wire wrapped around the length of the reservoir supplies the auxiliary power. The reservoir is wrapped with fiberglass insulation approximately 1.5-cm thick. An on/off controller with a $\pm 0.25°C$ deadband is used to regulate the auxiliary power and a thermistor is used as the feedback-temperature sensing element.

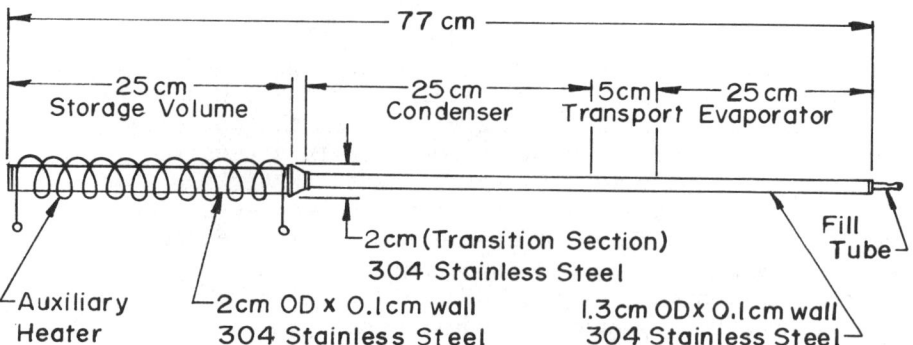

Fig. 6 Experimental model of electrical feedback controlled heat pipe.

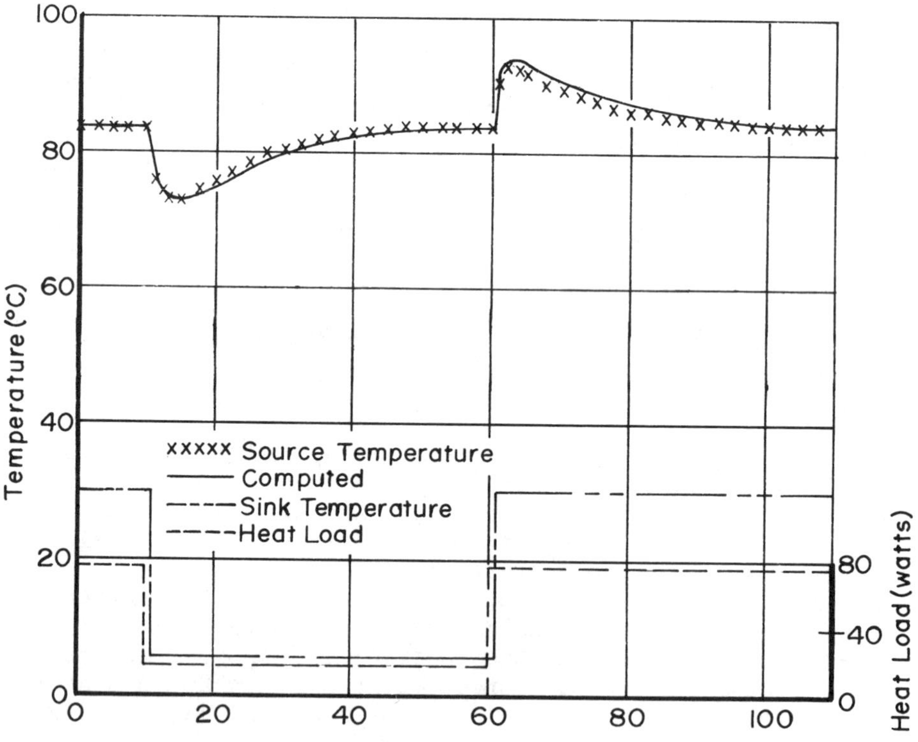

Fig. 7 Transient response of heat source with electrical feedback control.

Fiberglass tape was wrapped around the evaporator section of the heat pipe in order to increase the thermal resistance ($R_s$) between the heat source and the vapor. An aluminum cylinder, weighing 185 g and wrapped full length with a heater wire, is clamped around the heat pipe over the fiberglass tape to simulate a heat source. Two thermocouples and the control thermistor are attached to the outside diameter of the cylinder.

When conducting the tests, the heat-pipe system is inserted within a 5-cm o.d. copper tube surrounded by a water bath. Cooling of the heat pipe is affected by circulating water from the bath to copper fins brazed to the heat pipe along the 25-cm condenser section. This setup permits the storage reservoir and the heat pipe to see the same sink temperature and, at the same time, insulates the reservoir from the convective cooling of the condenser section, therein reducing auxiliary power requirements.

The nominal heat-source temperature control point had been chosen so that, for the particular gas charge, absolute control was achieved when the temperature of the storage reservoir at the high power/sink condition was essentially equal to the sink temperature. Similarly, at the low power/sink condition, the auxiliary power was just sufficient to achieve absolute control at the nominal source temperature. Thus, for a step change from high to low power and sink temperature, the auxiliary power was on throughout the entire transient. Conversely, the auxiliary power was off throughout the entire transient associated with the step decrease. This set of test conditions represents the limiting case since the total variation in reservoir temperature from high sink temperature to the system vapor temperature, or vice versa, must be realized in order to achieve control.

## B. Test Results

The transient response of the "simulated" heat source to simultaneous step changes from a low power/sink condition to a high power/sink condition and vice versa is shown in Fig. 7. Essentially, absolute control of the source temperature is achieved for the two step changes. This control was attained for variations in power ranging from 15 to 75 w and simultaneous variations in sink temperature from 5 to 30°C. The auxiliary power required to maintain the heat source at 84°C at the low power/sink condition was approximately 8.5 w.

The maximum overshoot of the heat-source temperature was 9°C, while the maximum undershoot was 11°C. The time for the heat-source temperature to settle to within 1°C of its final steady-state value was 29 min for the step increase and 36 min for the step decrease.

The difference between overshoot/undershoot and different recovery times for the two cases can be attributed to the effect of the vapor pressure on the response of the system. Although the response of the storage temperature is essentially identical for both cases, the initial change in vapor pressure in the storage volume is less in the case where the reservoir temperature begins to increase from the sink condition (i.e., step change from high power/sink to low power/sink). Consequently, the interface does not adjust as rapidly for the step change from high to low power/sink; therefore, the undershoot is greater than the overshoot which, in turn, leads to longer recovery times.

An optimum system is one for which the overshoot/undershoot and the response times are minimized along with the auxiliary power input. The most effective way of accomplishing this is to minimize the heat capacitance of the storage reservoir so that it will respond as rapidly as possible to the regulated auxiliary power. A detailed discussion of the transient performance of electrical feedback controlled heat pipes is given in Ref. 10.

The steady-state temperature distribution of the system is shown in Figs. 8 and 9. For the low power/sink condition, the gas-vapor interface is located at the beginning of the condenser section. The temperature of the storage volume is less than the vapor temperature at this condition, indicating that there is an overcharge of argon gas. The gradient across the condenser/reservoir transition region indicates conduction losses from the storage volume to the condenser section.

Except for a slight conduction effect from the condenser section to the storage volume, the reservoir is essentially at a uniform temperature less than 2°C above the sink temperature for the steady-state high power/sink case. For this condition, the active condenser length extends over approximately two-thirds of the total condenser.

The difference between the source temperature and the vapor temperature at the high power/sink condition is 10°C. Thus, the thermal impedance between the heat source and the

# FEEDBACK CONTROLLED VARIABLE CONDUCTANCE HEAT PIPES

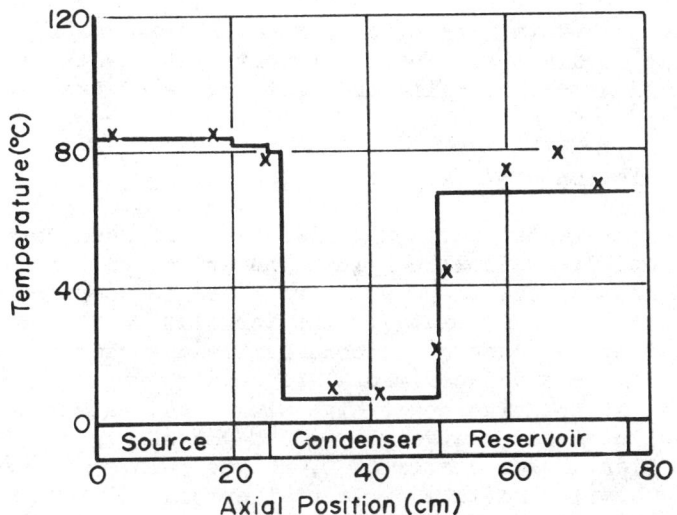

Fig. 8 Steady-state axial temperature distribution for low power/sink condition.

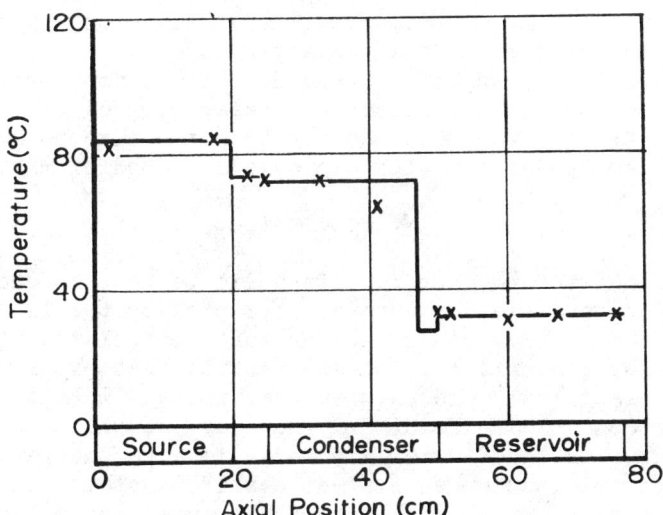

Fig. 9 Steady-state axial temperature distribution for high power/sink condition.

vapor ($R_s$) is 0.133°C/w. Consequently, an ideal variable-conductance heat pipe with no feedback ($V_{st} \to \infty$) would experience a ±5°C variation in source temperature under the same test conditions. The experiment system without feedback could have controlled the heat source at best to within ±8°C.

## C. Correlation of Test Data

The experimental data just discussed has been correlated using a one-dimensional nonlinear numerical solution. Calculated heat source temperature response is compared with experimental data in Fig. 7. The correlation was established by adjusting the mass of noncondensible gas charge such that the calculated and experimental reservoir temperatures are identical at the high power/high sink steady-state condition. The calculated gas charge was 0.0288 gms of argon compared to an experimental gas charge of 0.026 gms. This difference could be attributed to experimental error. Although a one-dimensional model and lumped elements are used to define the feedback controlled heat pipe system, very good correlation of the experimental data is obtained. This is due to the ability of the model to match the reservoir's temperature response.

Steady-state axial temperature distributions, which were calculated using the nonlinear analysis are compared with experimental data in Figs. 8 and 9. Except for conduction effects between the condenser and reservoir, close correlation of the data was obtained for both the high power/high sink and low power/low sink steady-state conditions.

## V. Conclusions

The analysis and test results indicate that a feedback controlled variable conductance heat pipe system is capable of much better heat source temperature control than that afforded by conventional thermal control heat pipes. Both active (electrical) and passive (mechanical) methods of feedback control are feasible. However, an electrical system has greater design flexibility and will give sharper temperature control than an equivalent passive system. Test results indicate that essentially absolute source temperature control can be achieved over a broad range of power and substantial changes in sink temperature with an electrical feedback controlled variable conductance heat pipe.

## References

[1] Ranken, W. A. and Kemme, J. E., "Survey of Los Alamos and Euratom Heat Pipe Investigations," *Proceedings of 1965 Thermionic Conversion Specialist Conference*, pp. 325-336.

[2] Feldman, K. T. and Whiting, G. H., "Application of the Heat Pipe," *Mechanical Engineering*, Vol. 90, Nov. 1968, pp. 48-53.

[3] Harbaugh, W. E. and Eastman, G. Y., "Experimental Evaluation of an Automatic Temperature Controlled Heat Pipe," *Proceedings of the Intersociety Energy Conversion and Engineering Conference*, Boulder, Colorado, August 1968.

[4] Sheppard, T. D., "The Heat Transistor, a Device for Thermal Control," The Bendix Corp., MT-14, p. 182.

[5] Marcus, B. D. and Fleishman, G. L., "Steady State and Transient Performance of Hot Reservoir Gas Controlled Heat Pipes," ASME 1970 Space Technology and Heat Transfer Conference, June 1970, Paper 70-HT/SPT-11 (this volume).

[6] Hinderman, J. D., Waters, E. D. and Kaser, R. V., "Design and Performance of Noncondensible Gas Controlled Heat Pipes," AIAA Sixth Thermophysics Conference, April 1971, Paper 71-420 (this volume).

[7] Edelstein, F. and Hembach, R. J., "The Design, Fabrication, and Testing of a Variable Conductance Heat Pipe for Equipment Thermal Control," AIAA Sixth Thermophysics Conference, April 1971, Paper 71-422.

[8] Bienert, W. "Heat Pipes for Temperature Control," Proceedings of Fourth Intersociety Energy Conversion and Engineering Conference, Washington, D. C., Sept. 1969, pp. 1033-1041.

[9] "Study to Evaluate the Feasibility of a Feedback Controlled Variable Conductance Heat Pipe," NASA CR-73475, Dynatherm Corporation, Sept. 1970.

[10] Bienert, W. B. and Brennan, P. J. "Transient Performance of Electrical Feedback Controlled Heat Pipes," SAE/ASME/AIAA Life Support and Environmental Control Conference, July 1971, Paper 71-Av-27.

DESIGN, FABRICATION, AND TESTING OF A VARIABLE CONDUCTANCE
HEAT PIPE FOR EQUIPMENT THERMAL CONTROL

F. Edelstein* and R. J. Hembach+

Grumman Aerospace Corporation, Bethpage, N. Y.

## Abstract

A variable conductance inert gas type heat pipe has been built to provide fine temperature control for spacecraft equipment. The pipe consists of a self-filling artery with a grooved wall capillary system that provides low evaporator-to-condenser temperature drops. Storage of the inert gas in a low temperature reservoir which communicates with the working fluid through the condenser eliminates the usual startup problems with these devices. Fabrication of the pipe emphasizes the importance of adequate cleanliness procedures. Latest test results are presented.

## I. Introduction

Current design practices for thermal control of electronic packages on board unmanned space satellites rely on thermal radiation between the heat source (electronics) and sink (space) using an intermediate radiating surface such as the vehicle outer skin. This arrangement is shown in Fig. 1a. Because of the fixed nature of the thermal coupling, wide ranges in equipment heat generation and/or environment heat loads will result in large equipment temperature variations, typically between 0 and $130°F$. When equipment or environment loads vary too widely, or when it is desired to maintain

---

Presented as Paper 71-422 at the AIAA 6th Thermophysics Conference, Tullahoma, Tennessee, April 26-28, 1971. The writers wish to express their thanks to Don Visceglie and Ed Leszak for their help in the machining and fabrication of the heat pipe, and to G. Knowles for design concept suggestions.

*Project Group Leader.
+Principal Investigator.

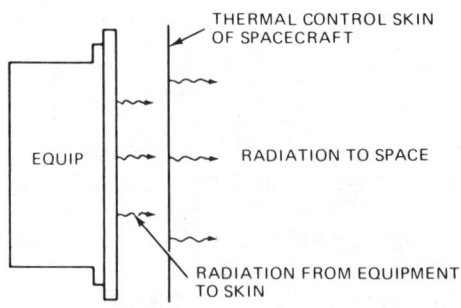
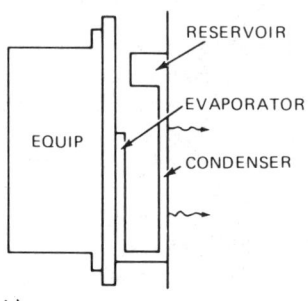

Fig. 1 Typical spacecraft equipment thermal control-plan view.

tighter temperature control, a variable conductance heat pipe (VCHP) may be used to provide near constant equipment temperatures. This is important because it assures improved reliability and reduced equipment cost for special temperature sensitive equipment. An illustration of this method of thermal control is shown in Fig. 1b. The VCHP employs a noncondensable gas, nitrogen, stored in a relatively cold reservoir to achieve the variable conductance feature.

Other means of achieving temperature control with heat pipes are covered in the literature,[1] as is the use of noncondensable gas for temperature control[2-5] and a description of the characteristics of a hot reservoir variable conductance heat pipe.[6]

The gas-controlled heat pipe has a number of advantages over existing mechanical devices (such as a louver), namely:
a) It is inherently simple in operation; since it contains no moving parts, it is a reliable and less expensive device;
b) The high conductance coupling between the equipment and sink which is typical of heat pipes allows the designer to

use less vehicle skin area for thermal control; and c) Turn-down ratios (or maximum to minimum load) are larger.

The design, fabrication, and performance results of a VCHP are presented in this paper. Application of a VCHP to provide temperature control for spacecraft electronic equipment was selected to evolve a practical design useful to thermal control engineers. A concurrent paper discusses the application of the VCHP to other areas of spacecraft thermal control, such as heat pipe radiators.[7]

## II. Design Requirements

As a basis for designing a variable conductance heat pipe, a number of design requirements were specified, not unlike those for typical electronic equipment on board the Grumman Orbiting Astronomical Observatory (OAO) satellite. The satellite[8] has an octagonal cross section 6-1/2 ft across flats; it is 10 ft high, and is designed for a 400-naut-mile Earth orbit. Of its 48 bays, approximately 75% contain electronic equipment. Environment heat loads can vary from direct solar radiation on one side of the spacecraft to planetary and reflected solar radiation on the other. Design requirements for integration of a VCHP in an equipment bay include:

a) Provision for equipment temperature control between approximately 70 and 80°F over the entire range of equipment heat generation and environmental load changes (current design techniques, Fig. 1a, provide temperature control between 0 and 130°F).

b) Heat pipe "shut-off" during low-load conditions to minimize survival heater power requirements.

c) Quick heat pipe response to step changes in load as well as startup after launch and long-term quiescence.

d) Heat pipe installed within existing spacecraft envelope without imposition of any new geometric or weight constraints. The distance between the evaporator (which attaches to the equipment) and the condenser (which attaches to the vehicle skin) is limited to approximately two inches.

e) Satisfactory operation of the heat pipe in zero-g and one-g field; the latter being important to insure valid ground testing.

f) Heat pipe reliability to provide the capability of operating under all mechanical loads, over the 2 yr life of the spacecraft.

### III. Design and Fabrication

Functional Components
---

The cold reservoir variable conductance heat pipe (VCHP) may be subdivided into five basic components as shown in Fig. 2. The components (in the direction of vapor flow) are the evaporator, a low-conductance section, the condenser, another low-conductance section, and the inert gas reservoir. When the pipe is closed (no load) the interface zone between the vapor and the inert gas lies at the end of Sect. (1) or in Sect. (2). When the pipe is open (full load) the interface zone is in Sect. (4). At partial loads, the interface will assume a position in the condenser Sect. (3) to maintain the evaporator temperature at a near constant value.

Each section of the VCHP has a specific function. The functions of the evaporator and condenser are self-explanatory, and are the same for any style of heat pipe. The first low-conductance section has a specific function of thermally isolating the evaporator and the equipment which is attached to it from the external sink or environment at no load conditions. It defines the low side of the turn down ratio for this device. The second low-conductance section is provided

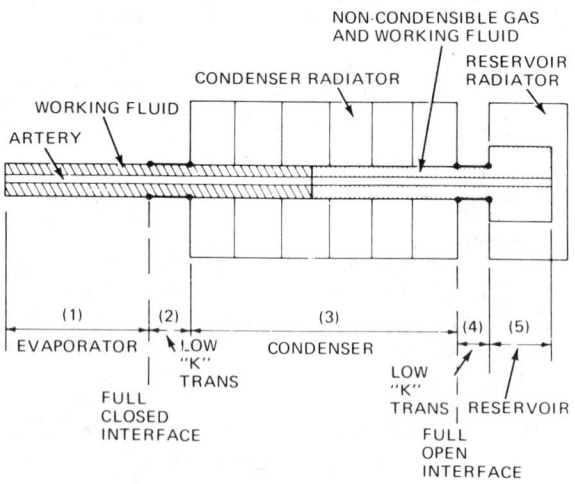

Fig. 2 Cold reservoir VCHP schematic with working fluid/gas interface within condenser.

to thermally isolate the inert gas reservoir from the condenser. It tends to minimize the reservoir temperature excursion for hot and cold case environmental conditions and permits use of a smaller volume reservoir. On the other hand, enough heat leakage into the reservoir must be provided to prevent freezing of the fluid in this section under cold case conditions. The function of the reservoir is to provide temperature control in the device. In particular, it determines the control tolerance of the heat pipe or the change in evaporator temperature that is required to move the interface from the beginning of the condenser (zero load) to the end of the condenser (full load). The reservoir normally contains inert gas as well as the liquid and vapor phases of the working fluid. The influence of the working fluid is diminished by maintaining the reservoir at temperatures below those of the evaporator, thereby resulting in reduced local vapor pressures. If the reservoir volume is large relative to the condenser, the evaporator temperature change is small. The reservoir may also be made to set the general temperature level of the device. Increasing the reservoir temperature or quantity of inert gas will raise the temperature level in the pipe with a very slight change in control tolerance.

Four fundamental requirements must be met within the heat pipe: a) wall capillary system in the evaporator, condenser and reservoir (not required in the low K sections) from which to evaporate and condense fluid; b) large capacity fluid return capillary system (or, in this case, the artery as indicated in Fig. 2); c) capillary communication system between the artery and every section of the capillary system (a); and d) vapor transport passage, continuous throughout the full length of the pipe.

Configuration and Construction

The foregoing functions and design fundamentals must be preserved in an actual application. One application for a cold reservoir pipe is shown in Fig. 3. This is an opened view of a characteristic equipment bay in OAO. The area of the condenser radiator is determined from its temperature and total radiant load requirement, with the latter being comprised of the maximum absorbed radiant flux (65 Btu/hr-ft$^2$) and the maximum electrical load (30 watts). The temperature of the condenser radiator is the same as that of the equipment (approximately 75°F), minus the following temperature drops:
a) conductance between equipment and evaporator wall;
b) evaporator wall to condenser wall (film coefficients); and
c) conductance between condenser wall and radiator.

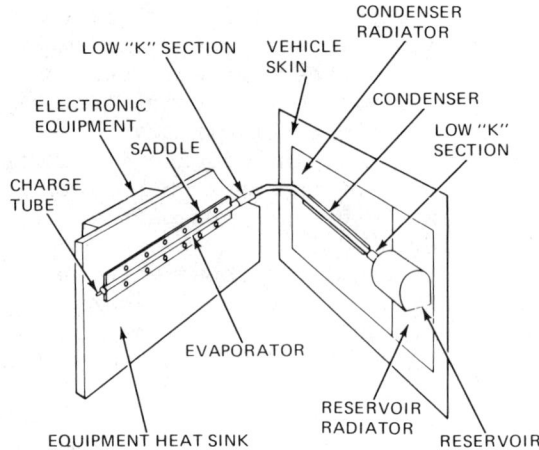

Fig. 3  Application of VCHP in equipment bay of orbiting astronomical observatory (OAO) shown open.

At a 30 w maximum equipment load, the temperature drop totaled approximately 5°F. The foregoing considerations fix the overall dimensions of the radiator panels as shown in Fig. 4. Evaporator length (18 in.) is matched with equipment length to minimize gradients. As mentioned before, the heat pipe design must also accommodate the remaining geometrical constraints of the spacecraft structure. The most difficult constraint was the 2 in. distance between the heat sink and skin panel planes (which are parallel). The distance limitation dictated the maximum size of the reservoir and, in addition, required that the artery capillary system (which must be continuous from the beginning of the evaporator to the end of the reservoir) be folded back on itself in a distance of two inches. Fig. 4 shows the resulting external configuration of the heat pipe.

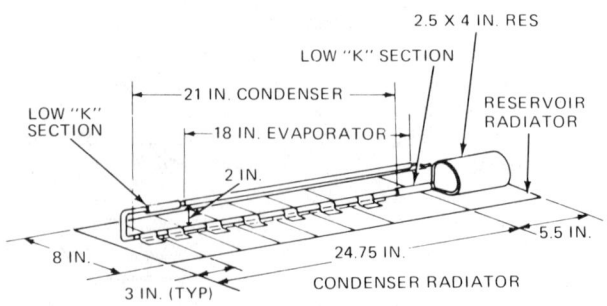

Fig. 4  Configuration of cold VCHP.

Note that the condenser radiator was constructed of 3 in. sections to aid in the experimental definition of the interface zone while it is traversing the condenser section.

The basic material selections were 6061 aluminum for all tubing, transitions and reservoir, and stainless steel mesh for the internal artery. The mode of assembly was the heliarc welding together of the five sections shown in Fig. 2 which were given real forms in Fig. 4. The assembly procedure is quite crucial with the last welding operations being performed after the "U" shaped artery is in place in the pipe. A special, cold helium-gas purge technique was developed to protect and cool the artery assembly during welding operations. All aluminum details not requiring welding were anodized when possible; items not anodized were alodined.

The internal configuration of the pipe must reflect the four fundamental requirements previously listed. The wall capillary in the condenser and evaporator sections was formed by a single-point tool, internal threading operation. The pitch in the evaporator section was 100 threads per inch, with a 20° included angle profile 0.007 in. deep. The pitch in the condenser section was 82 threads per in., with a 20° included angle profile 0.013 in. deep. The wall capillary system inside the reservoir was provided by furnace brazing 120-mesh aluminum screening to the 0.060 in. sheet material which was used to form the can and the end covers. A cylindrical loop of this same screening material was formed at the time of brazing to provide a means of capillary attachment to the artery. The details of the reservoir construction and capillary system are shown in Fig. 5. Departure from the ideal cylindrical shape of the reservoir was necessary to:

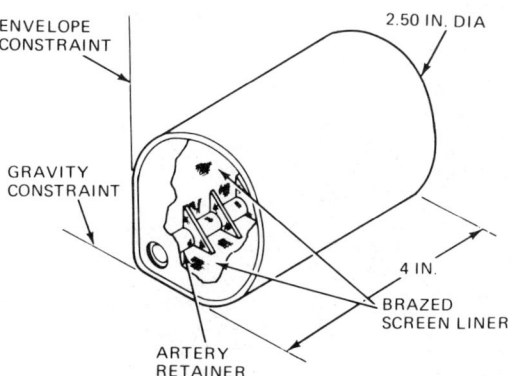

Fig. 5 Reservoir configuration.

a) allow ground testing without accumulation of excessive fluid in the reserovir; and b) minimize the penetration of the reservoir into the vehicle skin plane.

The large-capacity fluid return capillary system (or artery) is constructed of 100-mesh stainless steel spirally wound around a 1/16 in. solid rod with the space between wraps being set by 0.013 in. diam. stainless steel spacer wires. A cross section of the artery is shown in Fig. 6. Spacer wires are spot welded to the screen prior to the spiral winding of the artery. They provide the gap necessary to assure that the working fluid (ammonia) will "self-prime" or completely fill the artery in a one "g" field. This was visually demonstrated using a special glass tube table top experiment. The artery is encased in a 100-mesh stainless screen sleeve formed with a carefully soldered seam. (To realize the full pumping capacity of the mesh this outer sleeve must have no holes larger than the screen pore size). The artery is fabricated in three sections, the condenser-reservoir section, the evaporator section, and the transition section (all identical in cross section). The sections are mitered at 45° to form the 2 in. bend. The mitered ends are slipped into electroformed nickle elbows and held snugly in contact while the outer mesh sleeves of each section are soldered and sealed to the elbows. During this operation the double miter aluminum transition section of the pipe is slipped over the short section of the artery in preparation for final assembly. The mitered artery joints must be true within 0.013 in. (wire spacer size); this is checked by X-ray examination.

The capillary link between the artery and the evaporator wall capillary and the condenser wall capillary consists of 4 fiberglass strands, or legs 90° apart as indicated in Fig.

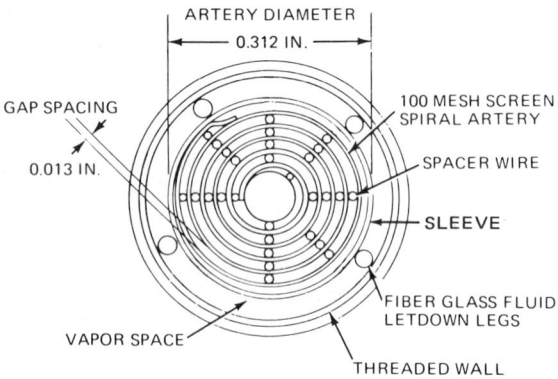

Fig. 6  Artery cross section.

6. Except for the two inch transition section the strands run the full length of the pipe from the beginning of the evaporator to the end of the reservoir. The OD of the strands exceed the gap (vapor space) between the artery sleeve OD and the pipe ID.

The low K section between the reservoir and condenser was formed using a neck down aluminum tube with an internal press-fit stainless steel sleeve. The other low K section was made using external fiberglass reinformcement layers in lieu of the steel sleeve. The attachment details between the condenser and radiator are shown in Fig. 7.

Cleaning procedures were as follows: a) wall capillary area: anodize, acetone pick, distilled water pick; b) stainless steel screen: passivation of raw materials, periodic reverse cleaning of screen assemblies, final degreasing; c) fiberglass: acetone soak, distilled water soak, and rinse; and d) final pipe assembly: $10^{-7}$ torr vacuum furnished by vac-ion pumping, with a 150°F bake-out of all sections for 24 hr.

The assembled pipe, after test instrumentation has been installed, as shown in Fig. 8 and 9.

IV. Design Considerations

Reservoir Sizing

The size of the reservoir is a function of the desired control band and reservoir temperature variation. It is determined by calculating the gas inventory at the hot and cold case operating conditions. For example, in the cold case it is assumed that the condenser is blocked and the interface is

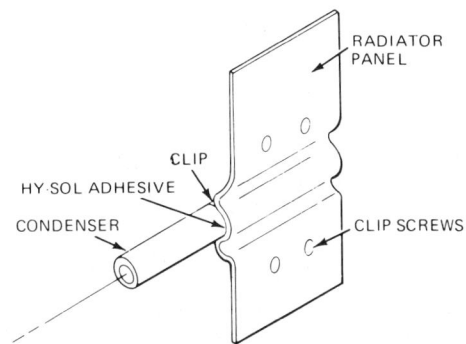

Fig. 7 Direct bonding of condenser to radiator panel.

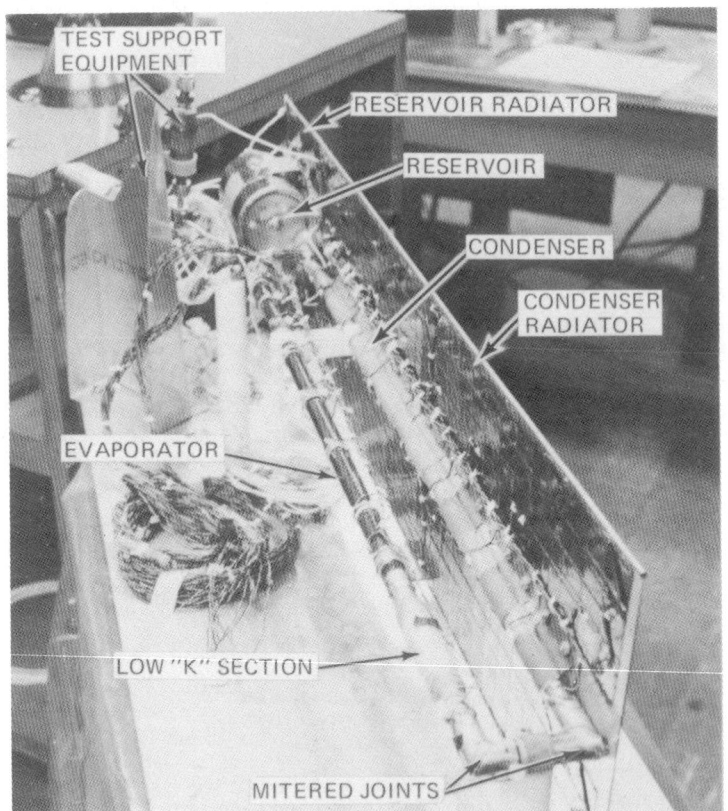

Fig. 8  Final VCHP assembly with test instrumentation.

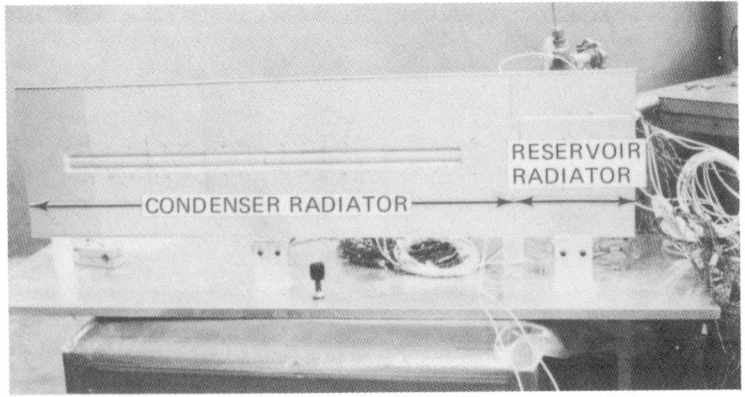

Fig. 9  Radiator view of assembled VCHP before painting.

## VCHP FOR EQUIPMENT THERMAL CONTROL

positioned between Sec. 1 and 2 in Fig. 2. The gas inventory is

$$M = \frac{1}{R} \sum_{i=2}^{5} \frac{P_i V_i}{T_i} \qquad (1)$$

In the hot case (denoted by prime) the heat pipe is fully operational with the interface positioned between Sec. 3 and 4. The same gas inventory is now given by

$$M = \frac{1}{R} \sum_{i=4}^{5} \frac{P'_i V_i}{T'_i} \qquad (2)$$

The partial pressure of the gas is the difference between the total pipe pressure and local vapor pressure of working fluid, i.e.,

$$P_i = P_{vl} - P_{vi} \qquad (3)$$

$$P'_i = P'_{vl} - P'_{vi} \qquad (4)$$

where: $M$ = mass of noncondensable gas, $R$ = gas constant, $P_i$ = partial pressure of gas in Sect. i, $P_{vi}$ = vapor pressure of working fluid in Sect. i, $P_{vl}$ = vapor pressure of working fluid in evaporator (i = 1), $V_i$ = total gas and vapor volume in Sect. i, $T_i$ = temperature in Sect. i (assumed isothermal), and

$$i \begin{cases} = 1 \text{ evaporator} \\ = 2 \text{ transition between evaporator and condenser} \\ = 3 \text{ condenser} \\ = 4 \text{ transition between condenser and reservoir} \\ = 5 \text{ reservoir.} \end{cases}$$

It has been assumed that the interface is sharp and the temperature is constant along a given section. In general, the temperature of each section will be known, thereby yielding the gas partial pressure from Eqs. 3 and 4. The solution of the reservoir volume, $V_5$, can be found from Eqs. 1 and 2.

$$V_5 = \frac{T_5 T'_5}{T_5 P'_5 - T'_5 P_5} \left[ \frac{P_2 V_2}{T_2} + \frac{P_3 V_3}{T_3} + \frac{P_4 V_4}{T_4} - \frac{P'_4 V_4}{T'_4} \right] \qquad (5)$$

The typical effect of evaporator control tolerance on reservoir size is shown in Fig. 10 for two working fluids, methanol and ammonia. Large control tolerances require small reservoir volumes which are not significantly affected by choice of fluid. Where tight temperature control is desired, choice of fluid is important to minimize reservoir size. The typical effect of variation of reservoir temperature on reservoir volume is shown in Fig. 11. It is seen that minimum volumes result when there is little or no change in reservoir temperature. As the variation in this temperature increases, large reservoir sizes are required particularly for ammonia which exhibits large changes in pressure for the temperature range of interest. Note that the characteristics shown in both Figs. 10 and 11 are typical since they depend on the absolute values of the evaporator and reservoir temperatures. The size of the reservoir selected for the engineering test model shown in Fig. 4 was maximized to provide the best possible evaporator temperature control for any given variation in reservoir temperature.

Passive and Active Control Modes

As previously mentioned, minimum reservoir volumes result when there is no change in reservoir temperature. However, in practice the reservoir temperature will vary or float as a function of spacecraft orientation and the ambient flux environment. This mode of operation (passive) results in larger volumes for the same control tolerance than the active mode where the reservoir, through a low-power thermostatically controlled heater, is held at constant temperature. This is

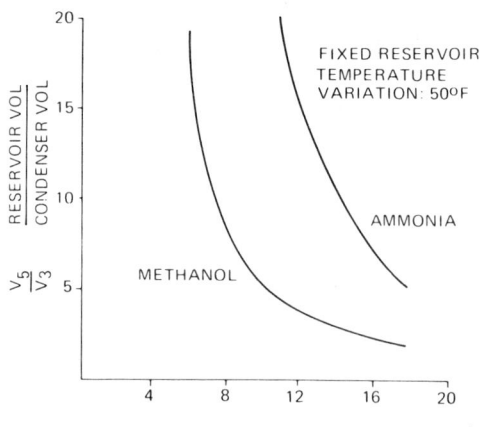

Fig. 10 Reservoir size vs control tolerance.

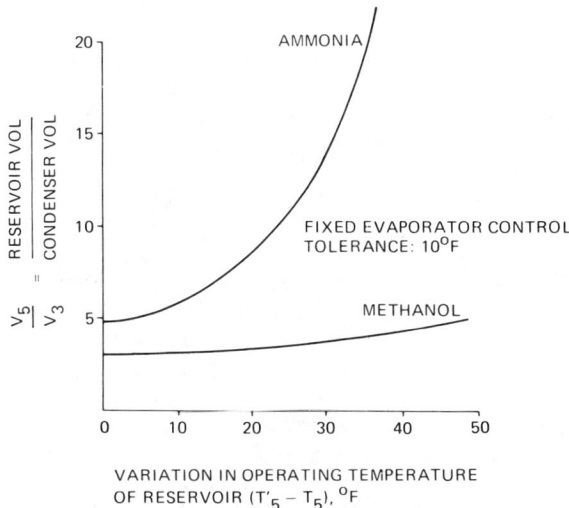

Fig. 11  Reservoir size vs reservoir temperature variation.

illustrated in Fig. 12 for ammonia. Where fine temperature control is desired in an ambient that has wide variations, the active control mode may be used. The tradeoff would involve heater power required to maintain constant temperature versus additional volume and weight of a larger reservoir volume.

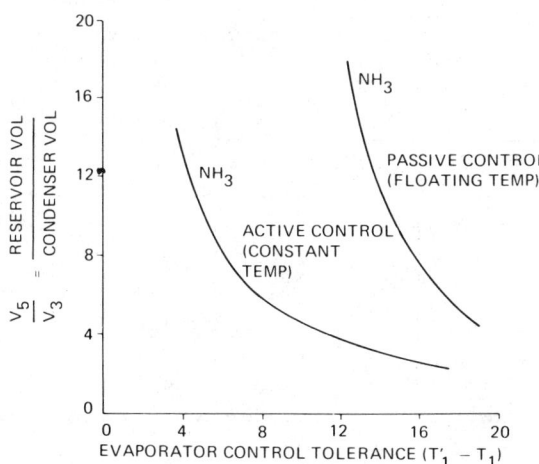

Fig. 12  Typical VCHP characteristic effects of reservoir temperature control on reservoir size.

Fluid Selection

Aside from chemical compatibility with other parts of the heat pipe, choice of working fluid depends on the features desired, such as temperature control, heat transport capacity, gravity sensitivity, operating environment, toxicity, etc. Four working fluids were evaluated for this VCHP configuration relative to some of these features. The results of this evaluation are given in Table 1.

Evaporator temperature control is based on a fixed reservoir volume relative to the condenser of 18.0. The passive control range is based on a given reservoir temperature variation of 50°F, while the active control band assumes a constant reservoir temperature at -43°F. Fluid pumping capacity was calculated from liquid and vapor viscous pressure losses in the configuration shown in Fig. 6.

Methanol provides the best temperature control characteristics; ammonia yields the highest heat transport capacity, and the best gravity characteristics. The gravity characteristic is important because it allows ground testing of the VCHP in orientations that do not have to be perfectly level. Note also that in the active control mode, evaporator temperature control is essentially constant and relatively insensitive to working fluid. Ammonia was selected as the working fluid because of its compatibility with the heat pipe materials as well as its superior heat transport capacity.

Table 1  VCHP working fluid evaluation

| Fluid | Evaporator temperature control, °F | | Max. fluid pumping capacity, w | Ability to act against gravity $a(\sigma/\rho)$ | Freezing Point °F |
|---|---|---|---|---|---|
| | Passive | Active | | | |
| Methanol | 72 - 78 | 74.0 - 76.0 | 71.6 | 1.0 | -144 |
| Ammonia | 69 - 81 | 73.3 - 76.7 | 161.4 | 1.4 | -104 |
| Acetone | 71 - 79 | 73.7 - 76.3 | 75.3 | 1.1 | -138 |
| Freon-21 | 70 - 80 | 73.5 - 76.5 | 35.6 | 0.5 | -211 |

[a] Surface tension divided by density relative to methanol.

## V. Testing

### Test Setup

The cold reservoir variable conductance heat pipe was instrumented with 48 thermocouples, plus a condenser radiator heater to simulate orbital flux loads; a reservoir radiator heater with the same function; a reservoir heater to demonstrate the active control mode feature; and an evaporator heater to simulate electronic box heat loads. A guard heater was added to the charging tube valve line to null the tube heat leak during test. In addition, a pressure transducer was installed at the evaporator end of the heat pipe. The pipe assembly, shown mounted on the testing plate in Figs. 8 and 9, was insulated with 25 layers of aluminized mylar in three basic isolated sections: the evaporator, condenser, and reservoir. The vacuum chamber selected for the test was 2 ft in diameter and 4 ft long and was connected to a 4 in. diffusion pumping system.

### Results

Figure 13 shows the steady-state temperature profile along the pipe for different heat loads applied to the evaporator. The condenser and reservoir radiators ($\epsilon = 0.85$) were heated to simulate an absorbed ambient heat flux of 65 Btu/hr-ft$^2$, or an equivalent sink temperature of 0°F. This corresponds to full solar impingement (440 Btu/hr-ft$^2$) on a panel whose absorptivity is 0.15. The reservoir temperature was allowed to float passively and remained relatively cold, even at an evaporator power level of 25 w due to the low conductivity section between it and the condenser. At the low-power level of 3 w the pipe was essentially shut off with the interface lying between the evaporator and condenser. The 3 w represent the sum of conduction losses from the evaporator to the condenser and radiation leakage through the insulation surrounding the evaporator. At 25 w the condenser was still partially blocked and not quite at its full load capacity. It is seen that at this power level the temperature drop between the evaporator and condenser wall was approximately 2°F. Also, the evaporator control range was between 66°F and 75°F for the power range of 3 - 25 w.

The effect of reservoir temperature on evaporator control range is shown in Fig. 14. The data from Fig. 13 is plotted along with data obtained at a higher reservoir temperature of 21°F - 26°F. It is seen that the evaporator temperature control range is unchanged but merely shifts upward by about 10°F when the reservoir temperature increases by approximately

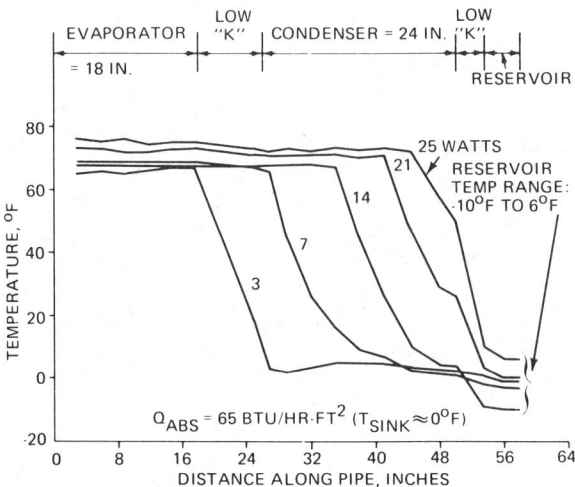

Fig. 13 Steady-state temperature profile along pipe - test results.

25°F. This effect is non-linear, due to the non-linear vapor partial pressure/temperature relation for ammonia. At a lower absolute reservoir temperature the same 25°F increase would result in a lower than 10°F increase in evaporator temperature. The dashed lines in Fig. 14 represent the predicted range between evaporator temperature at pipe full open and full closed positions, based on the reservoir size and gas weight introduced into the pipe. It is seen that these predictions were approximately 5°F lower than the data. This can be due to two reasons: inaccuracies in the amount of gas introduced, and the presence of a diffusion zone (instead of the sharp interface assumed in the calculations) which causes the pipe to behave as though it had more noncondensable gas than actually present. Note that the evaporator temperature can be controlled to remain absolutely constant or even decrease as power increases by adding a control scheme that will make the reservoir temperature decrease with increasing evaporator loads.

The turn down ratio, or maximum to minimum power at a relatively constant evaporator temperature, is probably higher than the 10 to 1 ratio indicated in Fig. 14. This is because the low-power insulation losses from the evaporator would be smaller in a real application.

During changes in evaporator power settings, it was noticed that the pipe achieved approximately 90% of its steady-state value within 3 min. after the change. Start up from a cold start with no power on the evaporator took approximately 15

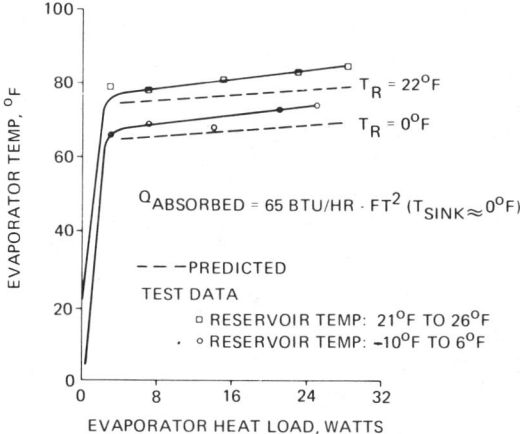

Fig. 14  Evaporator temperature control.

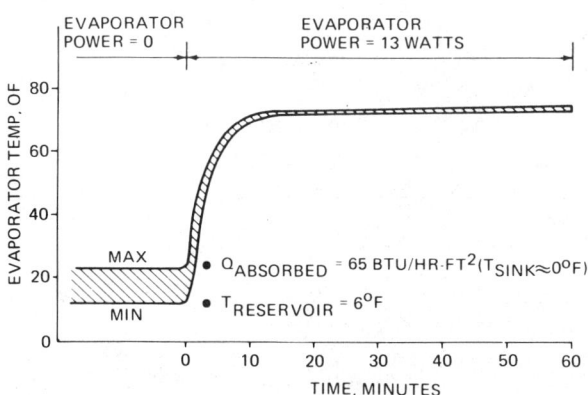

Fig. 15  Transient response - step change in load.

min. to achieve steady-state. This is seen in Fig. 15 where at zero time 13 w was applied to the evaporator. Previous to this time the heat pipe was unpowered (for a number of days) allowing the nitrogen to diffuse into the evaporator. Startup of this pipe is relatively fast compared to the long startup times with hot (unwicked) reservoir heat pipes.[6]

Remaining test effort will include: a) performance map over wide range of ambient flux conditions; b) diode characteristics of pipe; and c) characteristics of other working fluids.

## VI.  Conclusions

The following conclusions can be drawn from the work described in this paper: 1) A cold reservoir variable conductance heat pipe has been built that fulfills the design requirements for spacecraft electronic cooling, previously outlined; 2) depending primarily on ambient flux variations, active or passive control modes can be employed to provide equipment temperature control to $\pm 5°F$ or better; 3) the artery heat pipe can be mitered and spliced to allow installation in tight configurations; 4) the use of a low-conductivity section is effective in thermally isolating the reservoir from the condenser during conditions when the condenser is fully operational. This section also isolates the evaporator and condenser thereby minimizing losses when the pipe is shut off. Turn-down ratios of better than 10 to 1 can be achieved; 5) the pipe responds quickly, within minutes, to step changes in load as well as startup after long term quiescence and 6) interface predictions, using the sharp interface assumption, are generally adequate for engineering applications.

## References

[1] Shlosinger, A.P., "Heat Pipe Devices for Space Suit Temperature Control," CR-1400, 1968, NASA.

[2] Katzoff, S., "Heat Pipes and Vapor Chambers for Thermal Control of Spacecraft," AIAA Progress in Aeronautics and Astronautics: Thermophysics of Spacecraft and Planetary Bodies, Vol. 20, edited by G. B. Heller, Academic Press, New York, 1967, pp. 761-818.

[3] Eastman, G. Y., "The Heat Pipe - A Progress Report," RCA Corp., Lancaster, Pennsylvania.

[4] Turner, R. C., "The Constant Temperature Heat Pipe - A Unique Device for the Thermal Control of Spacecraft Components," AIAA Paper 69-632, San Francisco, Calif., 1969.

[5] Bienert, W., "Heat Pipes for Temperature Control," Proceedings: Fourth Intersociety Energy Conversion Engineering Conference, Sept. 1969.

[6] Marcus, B. D. and Fleischman, G. L., "Steady State and Transient Performance of Hot Reservoir Gas Controlled Heat Pipes," ASME Paper 70-HT/SpT-11.

[7] Roukis, J., Rogovin, J., and Swerdling, B., "Heat Pipe Applications to Space Vehicles," AIAA Paper 71-410, Tullahoma, Tenn., April 1971.

[8] Hemmerdinger, L. H., "Thermal Design of the Orbiting Astronomical Observatory," Journal of Spacecraft and Rockets, Vol. 1, No. 5, Sept.-Oct. 1964, pp. 477-483.

A VARIABLE CONDUCTANCE HEAT-PIPE
FLIGHT EXPERIMENT

J. P. Kirkpatrick[*]
NASA Ames Research Center, Moffett Field, Calif.

and

B. D. Marcus[†]
TRW Systems Group, Redondo Beach, Calif.

Abstract

A gas-controlled, variable conductance heat pipe, designated the Ames Heat Pipe Experiment (AHPE), has been qualified for flight aboard the Orbiting Astronomical Observatory (OAO-C) scheduled for launch in May 1972. The AHPE will provide temperature stability for the spacecraft's On-Board Processor (OBP) by maintaining the OBP platform/AHPE interface at 63 ±5°F for large variations in power dissipation and incident energy. The paper discusses the thermal boundary conditions imposed by the OAO-C spacecraft which made the AHPE a particularly difficult design problem; the selection of a "hot," nonwicked reservoir for the containment of the noncondensing control gas (nitrogen); the effect of mass diffusion on the reservoir design, fluid selection (methanol), and predicted thermal performance; and the influence of axial conductivity on radiator design. Also presented is the flight qualification and acceptance testing program which caused no significant change in AHPE performance. The results of

---

Presented as Paper 71-411 at the AIAA 6th Thermophysics Conference, Tullahoma, Tenn., April 26-28, 1971.
Work under NASA Contract No. NAS2-5503.
The authors wish to thank Stanford Ollendorf of the NASA Goddard Space Flight Center, whose active support made possible the OAO-C flight opportunity, and George Fleischman of TRW Systems Group for his contributions throughout the analytical and experimental phases of the AHPE development.
[*]Research Scientist.
[†]Manager, Heat-Pipe Projects, Material Science Department.

thermal performance tests conducted under simulated flight conditions are discussed to demonstrate that the predicted and actual thermal performance compare favorably. Although ultimate confidence must await flight proven performance, it is concluded that the behavior of gas-controlled heat pipes appears well enough understood, and that sufficient analytical design tools are available, to begin utilizing this technology effectively in spacecraft thermal control applications.

## Introduction

The development of variable conductance heat-pipe technology is becoming one of the most fascinating and potentially useful areas of heat-pipe endeavor. Several investigators[1-8] have identified methods by which a heat pipe can "adjust" its effective conductivity in response to heat-transfer requirements. To explore these methods in greater detail and to determine their usefulness in specific engineering applications, a technology-development program was initiated. As part of this program, a gas-controlled heat pipe has been built and qualified for flight aboard the Orbiting Astronomical Observatory (OAO-C) scheduled for launch in March 1972. This flight opportunity will demonstrate in a specific application the effectiveness of a variable conductance heat pipe in providing temperature stability for spacecraft equipment which experiences varying electronic duty cycles and changing thermal boundary conditions. This paper will discuss the thermal boundary conditions imposed by the OAO-C flight opportunity; the trade-off studies resulting in the selection of a "hot," nonwicked gas reservoir; detailed design considerations; flight-qualification and acceptance-testing programs; and finally, the thermal performance of the flight hardware during ground testing.

## OAO-C Flight Opportunity

The purpose of the variable conductance heat pipe/radiator, officially entitled "The Ames Heat-Pipe Experiment" (AHPE), is to provide temperature control for the OAO-C spacecraft's On-Board Processor (OBP) by regulating the heat transfer from the back of the OBP honeycomb equipment shelf to space (Fig. 1). Power dissipation from the OBP varies from about 10 to 30 w, and the energy incident on the Alzak-coated radiator ($\alpha/\epsilon$ = 0.17/0.75) varies as shown in Table 1. The radiator receives no direct insolation and the large amount of incident infrared energy is emitted from a nearby solar cell panel. It will be shown later that this large infrared flux is a dominant factor in the AHPE design.

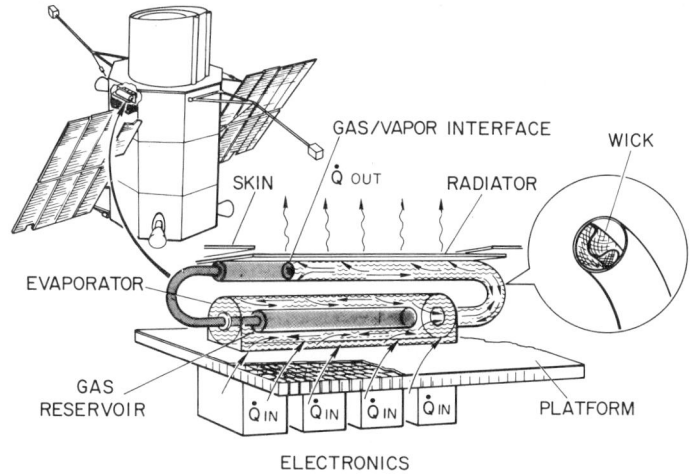

Fig. 1 Ames heat-pipe experiment (AHPE) on OAO-C.

Table 1 Incident fluxes on AHPE radiator[a]

|  | Solar | Infrared | | |
|---|---|---|---|---|
|  | Albedo | Earth | Panel | Total IR |
| MAX | 15.96 | 16.75 | 43.23 | 59.98 |
| MIN | 7.22 | 10.24 | 14.87 | 25.11 |

[a]Orbital average (Btu/hr-ft$^2$).

Without the AHPE, the conventional use of thermostatically controlled heaters and radiative coupling to space would result in an OBP platform temperature fluctuation from 0° to 140°F. A major experiment constraint was that, for any AHPE failure mode, the temperature of the OBP would not exceed these limits. This required a radiative heat-transfer path parallel to the AHPE which, at 30-w dissipation on the platform, allows only 22 w to be conducted through the heat pipe. Six watts are radiated directly to the radiator, and 2 w are radiated to the surrounding walls. Therefore, an AHPE performance goal was established to maintain the pipe's mating surface with the OBP platform at a nominal 65 ±5°F, for changes from minimum to maximum incident fluxes, and for power variations through the heat pipe up to a maximum of 22 w. Special concern for the minimum power through the heat pipe was not warranted because of the large amount of heat (about 8 w) being lost through the parallel radiative coupling.

Additional constraints were 1) the available volumetric envelope of 28 x 16 x 3 1/2 in.; 2) a requirement for meaningful testing in Earth's gravitational field; and 3) a scheduled delivery 11 months after contract go-ahead.

Variable Conductance Technique

Liquid-flow control, vapor-flow control, condenser blockage by excess liquid, and condenser blockage by noncondensing gas were evaluated as potential variable conductance techniques for this application. The details of this evaluation are beyond the scope of this paper and will be reported elsewhere. However, the results indicated that the most promising scheme for the AHPE was to employ noncondensable gas control.

The use of a noncondensing gas to vary the effective condenser area of a heat pipe in order to vary its conductance has been described previously in the literature.[4,5,8] Basically, the technique involves the addition of a fixed quantity of gas which does not condense at the lowest temperatures experienced by the heat pipe. The gas, since it cannot be wicked back to the evaporator region, is swept to the end of the condenser and any reservoir volume that has been provided, forming a gas plug (Fig. 2). This gas plug acts as a diffusion barrier to the flowing vapor and effectively "shuts off" that portion of the condenser which it fills. Consequently, variation in the length of this gas plug in response to changes in evaporator and/or

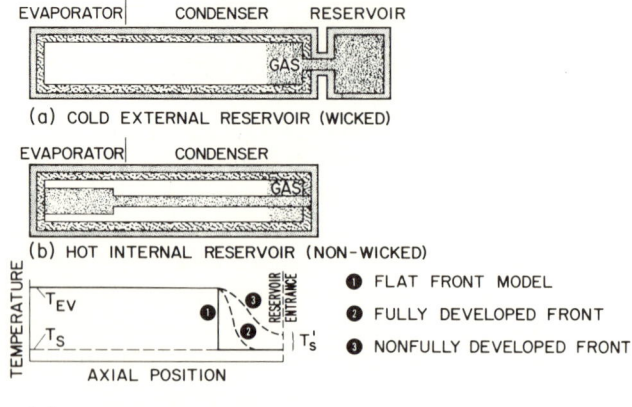

Fig. 2 Cold and hot reservoir heat pipes.

reservoir temperature represents a variation in active condenser length and, hence, in heat transfer from the system.

## Design Considerations

### Control Range

<u>Preliminary control analysis.</u> The AHPE performance goals call for rather close control of the evaporator temperature (±5°F) where the thermal environment both inside and outside of the spacecraft varies substantially. Because the operating temperature of a gas-controlled heat pipe varies with reservoir temperature,[4,8] and because there was no constant-temperature position in the OBP bay at which to mount the reservoir, a design analysis was performed for the two locations where its temperature could be determined. In one case (cold reservoir), the reservoir is located at the end of the condenser, so that its temperature depends on the effective space temperature ($T_s$) and fluctuates with variations in thermal environment (Fig. 2a). The second case (hot reservoir) places the reservoir inside the evaporator, so that its temperature range corresponds to the heat pipe's control range (Fig. 2b).

There exists a fundamental difference in these two approaches. The cold external reservoir must be wicked, or else vapor diffusing through the gas will condense in the reservoir and be lost to the wicking system. The partial pressure of vapor in the reservoir then will be the vapor pressure corresponding to its temperature. On the other hand, the hot internal reservoir must not be wicked, for its vapor pressure would then be equal to that in the evaporator (i.e., the total pressure) and it could contain no gas. Without wicking, the partial pressure of vapor in the reservoir is established by diffusion to and from the reservoir entrance (e.g., the end of the condenser) and hence, at steady-state conditions, corresponds to the temperature at this point ($T_s'$, Fig. 2c).

The basic principle in designing a gas reservoir for a desired control range is that the molar gas inventory in a given heat pipe remains constant for all operating conditions. Assuming an ideal gas mixture, the molar inventory for an element of pipe volume is simply

$$dn = (P_g/R_u T_g) dV \qquad (1)$$

where dn = number of moles of gas in the volume element dV, $P_g$ = partial pressure of gas in dV, $T_g$ = temperature of gas in dV, and $R_u$ = universal gas constant. Thus, for any given operating condition, one can obtain a value for the total molar inventory in the heat pipe by integrating Eq. (1) over its volume. To size the gas reservoir, such expressions are written for the two operating extremes; i.e., the full-on condenser at maximum thermal boundary conditions and the full-off condenser at minimum thermal boundary conditions, without specifying the reservoir volume. These expressions are then solved simultaneously for the molar inventory and reservoir volume, other pipe parameters being specified.

By using this approach, a parametric analysis was performed for preliminary AHPE specifications, comparing cold external vs hot internal reservoirs for ammonia and methanol working fluids. The calculations were performed for several possible combinations of maximum and minimum radiator effective space temperature ($T_s$) that might be obtained using various coatings on the back of the radiator. The results are shown in Fig. 3 for the cases where the back of the radiator is painted black (-60°F < $T_s$ < -2°F) and aluminized (-107°F < $T_s$ < -19°F).

The curves, which represent the required reservoir-to-condenser volume ratio ($V_r/V_c$) to achieve a particular evaporator control range for the specified conditions, show several interesting features: 1) For either working fluid, the internal hot reservoir allows much closer control than the external cold reservoir, and 2) for either reservoir design, methanol allows much closer control than does ammonia.

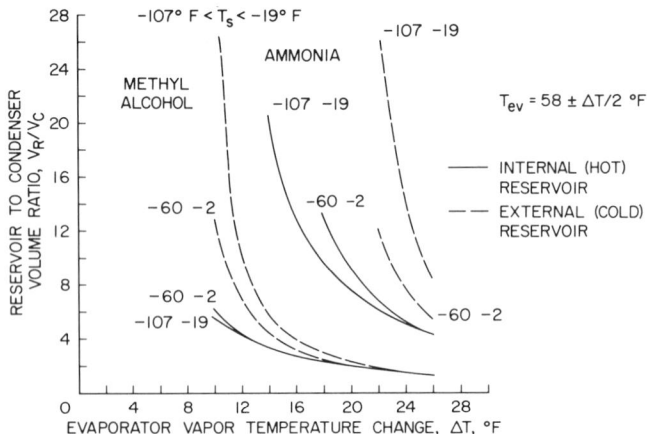

Fig. 3 Volume requirements for hot and cold reservoirs.

This behavior becomes clear when one considers the way in which sink-temperature variations affect control. As shown by Marcus and Fleischman,[8] the heat-pipe operating temperature is affected by changes in the reservoir-gas temperature and by variations in the partial pressure of vapor in the reservoir compared with the total pressure in the system. Thus, since the hot reservoir design minimizes reservoir-gas temperature fluctuations, it offers superior control. Also, for a given design approach, methanol offers better control because the change in vapor pressure over the specified ranges in effective space conditions is smaller compared with the total system pressure than for ammonia.

The results of Fig. 3 indicated that it was not practical to obtain the desired ±5°F control range ($\Delta T = 10°F$) using ammonia, the preferred fluid from a hydrodynamic point of view. Furthermore, even using methanol (the second best hydrodynamic fluid), it appeared that the only practical approach to achieving the desired control range was to use a hot reservoir design. Thus, a hot, internal reservoir pipe with methanol as the working fluid was selected for the AHPE.

Final control analysis. The preliminary analysis previously discussed was performed using the "flat-front" model for gas-loaded heat pipes. This model assumes that the interface between the active and inactive portions of the condenser is very sharp and that axial conduction in the pipe wall and radiator is negligible (Fig. 2c). Thus, the analysis presumes that the temperature in the "shut-off" portion of the condenser is everywhere equal to the effective sink temperature.

But it has been shown[8] that axial conduction is not negligible and leads to considerable spreading of the vapor-gas front. If, under conditions of high condenser utilization, this causes the temperature at the end of the condenser to rise above the sink temperature ($T_s' > T_s$), the partial pressure of the vapor in a hot reservoir increases, causing an increase in operating temperature of the pipe and a widening of the control range.

Since these effects are quantitative rather than qualitative, the simple flat-front model does permit useful preliminary design analyses and trade-offs as described in the last section. However, to perform the final design, it was necessary to treat the problem more rigorously. This was especially true for the AHPE in that, because of envelope constraints and active radiator-area requirements, only about 3 in. at the end of the condenser were available to develop the vapor-gas front and drop the temperature at the reservoir entrance ($T_s'$) low enough to minimize the partial pressure of vapor in the hot reservoir. To accomplish this, an analysis was formulated, based on a one-dimensional model which included 1) radiation to and from the finned condenser, 2) axial conduction in the walls, fins, and wicks, 3) binary mass diffusion between the vapor and gas, and 4) an approximate treatment of wick resistance which is accurate for high conductance wicks.

The governing equations were programed for numerical solution on a digital computer. This analysis and numerical solution are beyond the scope of this paper and will be reported elsewhere. However, the results of its application to the AHPE design are discussed below.

By studying the problem parametrically, it was found that the key variable affecting the length of the gas front, which was not constrained by other design considerations, was the axial conductance ($k_{eff}$) of the condenser tube and radiator fin. The effect of axial conductance on the calculated performance of the AHPE is shown in Fig. 4. This graph shows the variation of two parameters as a function of the effective axial thermal conductivity (total axial conductance referenced to the cross-sectional tube wall area). The left-hand ordinate represents the equilibrium temperature ($T_s'$) at the entrance to the reservoir (end of the condenser) for full power at maximum boundary

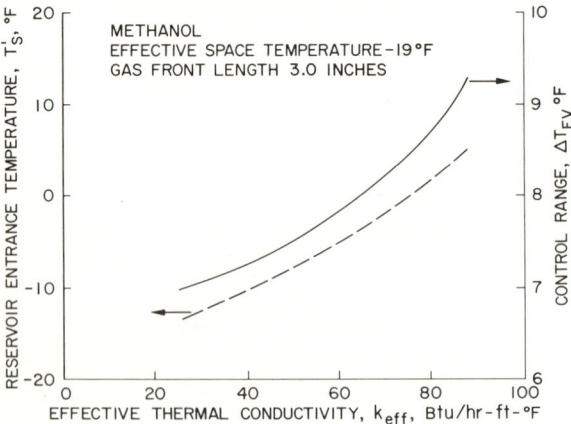

Fig. 4 Effect of axial conductivity on control range and reservoir-entrance temperature.

conditions. Thus, one sees that, even for relatively small values of $k_{eff}$, axial conduction causes the temperature at this point to rise above sink conditions; i.e., the front does not fully develop in 3 in.

As stated previously, the effect of this on hot reservoir pipes is to increase the partial pressure of vapor in the reservoir and widen the control range. This is clearly seen on the right-hand ordinate of the curve, which represents the variation in evaporator vapor temperature ($\Delta T_{EV}$) between operating extremes of the heat pipe. As axial conductance increases, $T_s'$ increases, which results in a broader control range ($\Delta T_{EV}$).

These calculations led to the conclusion that relatively small values of $k_{eff}$ were required to achieve the desired control range. To accomplish this it was necessary to segment the radiator by contructing it of individual fins so that its conductance was anisotropic. That is, it had a high conductance, perpendicular to the condenser tube to yield a high radiator effectiveness but a low conductance in the axial direction.

A nonsegmented radiator of the size used would have a $k_{eff}$ on the order of 2000 Btu/hr-ft-°F. By designing the last 3 in. of the radiator with 0.5-in. fins and 0.150-in. gaps at their roots, it was possible to reduce $k_{eff}$ to 41.3 Btu/hr-ft-°F in this critical region and establish an anticipated control range of 7.3°F. This range, however, is that of the evaporator vapor temperature. To it must be added the range of temperature drop into the evaporator (1.3°F). Thus, the total predicted variation in the saddle interface temperature was 8.6°, slightly better than the design goal of 10°F.

Diffusion-Controlled Transients

The selection of a hot, nonwicked reservoir involved a trade-off in terms of transient performance. Because the partial pressure of vapor within the reservoir is established by diffusion of vapor to and from the reservoir entrance, any changes in this parameter occur relatively slowly. With the understanding previously gained, it was possible to design the AHPE with a diffusion time constant of only about 2 hr by minimizing the length and maximizing the diameter of the reservoir feed tube. Experiments on the prototype unit, in which the heat pipe was overdriven (forcing vapor into the reservoir) and allowed to recover, substantiated the predicted transient response.

## Start-up with Liquid in the Reservoir

It has been demonstrated experimentally that the presence of liquid in the gas reservoir of a hot reservoir heat pipe gives rise to high pressure and temperature transients when the pipe is started. This is caused by the liquid in the hot reservoir vaporizing and displacing the gas. An upper bound for this phenomenon occurs when all the gas is forced into the condenser. A quantitative estimate for this condition can be obtained using the "flat-front" analysis discussed in Ref. 8. The results of such an analysis for the AHPE design are shown in Fig. 5. Evaporator temperature and pressure are plotted as functions of the input power to the heat pipe for start-up under maximum ($T_S = -2°F$) and minimum ($T_S = -60°F$) effective space temperatures. It is apparent from Fig. 5 that a transient overpressure of as much as 11 times the nominal design pressure (1.95 psia at 70°F) could result from a start-up with liquid in the reservoir. In the AHPE, these pressures pose no problem. However, this phenomenon would certainly be troublesome with an ammonia heat pipe, in which the vapor pressure would be 129 psia at 70°F.

Any liquid in the reservoir will slowly diffuse out due to the concentration gradient of vapor between the hot reservoir and its cold entrance. Thus, the heat pipe automatically will "rectify" itself; however, the rectification process occurs by diffusion and is relatively slow. Thus, to minimize such effects on the AHPE, the design included

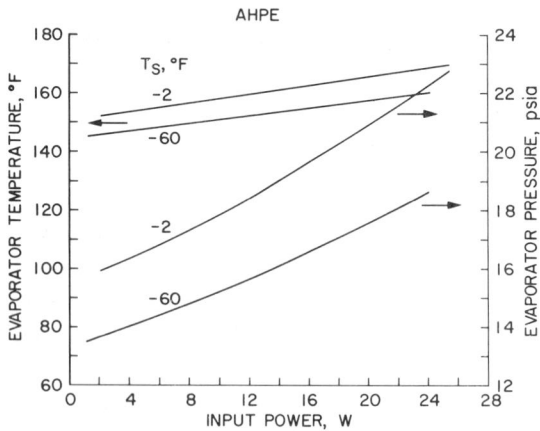

Fig. 5 Maximum evaporator pressure and temperature with liquid in the reservoir.

a perforated Teflon plug blocking the entrance to the reservoir, which serves to impede liquid from entering the reservoir while permitting the gas to pass freely.

Hydrodynamic and Thermal Design

The hydrodynamic and thermal design of the AHPE evolved through multiple iterations of the pertinent factors, resulting in the hardware configuration shown schematically in Fig. 6. The major design features and the considerations leading to their selection are as follows:

Methanol was selected as the working fluid from a control point of view, even though it is hydrodynamically inferior to ammonia at the operating temperatures involved.

The heat pipe had to operate in a 1-g as well as 0-g field to permit testing. Since the optimum wick structure for a heat pipe is a function of the g-field in which it operates, it was necessary to compromise 0-g capacity to provide sufficient 1-g performance.

The dimensional constraints on the available volumetric envelope required that the pipe be bent on a tight radius. This, and the desire to achieve a large reservoir-to-condenser volume ratio ($\sim$ 10:1) with a reasonable reservoir volume, dictated a relatively small-diameter condenser tube (7/16 in. o.d.). The consequent limitation on condenser

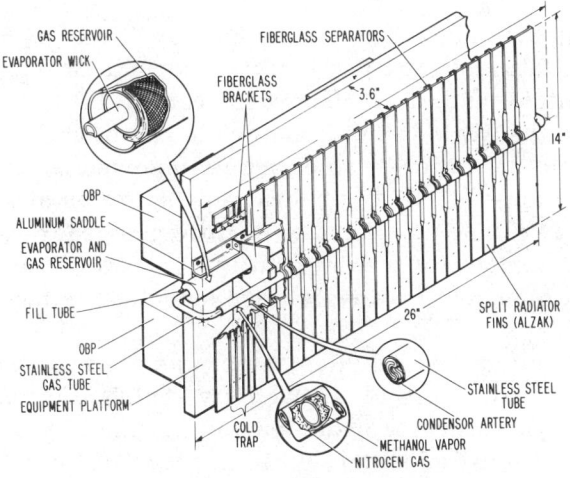

Fig. 6 AHPE hardware configuration.

wick flow area, in conjunction with the use of methanol as the working fluid, required the use of an artery-type condenser wick. Several types of arteries were considered, but the need for the artery to maintain its integrity when sharply bent led to the selection of a "filled" artery consisting of six layers of 40-mesh screen wrapped in a double layer of 94-mesh screen.

Because the heat pipe contains a noncondensible gas, which tends to promote nucleation in a superheated liquid, it was decided not to use arteries in the evaporator. Even though calculations indicated that nucleation would not occur at the anticipated superheat levels, the margin of safety was not large enough to risk vapor blockage of an artery. Thus, the primary evaporator wick consisted of multiple layers of 145-mesh screen.

To minimize temperature drops due to heat transfer into and out of the device, and thereby to maximize full-on conductance, the primary wicks were designed to occupy only half the circumference of the pipe, as shown in Fig. 6. A single layer of 150-mesh screen in the condenser and a double layer in the evaporator were provided to pump the liquid around the circumference of the pipe to and from the primary wicks carrying the axial flow. Heat transfer into and out of the pipe was principally through these thin wicks, which were sintered to the pipe wall to improve thermal conduction.

## Materials

In selecting the materials for the heat pipe, the principal criteria were weight, fabricability, availability, thermal conductivity (in that it effects control) and, most important, compatibility. Since methanol was to be the working fluid, materials compatibility—particularly in terms of gas evolution—was of prime importance. At the design temperature range (65 ±5°F) the heat pipe operates at relatively low pressure (1.44 - 1.96 psia), and very little gas generation would raise the operating temperature significantly. In view of the above criteria, the following principal materials were selected for the AHPE.

<u>Heat pipe and wicks.</u> Stainless steel was used for the pipe and wicks. It has a low thermal conductivity, which allowed developing the vapor-gas front over a short length of condenser; it is strong and available in thin-walled tubing (0.016 in.) for light weight; it is easily welded and sintered; and it is compatible with methanol, as demonstrated by life tests of subscale heat pipes.

Working fluid. Spectrophotometric grade methanol, selected for minimum water content, was used as the working fluid. Great care was taken in the process and fill operations to avoid contaminating the system with water, which could react with the stainless steel to liberate hydrogen.

Control Gas. Research grade nitrogen (99.999% purity), seeded with a similar grade helium as a leak detection aid, was used as the control gas. Nitrogen was used to match closely the molecular weight of methanol (28 vs 32) so as to avoid stratification of the vapor and gas in 1-g testing. Oxygen, which would yield a closer match, was deemed unsuitable because of its chemical reactivity.

Radiator and saddles. The radiator and saddles were fabricated from aluminum because of its high thermal conductivity and light weight. The radiator surface was Alzak-Type M1 to provide thermal radiation properties consistent with the rest of the OAO-C spacecraft. The cold trap region uses a finer segment than the main radiator to further reduce the effective axial conductance.

## Qualification and Flight Acceptance Tests

Tests were conducted on the qualification and flight units of the AHPE according to the sequence shown in Table 2.

Table 2  Test sequences

| Qualification | Acceptance |
|---|---|
| Initial leak | Initial leak |
| Initial functional | Initial functional |
| Vibration | Vibration |
| Post vibration functional | Post vibration functional |
| Post vibration leak | Post vibration leak |
| Temperature cycling | Temperature cycling |
| Thermocouple calibration | Thermocouple calibration |
| Post thermal cycle leak | Final functional |
| Shock | Thermal performance |
| Acceleration | |
| Final leak | |
| Final functional | |
| Thermal performance | |

By "seeding" the nitrogen control gas with helium, it was possible to perform leak tests by placing the entire AHPE in a vacuum system coupled to a standard helium leak detector. Measured leak rates, which did not increase with environmental exposure, were typically $3.5 \times 10^{-9}$ SCC He per sec for the qualification unit and $3.0 \times 10^{-9}$ SCC He per sec for the flight unit.

During functional testing in the ambient laboratory environment, the heat pipe was mounted with the centerline of the evaporator approximately 1/4 in. above the centerline of the condenser. This was done so that gravity would not aid the capillary return of liquid from condenser to evaporator. The unit was thermally insulated, except for the radiator, and was operated at consecutive power inputs of 10 and 30 w. The axial temperature profiles obtained were examined for any changes traceable to mechanical damage, such as wick separation from the tube wall or leaks through cracked welds that might have developed during the vibration, shock, and acceleration tests.

The time temperature history also was useful in determining start-up transients caused by any spillage of liquid into the gas reservoir during mechanical testing. The previbration functional test (Fig. 7) illustrates a start-up with liquid in the reservoir. During the preceding leak test the condenser had been oriented so that a constant liquid head had forced fluid through the perforated Teflon plug into the reservoir. Most of this liquid was removed just prior to the functional test by a manual rotation procedure, but the temperature "overshoot" on start-up indicates that a small amount of liquid remained.

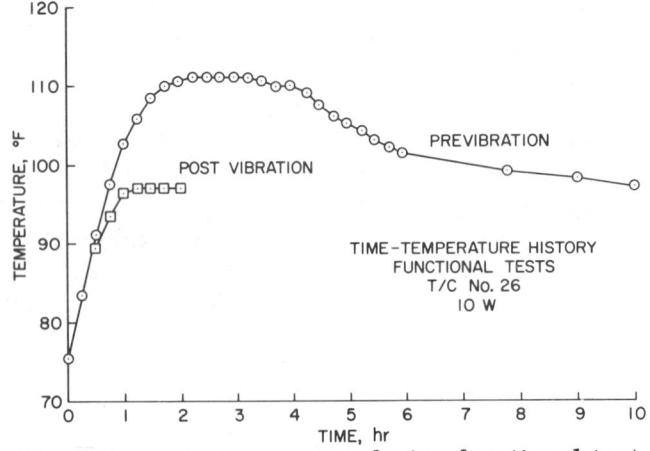

Fig. 7 Transient response during functional tests.

# VARIABLE CONDUCTANCE FLIGHT EXPERIMENT

The test was continued until the vaporized liquid had diffused back into the condenser. The AHPE was then vibrated according to the specifications of Appendix A and an immediate functional recheck performed. The postvibration start-up displayed no overshoot from excess vapor or liquid in the reservoir, thus demonstrating the effectiveness of the Teflon plug in preventing liquid from "splashing" into the reservoir.

The mechanical launch environments of vibration, shock, and acceleration were simulated on standard environmental-test equipment to the levels specified in Appendix A. Visible inspection and functional testing did not reveal any damage from these exposures. Temperature cycling was performed in a temperature chamber under dry nitrogen. Following cycling, the AHPE was allowed to soak at a constant temperature for several hours until the unt became isothermal. A thermocouple calibration was then performed.

## Thermal Performance Tests

Thermal performance tests were conducted to verify the predicted performance of the AHPE for the fluctuations in power dissipation and external boundary conditions anticipated during flight. Of particular interest were the control range and axial temperature profiles, the minimum and maximum pipe conductances, and the maximum heat-load capacity. Although these tests simulated the thermal energy incident on the radiator during flight, their purpose was to determine the performance of the AHPE itself with well-defined thermal boundary conditions, rather than attempting to simulate the complex interactions within the OAO-C/G-1 bay. A separate set of tests was performed for this purpose and will not be reported here.

The AHPE was mounted on the vertical supports so that the centerline of the evaporator was approximately 1/4 in. above the centerline of the condenser. Teflon washers were used to insulate the AHPE conductively from the support structures, and guard heaters were used to minimize heat losses through the thermocouple leads and from the back of the evaporator. All but the outboard surface of the radiator was insulated with approximately 20 layers of crinkled aluminized mylar. The albedo (S) and infrared energy (IR) incident on the radiator during flight were simulated by a shroud completely encompassing the radiator's field of view and heated to an effective space temperature ($T_S$) defined by the equation

$$\epsilon T_s^4 = \alpha S + \epsilon IR \qquad (2)$$

where $\alpha$ and $\epsilon$ are the solar absorptance and infrared emittance of the Alzak radiator. The heated shroud was lined with open-cell honeycomb and painted with Cat-a-lac black to yield a shroud emittance of nearly unity. Power to the evaporator was supplied by four flight-type foil heaters bonded to the sides of the evaporator saddle.

The test conditions were specified to map the heat pipe's performance under maximum and minimum external flux conditions. Each test condition was maintained until no temperature changed more than 2°F/hr, and generally less than 1°F/hr. All thermocouples and heater powers were continuously monitored by a digital data-logging system. Table 3 is a summary of the tests performed. The thermocouple numbers refer to the locations shown in Fig. 8.

Table 3  Thermal performance test summary

| T/C no.<br>Test | Shroud | | | | | Cold trap | | | | | | Main Radiator | | | | | | | | | | | Gas tube | Vap tube | Evaporator | | | |
|---|---|---|---|---|---|---|---|---|---|---|---|---|---|---|---|---|---|---|---|---|---|---|---|---|---|---|---|---|
| | 1 | 2 | 3 | 4 | 5 | 6 | 7 | 8 | 9 | 10 | 11 | 12 | 13 | 14 | 15 | 16 | 17 | 18 | 19 | 20 | 21 | 22 | 23 | 24 | 25 | 26 | 27 | 28 |
| Minimum external fluxes, w<sup>a</sup> | | | | | | | | | | | | | | | | | | | | | | | | | | | | |
| 2 | 59 | 63 | 63 | 60 | 60 | 55 | 56 | 57 | 56 | 56 | 54 | 55 | 56 | 56 | 56 | 56 | 56 | 54 | 53 | 47 | 46 | 44 | 42 | 21 | 59 | 58 | 58 | 56 |
| 10 | 60 | 61 | 61 | 59 | 61 | 53 | 54 | 55 | 54 | 53 | 51 | 50 | 50 | 49 | 48 | 49 | 47 | 42 | 27 | 01 | 20 | 50 | 41 | 60 | 61 | 60 | 59 | 60 |
| 22 | 62 | 61 | 61 | 61 | 61 | 50 | 50 | 52 | 49 | 47 | 44 | 41 | 33 | 30 | 26 | 26 | 19 | 1 | 50 | 54 | 56 | 59 | 38 | 62 | 63 | 63 | 62 | 62 |
| 30 | 63 | 61 | 62 | 62 | 61 | 14 | 6 | 7 | 18 | 34 | 42 | 52 | 55 | 57 | 58 | 59 | 58 | 56 | 57 | 56 | 58 | 61 | 17 | 65 | 71 | 66 | 74 | 66 |
| 40 | 62 | 61 | 62 | 61 | 63 | 18 | 28 | 42 | 51 | 51 | 52 | 56 | 58 | 60 | 60 | 61 | 60 | 58 | 59 | 59 | 60 | 63 | 19 | 67 | 73 | 68 | 78 | 69 |
| 50 | 61 | 61 | 60 | 61 | 60 | 58 | 65 | 77 | 73 | 71 | 73 | 75 | 77 | 79 | 80 | 80 | 79 | 77 | 79 | 77 | 78 | 76 | 63 | 89 | 96 | 89 | 105 | 90 |
| Maximum external fluxes, w | | | | | | | | | | | | | | | | | | | | | | | | | | | | |
| 2 | 3 | 2 | 3 | 2 | 42 | 16 | 15 | 17 | 14 | 12 | 9 | 6 | 5 | 4 | 4 | 4 | 3 | 4 | 1 | 6 | 8 | 17 | 8 | 62 | 63 | 62 | 62 | 61 |
| 10 | 2 | 2 | 2 | 2 | 42 | 16 | 15 | 16 | 13 | 11 | 7 | 3 | 1 | 0 | 1 | 2 | 3 | 7 | 23 | 52 | 57 | 61 | 8 | 64 | 65 | 63 | 64 | 63 |
| 20 | 2 | 2 | 2 | 2 | 43 | 11 | 8 | 11 | 5 | 0 | 6 | 18 | 26 | 32 | 42 | 59 | 61 | 60 | 62 | 62 | 62 | 64 | 5 | 66 | 68 | 67 | 67 | 66 |
| 30 | 2 | 2 | 3 | 9 | 42 | 32 | 40 | 53 | 59 | 59 | 60 | 64 | 66 | 67 | 68 | 68 | 67 | 68 | 67 | 67 | 67 | 70 | 32 | 72 | 75 | 72 | 75 | 73 |

<sup>a</sup> TEMPERATURE, °F

Of major interest to the thermal control engineer is the control characteristics or the relationship between evaporator temperature and power input. Figure 9 describes the measured performance of the evaporator temperature, thermocouple (T/C) No. 26, for various power inputs at maximum and minimum external flux conditions. A performance goal of 65 ±5°F for powers up to 22 w compares favorably with the measured value of 63 ±5°F. The lower nominal temperature is the result of intentionally providing a margin of safety for any temperature rise resulting from chemical gas evolution within the heat pipe. It is interesting to note the strong effect of the external

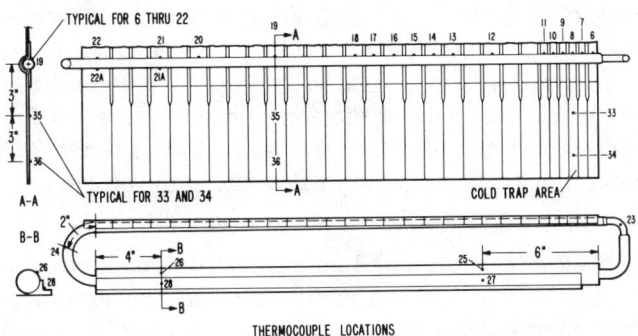

Fig. 8   AHPE thermocouple locations.

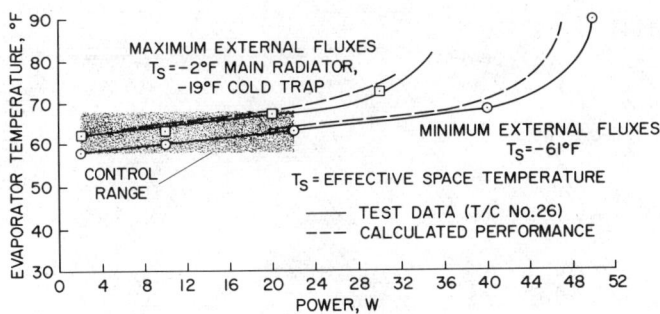

Fig. 9   AHPE control characteristics.

heat fluxes on the radiator. For the design control range (up to 22 w), the evaporator temperature change due to a power fluctuation of 20 w was 5°F, which is the same as the evaporator temperature fluctuation due to the change from minimum to maximum external flux conditions.

Shown as the dotted lines in Fig. 9 is the calculated performance obtained with the computer program of Ref. 7. Good correlation is achieved within the control range, with poorer agreement as the gas front moves into the cold trap region at the higher power levels. As the gas front influences the reservoir-entrance temperature (T/C No. 6), it also influences by diffusion the amount of vapor in the gas reservoir. This vapor diffusion and the associated temperature rise is a relatively slow process and may not have reached equilibrium before the tests were terminated. This would explain much of the difference between the calculated and measured evaporator temperatures.

The axial temperature profiles for each test are plotted in Figs. 10 and 11. The effect of the cold trap in "sharpening" the gas front is readily seen at the higher power levels. The reduced axial conductivity of the cold trap is demonstrated in Fig. 11 (20- and 30-w tests) by the distinct change in slope of the temperature profile. In Fig. 10 (2-, 10-, and 22-w tests), the slight increase in T/C's 6 and 7 is indicative of heat conduction from the evaporator through the reservoir feed tube.

If the heat pipe is envisioned as a variable conductor between the evaporator saddle ($T_e$) and the radiator at an average temperature ($T_r$), then it becomes appropriate to define an over-all conductance

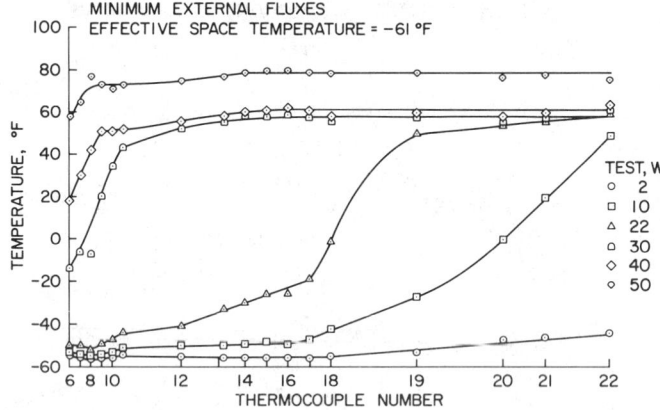

Fig. 10 Axial temperature profiles - minimum incident fluxes.

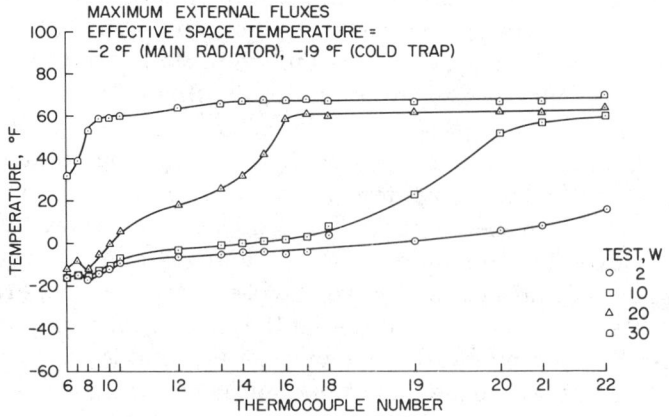

Fig. 11 Axial temperature profiles - maximum incident fluxes.

($K_o$) for the on and off conditions, where Q is the power conducted through the heat pipe.

$$K_o = Q/(T_e - \overline{T}_r) \tag{3}$$

By using Eq. 2 the data from Table 3 for the minimum flux/2-w test and the maximum flux/30-w test yield $K_o$ values of 0.0175 and 6.00 w/°F, respectively.

Another method of expressing variable conduction capability, more useful in spacecraft thermal control calculation, is to assume that the variable conductance heat pipe behaves like a thermal louver in regulating the energy radiated from the platform to space. For the AHPE we can define a variable effective emittance ($\epsilon_{eff}$) as follows:

$$\epsilon_{eff} = f(T_e) = \frac{Q}{A(\sigma T_e^4 - \sigma T_s^4)} \tag{4}$$

Figure 12 is a plot of $\epsilon_{eff}$ vs $T_e$ for the various tests performed.

The AHPE was hydrodynamically designed to pump 24 w with a 20% factor of safety during 1-g testing. Close examination of the data for minimum flux conditions shows that the temperature of the evaporator end furthest from the condenser end (T/C No. 25) begins to rise significantly above the evaporator end nearest the condenser (T/C No. 26) for powers of 30 w and greater. It therefore can be concluded that the onset of evaporator dryout occurred in the design region between 22 and 30 w.

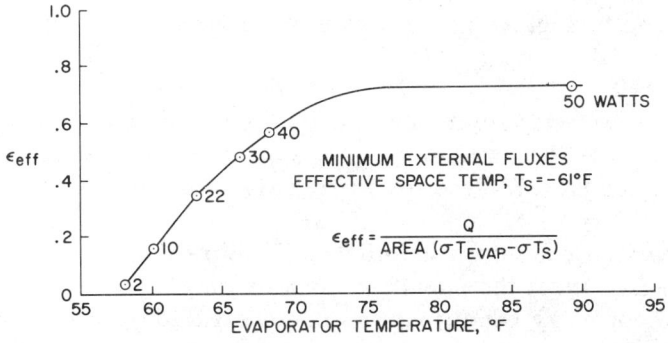

Fig. 12 Effective emittance as a function of evaporator temperature.

## Discussion and Conclusions

The use of a gas-controlled, variable conductance heat pipe is an attractive technique for passively regulating the temperature of spacecraft equipment. Under certain conditions extremely close temperature control is possible. However, the degree of control which can be achieved is a strong function of the operating temperature and the thermal boundary conditions involved in a given application. In this regard, the AHPE posed a particularly difficult design problem in that close temperature control was required while the effective radiator space temperature was both high and widely varying. Such conditions compromise control due to 1) variation of gas temperature within the gas reservoir and 2) variation of the vapor pressure in the reservoir.

Analytical and experimental examination showed that the effects of gas-temperature variations can be minimized by utilizing a hot non-wicked reservoir thermally coupled to the evaporator. However, under these circumstances careful consideration must be given to the diffusion-controlled transient phenomena associated with vapor and/or liquid in the reservoir.

Variations in reservoir vapor pressure can be minimized by a number of techniques:

1) For hot reservoir heat pipes a cold-trap can be used to lower the temperature at the reservoir entrance. For cold reservoir heat pipes similar techniques can be used to lower the reservoir temperature itself.

2) The axial conductance of the condenser/radiator should be minimized to achieve sharp vapor/gas fronts and to allow maximum radiator utilization.

3) The working fluid should be selected so that its vapor-pressure variation over the anticipated range of effective space temperatures is small compared to the total pressure in the pipe.

By careful consideration of these factors it was possible to design the AHPE to meet the specified control range ($\pm 5°F$), in spite of the adverse boundary conditions, as demonstrated by the favorable comparison of design and measured performance. It is concluded, therefore, that the behavior of gas-controlled heat pipes appears sufficiently well understood and that sufficient analytical design tools are available to begin to utilize effectively this technology in spacecraft thermal control applications. Ultimate confidence must, however, await flight proven performance.

Appendix A. Qualification and Acceptance Test Specifications

## VIBRATION
### Qualification (All Axes)

| Sinusoidal, 2 octaves/min | | Random, 4 min/axis | |
|---|---|---|---|
| Frequency | Level | Frequency | Level |
| 5-24 Hz ±2% | 1/2" DA ±10% | 15 Hz ±2% | 0.023 $g^2$/Hz ±10% |
| 24-100 Hz ±2% | 15 g Peak ±10% | 15-70 Hz ±2% | Linear increase |
| 110-2000 Hz ±2% | 7.5 g Peak ±10% | 70-100 Hz ±2% | 0.7 $g^2$/Hz ±10% |
| | | 100-400 Hz ±2% | Linear decrease |
| | | 400-2000 Hz ±2% | 0.045 $g^2$/Hz ±10% |

### Acceptance (All Axes)

| Sinusoidal, 4 octaves/min | | Random, 4 min/axis | |
|---|---|---|---|
| Frequency | Level | Frequency | Level |
| 5-20 Hz ±2% | 1/2" DA ±10% | 15 Hz ±2% | 0.010 $g^2$/Hz ±10% |
| 20-110 Hz ±2% | 10.0 g Peak -10% | 15-70 Hz ±2% | Linear increase |
| 110-2000 Hz ±2% | 5.0 g Peak ±10% | 70-100 Hz ±2% | 0.31 $g^2$/Hz ±10% |
| | | 100-400 Hz ±2% | Linear decrease |
| | | 400-2000 Hz ±2% | 0.02 $g^2$/Hz ±10% |

NOTE: ONE SWEEP FOR EACH FREQUENCY.

## SHOCK
### Qualification

| Direction | | Lead | No. shocks | Wave shape | Duration |
|---|---|---|---|---|---|
| Long. | +$X_c$ | 30g ±10% | 2 | | |
| | -$X_c$ | 30g ±10% | 2 | | |
| Lat. | +$Y_c$ | 15g ±10% | 2 | 1/2 Sine | 1 shock each axis 6 ms |
| | -$Y_c$ | 15g ±10% | 2 | | 1 shock each axis 12 ms |
| | +$Z_c$ | 15g ±10% | 2 | | |
| | -$Z_c$ | 15g ±10% | 2 | | |

## ACCELERATION LEVELS
### Qualification

| Direction | Load | Duration |
|---|---|---|
| +X axis | 11.5 g ±10% | |
| | | 4.5 min/axis |
| -X, ±Y, ±Z axes | 3.8 g ±10% | |

## TEMPERATURE LEVELS
### Qualification and Acceptance[a]

| Temperature | Cycles |
|---|---|
| -35°F ±5°F | |
| 140°F ±5°F | 12 |

[a]Total duration of test shall be at least 48 hr.

## References

[1]Katzoff, S., "Heat Pipes and Vapor Chambers for Thermal Control of Spacecraft," AIAA Progress in Astronautics and Aeronautics: Thermophysics of Spacecraft and Planetary Bodies, edited by G. B. Heller, Vol. 20, Academic Press, New York, 1967, pp. 761-818.

[2]Shlosinger, A. P., "Heat Pipes for Space Suit Temperature Control," Proceedings of the ASME Aviation and Space Conference, Beverly Hills, Calif., June 16-19, 1968, pp. 644-648.

[3]Anand, D. K. and Hester, R. B., "Heat Pipe Application for Spacecraft Thermal Control," Tech. Memo. DDC AD 662 241, NASA N68-15338, Aug. 1967, Washington, D.C.

[4]Bienert, W., "Heat Pipes for Temperature Control," Proceedings: Fourth Intersociety Energy Conversion Engineering Conference, Washington, D.C., Sept. 22-26, 1969.

[5]Turner, R. C., "The Constant Temperature Heat Pipe - A Unique Device for the Thermal Control of Spacecraft Components," AIAA Paper 60-632, San Francisco, Calif., June 1969.

[6] Hinderman, J. D. and Waters, E. D., "Design and Performance of Noncondensable Gas Controlled Heat Pipes," AIAA Progress in Astronautics and Aeronautics: Fundamentals of Spacecraft Thermal Design, Vol. 29, edited by J. W. Lucas, MIT Press, Cambridge, Mass., this volume.

[7] Edelstein, F., and Hemback, R. J., "The Design, Fabrication, and Testing of a Variable Conductance Heat Pipe for Equipment Thermal Control," AIAA Progress in Astronautics and Aeronautics: Fundamentals of Spacecraft Thermal Design, Vol. 29, edited by J. W. Lucas, MIT Press, Cambridge, Mass., this volume.

[8] Marcus, B. D. and Fleischman, G., "Steady State and Transient Performance of Hot Reservoir Gas-Controlled Heat Pipes," ASME Paper 70-HT/SPT-11, Space Technology and Heat Transfer Conference, Los Angeles, Calif., March 1970.

[9] Edwards, D. K., Fleischman, G. L., and Marcus, B. D., "User's Manual for the TRW Gaspipe Program," NASA CR 114306, April 1971.

# IV Thermal Design

APOLLO TELESCOPE MOUNT/THERMAL SYSTEMS UNIT:
CORRELATION OF PREDICTED DATA AND TEST RESULTS

Uwe Hueter*

NASA Marshall Space Flight Center, Huntsville, Ala.

and

J. Micheal Connolly† and Paul A. Christensen†

Martin-Marietta Corporation, Denver, Colo.

Abstract

The Apollo Telescope Mount (ATM) will be the first manned solar observatory to observe, monitor, and record the structure and behavior of the sun outside the earth's atmosphere. A full-scale thermal vacuum test was conducted to verify the ATM thermal design and the analytical techniques used to construct the thermal mathematical models. Methods were developed to mathematically simulate the test environment created. Post-test analysis revealed that proper handling of these test boundary conditions was vital to successful correlation of test results. Thermal design, thermal vacuum test philosophy, mathematical models used, and the correlation of analytical and test data are discussed herein.

---

Presented as Paper 71-433 at the AIAA 6th Thermophysics Conference, Tullahoma, Tenn., April 26-28, 1971. The authors gratefully acknowledge Mr. Howard Trucks for his valuable assistance and Mr. James McLane and his group for providing overall direction of the thermal vacuum test facility.
　*Aerospace Engineer, Astronautics Laboratory.
　†Engineer, ATM Thermal Unit.

## I. Introduction

The ATM is part of the Skylab, an earth-orbiting space station, scheduled to be launched by NASA in 1973. The Skylab orbital assembly (Fig. 1) includes the Saturn V Workshop, the Command Service Module (CSM), the Multiple Docking Adapter (MDA), the Airlock Module, and the ATM. The Skylab program intends to perform scientific investigations in earth orbit, to study long-duration space-flight effects on men and systems, and to collect information useful for future space programs. The ATM mission objective is to observe the sun outside the earth's atmosphere. High-resolution observations of the solar disk will provide data in the visible, ultraviolet, and x-ray regions of the electromagnetic spectrum to enable better understanding of physical processes occurring on the sun.

The ATM contains eight solar telescopes housed in a cylindrical canister that mounts to an octagonal rack through gimbal and roll rings. The rack structure also provides support for a sun shield, four x-shaped solar arrays, and various ATM electrical and mechanical components. Figure 2 shows the overall ATM configuration, excluding the solar arrays.

To verify the thermal design, a thermal vacuum test was conducted utilizing a full-scale unit, referred to as the ATM/Thermal

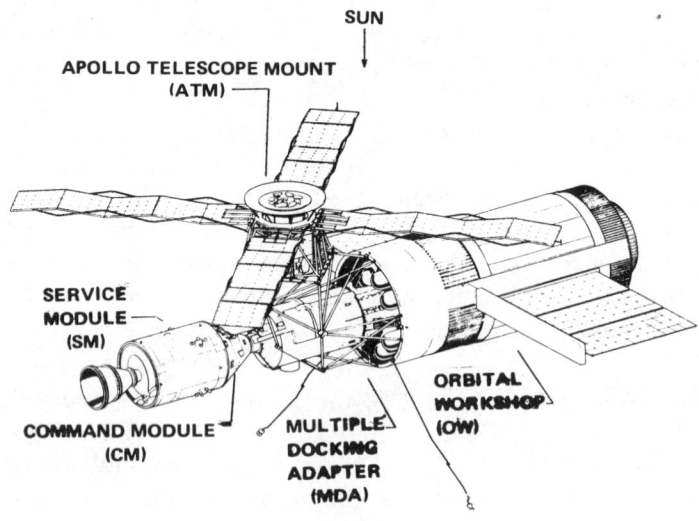

Fig. 1 Skylab orbital assembly.

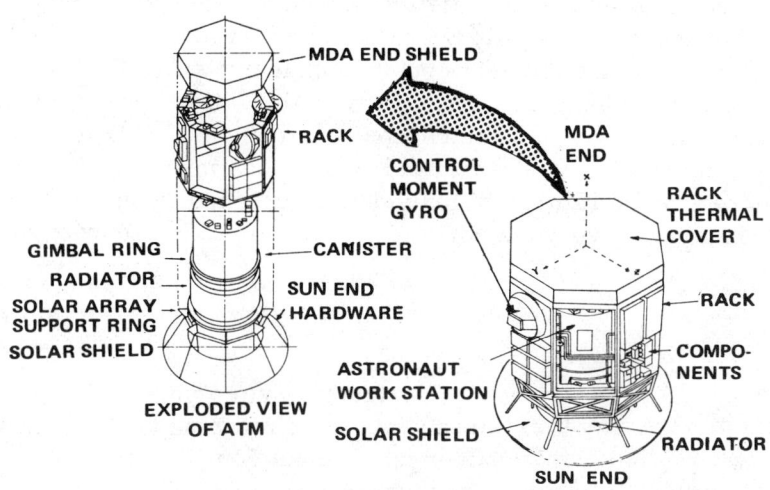

Fig. 2 Apollo telescope mount.

Unit (TSU) (Fig. 3). Most components on the test article were thermal simulators of the actual flight hardware. However, the canister active Thermal Control System (TCS) and the rack and canister structure and insulation were flight hardware. The TSU was subjected to the most severe thermal conditions expected during the ATM mission. This paper discusses the thermal design, thermal vacuum testing, mathematical models, and data correlation.

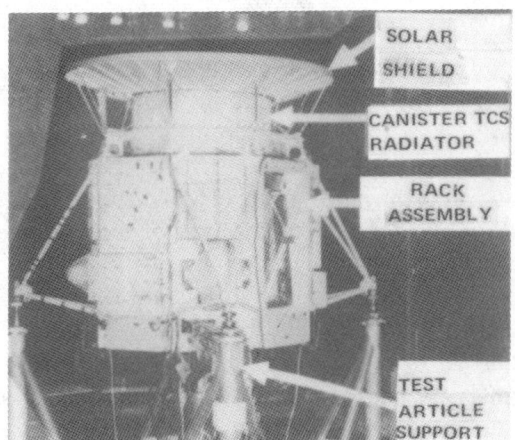

Fig. 3 Apollo telescope mount/ thermal systems unit.

## II. Thermal Design

The wide variation of thermal design requirements for the telescopes and equipment resulted in three distinct thermal designs. First, the rack components, generally operating over large temperature ranges, use passive thermal control supplemented by thermostatically controlled heaters. Second, an active TCS provides for the canister a nearly constant telescope environment. Third, electrical heaters and insulation and thermal control coatings closely maintain the correct temperatures of individual telescopes to reduce thermal distortions that otherwise adversely affect pointing accuracy and stability.

Rack

The 9- × 13-ft octagonal rack structure surrounds and supports the canister housing the telescopes. Also mounted to the rack are more than 100 electrical and/or mechanical components that compose the power, telemetry, and pointing and control systems. The 3000 w of heat dissipated by the components are rejected to space by passive methods consisting of surface coatings, thermal isolation mounts, insulation, radiation shields, and thermal coupling. The passive thermal design must maintain satisfactory component temperatures under all anticipated orbital conditions. Consequently, the analyses and tests consider hot and cold conditions established by the following guidelines: 1) all orbital environmental influences (e.g., solar, albedo, earth IR) are assumed to have a $\pm 3\sigma$ deviation; 2) the nominal $\alpha_s/\epsilon$ of 0.25/0.90 of the S-13G white paint is assumed to vary from 0.50/0.90 to 0.20/0.90 for hot and cold conditions, respectively; and 3) maximum and minimum component heat dissipation rates are considered.

The philosophy underlying the rack thermal control concept is first to eliminate any "hot case" operational conditions that would cause a component to exceed its maximum allowable temperature limit. For example, this may be accomplished by mounting the component to a radiating surface with good thermal contact. Second, components operating below their allowable minimum temperatures are 1) individually and/or collectively thermally isolated from the rack structure with fiberglass and titanium bracketry; 2) covered with thermal shields to reduce radiation heat losses; and 3) insulated until the predicted component temperatures approach their maximum allowable temperatures. Finally, supplemental, thermostatically controlled heaters are added to components that operate below their minimum temperature during the "cold case" and near their maximum temperature limit during the "hot case." Thus, rack thermal control design stresses maximum use of passive thermal control and minimizing heater power requirements. The following significant thermal requirements resulted:

1) All ATM surfaces exposed to the external environment have white surface coatings ($\alpha_s/\epsilon = 0.25/0.90$) to eliminate hot problems resulting from solar heating.

2) Thermal shields protect components with low-power density from cold environments. These shields are covered with various amounts of high-performance insulation to control the heat losses.

3) All major mounting panels, except those containing high-heat dissipating components such as the nickel/cadmium batteries, are isolated from the rack structure and covered with various amounts of high-performance insulation.

4) Components are located so that power distribution in major zones around the rack sides is reasonably uniform.

5) Components dissipating heat at high rates are mounted on external panels.

6) A rack-mounted sun shield prevents direct solar impingement on rack components.

Canister

The canister is a 135-in.-long, 86-in.-diam insulated cylinder that houses eight solar experiments, a fine pointing control system, several supporting electronic boxes, and an active thermal control system. The canister's primary function is to provide a uniform and nearly constant environment and to serve as a stable telesope-mounting platform.

The telescopes mount to an isolated structure (referred to as the spar) that divides the canister interior into four quadrants. The spar is supported by a girth ring that connects to the gimbal system. Thermal isolation of the spar must minimize spar temperature gradients that cause optical misalignment. Thermal isolation is achieved by use of low-conductance mounts between the telescopes/spar and the spar/girth ring connections. In addition, the entire spar is encapsulated by multilayer insulation and surrounded by a cylindrical canister that has black interior walls to minimize reflections and maximize radiative coupling between telescopes and cold plates. The entire canister is encapsulated with multilayer insulation to isolate the canister interior thermally from the external environment. Major canister heat shorts are six film retrieval, one telescope access and ten sun-end aperture doors, and the standoffs for the TCS radiator panels. The canister sun end has a 10-in. overhang to prevent solar energy impingement directly on the radiator panels mounted to the exterior sun-end sidewall.

The canister's interior walls are maintained at a nearly constant temperature by a closed-loop fluid, heat-transport system (Fig. 4). The heat dissipated by the telescopes is absorbed by the canister wall and cold plates, transferred to the radiator by the fluid loop, and then rejected to the external environment. The TCS design must maintain the temperature of the fluid entering the canister cold plates within 50° ±3° F and assure that the fluid temperature rise in the canister does not exceed 5° F with a maximum telescope heat output of 500 w. The TCS contains 16 cold plates, 4 radiator panels, a pump package, an accumulator, an in-line heater, a mixing valve, a fluid filter, associated valves, and connective tubing. The working fluid is methanol/water (80/20 wt %) with a nominal system flow of 850 lb/hr. The fluid exits the pump and is filtered and split into two parallel paths, radiator and heater legs. A mixing valve proportions flow through the legs and is positioned by signals from an Electrical Control Assembly (ECA) that monitors the canister inlet

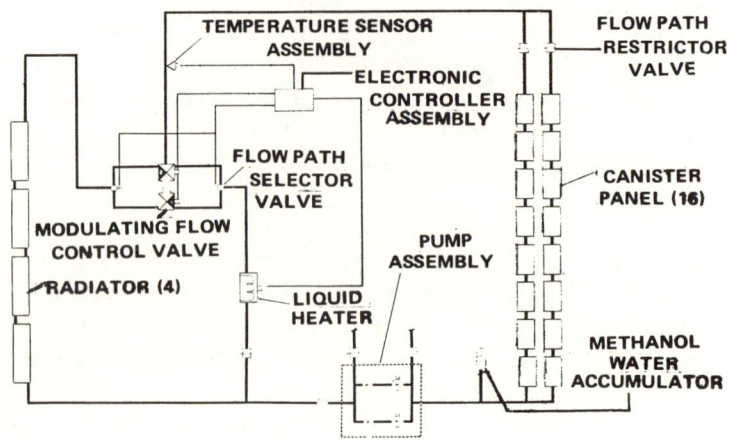

Fig. 4 Canister thermal control system.

temperature. The ECA also energizes a TCS heater if the mixed temperature drops below 47.7°F. The flow splits into equal flow rates before entering the cold plates and flows in parallel through the two canister halves. Flow leaving the canister halves mixes and then passes an accumulator before entering the pump.

The thermal design of the telescope assemblies must satisfy the telescopes' pointing stability requirements and maintain the focal length within specified tolerances. To meet these objectives, temperature gradients in the telescopes must be minimized. The following thermal control techniques were implemented in the telescopes thermal designs: 1) thermal isolation mounts to minimize heat conduction between the telescopes and the mounting spar, thus forcing radiation heat transfer to control the energy exchange between the telescopes and their heat sink, the TCS cold plates; 2) thermal coatings and insulation are utilized as a means of controlling the different heat dissipation rates; 3) use of high-conductance material in construction of the instrument housing reduces gradients; and 4) thermostatic heaters correct thermal disturbances. Heater systems for the telescopes include standoff heaters and integral heaters. The standoff heater concept[1] utilizes a low-capacitance, thermostatically controlled radiation shield over the case of the instrument. Because of the low capacitance of the shield, the temperature of the shield cycles rapidly enough so that the instrument, with its higher thermal capacity, remains at a relatively constant temperature. These are on/off-type heaters and have a fixed set point temperature. Each heater panel activates 0.5°F below and deactivates 0.5°F above its set point temperature. The integral heaters surround the instrument housing and are physically attached to it. The integral heaters utilize either an on/off type or a proportional type of power control. Two telescopes are passively controlled. These telescopes have a very low heat dissipation rate that allows the use of a gold coating whose low emissivity, in effect, isolates the telescopes from external temperature variations, thus elminating the need for an active thermal system.

## III. Thermal Vacuum Testing

Because of the thermal design complexity and the stringent telescope thermal control requirements, a comprehensive test program was conducted, with the TSU test being the last of a series before thermal vacuum accpetance testing of the ATM prototype and flight vehicles. Since the prototype and flight ATM's will be equipped with only limited thermal instrumentation, the TSU test constitutes the prime source of detailed thermal behavior data on the fully assembled ATM. The objectives of the TSU test were 1) verification of thermal design and operation of the ATM, 2) collection of test data for verification of the analytical techniques used to construct the ATM thermal models, 3) determination of any significant thermal problems that could adversely affect the success of the ATM program in subsequent testing and flight, and 4) development of shipping, handling, and testing techniques for flight hardware.

The test facility, Chamber A, at the NASA Manned Spacecraft Center, Houston, Tex., is a stainless-steel chamber 65 ft in diameter and 120 ft high. The chamber vacuum system is a combination of mechanical and diffusion pumps and a 20°K cryopump employing gaseous helium. The chamber can pump down to $1.0 \times 10^{-6}$ t in 20 hr. The interior of the chamber is lined with black, nitrogen-cooled, heat sink panels that operate at approximately 90°K. The chamber is equipped with a top solar simulator. The solar simulation system for Chamber A consists of 19 solar simulator modules mounted externally on top of the chamber. The simulator modules are an on-axis carbon arc system with a carbon arc burner assembly. The carbon-arc-type solar simulators have a wavelength range from 0.25-3.0 $\mu$. The simulators irradiate the vehicle through penetrations in the chamber top with an intensity controllable in the range from 60-140 w/ft$^2$ over a 13-ft diam. The test facility can provide a simulated space environment of vacuum, LN$_2$ temperature heat sink, and solar heat flux on the uper surface of the test article.

Predicted heat fluxes on the sides and MDA end of the test article from solar, earth albedo, earth emission, and the infrared radiation interchange with other surfaces of the orbital assembly were provided by an IR cage,[2] consisting of approximately 250 IR lamps mounted on a structural cage surrounding the TSU. The proper intensity of the heat flux on the test article was achieved by a closed-loop control system. The ATM/TSU was divided into 25 zones for heat flux simulation. These zones included 8 rack zones, 16 radiator zones, and 1 zone for the MDA end of the TSU. Alzak baffles reduced lamp overlap to adjacent zones and insured proper heat flux levels and uniformity.

The ATM/TSU test program[3] provided a series of test runs to simulate hot and cold thermal conditions, as previously discussed, for the various phases of the mission (orbital insertion, orbital operation, orbital storage, and earth-pointing modes). Orbital environment variations and surface coating changes were simulated by varying the total energy applied by the environmental simulators. A cold soak test was also conducted with all power off to determine the thermal capacitance of the components from their cooldown rates.

The test program was conducted in two phases. The first phase, or chamber pumpdown, was solely for calibrating the IR cage. The second phase, or pumpdown, consisted of 13 different tests, both steady state and transients, simulating various extreme thermal conditions on the test article.

The size of the test chamber placed certain constraints on the test article. The solar array wings and the remaining portions of the orbital assembly could not be included in the test configuration. Therefore, the effects of these surfaces on the ATM were simulated by the IR cage. An attempt to simulate the heat flux on all surfaces of the ATM was considered not practical because of the complex geometry of the test article. Therefore, the IR cage was designed to produce the desired absorbed heat flux at a plane on the vehicle surface parallel with the plane of the IR cage. Thus, heat fluxes on the sides of the exposed components were not simulated. The TSU test instrumentation consisted of various types of measurements (e.g., temperature, pressure, heat flux, voltage, currents, and power). Approximately 1300 test article measurements[4] were being recorded, of which 1100 were being displayed in real or near-real time during the test for monitoring purposes. The test data output was in digital tabulation, plots, and magnetic tape.

## IV. Thermal Mathematical Models

Thermal design of the ATM used thermal mathematical models to predict temperature levels and gradients. One of the primary TSU test objectives was to verify these models by predicting ATM thermal performance during thermal vacuum tests, thus assuring the ability of these models to predict thermal performance during actual orbital operations. Techniques used in model development and utilization with emphasis on thermal vacuum testing are discussed in this section.

Complexity, size, and thermal requirements of the ATM dictated extremely large nodal mathematical models to obtain proper data resolution for thermal evaluation. Thus, methods limiting model size without loss of data resolution had to be developed. Definable boundaries were identified which allowed the use of several models rather than one large model. Methods were developed which allowed reduction of the thermal network and simulation of the IR cage flux without actually modeling the lamps.

Reducing the radiation network size in the models was important, since the major portion of the thermal network is radiation conductors. A technique to reduce the number of radiation conductors in the gray-body network[5] was systematically to eliminate small gray-body radiation conductors based on their percent of energy contribution. The total network energy was conserved by summing the eliminated conductors to a pseudoradiation node that represents a conductor-weighted average temperature of the surrounding nodes. The radiation network in one model, for example, was reduced in this manner by approximately 50% without any appreciable loss in accuracy.

Another technique used to reduce model size was the manner in which the heat flux was applied to the model to simulate test conditions. Imposed fluxes on the TSU during the test could not exactly simulate the actual orbital environment. Therefore, an attempt was made to simulate test fluxes in the mathematical models closely by modeling the IR cage. Heat flux simulation was achieved in the model by calculating the view and gray-body factors from the IR lamps to each external node and to the control calorimeter for each IR zone. Then, the ratio of the gray-body form factor from the lamps to a node and from the lamps to the control calorimeter was multiplied by the absorbed flux of the calorimeter to determine the flux per unit area for each node. Thus, the radiation network between the lamps and the test article surfaces was not included as part of the model, since only the gray-body form factor ratios were needed to determine the flux on each node.

The sequential procedure for performing a thermal analysis with the ATM thermal mathematical models is as follows. Because of model size limitations, three models are utilized to obtain the total absorbed heat flux on the ATM surfaces. All three models mathematically simulate the entire Skylab Orbital Assembly. The first two models, having 140 nodes each, calculate fluxes on the ATM caused by the external environment (e.g., solar, albedo, and earth IR). One model calculates detailed fluxes on the rack structure; whereas, the other model computes detailed fluxes on the TCS radiator. These models are operational on the Martin Thermal Radiation Analyzer Program (MTRAP) which is a version of the Lockheed Orbital Heat Rate Package (LOHRP). Outputs of these two models are combined to attain the total absorbed fluxes on approximately 628 surfaces. These external absorbed fluxes define boundary conditions for the rack model and partially define boundary conditions for the canister model. The 1100-node rack model calculates the rack-mounted component temperatures, rack structural temperatures, and canister exterior insulation temperatures. These rack and canister temperatures are used as boundary conditions in an MDA end model, consisting of 200 nodes, which calculates the component temperatures on the MDA end of the canister. These component temperatures plus the canister insulation temperatures calculated by the rack model are then used as boundary conditions for the canister model. The 1300-node canister thermal model calculates the TCS cold plate, canister structure, and telescope temperatures. All temperatures are calculated by utilizing Martin's Improved Thermal Analyzer (MITA), a modified version of the Chrysler Improved Numerical Differencing Analyzer (CINDA).

Considerable amounts of computer time are required to run these models, because of the large nodal breakdown and the nonlinearity of the thermal network. The overall canister model, for example, has a total of 11,215 conductors of which 9800 are nonlinear or radiation conductors. The machine time for a steady-state solution for each model required approximately 10 min on a third-generation computer (e.g., CDC6600). The ratio of real, or orbit, time to computer time is approximately 5/1 for the rack and canister model.

## V. Data Correlation

Analytical and test data were correlated to verify the modeling techniques used and to update the mathematical models to make them representative of the physical system. Correlation procedures and results are described in the following paragraphs.

The initial effort for the post-test model correlation was to update the simulation of the chamber environment. One of the pretest assumptions was that all fluxes incident on the test article surface were emitted from the IR lamps; i.e., the background flux was zero. This assumption was known to be in error before the test. However, rather than make an estimate of the background level in the chamber, it was better for test-monitoring purposes to know that all predicted temperatures were cold-biased. In order to simulate the chamber environment properly, it was necessary in the post-test analysis to divide the measured flux into lamp and background fluxes. The background flux was determined from IR lamp calibration data and was utilized to calculate equivalent environment temperatures for each IR zone. A hemispherical enclosure, at the equivalent environment temperature, surrounding each IR zone was used in the mathematical model to simulate the background flux. The difference between total measured and background flux for each zone was considered as lamp flux and was then handled in the mathematical model as described in the previous section.

Initial verification of the test environment modeling techniques was achieved with power-off, steady-state test data. These tests were chosen for environmental correlation because they eliminate two possible sources of errors: capacitance and component power dissipation rates. The results showed that 98% of the 640 correlations on the rack were within ±10° F, which indicated the acceptability of environmental modeling techniques. Following the component power-off data correlation, an attempt was made to correlate the steady-state test data with the component powers activated. Although it had appeared during the component power-off correlation that the thermal networks were correct, the component power on test data indicated that several assumptions in the models were not valid. Detection of these errors during the component power-off tests was not possible because of the temperature uniformity of the test article.

The models, after correction, appear to correlate the data reasonably well and can be considered adequate steady-state analytical tools. Model accuracy is supported by the data presented in Figs. 5 and 6, which show the percent of steady-state test correlations for different temperature deviations for the rack and canister measurements. These figures represent approximately 600-node/ measurement comparisons for the rack and 150-node/measurements for the canister. The measurement accuracy, because of the data reduction method, was ±3.1° F for the rack measurements and ±0.31° F for the canister measurements. The data represented in Figs. 5 and 6 show that 48.5% of the rack measurements and 12.2% of the canister measurements fall within their measurement accuracy tolerance. Also, Figs. 5 and 6 show that 93% of the rack and canister measurements fall within ±10° and ±5° F temperature bands.

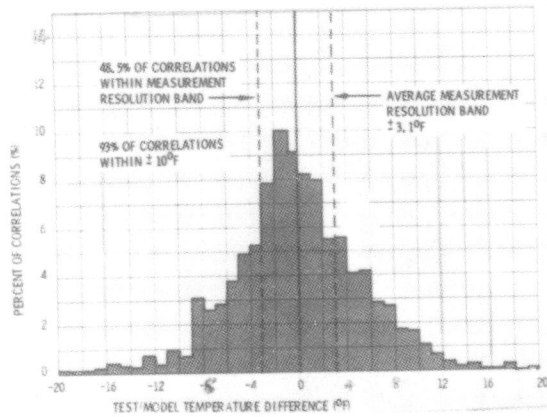

Fig. 5 Rack model/test data correlation distribution.

The next step in the data correlation was to adjust the capacitance of the model by using the data from the cold soak test to eliminate power and flux simulation uncertainties. The cold soak test consisted of preconditioning the test article to approximately 70°F. After the test article had reached a fairly uniform 70°F temperature, the IR cage and all test article power were turned off and the test article allowed to cool. Caution was exercised to insure that background flux was properly simulated, since the IR cage was used to precondition the test article before beginning cold soak. If the background flux is not accurately simulated in cold soak correlations, the difference in slope between the test and analytical temperatures may be interpreted as a capacitance mismatch. Upon completion of the cold soak test correlations, the correlations for the transient orbital tests were made. Additional model corrections were not required in the transient correlation, and the accuracy of these correlations compared to those for the steady-state correlations. Figures 7 and 8 show typical transient correlations for the rack and canister.

The results of the correlation procedures are shown in Figs. 9 and 10. Figure 9 shows the pre-test and post-test correlation accuracy for the rack measurements, whereas Fig. 10 shows the same results for the canister measurements. If the models had correlated perfectly with the test data, a 45° line beginning at the origin would have been shown. The adjustments of the models based on the test results reduced the correlation error band by approximately 50% and increased the models accuracies so that predictions for the rack and canister can now be achieved within a temperature band of ±10° and ±4°F.

In general, the ATM/TSU thermal vacuum test demonstrated the adequacy of the thermal design and provided sufficient test data[6] to correct and verify the mathematical models. The design philosophy

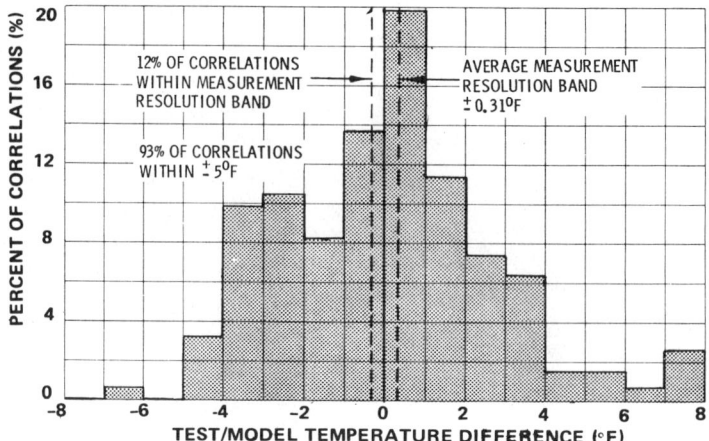

Fig. 6 Canister model/test data correlation distribution.

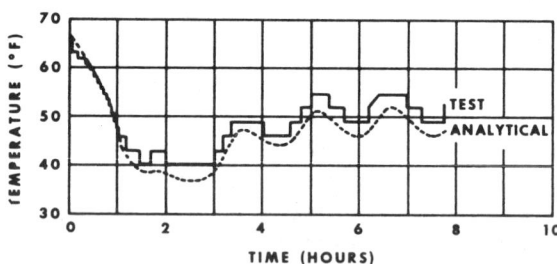

Fig. 7 Typical rack component data correlation.

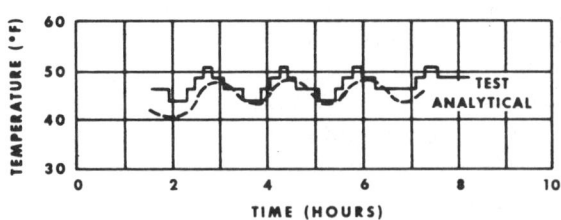

Fig. 8 Typical canister component data correlation.

to assure no hot problems was proved, since no high-temperature, out-of-limits conditions were encountered. Twelve electronic components were discovered to have cold problems during testing; however, four of these problems were attributed to test anomalies. The remaining eight components required design fixes on the prototype and flight units. In the canister, six of the eight telescopes were out of limits at some time during the tests; however, a full evaluation of telescope thermal performance on telescope data-taking capability is still in progress. Also, conditions attributed to test anomalies have not been fully evaluated to date.

## VI. Conclusions

The TSU test was a successful and useful part of the ATM program. Items that attributed to the test success were heat flux simulation and real-time data-monitoring capabilities. Good heat flux simulation established well-defined boundary conditions for analytical models which contributed, to a large degree, to satisfactory data correlation. Real-time monitoring of approximately 1100 measurements, supported by pretest temperature predictions, enabled test anomalies to be discovered almost immediately and corrected before costly test time was wasted. Tests such as the cold soak and steady-state tests, whose primary objectives were to verify the mathematical model, also assisted model verifications.

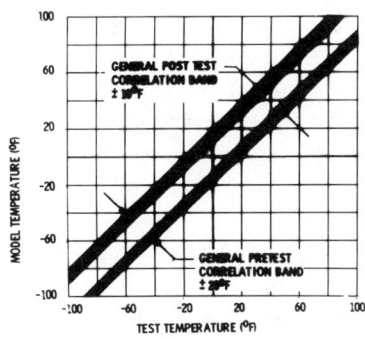

Fig. 9 Rack model/test data correlation.

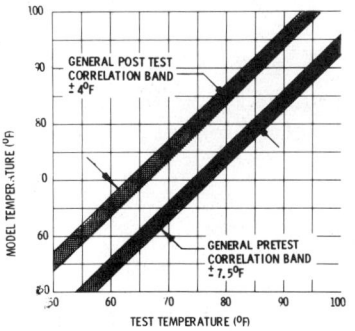

Fig. 10 Canister model/test data correlation.

Two problems hindered the use of test data for model correlation. Determination of exact component power dissipation rate was difficult because facility limitations required grouping of several components power under a single measurement. Consequently, the exact power dissipation of each component could not be determined and resulted in erroneous component power dissipation rates. Another limitation was lack of instrumentation to define temperature gradients adequately in various areas. Instrumentation limitations and thermocouple location definition before completion of the mathematical models were the underlying reasons for being unable to measure all desired temperatures.

The data correlation was judged to be good, generally within $\pm 10°F$ for the rack and $\pm 4°F$ for the canister. Nearly half (48.5%) of the rack correlations were within the resolution or recorded accuracy band of the test data. Only 12.2% of the canister measurements fell within the recorded accuracy band of $\pm 0.313°F$. However, over half of the canister measurements fell within a $\pm 2°F$ resolution. The average measurement correlation accuracy for the overall ATM was approximately ±3°F. A major factor in achieving good correlation was the unique mathematical simulation of the flux environment. The technique of treating the total measured flux as originating partially from both a directional (IR lamps) and a hemispherical (background radiation) source instead of from a strictly hemispherical source changed the temperature level of the models by as much as 15° to 20°F.

As a result of thermal vacuum test, the models have been corrected and verified. Errors between the test data and model predictions have been reduced by approximately 50%, thus increasing the confidence level for using models for flight predictions. In general, considering the ATM complex geometry, the size of the thermal mathematical models, and the many test variables, the predicted and test data agreed with exceptional accuracy.[7]

## References

[1]Regenbrecht, D. E., "Thermal Control of Four Apollo Telescope Mount Instruments," Joint National Meeting of the American Astronautical Society and Operations Research Society, Denver, Colo., June 1969.

[2]Bachtel, F. D. and Loose, J. D., "Design and Control of an Orbital Heating Simulator," Paper 71-432, AIAA 6th Thermophysics Conference, Tullahoma, Tenn., April 26, 1971.

[3]"Apollo Telescope Mount Thermal Vacuum Test Plan for Thermal Systems Unit," Doc. 50M74742, April 3, 1970, NASA.

[4]"Apollo Telescope Mount Instrumentation Program and Component List (Thermal Systems Unit)," Doc. 50M02479 (Rev. C), July 8, 1970, NASA.

[5]Chapter, J. J., "A Simplification Technique for Thermal Radiation in Grey Enclosures," Journal of Spacecraft and Rockets, Vol. 7, No. 12, Dec. 1970, pp. 1476-1478.

[6]"Apollo Telescope Mount Thermal System Unit Test Thermal Evaluation — Data Book," Rept. ED-2002-1245, Jan. 31, 1971, Martin-Marietta Corp., Denver, Colo.

[7]"Apollo Telescope Mount Thermal System Unit Test — Final Report," Rept. ED-2002-1174-2, Jan. 31, 1971, Martin-Marietta Corp., Denver, Colo.

# RADIATIVE, ABLATIVE, AND ACTIVE COOLING THERMAL PROTECTION STUDIES FOR THE LEADING EDGE OF A FIXED-STRAIGHT WING SPACE SHUTTLE

A. V. Gomez* and C. G. Johnston[†]
TRW Systems Group, Houston, Texas

and

D. M. Curry[‡]
NASA Manned Spacecraft Center, Houston, Texas

## Abstract

Tradeoff comparisons of thermal protection system weights for entry trajectories with different maximum cross-range capability are presented as a function of the number of entry missions (without refurbishment) for the wing leading edge of a fixed-straight wing space shuttle configuration. The range of radiative and ablative heat shield materials investigated included pyrolytic materials, graphite, and refractory metals (e.g., Be, C, Nb, Mo, Ta, W, etc.). For the actively cooled systems, regenerative and transpiration cooling design concepts were investigated. The choice of coolants were hydrogen, helium, oxygen, and water. An extensive summary of the results with emphasis on the radiative and ablative systems is presented.

## Nomenclature

| | |
|---|---|
| $a_H, a_M$ | = correlation function constants |
| B | = blowing parameter |
| c | = mass fraction |
| $c_p$ | = specific heat at constant pressure |
| $c_{p,mono}$ | = specific heat for an equivalent monatomic species, $[= 5/2(R/\mathcal{M})]$ |
| e | = constant, = 2.718 |
| F | = heating factor |
| g | = conductance $g = \rho_e u_e St$ |

---
Presented as Paper No. 71-445 at the AIAA 6th Thermophysics Conference, Tullahoma, Tenn., April 26-28, 1971.
*Staff Engineer.
[†]Engineer.
[‡]Engineer.

| | | |
|---|---|---|
| $G$ | = | correlation function, $G = g/g^*$ |
| $h$ | = | static enthalpy |
| $H_r$ | = | recovery enthalpy |
| $j$ | = | mass diffusion flux |
| $K$ | = | thermal conductivity |
| $\ell$ | = | reference length |
| $Le$ | = | Lewis number |
| $\mathcal{M}$ | = | molecular weight |
| $\dot{m}$ | = | mass flux per unit area |
| $p$ | = | pressure |
| $\dot{q}$ | = | heat transfer per unit area |
| $Q_{oxid}$ | = | heat of oxidation |
| $Q_{subl}$ | = | heat of sublimation |
| $\vec{r}$ | = | spatial location vector |
| $R$ | = | universal gas constant |
| $Re$ | = | Reynolds number |
| $R_h$ | = | chemical energy recovery factor |
| $St$ | = | Stanton number |
| $t$ | = | time |
| $T$ | = | temperature |
| $v$ | = | velocity component normal to the wall |
| $V$ | = | bulk flow velocity |
| $X$ | = | curvilinear coordinate parallel to the wall |
| $Y$ | = | coordinate normal to the wall |
| $\alpha$ | = | wing angle of attack |
| $\alpha_T$ | = | thermal diffusivity |
| $\varepsilon$ | = | gray-body radiation emissivity coefficient |
| $\mu$ | = | absolute coefficient of viscosity |
| $\nu$ | = | number of gas-phase species |
| $\rho$ | = | density |
| $\rho_R$ | = | gray radiation wall reflectance coefficient |
| $\sigma$ | = | Stefan-Boltzmann constant |

Subscripts

| | | |
|---|---|---|
| BL | = | boundary layer |
| c | = | conduction |
| c | = | wall or char |
| cw | = | cold wall |
| D | = | diffusion rate limited oxidation regime |
| e | = | equilibrium |
| e | = | outer edge of boundary layer |
| g | = | gas |
| H | = | heat transfer |
| HS | = | heat shield |
| M | = | mass transfer |
| P | = | product species |

| | | |
|---|---|---|
| r | = | reactant species |
| ref | = | reference |
| R | = | reaction rate limited oxidation regime |
| R | = | radiation |
| s | = | evaluated at the S surface |
| S | = | sublimation ablation regime |
| SL | = | sea level |
| t | = | transferred state |
| T | = | transition ablation regime |
| u | = | evaluated at the U surface |
| w | = | wall |
| $\phi$ | = | oxidizer species |
| $\infty$ | = | freestream conditions |

Superscripts

| | | |
|---|---|---|
| b | = | binary |
| I | = | inert species |
| s | = | solid-phase species |
| * | = | evaluated in the limit of zero mass injection |

## I. Introduction

Space shuttle orbiter vehicle concepts adopting a fixed-straight wing configuration present the thermal protection system designer with a critical design problem: protection of the leading edge of the wing during the atmospheric entry flight phase. A number of techniques to protect the leading edge adequately are feasible within current technology. Radiative heat shields and active cooling thermal protection systems have been identified as possible solutions for reusable space shuttle configurations. Ablative heat shields, although simpler solutions, do not always fit exactly the reusability criterion because of the need of ablator replacement after one or a limited number of entry missions. Radiative and/or ablative heat shields are limited by the material's radiative properties, oxidation chemistry, and melting temperature, which may severely restrict the space shuttle entry trajectory. This results in an undesirably limited crossrange capability. Actively cooled heat shields do not place such a limitation on the shuttle design. Under this category, regenerative cooling and transpiration cooling system concepts must be considered. In regenerative cooling, thermal protection is obtained by forced convection of coolants at the critical areas of the space shuttle. Heat is carried away by the coolant from the hot spots to other areas of the vehicle, where it may be dumped, radiated away, or soaked in, i.e., heat sink. For transpiration cooling, the coolant is injected into the external flow at the critical surface areas where thermal protection is required. Thermal protection is achieved in two ways:

first, by the heat absorbed by the coolant as it flows through the porous wall material, and second, by considerably altering the external flow near the wall (the boundary layer) in a manner that results in a substantial decrease in the convective heat-transfer rate to the wall.

The methods for predicting the surface degradation or recession of radiative and ablative thermal protection systems were either purely empirical methods or semitheoretical methods.[1,2] The empirical methods considered were based on heat- and mass-transfer recession rate correlations based on arcjet tests for candidate heat shield materials that were conducted in the range of application of pressure and temperature. The theoretical methods were based on multicomponent mass injection heat- and mass-transfer correlations of the boundary layer coupled with the wall chemical kinetics.[3,4] In the latter case, only published chemical kinetic constants[5-29] for candidate wall materials based on experimental data in the approximate range of applications of pressure and temperature were used. The theoretical methods properly accounted for the wall diffusion fluxes at the wall due to the injection of foreign species into the external flow. Theoretical predictions for selected materials for which arcjet testing data were available were compared and found to be in good agreement with the experimental results. The methods of analyzing the various heat-shield materials considered varied depending on the melt temperature and oxidation chemistry. Generally they may be categorized as 1) oxidation controlled - when their melting temperature is found to be much grater than the radiative equilibrium temperature calculated for the convective heat-transfer rate, 2) simple sublimers - when the melting temperature is lower than the radiative equilibrium temperature, and 3) pyrolytic ablators - where the material decomposition into pyrolysis gas and char occurs in depth. In all cases, the heat of ablation was determined by considering as many species as possible for which thermochemical data were available. Thermodynamic properties of species were obtained from JANAF tables and other sources[30-36] and used in the form of polynomial curve fits to the tabulated data. The viscosities and thermal conductivities of the pure components were computed in the manner recommended by Hirschfelder, et al.[37] The Lennard-Jones interaction potential was assumed and the required parameters obtained from Svehla[38] and Hochstim.[39] The mixture viscosity was calculated following the method of Wilke[40] and the mixture thermal conductivity according to Mason and Saxena.[41] The multicomponent binary diffusion coefficients of species were calculated according to the bifurcation approximation.[42]

For regenerative cooling, the wing leading edge was assumed constructed with small orifices or passages where the coolant flows. The flow in these passages was assumed to be either laminar or turbulent, depending on the Reynolds number, and in the liquid or gaseous phase, depending on the inlet conditions. Viscous dissipation (heating) and the wall surface reradiation were included in the analysis. To obtain solutions, the wall (skin) was assumed to be at the equilibrium temperature from which the coolant temperature gradient, pressure gradient, and velocity gradients were calculated. The procedure followed for obtaining solutions was based on finite difference numerical analysis, and iterations were performed until a satisfactory solution was obtained. The criterion for determining sa satisfactory solutions depended on the maximum wall temperature desired and on the fact that the flow Mach number cannot exceed unity.

Transpiration cooling is provided by gas or liquid injection through a porous wall or matrix of either connected open porosity or a simulation of this by many small discrete holes or capillary tubes distributed over the wall. In this manner, a continuous distribution of coolant injection to the exposed surface is achieved which may be varied according to the heating requirements by varying the permeability and the thickness of the matrix. Because of this fact, this system features the most efficient utilization of the injectant. The advantages of transpiration cooling over other thermal protection systems have been demonstrated in many studies completed in recent years. These studies invariably considered the flow through the porous wall separate from the injection heat-blocking effects. Therefore, attention has been directed to the solution of the coupled problem (matrix-flow/heat-blocking effects) with the objective of determining the over-all system performance and injection control requirements.[1,2] Analytical predictions for the coupled problem showed that the characteristic behavior of both gas and liquid coolants exhibits a region of unstable flow where injection control, by varying the inlet pressure, is ineffective. The matrix temperature and the injectant properties play an important part with respect to the injection control. When the mass injection rates are reduced sufficiently, injection control is lost. In this region, the outer-wall boundary conditions and the heat stored in the matrix govern the flow rate regardless of the backface pressure. This condition is akin to ablation, where the outer wall conditions govern the mass injection into the boundary layer. The phenomenon was found to be characteristic of gas as well as liquid injection. For liquids, the unstable region was encountered when a vapor phase change occurs. The prediction of the onset of the critical region was included in the analysis.

The purpose of the study was to predict the performance of the preceeding thermal protection system concepts and compare the resulting system weights for a range of entry aerothermodynamic environmental conditions representative of space shuttle entry missions. In the present paper, a summary of the numerical results, analytical assumptions, weight comparisons, and the final conclusions for each system are presented with emphasis on the radiative and ablative systems.

## II. Analysis

### Boundary-Layer Mass and Energy Fluxes

Consider the wall mass and energy flux balance presented in Fig. 1. In the figure, a small control volume is presented that encloses the wall surface and is bounded by the mathematical surfaces S and U. These surfaces are assumed to lie very close to the wall so that the control volume width is negligibly small in comparison with the characteristic dimensions of the boundary layer, i.e., the boundary-layer thickness. At the S surface, the wall gas-phase properties are defined. Some properties are continuous across the wall, e.g., T and $\dot{m}$, whereas others are discontinuous, e.g., h and c. In this manner, the boundary layer and heat shield mass and energy fluxes may be precisely defined, even though chemical reactions are occurring at the physical surface. The temperatures at the S, U, and wall surfaces are assumed to be equal. The temperature is a characteristic thermodynamic quantity that is uniform within the control volume and is dependent on the heat-shield time-integrated heat input. The boundary-layer energy fluxes at the S surface relevant to the boundary-layer analysis are given by 1) the diffusive energy flux to the wall (also called the convective heat-transfer rate to the wall):

$$\dot{q}_{BL} = \left[ \dot{q}_c - \sum_{i=1}^{\nu} j_i h_i \right]_s = \left[ \dot{q}_c - \sum_{i=1}^{\nu} (\dot{m}_i - \dot{m}) h_i \right]_s \qquad (1)$$

and 2) the convective energy flux from the wall

$$\dot{m} h_{w,s} = \left[ \rho v \sum_{i=1}^{\nu} c_i h_i \right]_s \qquad (2)$$

These definitions result from the treatment of the governing boundary-layer equations in terms of mass average velocities. The conduction heat flux rate into the heat shield evaluated at the U surface can be written as

$$\dot{q}_{HS} = \dot{q}_{c,s} + \dot{m} Q_{oxid} - \dot{m}_c Q_{sub1} + \dot{q}_R (1 - \rho_R) - \epsilon \sigma T_w^4 \qquad (3)$$

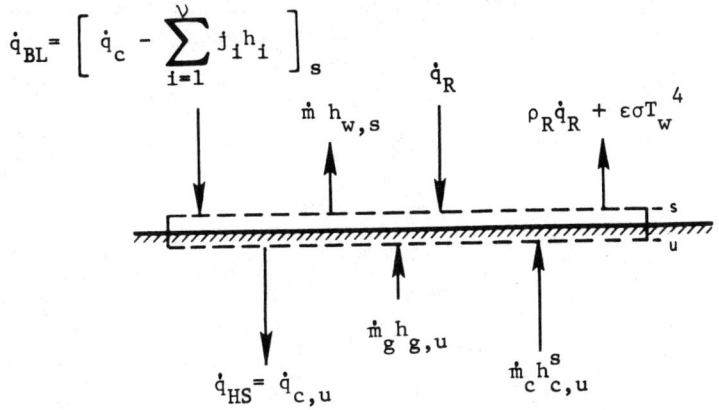

Fig. 1 Wall interface mass and energy balance.

The wall heat of oxidation and heat of sublimation due to chemical reactions that occur within the control volume are obtained from the preceding relations. They are given by

$$Q_{oxid} = (1/\dot{m}) \left[ \dot{m}_g h_g + \dot{m}_c h_c \right]_u - \left[ \sum_{i=1}^{\nu} \dot{m}_i h_i \right]_s \tag{4}$$

$$Q_{subl} = \left[ h_c - h_c^s \right]_u \tag{5}$$

The most refined method available for predicting the boundary-layer heat- and mass-transfer fluxes to the wall involves the coupling of a numerical solution procedure for the boundary-layer flow conservation equations to a finite-difference solution procedure for the indepth response of the heat shield. Such an approach is not feasible for performing engineering calculations over a complete entry trajectory because of the excessive computer time required and the inherent unreliability of such a complex computer program. Thus, current practice involves a finite-difference solution of the equations governing the indepth response of the heat shield but uses simple correlation formulas to describe the convective heat- and mass-transfer fluxes to the wall. These formulas are in the form of standard correlations for high-temperature boundary-layer flows, with correction factors to account for the reduction of heat and mass transfer due to surface mass addition. The most extensive computations of these correlations are those of Gomez et al.[3] and Mills et al.[4] which include multicomponent diffusion effects and chemical reactions between injectant and air species. These correlations with minor modifications[1] have been adopted for the present analysis. The correlation equations used are:

Heat Transfer

$$\dot{q}_{HS} = \dot{q}_{c,u} = \dot{q}_{c,u}^{I} R_h + \dot{q}_R (1 - \rho_R) - \varepsilon \sigma T_w^4 \tag{6}$$

$$\dot{q}_{c,u}^{I} = \left[\dot{q}_c - \sum_{i=1}^{\nu} \dot{m}_i h_i\right]_s^{I} + \left[\dot{m}_g h_g + \dot{m}_c h_c^s\right]_u \tag{7}$$

$$\dot{q}_{c,s}^{I} = \dot{q}_{c,s}^{*} G_H \tag{8}$$

$$G_H = \frac{q_{c,s}^{I}}{q_{c,s}^{*}} \prod_{i=1}^{\nu} \frac{a_{H_i} c_{i,t} B_H}{e^{(a_{H_i} c_{i,t} B_H)} - 1} \tag{9}$$

$$a_{H_i} = 1.25 \left\{ \left(\mathcal{M}_{air}/\mathcal{M}_i\right)^{0.4} \left(c_{p_i}/c_{p_i,mono}\right)^{0.3} \right\} \tag{10}$$

$$B_H = \dot{m}/(\rho_e u_e St_H^{*}) = -\frac{\dot{m}}{q_{c,s}^{*}/\left(H_r - h_{(T)}\right)_{air}} \tag{11}$$

$$\dot{q}_{c,s}^{*} = \dot{q}_{cw(T_{ref})} \left[\left(H_r - h(T_w)\right)/\left(H_r - h(T_{ref})\right)\right]_{air} \tag{12}$$

Mass Transfer

$$\dot{m}_i = \dot{m} c_{i,s} + j_{i,s} = \dot{m} c_{i,t} \tag{13}$$

$$g_{M_i} = g_{M_i}^{*} G_{M_i} = - j_{i,s} / (c_{i,e} - c_{i,s}) \tag{14}$$

$$G_{M_i} = \left(a_{M_i} B_{M_i}\right) / \left(e^{(a_{M_i} B_{M_i})} - 1\right) \tag{15}$$

$$a_{M_i} = \left(\sum_{i=1}^{\nu} c_{i,t} a_{M_i}^{b}\right) \left[a_{M_i}^{b} / \sum_{i=1}^{\nu} c_{i,t} a_{M_i}^{b}\right]^{1/3} \tag{16}$$

$$g_{M_i}^{*} = \rho_e u_e St_H^{*} Le_i^{0.72} \tag{17}$$

where

$$B_{M_i} = \dot{m}/g_{M_i}^{*} \qquad a_{M_i}^{b} = 1.462 \left(\mathcal{M}_{air}/\mathcal{M}_i\right)^{0.922} \tag{18}$$

The quantity $\dot{q}_{c,u}^I$ introduced in Eq. (7) is the effective conduction heat flux into the heat shield evaluated at the U surface in the absence of chemical reactions between injectant and air species in the boundary-layer flow and negligible radiation fluxes at the wall. The factor $R_h$, which is called the chemical energy recovery factor, scales the effect on the conduction heat flux to the heat shield due to these chemical reactions. $R_h = 1$ implies inert injectant species in the boundary layer, but it does not exclude the presence of chemical reactions at the wall interface. The quantities $g_{M_i}$ and $c_{i,t}$ introduced in Eqs. (13) and (14) are the usual definition of mass-transfer conductance and transfer state mass fraction for the ith species which are derived in the Spalding[43] formulation of the steady-state mass-transfer problem. The correlation of mass-transfer and heat-transfer conductances in the limit of zero mass injection (Eq. 17) introduces the effective binary Lewis number $Le_i$ for injectants into air.

The cold-wall heating rate $\dot{q}_{cw}$ is determined using either wind-tunnel tests for the wing and the complete space-shuttle configuration or using current aerodynamic heating two-dimensional flow theory. The empirical criterion is based on the definition of heating factors which may be assumed invariant with entry time. These factors are the ratios of the wind-tunnel heating rates, corrected for cold-wall effects (i.e., $T_{ref} = 140°F$) and scaled to the full-size vehicle, normalized by the heating rate to a 1-ft-radius sphere in the same heating environment. Accordingly, using the Detra et al. correlation,[44]

$$\dot{q}_{ref} = \frac{17600}{\sqrt{1 \text{ ft}}} \sqrt{\frac{\rho_\infty}{\rho_{SL}}} \left(\frac{V_\infty}{26000 \text{ FPS}}\right)^{3.15} \left(\frac{H_r - h(T_{ref})}{H_r - h(300° K)}\right)_{air} \quad (19)$$

$(BTU/ft^2\text{-sec})$

$$\dot{q}_{cw} = \dot{q}_{ref} F(\vec{r}/\ell) \quad (20)$$

## Radiative and Ablative Heat Shields

The performance of radiative and ablative thermal protection systems is governed by the coupled effects of the flow next to the wall (the boundary layer) and the wall-air species chemical reactions at the wall. The surface degradation or recession rate depends on the time rates about which the wall species are being consumed by the chemical reactions at the wall interface. These include oxidation reactions and the sublimation of solid-phase species. The availability of oxidizing species at the wall depends on the mass-transfer conductance properties of the boundary-layer flow which are affected by the injection at the wall of the products of ablation into the external flow. The character and time rate of the chemical reactions at the wall are also affected by the pressure and the temperature. The dependency of the wall mass recession rate on the pressure, temperature, and the availability of oxidizing species categorizes the problem into four distinct separate ablation regimes, which, in the order of increasing temperature, are designated as 1) reaction rate limited regime, 2) transition regime, 3) diffusion rate limited regime, and 4) sublimation regime.

The reaction rate limited regime is characterized by nonequilibrium oxidizing chemical reactions of the wall solid-phase species. The character and number of chemical reactions change with temperature and pressure. For most ablators at relatively low temperature ($T_w < 1000°C$), the solid-phase wall species combine with the external flow oxygen species to form solid phase oxides. The rates of formation of the oxides depend on the diffusion of oxygen species through the solid oxide film that coats the surface. Increasing the wall temperature or decreasing the pressure leads to the formation of volatile oxides that result from the vaporization of the solid oxide film. Eventually, at more elevated wall temperatures ($T_w > 1000°C$), the solid oxide species vaporize at the same rate as they are being formed, which exposes the wall species to the external oxidizing environment. The surface degradation rate is first characterized by a weight gain corresponding to the formation of the solid oxide film on the surface at low temperatures followed by a weight loss rate or recession rate at the higher temperatures. For fixed wall temperature and pressure, the equations that best describe the surface degradation rate as a function of exposure time are parabolic in the lower temperature range (the weight gain rate decreases with time) and linear at the higher temperature range (constant weight gain or constant weight loss rate). The dependency on pressure and temperature for the wall degradation rate in the elevated temperature range is generally well described by Arrhenius-type equations. From the preceding discourse, it is obvious that no single equation or single set of kinetic constants can be used to describe the wall surface degradation rate accurately over a wide range of temperature and pressure. Instead, the depth in which this regime is to be analyzed depends on its relative importance with respect to the other wall surface degradation regimes that must be considered in a given entry mission problem.

For the transition regime, the mass fractions and partial pressures of oxidizing species at the wall interface are less than those for the limiting case of no oxidizing chemical reactions. The mass rate of arrival of oxidizing species depends on the interdiffusion of species in the boundary-layer flow at the wall. Therefore, the exact analysis of the wall surface degradation problem is more difficult than the reaction-rate- or the diffusion-rate-limited oxidation problems when either is treated separately. As a consequence, an approximation for the wall surface degradation rate based on the reaction-rate- and diffusion-rate-limited oxidation solutions treated separately is generally used.

In the diffusion-rate-limited regime, the oxidation chemical reactions at the wall are controlled exclusively by the mass rate of arrival at the wall interface of oxidizing species. This rate depends on the interdiffusion of species in the boundary-layer flow which is affected by the injection of ablation products at the wall. The analytical problem is indeed difficult if more than one product species react with air species in the boundary layer in the vicinity of the wall. The problem, however, becomes relatively simple if the number of product species is limited to one, and the product-air species chemical reactions are neglected. In this instance, calculations on the wall surface degradation rate can be performed readily with the aid of the boundary-layer multicomponent injection correlations Eqs. (13-18).

The sublimation regime is characterized by the vaporization of the wall solid phase species which may occur in the presence of oxidation chemical reactions at the wall interface. The number of species formed at the wall may be large, and the chemical reactions may be equilibrium or non-equilibrium. However, for most applications, the assumption of equilibrium chemistry at the wall interface greatly simplifies the problem and was found to provide good accuracy for most engineering purposes.[4] Nevertheless, the problem of analytically coupling the boundary-layer flow with the wall interface thermochemistry still remains a formidable problem. Fortunately, the problem is tractable to numerical analysis, and computer codes[2] are available to perform these types of calculations. On the other hand, a very simple and naive treatment of the problem is possible if the oxidation reactions at the wall are neglected and if the sublimation of wall species is restricted to a single molecular species. In this instance, the surface degradation rate is proportional to the convective heat-transfer rate to the wall divided by the heat of vaporization of the wall solid-phase species.

Five separate analytical procedures were developed for determining the heat-shield performance which are applicable to the range of materials considered. The analytical procedures are as follows:

1) Radiative heat shields: In the absence of ablation ($T_{wall} \ll T_{melt}$), the wall temperature was determined to correspond closely with the equilibrium radiation temperature for a wide range of heat shield materials of finite thickness ($\tau \leq 0.15$ inch, for thicker skins, the temperature at the wall and corresponding convective heat flux are smaller).[45] Neglecting the wall incident radiation, which is negligibly small in comparison with the boundary-layer convective heat-transfer rate, the wall interface energy balance yields

$$\dot{q}_{cw(T_{ref})} \left[ \left(H_r - h(T_e)\right) / \left(H_r - h(T_{ref})\right) \right]_{air} = \epsilon \sigma T_w^4 \qquad (21)$$

$T_e$ can be readily solved for by the method of successive approximations ($\epsilon = 0.80$ was used in the analysis).

2) Simple sublimers: In the presence of ablation, the heat of oxidation, the heat of sublimation, and the boundary-layer heat-blocking effects due to mass injection must be included in the wall interface energy balance. For a simple sublimer, the wall temperature was assumed constant and equal to the melting temperature of the heat-shield material ($T_w = T_{melt}$). If the change of phase of the wall material is the only chemical reaction considered, $Q_{oxid} = 0$ and $R_h = 1.0$. The wall interface energy balance yields

$$\dot{q}_{HS} = \dot{q}_{c,s}^* \left[ \frac{a_H B_H}{e^{a_H B_H} - 1} \right] - \dot{m}_c Q_{subl} + \dot{q}_R (1 - \rho_R) - \epsilon \sigma T_w^4 \qquad (22)$$

The mass-transfer correlations provide the information required to determine the heat-blocking weighting constant $a_H$. For the simple sublimer, the trivial solution for the transferred state mass fractions of $c_{i,t} = 1.0$ for the single injectant species and $c_{i,t} = 0$ for the air species applies. Then, using Eq. (10), $a_H$ may be evaluated. Neglecting the heat capacitance of the heat-shield material and the incident radiation ($q_{HS} = q_R = 0$), the blowing parameter $B_H$ may be determined by the method of successive approximations. The surface degradation rate is obtained from $B_H$ using Eq. (11).

3) Reaction rate limited oxidation: It has been shown that all metal-oxygen reactions are first-order complex chain reactions[5] whose rate of attack to the metal surface is described by

$$\dot{m}_R = \dot{m}_c = k_1 (p_{O_2})^z e^{-\frac{\Delta E}{RT}} \qquad (23)$$

where $k_1$ is a constant, $PO_2$ is the pressure of oxygen at the metal surface, and $\Delta E$ is the activation energy. Since the exponent z lies between zero and unity, linear kinetics are implied. Finite difference numerical solutions for the heat-shield response using Eq. (23) are readily obtainable. For example, at a given entry trajectory time, the wall temperature that is a function of the time-integrated conduction heat flux into the heat shield is determined from the previous time heat-shield indepth response solution. Then, the ablation rate $\dot{m}$ can be determined. The conduction heat flux into the heat shield at the selected time can be calculated next using the wall interface energy balance and the aerodynamic heating correlations. As a first order approximation, the equilibrium radiation temperature Eq. (21) was used to determine the ablation mass loss rate.

4) Diffusion rate limited oxidation: In the diffusion-rate-limited or diffusion-controlled oxidation regime, the reactant species (oxygen) mass fractions at the wall are depleted by the surface chemical reactions ($c_{r,s} = 0$). The total mass fluxes of reactants (0 and $O_2$) to the wall are determined using the mass-transfer correlations. Using Eqs. (13) and (14),

$$\dot{m}_{(\phi)} = \sum_{r=1}^{R} \dot{m}\, c_{r,t} = -\sum_{r=1}^{R} c_{r,e}\, g_{M_r} \qquad (24)$$

where $\dot{m}_{(\phi)}$ is the elemental mass flux of oxygen to the wall. The net mass flux of products of oxidation reactions injected into the boundary layer must satisfy the conservation of elemental oxygen at the wall given by

$$\dot{m}_{(\phi)} + \sum_{j=1}^{R} \mu_{\phi,j}\, \dot{m}_j = 0 \qquad (25)$$

and the elemental flux of surface solid-phase species s which is being depleted by the oxidation process is

$$\dot{m}_D = \dot{m}_{(c^s,u)} = \sum_{j=1}^{J} \mu_{s,j}\, \dot{m}_j \qquad (26)$$

In these expressions, the quantities $\mu_{\phi,j}$ and $\mu_{s,j}$ are formation matrices that specify the mass concentration of $\phi$ and $c^s$ present in the jth product series. For determining the surface mass ablation rate, the mass-transfer conductances of reactant species (oxygen) must be evaluated. These conductances will be affected by the magnitude of the total mass injection rate into the boundary layer. For simple ablators, the total mass injection rate is equal to the surface mass ablation rate. In the presence of pyrolysis products gas injection (i.e., pyrolytic ablators), the reactant species at the outer edge of the boundary layer must diffuse toward the wall against the larger mass average velocity flux of pyrolysis and ablation products. This results in a reduced mass rate of oxidizer species available to enter into the surface oxidation chemical reactions and less surface degradation in comparison with the simple ablator limiting case. The calculation of the species mass-transfer conductances was performed by the method of successive approximations. In the first approximation, the surface degradation rate may be neglected in calculating the mass-transfer conductance of the oxidizer species. In the successive approximations, the preceding calculation of the surface mass ablation rate is used, together with the appropriate transferred state mass fractions. Generally, two or three iterations suffice to obtain adequate engineering accuracy. As an illustration, computations performed for the ablation of carbon (graphite) and tungsten using the present method are presented and compared with the published date of Scala[46] and Gilbert[47] in Table 1.

5) Transition regime: For the transition regime, the relation suggested by Scala[46] given by

$$\dot{m}_T = \left[ \dot{m}_R^{-2} + \dot{m}_D^{-2} \right]^{-1/2} \tag{27}$$

is considered satisfactory. The mass fluxes $\dot{m}_R$ and $\dot{m}_D$ are obtained from Eqs. (23) and (26), respectively.

6) Complex ablator (e.g., pyrolytic ablators): The boundary-layer thermochemistry was represented by heat- and mass-transfer correlations, together with a zero mass injection aerodynamic heating calculation. The coupling between the boundary layer and the heat-shield response was achieved via an open system surface thermochemistry computer code.[4] The principal advantages realized over all other ablator numerical solutions procedures are a) in evaluating the aerodynamic heat blockage, the contribution of injectant species and outer edge reactants species are considered, b) the surface recession rate is calculated based on the generalized surface thermochemistry model (both forward and backward chemical reaction rates are properly accounted for), and c) the surface energy balance that yields the heat shield input conduction flux is calculated with greater accuracy (in this instance, the heat of oxidation and heat of sublimation were calculated considering as many as 30 molecular species). A detailed description for the complex ablator analysis is presented in References 2 and 4.

Table 1 Diffusion rate limited oxidation regime.

|  | Carbon $C + 1/2\ O_2 \rightarrow CO$ | | Tungsten $W + O_2 \rightarrow WO_2$ |
|---|---|---|---|
|  | Present work | Scala[46] | Present work |
| $c_{CO,s}$ or $c_{WO_2,s}$ | 0.3554 | 0.3498 | 0.0154 |
| $c_{C,s}$ or $c_{W,s}$ | 0.1524 | 0.15 | 0.0131 |
| $B'_{M_i} = B_{M_i}/G_{M_i}$ for $O_2$ | 0.1751 | 0.175 | 0.0070 |
| For $3W + 9/2\ O_2 \rightarrow W_3O_9$ | $\dot{m}(W)/\dot{m}(C) = 4.693$ Present work | | |
| $C + 1/2\ O_2 \rightarrow CO$ | " $= 5.11$ Gilbert[47] | | |

## Actively Cooled Heat Shields

For regenerative cooling, the wing leading edge was assumed constructed with small orifices or passages where the coolant flows. Since the skin was assumed thin, the time-dependent heat storage capacity is negligible, and the skin or wall temperature corresponds to the equilibrium temperature. The coolant flow in the orifices was assumed to be parallel to the wing chord planes. Steady state, variable property pipe flow theory was assumed satisfactory to determine the energy heat transfer rates from the wall to the coolant. For transpiration cooling, one dimensional, single-phase, variable property flow analysis was assumed in describing the flow of coolant in the porous wall. This assumption is considered satisfactory since the wall thickness is very small in comparison with the dimensions of the chord of the leading edge. In the analysis, light molecular weight gaseous coolants were assumed to be stored in liquid form in cryogenic storage tanks. Before delivery to the wing leading edge backface manifold, gaseous coolants were assumed to undergo an irreversible heat addition process where the liquid-vapor phase change occurs. The coolant delivery system was designed as a dual system with cross feed. The delivery system weight was calculated including storage tanks, pumps, heaters or heat exchangers, piping, controls, and the leading edge skin. The details of the analysis for the active cooling systems are presented in References 1 and 2.

## III. Results and Discussion

### Entry Environments

The prediction of the aerodynamic heating environment for the wing leading edge of the space shuttle at small and large angles of attack in the hypersonic-orbital entry speed regime presents a difficult problem. The difficulties encountered originate primarily from two sources: first, at large angles of attack, the flowfield behind the bow shock wave on the exposed wing surface is predominately three-dimensional, and, therefore, crossflow effects cannot be neglected; second, early in the entry trajectory, the boundary-layer or viscous layer thickness in the vicinity of the leading edge sonic line approaches the magnitude of the thickness of the flow between the wall and the bow shock. Under the latter condition, the heating rate to the wall is augmented by vorticity interactions that arise due to the curvature of the bow shock and the vorticity swallowing at the outer edge of the boundary layer. Because of these difficulties, three approximate analytical procedures which bound the problem of calculating the convective heating rate distributions to the exposed surface of the wing of the space shuttle were investigated. The first two procedures designated "2-D Theory" and "2-D Empirical" are based on the following assumptions: 1) the flow over the wing is two-dimensional, i.e., negligible crossflow effects, and 2) the fuselage bow shock interference effects on the convective heating to the wing are excluded. The two-dimensional theory analysis is based on theoretical solutions for the shock layer and the boundary-layer flows where the vorticity interaction effects on the convective heating are calculated as first-order corrections to the no-interaction solutions. The two-dimensional empirical analysis is based on two-dimensional flow empirical data obtained at NASA, together with small corrections derived from theoretical solutions that account for the vorticity interaction effects. The third analytical procedure, which was designated "3-D empirical," is based on wind-tunnel test data for the complete space-shuttle configuration. The "3-D empirical" heating environment calculation procedure is based on three-dimensional wind-tunnel tests for the complete space-shuttle vehicle at a wing angle of attack of $60°$. (Comparable test data was not available for the smaller-wing angle-of-attack attitudes.) Therefore, the flow over the wing was three-dimensional, and the crossflow effects, fuselage bow shock interference, etc., are automatically included. The heating distributions were obtained using the heating factors determined in ground tests and the reference heating calculated in the manner shown by Eqs. (19) and (20). Figure 2 shows a comparison of the heating factors for the three heating calculation procedures used in this investigation. The prescribed wing geometry used in the investigation is presented in Table 2.

Three representative entry trajectories for a 25,000-lb payload space-shuttle orbiter with a fixed-straight wing config-

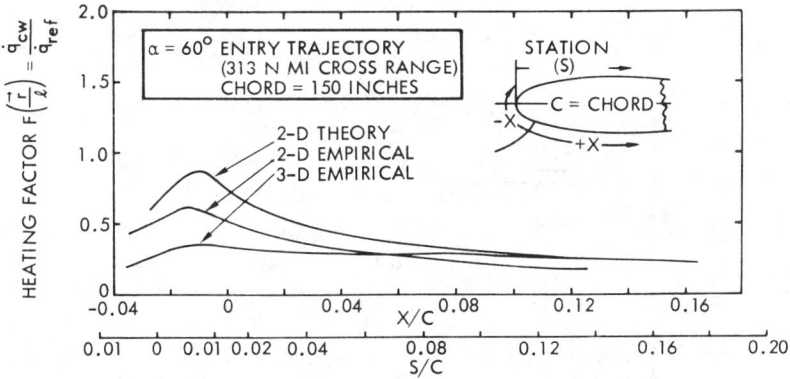

Fig. 2 Heating factor predictions.

uration were selected for analysis (Fig. 3). During entry, the wing angle of attack is held constant at 60°, 45°, or 22.5°, which provides for a crossrange capability of 313, 624, and 1196 naut. miles for each entry trajectory, respectively. Figure 4 gives a comparison of the peak radiation equilibrium temperatures around the leading edge for three wing span locations using the two-dimensional empirical heating method and entry profile for the α = 60° entry trajectory. Altitude and velocity histories for the three entry trajectories are shown in Fig. 3.

### Radiative and Ablative Heat Shields

The application of radiative and/or ablative heat-shield thermal protection system concepts for the wing leading edge of a fixed-straight wing space-shuttle configuration was investigated for representative shuttle entry missions. Calculations were performed for eight heat-shield materials for which adequate experimental data were available to describe the oxidation kinetics at environmental conditions of pressure and temperature similar to those encountered in the space-

Table 2  Wing leading edge dimensions.

| Quantity | Dimension | Quantity | Dimension |
| --- | --- | --- | --- |
| Wing area | 1175.00 ft$^2$ | Leading edge sweep | 14.00° |
| Wing span | 90.43 ft | Leading edge exposed area | 133.45 ft$^2$ |
| Root chord | 19.21 ft | Leading edge exposed span | 70.25 ft |
| Tip chord | 6.78 ft | Exposed root chord | 16.39 ft |
| Airfoil section | NACA 0012-64 | Leading edge chord | 15% chord |

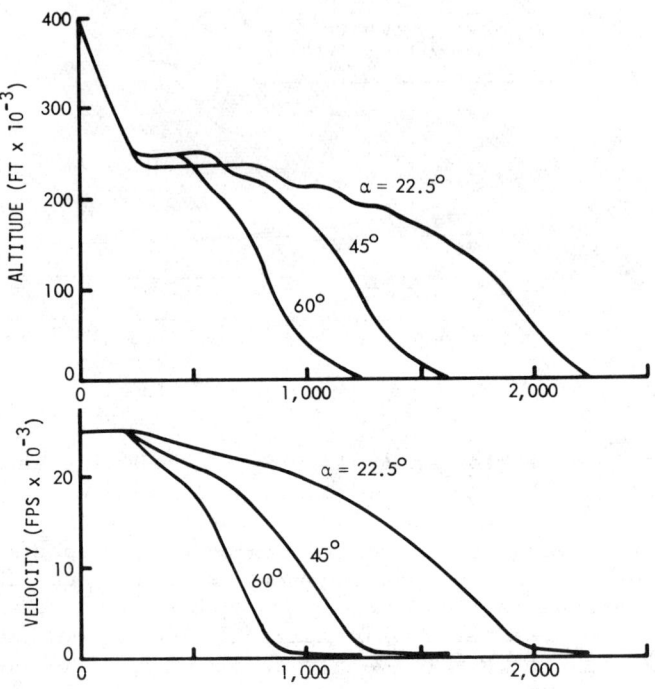

Fig. 3  Entry trajectories.

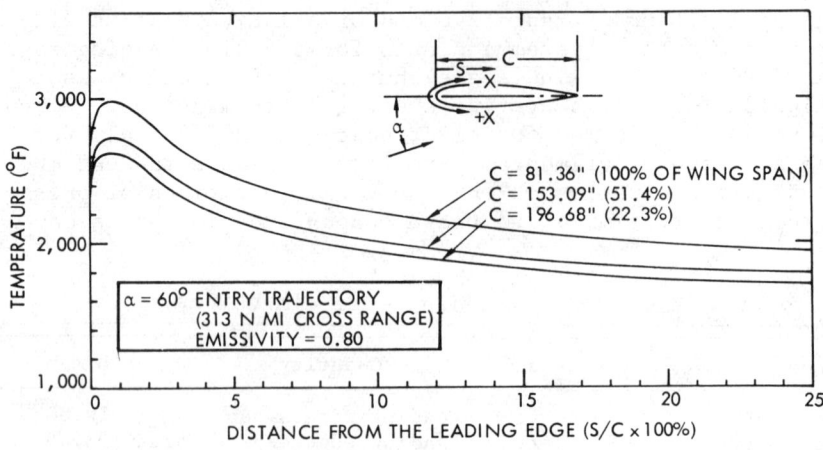

Fig. 4  Wall equilibrium radiation temperature predictions, heating criterion: "2-D Empirical".

shuttle entry missions. The materials for which calculations were performed may be categorized as follows:

High-temperature metals--- beryllium, molybdenum.
Refractory metals--------- tungsten, rhenium.
Carbonaceous materials---- ATJ graphite, LTV RPP(U)$^{48}$ unprotected†, LTV RPP(P)$^{48}$ protected†.
Pyrolytic materials------- AVCOAT 5026-39 (Apollo material)

The accuracy of the predictions of the wall surface degradation for selected heat-shield materials is dependent on the availability of quantitative experimental information in the range of application of temperature and pressure and for the same oxidizing atmosphere. Table 3 presents a review of the data sources used for high temperature of refractory metals. The range of interest in the present study is prescribed by environmental conditions at near peak heating for the space-shuttle entry trajectories. These environmental conditions are summarized in Table 4.

The performance of the heat-shield design concepts considered in this study can be evaluated best by comparing the total ablation mass loss and the maximum recession depth at the end of a selected mission. These quantities are dependent on four design or environmental factors, which are 1) the choice of heat-shield materials, 2) the entry trajectories considered, 3) the analytical procedure followed in determining the convective heating to the exposed surface of the wing leading edge, and 4) discrepancies on the published experimental data given by more than two sources for the same material. A comparison of the ablation mass loss and the maximum recession depth at the end of the entry missions led to the following qualitative evaluation of materials:

1) Unprotected refractory metals were found to provide satisfactory thermal protection for the space shuttle for all of the entry missions investigated. Of the three refractory metals calculated, the mass ablation loss for molybdenum was the smallest. The performance of tungsten and rhenium was about equal.

2) The recession depth per entry mission for the LTV RPP (P) protected heat-shield material was found to be negligibly small or comparable with the recession for refractory metals for all of the entry missions. Since the LTV RPP (P) material is lighter, the corresponding value of the ablation mass loss is at least an order of magnitude smaller than for the refractory metals.

3) The maximum recession per entry mission for ATJ graphite and beryllium materials was found to be of the order of 0.10

---

† Laminated carbon-carbon.

Table 3 Reaction rate limited oxidation data sources

| Reference & date | Ablator | Pressure $O_2$ torr[a] | Temperature °C |
|---|---|---|---|
| Ong[5] (1962) | W | 1-15,000 | 500-1300 |
| Perkins et al.[6] (1962) | W | $10^{-3}$-10 | 1300-3350 |
| Bartlett et al.[7] (1964) | W | $10^{-3}$-760 | 1300-3170 |
| Gulbransen et al.[8] (1964) | W | 2-100 | 1150-1615 |
| Rosner and Allendorf[9] (1964) | Mo and W | 1-$10^3$ | 800-1450 |
| Etemad and Holly[10] (1965) | W alloys | ~20,000 | ≤6000 |
| Kofstad[11] (1966) | Review | $10^{-2}$-$10^2$ | 400-1000 |
| Metzger et al.[12] (1967) | ATJ-Graphite | $10^2$-$10^4$ | 1000-3700 |
| Gulbransen et al.[13] (1967) | Re and Mo | $10^{-2}$-50 | 600-1600 |
| Gulbransen et al.[14] (1967) | Re-Ti alloys | 1-10 | 800-1400 |
| Charlot et al.[15] (1967) | Mo-Ta & Ni-Cr-Fe | ~15,000 | <1150 |
| Engdahl et al.[16] (1967) | Coatings for W | ~700 | ≤2000 |
| Van Orsdel et al.[17] (1967) | Nb and W | ... | ≤1400 |
| Walsh et al.[18] (1967) | W | 0.1-11 | 1750-3000 |
| Berman[19] (1967) | Porous W | ~$10^{-6}$ | <1150 |
| Babitzke et al.[20] (1968) | Ta and Nb | ~200 | 1000-1200 |
| Kolodney and Graff[21] (1968) | Silicide coatings for Nb and Ta | ≥760 | 300-1200 |
| Cassuto and Mihe[22] (1968) | W and Re | $10^{-6}$-$10^{-4}$ | 1150-2250 |
| Sama and Priceman[23] (1969) | Silicide coatings for Ta and W | ~200 | <1600 |
| Felten[24,25] (1969) | Nb-Ti alloys | 70-400 | 650-1000 |
| Watson[26] (1969) | Zr-Nb & Zr-Ti Review | ~160 | 400-800 |
| Gulbransen[27] (1970) | | 9-76 | 700-1600 |
| Babitzke et al.[28] (1970) | Nb, Hf, & W alloys | ~200 | 1000-1200 |
| Beryllium Conference[29] (1970) | Be | ... | <1000 |

[a] 1 torr = 0.0013157 atm.

Table 4 Heating environments for the space shuttle[a]

| Entry trajectory | $\alpha = 60°$ | $\alpha = 45°$ | $\alpha = 22.5°$ |
|---|---|---|---|
| Pressure, torr | 10.6-14.21 | 6.33-13.58 | 3.12-27.4 |
| Temperature, °C | 1163-1650 | 1177-1849 | 994-2284 |

[a] chord = 150 in and heating criterion is "2-D Empirical".

to 1.0 in. for the range of entry missions considered. Therefore, these materials are not suitable for multiple entry use without refurbishment.

4) The AVCOAT 5026-39 pyrolytic material (Apollo material) was found unsuitable for multiple reuse as the wing leading edge space shuttle thermal protection.

5) The oxidation resistance of refractory metals may be improved by the use of alloying elements. However, when different materials or alloys are used at different locations on the exposed surface of the wing, catalytic discontinuities will result. The presence of these discontinuities may negate the advantages gained by increasing the convective heating downstream of the wing leading edge.[49]

The different analytical procedures that were used in determining the convective heating to the wing surface had a negligible effect in the preceding qualitative evaluation of the heat-shield materials. Although differences as large as a factor of five with respect to the ablation mass loss or total recession resulted, differences of orders of magnitude larger due to the choice of heat-shield materials dominated the problem. For a reusable space-shuttle concept employing radiative and/or ablative heat shields, the parameters that must be weighted against the desired crossrange capability are 1) the maximum number of entry missions possible without refurbishment of the heat shield, and 2) the initial heat-shield weight. The maximum number of entry missions possible is dependent on the maximum permissible recession sum, i.e., the maximum recession per entry mission multiplied by the number of entry missions. Actually, the recession increment per entry mission will vary depending on the thickness of the heat shield that scales the heat capacity or heat-sink effects. The recession increment will be largest for the last entry mission when the thickness is smallest. Therefore, the assumption made is conservative. The analysis did not consider the effects of reuse on the basic materials' chemistry and thermophysical properties, e.g., gaseous diffusion into the wall surface, surface radiative emissivity changes, strength, ductility, etc. In the quantitative engineering analysis of multiple mission use without refurbishment of heat-shield materials, it was assumed that 1) 50 entry missions or more are equivalent to a fully reusable heat shield (100% reusability); 2) the maximum permissible recession sum for the heat-shield materials is 0.50 in.; and 3) refractory metals of approximately equal molecular weights will provide approximate equal performance, e.g., rhenium and tungsten.

On this basis, a total of 14 different materials were evaluated for their suitability to the solution of the reusable space-shuttle thermal protection problem. The results of this evaluation are summarized on Table 5. Note that, on the basis of the performance of molybdenum, it was concluded that columbium and zirconium would provide equivalent thermal protection. This naive approach, which is supported by qualitative evaluations of heat-shield materials found in the literature (Refs. 13, 20, and 50, in particular), was followed because of the lack of experimental data for the reaction-limited regime in the environmental range of interest.

Actively Cooled Heat Shields

Calculations for the regenerative cooled heat shields performance and the coolant delivery system weights were completed for three coolant species and for the same entry trajectories. The coolants considered include gaseous hydrogen, helium, and oxygen, which were assumed stored in the liquid phase in cryogenic storage tanks. Other factors that strongly affected the overall system weight were varied over a selected range. These include: the wing leading edge material maximum operating temperature, the speed of the flow in the leading edge coolant passages, and the procedures followed in predicting the convective heat-transfer rate to the wing.

The main conclusions drawn were as follows:
1) Light-molecular-weight gaseous species such as hydrogen and helium are the best coolants.
2) The maximum operating temperature of the wing leading edge material has a significant effect on the regenerative cooled heat shield coolant expenditure and the coolant delivery system weights.
3) The speed of the flow in the wing leading edge coolant passages must be large in order to enhance the cooling efficiency, which leads to lighter heat shield designs. Inlet Mach numbers of the order of 0.3 to 0.5 were found to provide optimum results.
4) Differences of a factor of 2 between the two-dimensional and three-dimensional empirical predictions of the convective heating for the $\alpha = 60°$ entry trajectory resulted in differ-

Table 5  Radiative and ablative heat shields performance[a]

| Heat Shield Materials | Molecular Weight | % Reusability[b] | | | Evaluation Reusable Material? |
|---|---|---|---|---|---|
| | | $\alpha = 60°$ | $\alpha = 45°$ | $\alpha = 22.5°$ | |
| Beryllium(Be) | 9.013 | 15.6 | 6.5 | 1.6 | No |
| ATJ-Graphite(C) | 12.01 | 44.8 | 22.2 | 7.4 | No |
| Titanium(Ti) | 47.9 | ≈ 44.8 | ≈ 22.2 | ≈ 7.4 | No |
| Vanadium(Va) | 50.95 | ≈ 44.8 | ≈ 22.2 | ≈ 7.4 | No |
| Zirconium(Zr) | 91.22 | ≈ 100.0+ | ≈ 100.0+ | ≈ 47.2 | Yes |
| Columbium(Nb) | 92.91 | ≈ 100.0+ | ≈ 100.0+ | ≈ 47.2 | Yes |
| Molybdenum(Mo) | 95.95 | 100.0+ | 100.0+ | 47.2 | Yes |
| Hafnium(Hf) | 178.6 | ≈ 100.0+ | ≈ 100.0+ | ≈ 11.7 | Yes |
| Tantalum(Ta) | 180.95 | ≈ 100.0+ | ≈ 100.0+ | ≈ 11.7 | Yes |
| Tungsten(W) | 183.86 | 100.0+ | 100.0+ | 11.7 | Yes |
| Rhenium(Re) | 186.22 | 100.0+ | 100.0+ | 66.6 | Yes |
| LTV-RPP (U)[48] unprotected[c] | 22 | 13.0 | 6.2 | 1.9 | No |
| LTV-RPP (P)[48] protected[c] | 22 | 100.0+ | 100.0+ | 59.2 | Yes |
| AVCOAT 5026-39 (Apollo Material) | 22 | 1.0 | 0.8 | 0.4 | No |

[a] Heating criterion is "2-D Empirical".
[b] % Reusability = $\left( \dfrac{\text{Recession sum for 50 entry missions}}{1/2 \text{ in}} \right)^{-1} \times 100\%$.
[c] Laminated carbon-carbon.

ences of 1.7 to 3.6 for the calculated complete heat shield system weights for hydrogen and helium coolants, respectively.

For transpiration-cooled heat shields, the zero mass injection convective heating to the exposed wing surface is many times smaller than for other entry vehicle applications for which transpiration cooling systems have been investigated. Under these conditions, the major part of the convective heating to the wing surface is blocked by the mass injection effects, and the remainder, which is very small, is conducted into the heat shield. The amount of heat blocked ranges from 96% to 100% for representative space-shuttle entry trajectories. For this range of applications, the heat-and-mass transfer correlations on which the analysis for the heat blocking effects is based do not provide accurate predictions.

As a direct result, the magnitudes of the calculated mass injection rates, although very small relative to the flow in the porous wall, are considered sufficiently large to cause the separation of the flow over the wing leading edge. Therefore, the mass injection rates and the spatial-time integrated coolant expenditure predictions using the present analytical procedure are considered to be very conservative.

The main conclusions drawn from the transpiration-cooled system studies were as follows:
1) For gaseous injection through the porous wall, the use of light-molecular-weight gases as coolants provides for the most effective transpiration designs.
2) For liquid injection through the porous wall, coolants that flow in the matrix in the liquid phase and vaporize at the wall interface were found to be many times more effective than gaseous coolants.
3) The variation of the permeability and thickness of the porous wall resulted in system weight savings of the order of two to three times in comparison with constant-thickness and constant-permeability designs.
4) The variation of the wing entry attitude (i.e., angle of attack) which provides for different maximum crossrange capabilities had a significant effect on the transpiration cooling system weight.
5) The different analytical procedures used in determining the convective heating rate to the exposed wing surface had a minor effect on the transpiration cooling systems weights.
6) The matrix (i.e., the wing leading edge skin) maximum permissible operating temperature determines the entry times at which the transpiration cooling system operation is started and terminated. At operating temperatures below and up to 2000°F, the dependence of the transpiration cooling system weight on the operating temperature of the matrix was found to be negligible.
7) The porous wing leading edge material is a major contributor to the transpiration cooling system overall weight, i.e., $\sim$ 40% to 70%.

## Thermal Protection System Concepts Comparisons

The evaluations of each thermal protection system concept for application to the space-shuttle entry missions were treated individually in the preceding sections. The basis for selecting the most effective system concept for a selected entry mission or missions is dependent on the comparison of the following system characteristics: 1) weight - the system over-all weight penalty, 2) reusability - the number of entry missions possible without refurbishment of the heat shield,

3) reliability - failsafe concept, and 4) cost - system development, testing, and hardware manufacturing costs.

The effectiveness of the thermal protection system concepts in terms of the over-all system weights and the reusability characteristics may be evaluated by comparing the individual system concept performance results. By definition, the active cooled systems are fully reusable, regardless of the number of entry missions. Radiative and ablative systems are considered reusable up to the maximum tolerable limit of surface degradation, which is called the maximum recession sum (i.e., the maximum recession sum = the sum of the maximum surface recession for each entry mission $\sim$ the recession per entry mission times the number of entry missions).

For engineering purposes, the maximum recession sum limit has been assumed to be 1/2 in. On this basis, the performance comparisons for the various thermal protection system concepts are presented in Figs. 5 and 6. In the first figure, the variation of the over-all system weights is given as a function of the number of entry missions for each trajectory investigated. In the second figure, the information presented in the preceding figure has been crossplotted and extrapolated in order to illustrate the variation of the system weights with respect to the entry trajectory maximum crossrange capability. Assuming 50 entry missions without refurbishment of the heat shield to be equivalent to a fully reusable system, an evaluation of the various system concepts led to the following conclusions:

1) For a maximum crossrange capability of 800 naut. miles or less, radiative and ablative thermal protection systems using unprotected refractory metals or protected carbonaceous heat-shield materials provide for the lightest system weights.

2) For a maximum crossrange of up to 1000 naut. miles, regenerative cooled and transpiration-cooled systems, although heavier than ablative systems by a factor of $\sim$ 10, have comparable system weights.

3) For a maximum crossrange capability of up to 1100 naut. miles, the LTV RPP(P) protected material provides for the lightest system weight.

4) For a maximum crossrange capability larger than 1100 naut. miles, the transpiration-cooled system provides for the lightest system weights.

5) For a maximum crossrange capability in excess of 1300 naut. miles, the thermal protection system weights become excessive, i.e., over 20% of the nominal payload (over 5000 lbm) for all of the system concepts considered.

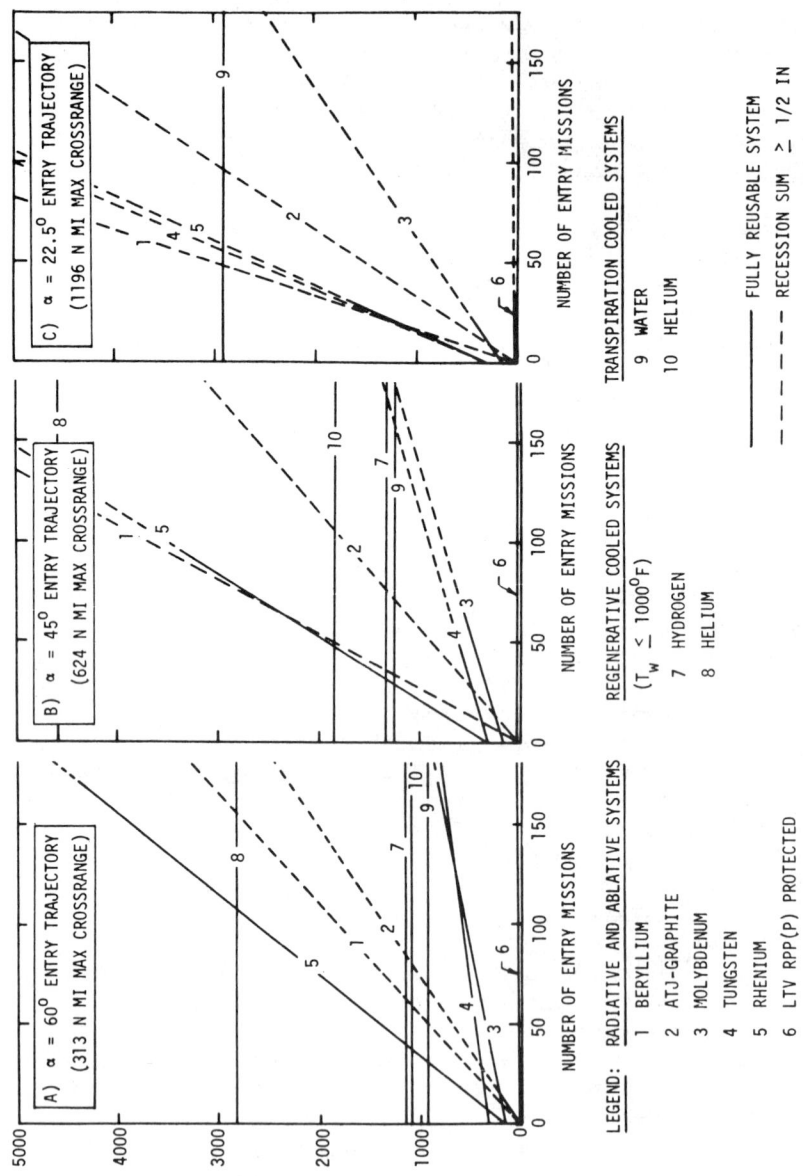

Fig. 5 Thermal protection systems weight comparisons, heating criterion: "2-D Empirical".

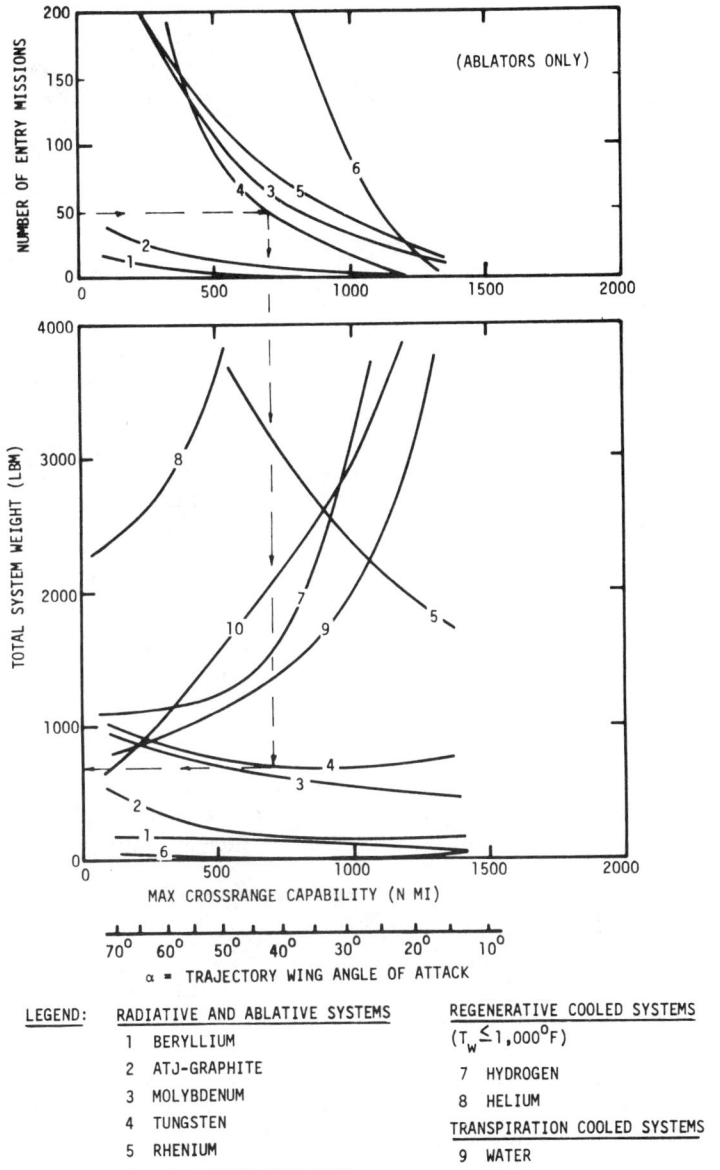

Fig. 6 Thermal protection systems weight comparisons, ablators recession sum is 1/2 in, and heating criterion: "2-D Empirical".

## IV. Concluding Remarks

The development of low-cost, high-reliability, lightweight and fully reusable entry vehicles requires the use of sophisticated, accurate numerical analysis design tools in order to perform meaningful tradeoff studies. Advanced analytical techniques have been used to predict the performance of a number of leading thermal protection system concepts and to compare the resulting weights for a range of entry aerothermodynamic environmental conditions for a fixed-wing space shuttle. The thermal-protection system concepts considered were radiative and/or ablative, regenerative cooled, and transpiration-cooled systems. The results of this investigation have indicated that unprotected refractory metals can be successfully used for up to 50 entry missions without refurbishment. The use of active cooling systems becomes competitive for cross-ranges in excess of 1100 naut. miles. Although the results presented are for a conventional fixed-wing configuration, the conclusions drawn are applicable to other wing configurations (e.g., delta wings) provided that the heating loads are not too different.

## References

[1] Gomez, A. V., Curry, D. M., and Johnston, C. G., "Radiative, Ablative, and Active Cooling Thermal Protection Studies for the Leading Edge of a Fixed-Straight Wing Space Shuttle," Paper 71-445, April 1971, AIAA.

[2] Gomez, A. V., "Radiative, Ablative, and Active Cooling Thermal Protection Studies for the Leading Edge of a Fixed-Straight Wing Space Shuttle," Rept. 17618-H076-R0-00, Dec. 1970, TRW.

[3] Gomez, A. V., Mills, A. F., and Curry, D. M., "Correlations of Heat Transfer for the Stagnation Region of a Reentry Vehicle with Multicomponent Mass Addition," Paper 70-HT/SpT-21, Space Technology for the 70's, Part 2, Sept. 1970, American Society of Mechanical Engineers.

[4] Mills, A. F., Gomez, A. V., and Strouhal, G., "The Effects of Gas Phase Chemical Reactions on Heat Transfer to a Charring Ablator," Journal of Spacecraft and Rockets, Vol. 8, No. 6, June 1971, pp. 618-625.

[5] Ong, J. N., Jr., "Oxidation of Refractory Metals as a Function of Pressure, Temperature, and Time: Tungsten in Oxygen," Journal of the Electrochemical Society, Vol. 109, No. 4, April 1962, pp. 285-288.

[17] Van Orsdel, J., Bartlett, E. S., and Barth, V. D., "Recent Developments Refractory Metals (Cb, Ta, Mo, W)," AD82-2460, Nov. 1967, Defense Metals Information Center, Battelle Memorial Institute, Columbus, Ohio.

[18] Walsh, P. N., Quets, J. M., and Graff, R. A., "Kinetics of the Oxygen-Tungsten Reaction at High Temperatures," Journal of Chemical Physics, Vol. 46, No. 3, Feb. 1967, pp. 1144-1153.

[19] Berman, D., "Porosity Effects on the Oxidation of Tungsten," Journal of Applied Physics, Vol. 38, No. 2, Feb. 1967, pp. 780-783.

[20] Babitzke, H. R., Oden, L. L., and Kelley, H. J., "Columbium and Tantalum Alloy Development," Rept. 7211, 1968, Bureau of Mines, U. S. Govt. Print. Office, Wash., D. C.

[21] Kolodney, M. and Graff, R. A., "Fundamentals of the Oxidation Protection of Columbium and Tantalum," Semiannual Rept. 5, April 1968, City College Research Foundation, University of New York; also NASA CR-94582.

[22] Cassuto, A. and Mihe, J. P., "Oxidation of Tungsten and Rhenium at Low Pressure," Academy of Sciences, Paris, C. R., Series C, Vol. 266, 1968, pp. 863-866.

[23] Sama, L. and Priceman, S., "Oxidation Resistant Coatings for Tantalum and Tungsten," Paper W 9-21.3, March 1969, American Society for Metals, Westec Meeting, Los Angeles, Calif.

[24] Felten, E. J., "The Interaction of the Alloy Niobium$^{-25}$ Titanium with Air, Oxygen and Nitrogen, Part I: The Unusual Oxidation Behavior of Nb$^{-25}$Ti at $1000°C$," Journal of Less-Common Metals, Vol. 17, 1969, pp. 185-197.

[25] Felten, E. J., "The Interaction of the Alloy Niobium$^{-25}$ Titanium with Air, Oxygen and Nitrogen, Part II: The Reaction of Nb$^{-25}$Ti in Air and Oxygen Between $650°$ and $1000°C$," Journal of Less-Common Metals, Vol. 17, 1969, pp. 199-206.

[26] Watson, R. D., "On the Oxidation of Zirconium Alloys in Air and the Dimensional Changes with Oxidation," Rept. 3375, June 1969, Atomic Energy of Canada Ltd.

[27] Gulbransen, E. A., "Thermochemistry and the Oxidation of Refractory Metals at High Temperature," Corrosion-Nace, Vol. 26, No. 1, Jan. 1970, pp. 19-28.

[6]Perkins, R. A., Price, W. L., and Crooks, D. D., "Oxidation of Tungsten at Ultra-High Temperatures," Rept. LMSD 6906298 Nov. 1962 (N63-18738), Lockheed.

[7]Bartlett, R. W., "Tungsten Oxidation Kinetics at High Temperature," Proceedings of the 3rd Conference, The Performance High Temperature Systems, Vol. 1, Dec. 1964, pp. 79-87.

[8]Gulbransen, E. A., Andrew, K. F., and Brassart, F. A., "Kinetics of Oxidation of Pure Tungsten, $1150°-1615°C$," Journal of the Electrochemical Society, Vol. 111, No. 1, J 1964, pp. 103-109.

[9]Rosner, D. C. and Allendorf, H. D., "Kinetic and Aerodyn Aspects of the Oxidation of Metals by Partially Dissociat Oxygen," TN-61, May 1964, AeroChem Research Lab., Inc., Princeton, J. J.

[10]Etemad, G. A., "Oxidation and Mechanical Performance o Tungsten at High Temperatures and High Pressure," AIAA J nal, Vol. 4, No. 9, Sept. 1966, pp. 1543-1548.

[11]Kofstad, P., High-Temperature Oxidation of Metals, Cha 6, and 7, John Wiley & Sons, Inc., N. Y., 1966.

[12]Metzer, J. W., Engel, M. J., and Diaconis, N. S., "Ox and Sublimation of Graphite in Simulated Reentry Enviro ments," AIAA Journal, Vol. 5, No. 3, March 1967, pp. 45

[13]Gulbransen, E. A., Andrew, K. F., and Brassart, F. A. "Studies on the High Temperature Oxidation of Molybdenu sten, Niobium, Tantalum, Titanium, and Zirconium," AD65 April 1967, Westinghouse Electric Corp.

[14]Gulbransen, E. A. and Brassart, F. A., "Oxidation of and a Rhenium-8% Titanium Alloy in Flow Environments o Pressures of 1-10 Torr and at $800°-1400°C$," Journal of Common Metals, Vol. 14, 1967, pp. 217-224.

[15]Charlot, L. A., Thiede, R. A., and Westerman, R. E. sion of Superalloys and Refractory Metals in High Tem Flowing Helium," VNWL-SA-1137, March 1967, Batelle Me Institute, Pacific Northwest Lab., Richland, Wash.

[16]Engdahl, R. E., Bedell, J. R., and Kroha, C. E., Jr Technique being Developed for Protecting Refractory M High Temperature," Materials Protection, Vol. 6, Oct. pp. 49-51.

[28] Babitzke, H. R., Oden, L. L., and Kelley, H. J., "Columbium Alloy Development with Boron, Hafnium, and Tungsten," Rept. 7388, June 1970, Bureau of Mines, U. S. Department of the Interior.

[29] "Proceedings of the Beryllium Conference," Rept. NMAB-272, July 1970, Vol. 1, National Research Council, National Academy of Science, National Academy of England, Washington, D. C.

[30] Joint Army, Navy, Air Force Thermochemical Panel, "JANAF Thermochemical Tables," U. S. Air Force Contract F04611-67-C0009, Dec. 1964 (last rev. 1966).

[31] "JANAF Thermochemical Tables," Dec. 1960 and supplements to date, Thermal Lab., DOW Chemical Co., Midland, Mich.

[32] D'Amur, I. and Mason, E. A., "Properties of Gases at Very High Temperatures," *Physics of Fluids*, Vol. 1, No. 5, Sept.-Oct. 1958, pp. 370-383.

[33] Duff, R. E. and Bauer, S. H., "Equilibrium Composition of the C/H System at Elevated Temperatures," *Journal of Chemical Physics*, Vol. 26, April 1962, p. 1754.

[34] Shick, H. L., "Thermodynamics of Certain Refractory Compounds," *Procedures of the 3rd Conference, The Performance of High Temperature Systems*, Vol. 1, Dec. 1964, pp. 9-17.

[35] Shick, H. L., *Thermodynamics of Certain Refractory Compounds, Vol. 2: Thermodynamic Tables, Bibliography, and Property File*, Academic Press, New York, 1966.

[36] Wicks, C. E. and Block, F. E., "Thermodynamic Properties of 65 Elements - Their Oxides, Halides, Carbides, and Nitrides," Bull. 605, 1963, Bureau of Mines, U. S. Government Printing Office, Washington, D. C.

[37] Hirschfelder, J. O., Curtiss, C. F., and Bird, R. B., *Molecular Theory of Gases and Liquids*, 2nd ed., John Wiley & Sons, New York, Nov. 1965.

[38] Svehla, R. A., "Estimated Viscosities and Thermal Conductivities of Gases at High Temperatures," TR R-132, 1962, NASA.

[39] Hochstim, A. R., "Equilibrium Compositions, Thermodynamic and Normal Shock Properties of Air with Additatives," Zph-122, Vol. 1, General Dynamics, Convair.

[40] Wilke, C. R., "A Viscosity Equation for Gases Mixtures," Journal of Chemical Physics, Vol. 18, No. 4, April 1950, pp. 517-519.

[41] Mason, E. A. and Saxena, S., "Approximate Formula for the Thermal Conductivity of Gas Mixtures," Physics of Fluids, Vol. 1, No. 5, 1958, pp. 361-369.

[42] Kendall, R. M., Barlett, E. P., Rindal, R. A., and Moyer, C. B., "An Analysis of the Coupled Chemically Reacting Boundary Layer and Charring Ablator, Part IV," CR-1063, June 1968, NASA.

[43] Spalding, D. B., "A Standard Formulation of the Steady Convective Mass Transfer Problem," International Journal of Heat and Mass Transfer, Vol. 1, 1960, pp. 192-207.

[44] Detra, R. W., Kemp, N. H., and Riddell, R. F., "Addendum to Heat Transfer to Satellite Vehicles Reentering the Atmosphere," Jet Propulsion, Vol. 27, Dec. 1957, p. 1256.

[45] Miller, B., "Thermal Analysis of the Proposed NASA/MSC Shuttle Wing Design," Rept. 69.4352.20-6, July 19, 1969, TRW.

[46] Scala, S. M. and Gilbert, L. M., "Sublimation of Graphite at Hypersonic Speeds," AIAA Journal, Vol. 3, No. 9, Sept. 1965, pp. 1635-1644.

[47] Gilbert, L. M., "The Hypersonic, Diffusion-Controlled Oxidation of Tungsten," R67SD38, July 1967, General Electric Missile and Space Division.

[48] While, D. M., "Development of a Thermal Protection System for the Wing of a Space Shuttle Vehicle, Phase I Final Report," Rept. T143-5R-00044, Feb. 1971, Vought Missiles and Space Corp.

[49] Sheldahl, R. E. and Winkler, E. L., "Effect of Discontinuities in Surface Catalytic Activity on Laminar Heat Transfer in Arc-Heated Nitrogen Streams," TN D-3615, Sept. 1966, NASA.

[50] Krochmal, J. J., "4000°F Oxidation Resistant Thermal Protection Materials," Paper 660659, Oct. 1966, Society of Automotive Engineers Aeronautic and Space Engineering and Manufacturing Meeting, Los Angeles, Calif.

# SPACE STATION ENVIRONMENTAL THERMAL CONTROL SYSTEM DEFINITION

Joseph C. Cody[*]

NASA George C. Marshall Space Flight Center, Huntsville, Ala.

and

R. M. Byke[†] and A. T. Stell[††]

McDonnell Douglas Astronautics Company-West,
Huntington Beach, Calif.

## Abstract

The methods used to select and define an environmental thermal control system (ETCS) for a 12-man space station are presented. The ETCS was defined to meet the space station program requirements and to require minimum resources for design, development, and qualification to support a 1977 launch. The design consists of an exterior radiator/fluid system with regenerative temperature control interfacing with an interior heat-transport water loop. The adequacy of the radiator design was verified by a parametric computer analysis that considered thermal coating degradation, vehicle attitude, docked module blockage, and variation in the solar constant, albedo, and earth emission.

## Introduction

The basic element of the space station is a core module 33 ft in diameter and 50 ft long containing two pressurizable compartments,

---

Presented as Paper 71-435 at the AIAA 6th Thermophysics Conference, Tullahoma, Tenn., April 26-28, 1971.

[*]Scientific Assistant to Chief, Propulsion & Thermodynamics Division, Astronautics Laboratory.
[†]Senior Staff Engineer.
[††]Senior Engineer/Scientist.

each with two decks (Fig. 1). Two of the decks are devoted to laboratories that support the experiment program. Each of the remaining two decks contains an operations area and living quarters for six men. Either of the pressurizable compartments could accommodate the entire 12-man crew for extended time periods should the need arise. Access between the decks is provided by a central tunnel 10 ft in diameter which can also be used as an emergency shelter. The core module also serves as a support station for permanently attached experiment and cargo modules and "free-flying" experiment modules that leave and return to the core module as necessary.

The core module will be launched into a 55° inclination, 200-300-naut-mile-altitude earth orbit by a Saturn-derivative launch vehicle. The mission lifetime will be 10 yr, with resupply of the space station to occur every 3 mo by an advanced logistics vehicle (space shuttle). Consequently, subsystems will be designed for 10-yr operation with in-flight maintenance, repair, or replacement. Subsystem operational requirements will impose a minimum of orientation restrictions on the space station.

The challenging problems in thermal control system design and integration derive from the combination of complex structural

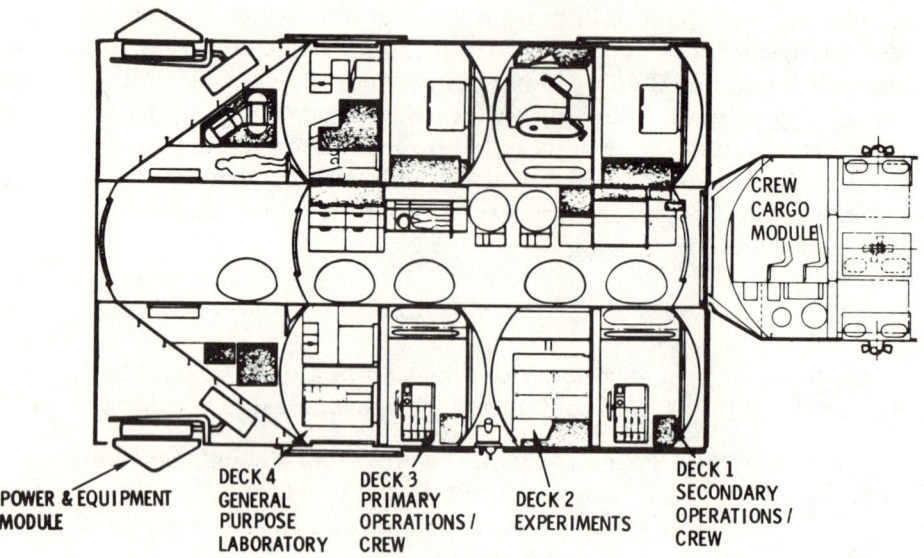

Fig. 1  Space station core modules.

geometry, long mission time, extremely variable exterior thermal environment, requirements for manned habitation and equipment thermal control, and docked module and other subsystem interfaces. For example, the 55° launch inclination actually results in a condition in which the space station radiator must be designed to operate in a range of $\beta$ angles of $\pm 78.5°$. Many of these problems were encountered in the Apollo and Skylab programs. However, the flexibility and life requirements for space station increase their complexity.

The estimation of the adsorbed heat flux to the ETCS radiator, located integrally with the meteoroid protection bumper (Figs. 2 and 3) presents a complex problem in radiation heat exchange between the radiator and surrounding docked modules. Parametric investigations were necessary to establish the worst design points for the ETCS radiator. The major parameters involved were thermal coating degradation, vehicle orientation and orbital position, docked module configuration, and variations in the orbital heating environment (solar constant, earth albedo, earth emission).

Two fluid loops are provided for the ETCS: an external radiator loop consisting of F-21 fluid and an internal water loop for atmosphere conditioning and interfacing hardware thermal control. The

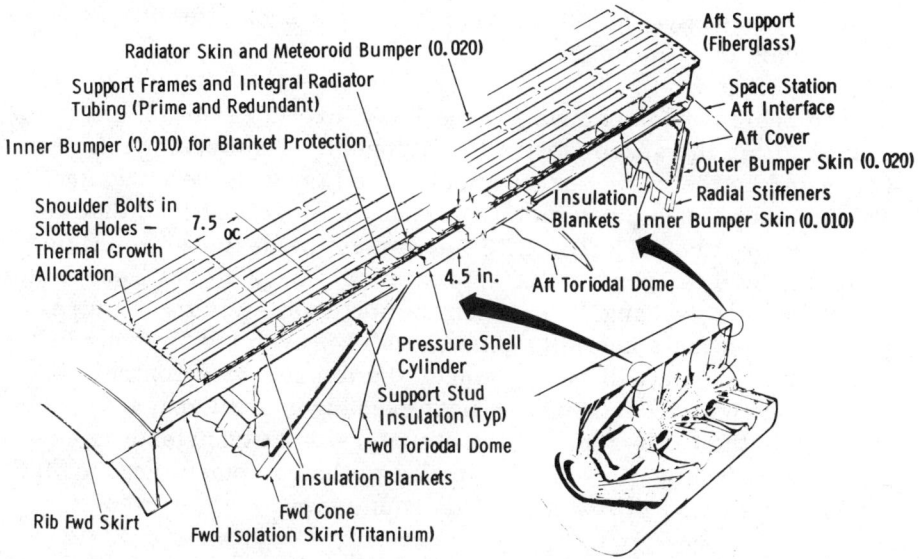

Fig. 2 Space station structure details.

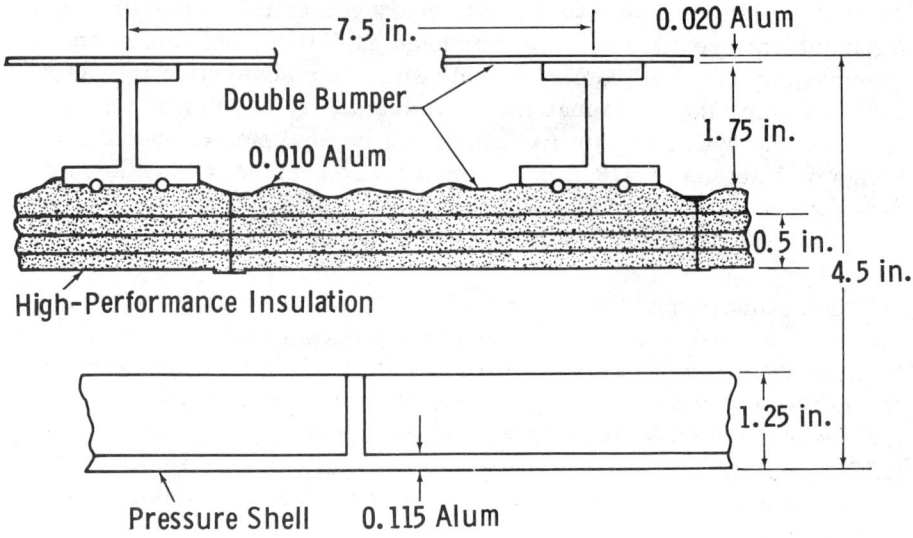

Fig. 3 Section through outer shell.

two-fluid system is utilized because it allows the choice of optimum fluids for heat rejection and heat collection. It also provides a significant safety advantage in that water, a nontoxic, non-flammable fluid, is used in internal occupied areas. Redundancy was also designed in these systems to provide maximum protection for the crew in the event of failures.

The design requirements for the ETCS are to reject the heat generated within the space station by equipment, the metabolic energy generated by the crew, and the heat gained through the structure. Table 1 shows how these factors produce the net heat load that the radiator must reject. The ETCS also must be designed to provide an outlet fluid temperature of approximately 39° F to the freon/water interface heat exchange. This temperature satisfies the requirements for adequate humidity control while preventing freezing in the water circuit. As will be explained, the temperature control is accomplished with a regenerative heat exchanger and thermal capacitors. For the trade studies, the structural heat transfer was assumed to be negligible when compared to the equipment heat load. As will be shown, this was a valid assumption.

To identify an ETCS for the space station which satisfies the objectives of high reliability and minimum design and development

Table 1  Factors producing net heat load

| | |
|---|---|
| Bus power | 24.50 |
| Internal line and conditioning losses | 4.08 |
| Battery charging inefficiency | 1.17 |
| Partial shunt regulator | |
| Parasitic dissipation | 0.57 |
| Total electrical power to space station bus | 30.32 |
| Power consumed by water electrolysis | (2.32) |
| Net electrical heat dissipated to EC/LS[a] | 28.00 |
| Thermal dissipation to EC/LS | 3.60[b] |
| Metabolic dissipation to EC/LS | 1.60 |
| EC/LS design heat load | 33.20 (kw) |

[a] Environmental control and life support.
[b] Isotope waste heat utilized by EC/LS system.

costs and which meets performance requirements, trade studies were made to evaluate the merits of several potential concepts. The major trades performed and the rationale developed are explained below.

## System Tradeoff Studies

Because high solar flux occurs on only one side of the space station at a time, it was clear that segmentation of the ETCS radiator would permit higher heat rejection capability. However, many radiator segments present a complex radiator fluid loop control problem that reduces reliability. A compromise was to break the radiator into two segments, each covering 180° of the space station circumference (Fig. 4) and connected by fluid circuitry. This arrangement offers several alternatives for further investigation (Fig. 5), which are: 1) flow modulation, splitting the flow to two panels such that the bulk of the flow is directed to the colder panel; 2) mixing, providing fluid or thermal mixing of the two outlet circuits to average the hot and cold temperatures; and 3) flow reversal, providing the capability of reverse flow in each panel such that flow can be directed away from the hot environment. Flow reversal can be combined with mixing, as shown.

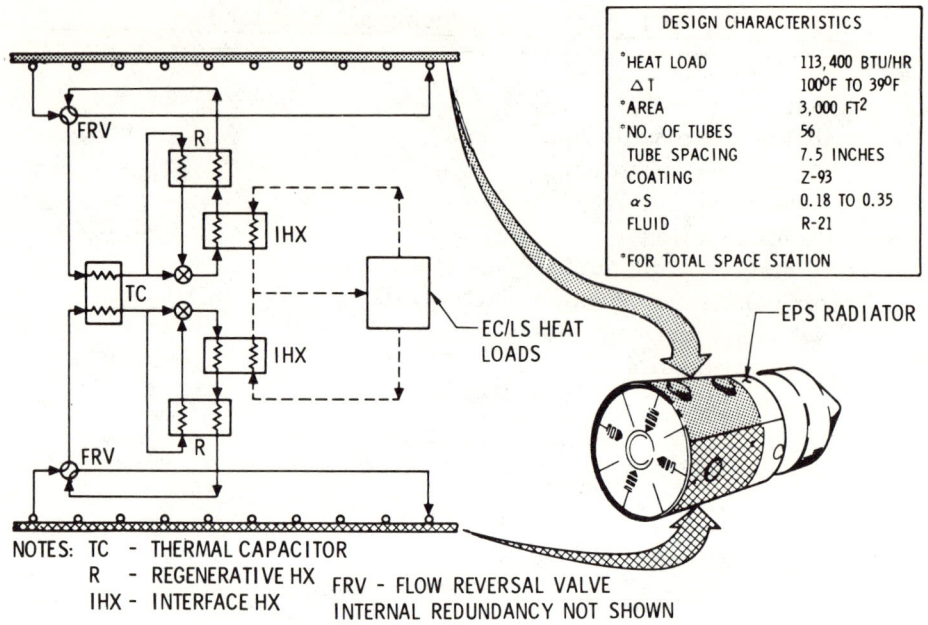

Fig. 4  EC/LS segmented radiator design.

To assess radiator performance, a transient radiator computer program and orbital heat flux program were used to evaluate the three previously described radiator concepts for $\beta$ angles (sun angles) of $0°$ and $78.5°$ and selected space station altitudes. As a result of the computer runs, flow reversal with mixing was found to yield the maximum heat rejection capacity. This concept then was selected for the ETCS. Thermal mixing by means of a heat exchanger was selected over fluid mixing because it eliminates flow maldistribution problems and provides some system redundancy. The results for the selected concept for the worst-case vehicle attitudes are shown in Fig. 6. A comparison of these results shows that the $78.5°$ $\beta$ case is a more severe design condition than the $0°$ $\beta$ case. For the extreme value of $\alpha/\epsilon$ assumed, the EC/LS temperature requirement cannot be met for $\beta = 78.5°$. Furthermore, it was concluded that a thermal capacitor would be required to maintain the average radiator fluid temperature below the design point of $39°$ F in the $0°$ $\beta$ case.

Four methods of radiator fluid temperature control were evaluated. These are shown in Fig. 7. A measure of performance of these control concepts is the minimum heat load required for the

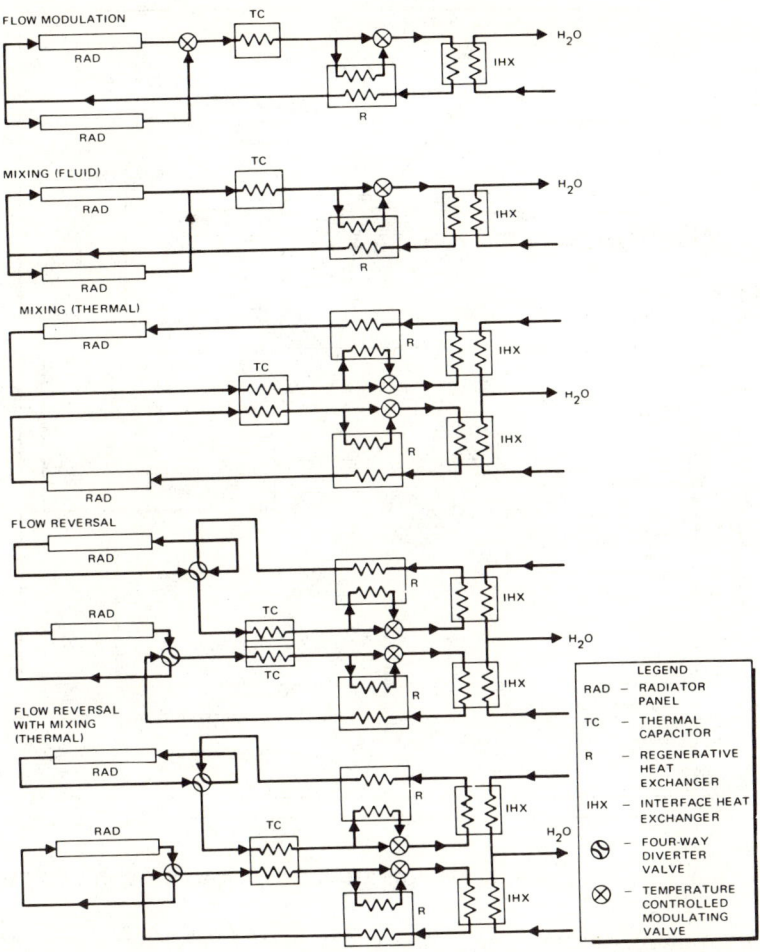

Fig. 5   Radiator segmentation concepts.

control system to satisfy its temperature control requirement. Expressed as a fraction of the design heat load, this attainable heat load is tabulated in Fig. 7 for each of the four concepts and is based on exposure of the radiators to deep space.

As indicated, the stagnation system offers appreciable performance advantages over the convenient bypass and regenerative systems. To assess whether these performance advantages warrant the design complexity and cost associated with the stagnation systems, an examination of the required performance levels was made. The results of this examination are shown in Fig. 8. Since the Isotope

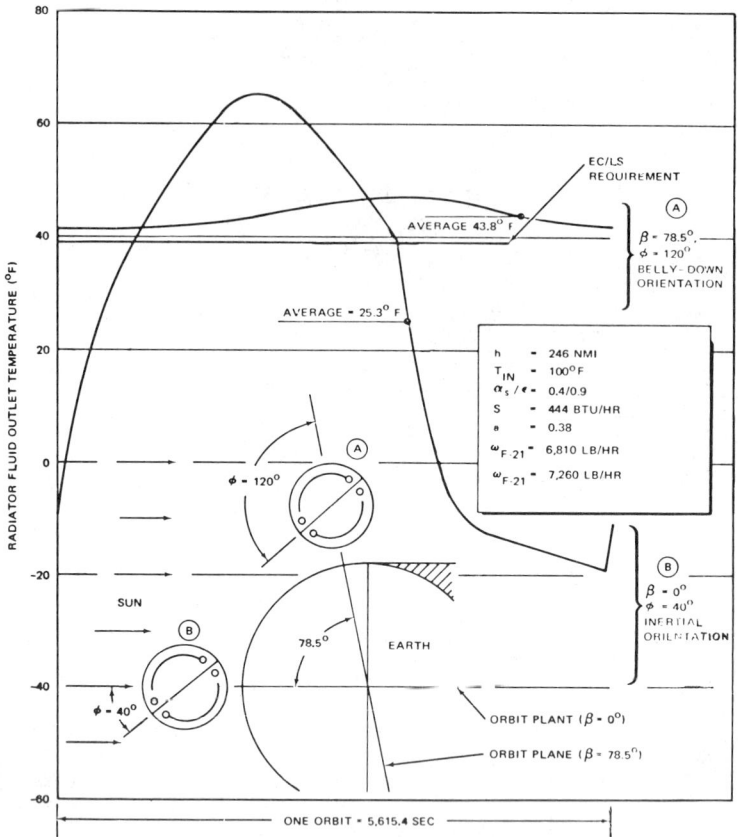

Fig. 6 Effect of space station attitude on radiator performance.

Brayton power system generates a constant load, the design heat load ratio for the space station was conservatively estimated to be 0.5. As shown in Fig. 8 bypass control will allow F-21 to freeze at heat load ratios less than 0.7. If regenerative control is used, F-21 will not freeze until the heat load ratio is less than 0.2. Thus, for the design heat load ratio of 0.5, bypass control is unsatisfactory, whereas regenerative control has considerable margin for not allowing freezing. Because of this design adequacy and the minimization of design complexity and cost, the regenerative control system was selected.

The basic radiator system resulting from the heat rejection and control system trades is shown schematically in Fig. 9. The system is comprised of two separate circuits, one for each radiator segment. Outlet radiator temperatures are thermally mixed in a

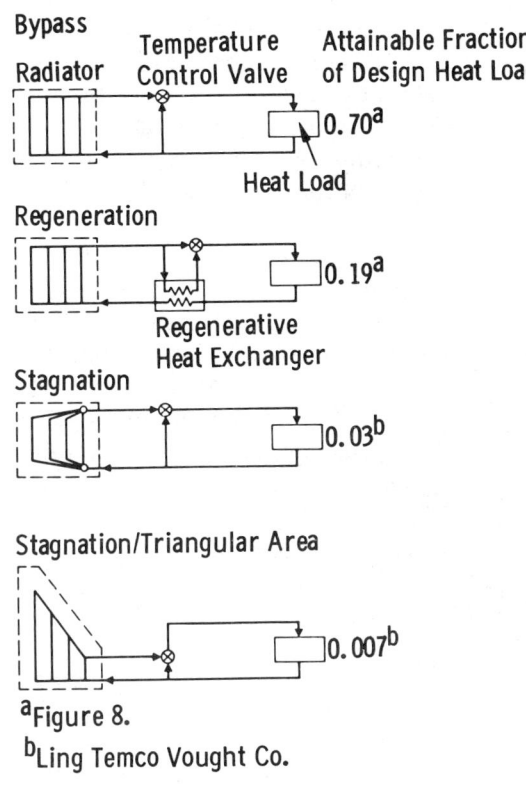

Fig. 7 Radiator control concepts.

[Labels in figure: Bypass Radiator, Temperature Control Valve, Attainable Fraction of Design Heat Load, 0.70[a], Heat Load, Regeneration, 0.19[a], Regenerative Heat Exchanger, Stagnation, 0.03[b], Stagnation/Triangular Area, 0.007[b]]

[a] Figure 8.
[b] Ling Temco Vought Co.

heat exchanger that also incorporates a phase change material for thermal storage. Each circuit incorporates a regenerative control system and an interface heat exchanger through which space station heat loads are transferred to the radiator circuits.

Two independent EC/LS systems are provided, one in each of two independently pressurized compartments of the space station. Each compartment has the capability of supporting 12 men, and each provides cooling for the electronic equipment in its respective compartments. If the heat load were equally distributed to these systems, the obvious solution would be to provide two independent radiator systems, each using half of the available radiator area. Unfortunately, unequal load sharing can occur, and the full radiator area must be used to reject the combined heat load.

Four fundamental solutions to this design requirement have been investigated and are shown schematically in Fig. 10. For simplicity, a single radiator is shown for each concept; however, each may be expanded to show the selected two-segment configuration or any number of additional circumferential or longitudinal radiator segments. The four concepts are described as follows:

1) Independent circuitry: Completely independent water and F-21 circuits are employed for the two EC/LS systems. Full radiator area is used by alternate tubes connected to the independent circuits. In order to satisfy unequal EC/LS system load sharing, water pumps, freon pumps, and interface heat exchangers are

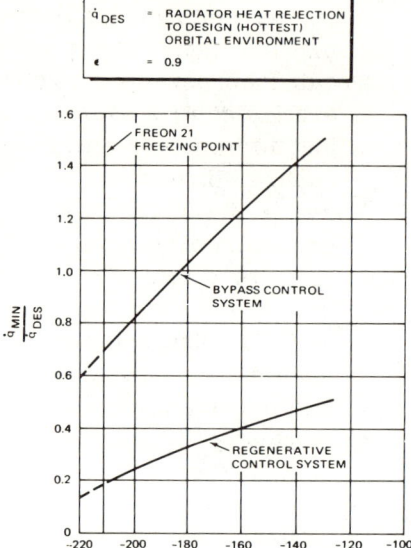

Fig. 8 Radiator control system performance.

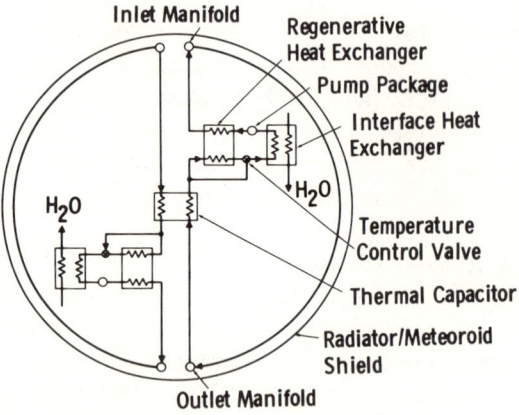

Fig. 9 Radiator system configuration

designed for the maximum single EC/LS system heat load. Variable-speed water and freon pumps are provided to maintain high heat rejection temperature levels independently of load sharing.

2) Common water circuit: Pumps and heat exchangers designed for the total heat load are used, and modulating valving is employed to satisfy unequal EC/LS system load sharing and yet maintain high heat rejection temperature levels. Although it is quite simple, this system possesses the fundamental disadantage that a water circuit failure could incapacitate both EC/LS systems.

3) Common radiator circuit: The EC/LS systems are separated, but similar flow modulation valving in the freon circiuts is required, together with variable speed water pumps to satisfy unequal EC/LS system load sharing while maintaining high heat rejection temperature levels. Interface heat exchangers and water pumps must also be designed for the maximum single EC/LS system heat load. A single radiator circuit failure could affect both EC/LS systems. However, in this case increased radiator segmentation could mitigate the effect. For example, if four radiator circuits were provided, loss of one would decrease the capacity of each EC/LS system by 25%.

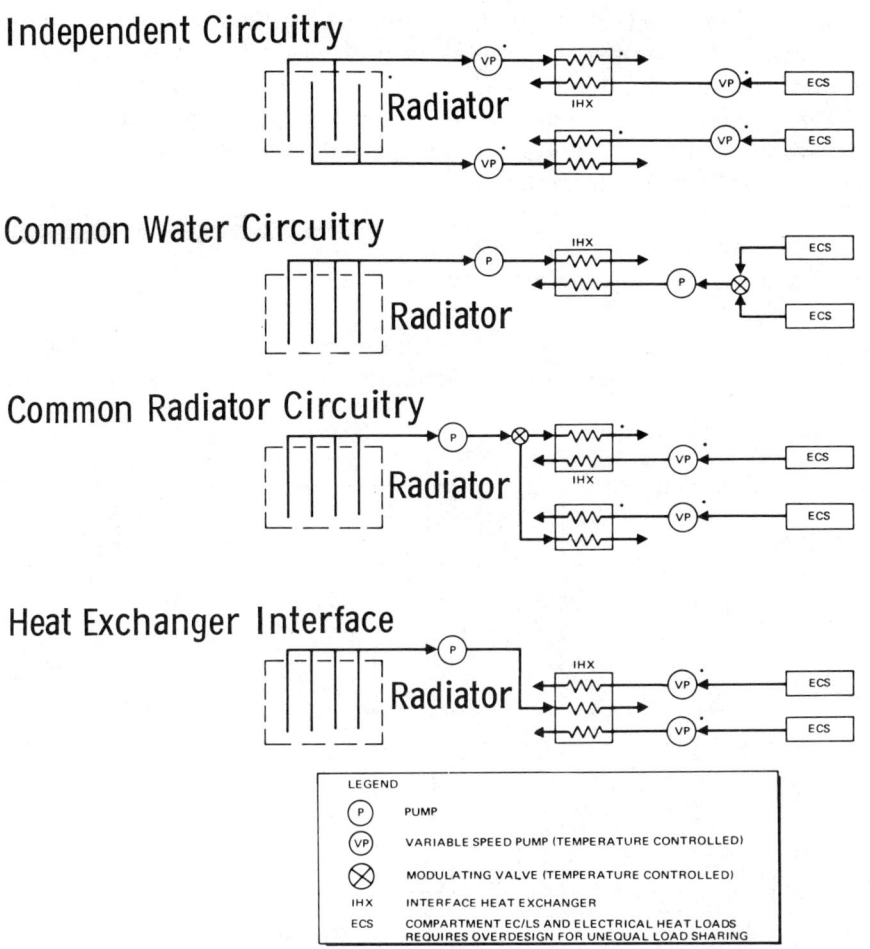

Fig. 10   Thermal control circuitry concepts.

4) Heat exchanger interface: The two water circuits and the freon circuit are linked in an interface heat exchanger. Water pumps must be designed for the maximum single EC/LS system heat load and must be variable speed to maintain the temperature levels necessary for radiator performance. As with concept 2, a single failure (i.e., both water passages of the heat exchanger) could incapacitate both EC/LS systems. However, because the location of such a failure is known and limited, proper sensing and isolation techniques should eliminate this problem. Again, several radiator segments could mitigate the effect of a freon circuit failure.

Concept 4 was selected. This system minimizes component overdesign and also minimizes complexity in the less maintainable freon circuits. As mentioned, the heat exchanger failure problem is solvable. Figure 11 illustrates how this selected circuitry concept is integrated with the selected radiator systems. This figure also reflects two additional design decisions:

1) Provision of redundant freon circuitry with radiator tubes spaced alternately with primary tubes.

2) Division of the radiator circuits into two longitudinal segments. This provides some additional redundancy, simplifies manufacturing,

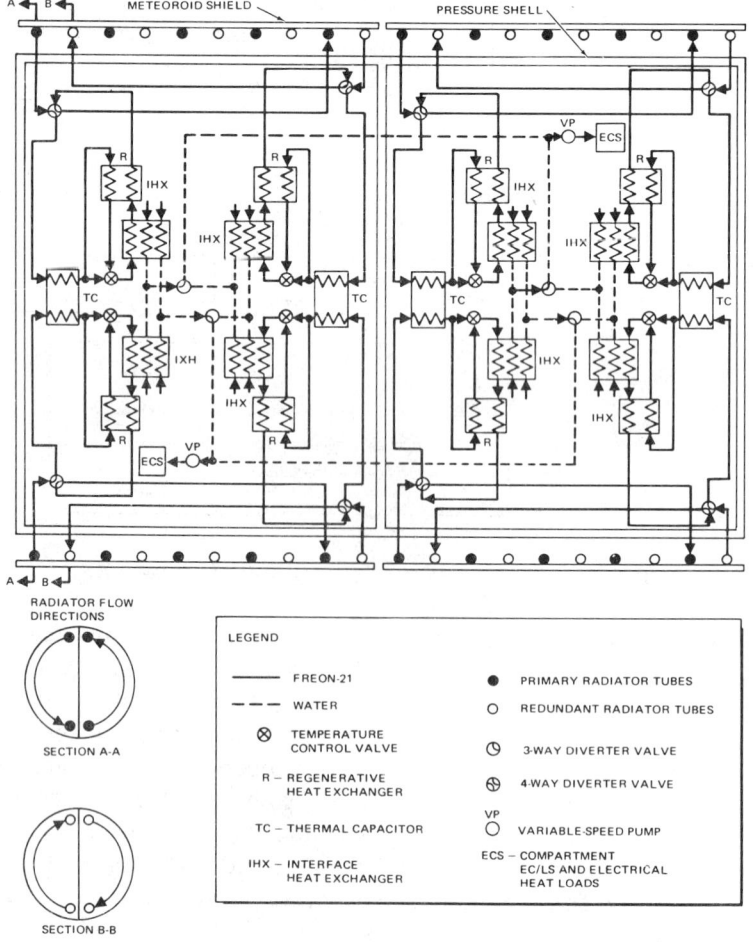

Fig. 11   Space station thermal control system schematic.

and decreases the probability of a docking accident incapacitating the entire radiator.

A complete discussion of the results of the thermal control tradeoff studies is reported in Ref. 1.

### Thermal Analyses and Verification of Concepts

At this point, an ETCS concept has been identified for further definition leading to a preliminary design. The major variables yet to be considered are the structural heat leak, docked module effects, and sensitivity of the system to thermal coating degradation.

The space station thermal design, which is similar in concept to that of a Thermos bottle, is intended to minimize heat transfer between the core module and the external environment and to control the termperature of the internal pressure shell walls. The former function minimizes the design range of the ETCS radiator, and the latter precludes atmosphere condensation and excessive temperature levels on the pressure shell wall.

The materials selected to effect this Thermos bottle design are summarized in Table 2. Figure 3 shows the integrated wall design of the structural insulation system. The high-performance insulation used between the integral meteoroid shield/radiator and the pressure wall consists of two overlapping blankets. Each blanket consists of 25 layers of double-aluminized mylar (DAM) separated by dacron net spacers with beaded dacron net face sheets that rigidize the blanket. The DAM is perforated to permit venting during boost.

The pressure level must be on the order of $10^{-3}$-$10^{-4}$ t if the insulation is to function effectively. With proper prelaunch purging, the orbit time required for the insulation pressure to reach $10^{-4}$ t is less than 1 hr; therefore, no initialization problems exist. Leakage of atmosphere into the blanket is not expected to be a problem because the expected leak rate will be about 2 scc/day/100 in. of seal length. However, if a meteoroid penetration causes a localized leak, the insulation effectiveness can be degraded significantly. The structure is designed for a 0.99 probability of no meteoroid penetration of the pressure shell, and provisions are made to detect and repair leaks caused by meteoroid penetrations. To prevent a localized 2 lb/day leak rate (design value) from causing insulation

Table 2  Summary of thermal control techniques

| Structure | Material | Purpose |
| --- | --- | --- |
| Radiator/meteoroid shroud | Two aluminum alloy bumpers | Eliminates HPI meteoroid punctures. Very low probability of radiator tube puncture (1/10-yr mission) |
| Core module insulation | DAM/dacron HPI (0.5-in. blanket) | Provides "Thermos bottle" effect for habitation compartments (side and aft areas) |
| Insulation on forward conical | High-temp. HPI such as multilayer Kapton | Provides Thermos bottle effect for habitation compartments (forward areas) |
| Forward thermal isolator | Titanium alloy (0.075 × 30 in.) | Isolates the hot Brayton radiator from the forward joint |
| Aft thermal Isolator | Titanium alloy (0.16 × 30 in.) | Isolates the aft meteoroid shield from aft joint. |
| Forward meteoroid shroud fairing | Fiberglass | Isolates EC/LS and Brayton radiators |
| Meteoroid shroud and midspan spacers | Fiberglass and polyimide materials | Low-conductivity support structure reduces heat leaks to a minimum, consistent with structural requirements |

Table 2 Summary of thermal control techniques (Cont'd)

| Structure | Material | Purpose |
|---|---|---|
| View port heaters | $0.2$ w/in.$^2$ imbedded resistance heaters | Eliminates dew point problems for cold case |

degradation and consequent core module wall condensation, adequate venting area must be provided for the leakage to escape through the meteoroid shield to space. This requires 90-900 in.$^2$ of venting area for insulation pressure levels of $10^{-3}$ and $10^{-4}$ t, respectively. Because the boost vent area necessary to maintain 1-1.5 psid positive annular pressure is 20-30 in.$^2$, holes under ejectable fairings are provided to acquire the additional vent area after orbit is achieved. To mitigate the leakage problem further, potential leakage paths such as hatch seals are vented directly to space and thus do not degrade insulation within the annulars.

The performance of the structural thermal design is summarized in Table 3. When compared to the radiator design heat load of 33 kwt, the total hot and cold case heat leaks (0.55 and 0.66 kwt, respectively) are quite small. In addition, the pressure shell temperature ranges from 64.5° - 85° F, which is well above the 58° F maximum dew point and well below the 105° F maximum touch temperature. Although heating will be required for view port warming (approximately 0.2 w/in.$^2$) and for atmosphere warmup, it may be generally concluded that the space station thermal design satisfies the Thermos bottle design goal.

ETCS Radiator Concept Verification

To verify the performance of the ETCS radiator for worst design conditions and maintain a manageable analysis effort, the thermal control analyses matrix shown in Table 4 was developed. From this matrix, it is seen that the influences of surface coating, orbital heating, docked configuration, vehicle orientation, sun angle $\beta$, and vehicle roll on the radiator were investigated. Figure 12 shows the relationship of the vehicle to the sun and earth for the sun and roll angles.

Table 3 Summary of heat leaks

| Space Station structure | Hot case $Q_{in}$, Btu/hr | Cold case $Q_{out}$, Btu/hr |
|---|---|---|
| A) EC/LS radiator section | | |
| Heat through HPI[a] | 31 | 1100 |
| Heat through docking ports | 41 | 210 |
| Heat through support channels | 2 | 63 |
| Heat through midspan spacers | 1 | 30 |
| Heat through other cutouts | 41 | 210 |
| B) Aft meteoroid shield | | |
| Heat through HPI[a] | 113 | 316 |
| Heat through titanium isolator | 109 | 301 |
| C) Conical structure | | |
| Heat through HPI[b] | 1260 | 0 |
| D) Forward joint | | |
| Heat through titanium isolator | 309 | 0 |
| Total | 1907 | 2230 |

[a] Based on thermal conductance value of 0.002 Btu/hr-ft$^2$-°F.
[b] Based on thermal conductance value of 0.003 Btu/hr-ft$^2$-°F.

To evaluate the preceding parameters properly, a detailed computer model of the space station docked configuration was developed to evaluate the adsorbed heat flux to the ETCS radiator. A nodal breakdown of the vehicle surface for the heat flux model is shown in Fig. 13. The nodal breakdown consisted of four nodes in the lateral direction (spacecraft x-axis) for every 22.5° in the circumferential direction.

The heat flux model evaluates the gray-body view factors for each radiator node and each node's view to the heat sink of space. This nodal evaluation is performed for each 20° increment of the orbit. This procedure, coupled with the exterior thermal environment

Table 4 Thermal control analysis case matrix

| Set No. | Case Number | Orbital Heating Constants 2 (per TM X-53865) | Surface Coating Properties $\alpha_s$ | Surface Coating Properties $\epsilon$ | Docked Module Configuration | Orientation | Sun Angle $\beta$ | Roll Angle $\phi$ | Remarks | |
|---|---|---|---|---|---|---|---|---|---|---|
| I | 1<br>2<br>3 | Maximum | 0.2<br>0.3<br>0.4 | 0.9 | 7 | Horizontal | 78.5° | 90° | Heating<br>Docked modules<br>Roll angle | Worst case |
| II | 4<br>5<br>6 | | 0.2<br>0.3<br>0.4 | | → | → | → | 0° | Heating<br>Docked modules<br>Roll angles less severe | Worst case |
| III | 7<br>8<br>9 | | 0.2<br>0.3 | | 1 (End cargo module) | → | → | 90° | Heating<br>Roll angle<br>Docked modules less severe | Worst case |
| IV | 10<br>11<br>12 | Nominal | 0.2<br>0.3<br>0.4 | | 7 | → | → | → | Roll angle<br>Docked modules<br>Heating less severe | Worst case |
| V | 13<br>14<br>15 | Maximum | 0.2<br>0.3<br>0.4 | | → | Solar | 0° | 0° | Heating<br>Docked modules<br>Roll angle | Worst case |
| VI | 16<br>17<br>18 | Maximum | 0.2<br>0.3<br>0.4 | | 7 (Rolled 180° from Set 1) | Horizontal | 78.5° | -90° | Heating<br>Roll angle<br>Docked modules<br>Orientation | Worst case<br>Less severe |
| VII | 19<br>20<br>21 | Maximum | 0.2<br>0.3<br>0.4 | | 1 (End cargo module) | → | → | 0° | Heating<br>Roll angle | Worst case<br>Less severe |
| VIII | 22<br>23<br>24 | Nominal | 0.2<br>0.3<br>0.4 | | → | → | → | 90° | Roll angle<br>Heating | Worst case<br>Less severe |

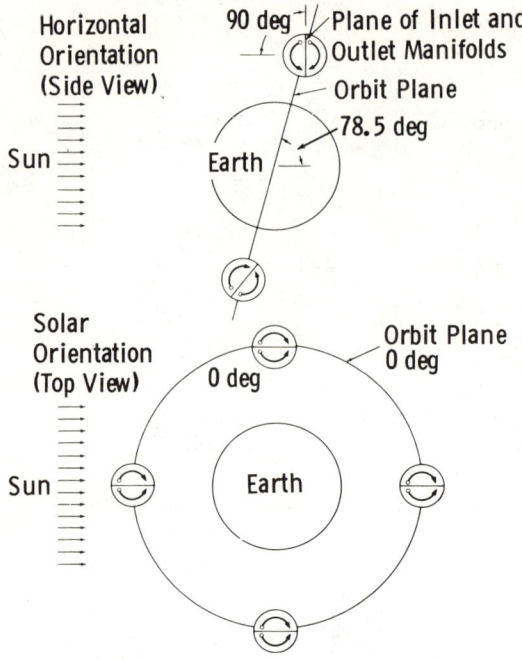

Fig. 12  Vehicle sun and roll angle orientation.

(solar constant, albedo, earth emission), provides the input data required for the thermal analysis computer model of the ETCS radiator. For this analysis, the radiator surface was divided into four nodes in the lateral direction for every 5.6° in the circumferential direction. The radiator thermal model is then used to evaluate the performance of the ETCS radiator.

Typical transient analyses results are shown in Fig. 14. Similar results were obtained for the other conditions specified in Table 4. The outlet temperature of the radiator can, in theory, be maintained below 39°F with a phase change material so long as the orbital average radiator outlet temperature is less than 39°F.

A summary of transient temperature data averaged over the orbit as a function of solar absorptance ($\alpha_S$) is shown in Fig. 15 for a space station with seven modules docked. Similar results were obtained for the other conditions specified in Table 4. For nominal heating conditions (nominal values of solar constant, albedo, earth thermal emmission), radiator outlet temperature is less than 39°F for solar absorptivities less than 0.38. Results of these analyses show that the ETCS radiator performance is extremely sensitive to $\alpha_S$ variation and, to a lesser degree, the influence of docked modules.

A complete discussion of the thermal analyses performed to verify the radiator design is reported in Ref. 3.

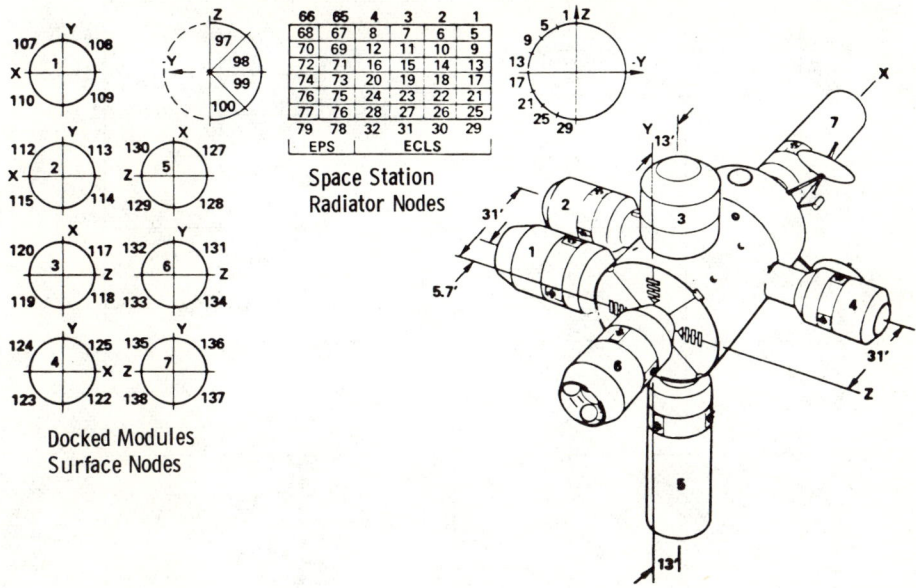

Fig. 13  Orbital heating model for space station.

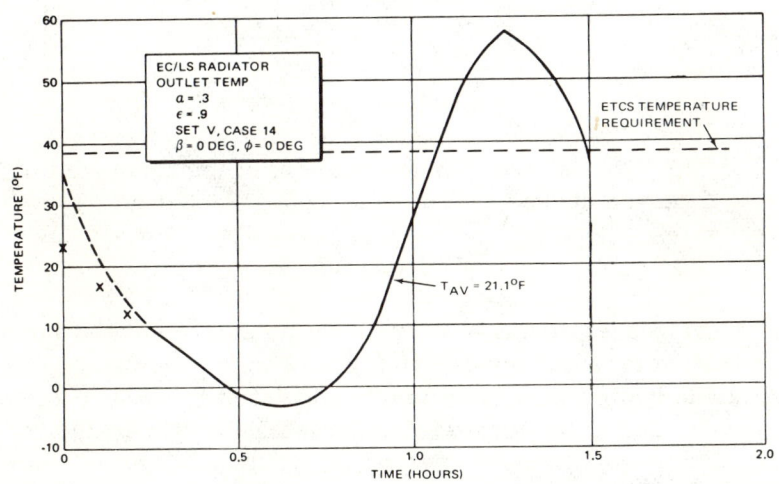

Fig. 14  Typical results of transient analyses.

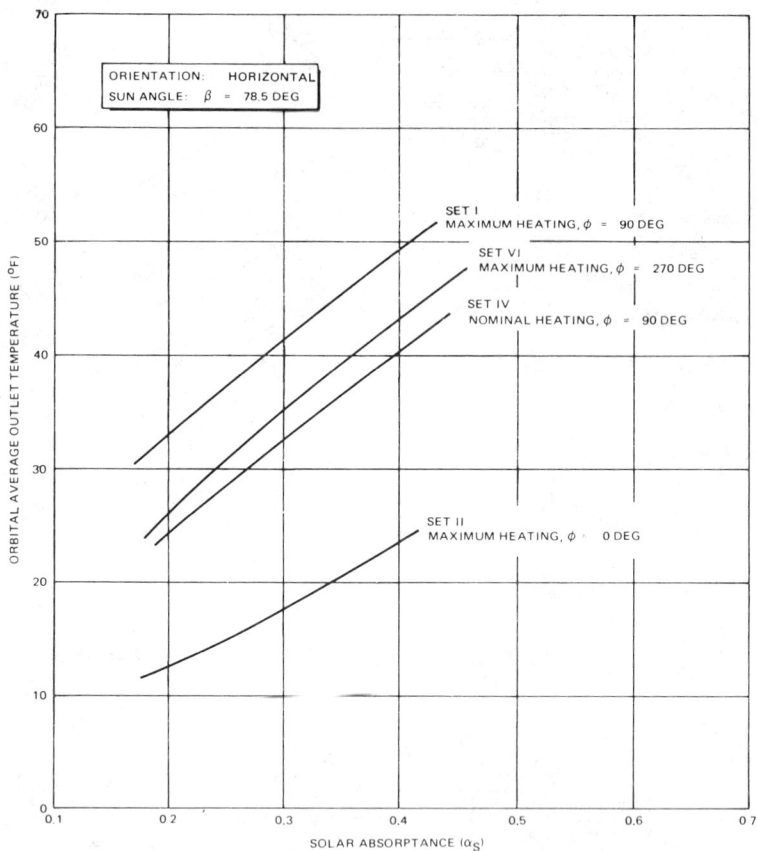

Fig. 15   EC/LS radiator average outlet temperature for seven docked modules and beta = 78.5 deg.

## Conclusions

The described approach — to arrive at an environmental thermal control system preliminary design for a 12-man space station — is considered to be optimum from the standpoint of manpower and cost requirements and in the level of detail required to generate a reliable design.

The results have shown that the reliable design of a thermal control system for large complex structures such as space stations requires complex detailed computer analyses early in the design phase to establish the most significant variables influencing the design, and the magnitude of their influence.

Since the thermal control system is extremely sensitive to the space station configuration, and in many cases is an integral part of the structure, accurate analyses of the thermal systems are required before a configuration can be considered feasible. In many cases, this is true for interfacing subsystems such as the power subsystem.

## References

[1] Report on Space Station Definition, Vol. V, Subsystems, Book 2, Crew Systems, MSFC DRL-160 Line Item 8, Contract NAS 8-25410, July 1970.

[2] Weidner, D. K., ed., "National Environment Criteria for the NASA Space Station Program," NASA TM X-53865, October 31, 1969, Marshall Space Flight Center, Alabama.

[3] Report on Selected Update Tasks for Baseline Space Station, Vol. II, "Thermal Control," MSFC-DRL-231, Line Item 8, Contract NAS8-25410, March 1971.

# Index to Contributors to Volume 29

Abu-Romia, M. M.
POLYTECHNIC INSTITUTE OF BROOKLYN
401

Anderson, D. L.
NASA AMES RESEARCH CENTER
205

Andres, R. J.
HUGHES AIRCRAFT COMPANY
107

Ashford, N. A.
IIT RESEARCH INSTITUTE
3

Basiulis, A.
HUGHES AIRCRAFT COMPANY
431

Berggren, C. C.
HUGHES AIRCRAFT COMPANY
33

Beverly, W. D.
THE BOEING COMPANY
167

Bienert, W. B.
DYNATHERM CORPORATION
463

Blair, Paul M., Jr.
HUGHES AIRCRAFT COMPANY
107, 221

Brennan, P. J.
DYNATHERM CORPORATION
463

Byke, R. M.
McDONNELL DOUGLAS ASTRONAUTICS COMPANY—WEST
579

Carroll, W. F.
JET PROPULSION LABORATORY
221

Chaabane, Ghassane M.
UNIVERSITY OF OKLAHOMA
53

Chin, Jin H.
LOCKHEED MISSILES & SPACE COMPANY
333

Christensen, Paul A.
MARTIN-MARIETTA CORPORATION
531

Cody, Joseph C.
NASA MARSHALL SPACE FLIGHT CENTER
579

Connolly, J. Micheal
MARTIN-MARIETTA CORPORATION
531

Cunningham, B. E.
NASA AMES RESEARCH CENTER
205

Curry, D. M.
NASA MANNED SPACECRAFT CENTER
547

Dahms, R. G.
NASA AMES RESEARCH CENTER
205

Dunn, J. C.
BELL TELEPHONE LABORATORIES INC.
319

Edelstein, F.
GRUMMAN AEROSPACE CORPORATION
487

Filler, M.
HUGHES AIRCRAFT COMPANY
431

Fischer, W. D.
UNIVERSITY OF ILLINOIS
269

Francis, John
UNIVERSITY OF OKLAHOMA
53

Gillette, R. B.
THE BOEING COMPANY
167

Gilligan, J. E.
IIT RESEARCH INSTITUTE
3

Goble, R. G.
MARTIN-MARIETTA CORPORATION
349

Gomez, A. V.
TRW SYSTEMS GROUP
547

Hamberg, O.
THE AEROSPACE CORPORATION
137

Hembach, R. J.
GRUMMAN AEROSPACE CORPORATION
487

Hering, R. G.
UNIVERSITY OF ILLINOIS
69, 243, 269

Hinderman, J. D.
McDONNELL DOUGLAS ASTRONAUTICS COMPANY
445

Houchens, A. F.
OAKLAND UNIVERSITY
243

Hueter, Uwe
NASA MARSHALL SPACE FLIGHT CENTER
531

Jacobs, S.
NASA MANNED SPACECRAFT CENTER
221

Johnston, C. G.
TRW SYSTEMS GROUP
547

Kaser, R. V.
UNIVERSITY OF OKLAHOMA
445

Kirkpatrick, J. P.
NASA AMES RESEARCH CENTER
463, 505

Kosowski, N.
GRUMMAN AEROSPACE CORPORATION
417

Kosson, R.
GRUMMAN AEROSPACE CORPORATION
417

Leger, L. J.
NASA MANNED SPACECRAFT CENTER
221

Levin, H.
HUGHES AIRCRAFT COMPANY
33

Levy, E. K.
LEHIGH UNIVERSITY
383

Linford, R. M. F.
McDONNELL AIRCRAFT COMPANY
123

Linton, R. C.
NASA MARSHALL SPACE FLIGHT CENTER
153

MacGregor, R. K.
THE BOEING COMPANY
87, 361

Marcus, B. D.
TRW SYSTEMS GROUP
505

Nickell, R. E.
BELL TELEPHONE LABORATORIES INC.
319

Peffley, W. M.
HUGHES AIRCRAFT COMPANY
33

Pieroway, Chesley
UNIVERSITY OF OKLAHOMA
53

Pogson, J. T.
THE BOEING COMPANY
87

Smith, E. C.
HUGHES AIRCRAFT COMPANY
107

Smith, T. F.
UNIVERSITY OF ILLINOIS
69

Smolak, George R.
NASA LEWIS RESEARCH CENTER
189

Stell, A. T.
McDONNELL DOUGLAS ASTRONAUTICS COMPANY—WEST
579

Steube, K. E.
McDONNELL AIRCRAFT COMPANY
123

Stevens, N. John
NASA LEWIS RESEARCH CENTER
189

Tomlinson, F. D.
THE AEROSPACE CORPORATION
137

Waters, E. D.
McDONNELL DOUGLAS ASTRONAUTICS COMPANY
445

Yovanovich, M. Michael
UNIVERSITY OF WATERLOO
289, 307

Zentner, R. C.
THE BOEING COMPANY
87

Zerlaut, G. A.
IIT RESEARCH INSTITUTE
3